W9-AOB-317

Ecological Dynamics of Tropical Inland Waters

Lakes and rivers of the tropics are rich with variety and human relevance, yet do not figure prominently in surveys of general freshwater biology and limnology. The fruits of their scientific exploration are largely embodied in regional and specialist descriptions and analyses. In this book the authors take a generalized view, on a world-wide scale, that is dynamic and quantitative in outlook. They set out to integrate events and processes under tropical conditions, not only geographically but also within a continuum of physics, chemistry and biology. The volume contains numerous illustrations and detailed documentation of literature. Together the two authors have gathered experience from several tropical countries over three to four decades. They provide a foundation that will be of value to all who work with tropical inland waters, with interests ranging from water quality to fisheries. The volume will also have an appeal to those researchers, teachers and students in limnology and freshwater biology everywhere who are curious about the tropical implication and application of their subject.

JACK TALLING is a Fellow of the Royal Society of London and an Honorary Research Fellow at the UK's Freshwater Biological Association.

JACQUES LEMOALLE is a Research Scientist at ORSTOM (Institut français de recherche scientifique pour le développement en coopération) in France.

ECOLOGICAL DYNAMICS OF TROPICAL INLAND WATERS

JACK F TALLING

JACQUES LEMOALLE

PUBLISHED BY THE PRESS SYNDICATE OF THE UNIVERSITY OF CAMBRIDGE
The Pitt Building, Trumpington Street, Cambridge CB2 1RP, United Kingdom

CAMBRIDGE UNIVERSITY PRESS
The Edinburgh Building, Cambridge CB2 2RU, UK http://www.cup.cam.ac.uk
40 West 20th Street, New York, NY 10011-4211, USA http://www/cup.org
10 Stamford Road, Oakleigh, Melbourne 3166, Australia

© Jack Talling, Jacques Lemoalle and ORSTOM 1998

This book is in copyright. Subject to statutory exception
and to the provisions of relevant collective licensing agreements,
no reproduction of any part may take place without
the written permission of Cambridge University Press

First published in 1998

Printed in the United Kingdom at the University Press, Cambridge

Typeset in Times 10/13pt, by Keyword plc, Wallington, Surrey

A catalogue record for this book is available from the British Library

Library of Congress Cataloguing in Publication data

Talling, J. F. (John Francis)
Ecological dynamics of tropical inland waters / Jack F. Talling,
Jacques Lemoalle.
p. cm.
Includes bibliographical references and index.
ISBN 0 521 62115 1 (hb)
1. Lake ecology–Tropics. 2. Stream ecology–Tropics.
3. Freshwater biology–Tropics. 4. Limnology–Tropics.
5. Biogeochemistry–Tropics. I. Lemoalle, J. (Jacques) II. Title.
QH84.5.T35 1998
577.63′0913–dc21 97-51548 CIP

ISBN 0 521 62115 1 hardback

Contents

Preface

This book has its origin in several ventures involving Franco-British collaboration over the past 30 years. One main focus was African shallow lakes, with intensive fieldwork during the late 1960s and early 1970s within the International Biological Programme and – later – a pan-African overview. Our fascination with tropical freshwater science developed from this and other experience in Africa.

We have felt, however, that regional experience and regional description have most reward when integrated into a generalized science. The stage then becomes the entire tropics and the specific a special case of the general. Such integration we have tried to provide here. The subject is developed using comparative examples. It is founded upon a resolution into fluxes and flux-interactions, using quantities of energy, water and chemical elements as 'common currencies'. Biological activities fit within, and participate in, these circulations. We then take up the consequences of time-variability at different frequencies and at various levels of organization. In the background is the question of tropical distinctiveness.

The book, is, therefore, organized around dynamic themes. We hope that it will be of value to those in tropical countries with scientific and practical interests in inland waters; also to those working at higher latitudes who would like to obtain a fuller perspective of their subject. For reference by both groups we have provided a detailed biography of the rather scattered literature.

Our venture has received help from numerous sources. Much early inspiration came from co-workers in the field, including our late colleagues Julian Rzóska and Leonard Beadle. We are grateful to others who have offered valuable comments and advice. They include Eddie Allison, Mary Burgis, Rob Hart, Xavier Lazzaro, Stephen Maberly, Jean Pagès, and Ed Tipping. Most of all, we have benefited from constructive criticism of the entire text by Geoffrey Fryer, Rosemary Lowe-McConnell and Roger Pourriot. Our two parent organizations, the Freshwater Biological Association (FBA) at Windermere and ORSTOM at Montpellier, have provided crucial support throughout. We are indebted to the FBA for use of its fine library facilities; from these much of the literature cited here can be obtained at moderate cost, under some conditions (excluding entire books), by photocopies supplied by post (the Document Delivery Service). Jack Talling is grateful to ORSTOM for financial support that enabled him to work at Montpellier. Also important was assistance with typing; here we are especially indebted to Kirsty Ross at Windermere and Marie-Christine Pascal at Montpellier. Funding from the Royal Society, and the professional skill of Kilian McDaid, enabled the redrawing of many text-figures. Finally the friendly help and advice of Alan Crowden and Maria Murphy of Cambridge University Press did much to smooth the path to final publication.

Jack F. Talling
Freshwater Biological Association
Windermere
Cumbria
UK

Jacques Lemoalle
ORSTOM
Montpellier
France

Acknowledgements (text-figures)

We are indebted to the following publishers and others for their kind permission to reproduce from published text-figures. The figures listed, in conjunction with the references, give the attribution to authors, journals or books, and bibliographic details.

Academic Press, London – Figs. 5.34, 5.39*a*

Akademie Verlag (now Wiley – VCH Verlag), Berlin – Figs. 5.7, 5.18, 5.20

American Fisheries Society, Bethesda – Fig. 3.47

American Geophysical Union, Washington DC – Fig. 2.27

American Society for Limnology & Oceanography – Figs. 2.5, 2.11, 2.12, 2.14, 2.32, 3.24, 3.26, 3.30, 4.17, 4.19

Asociacion Venezolana para el Avance de la Ciencia, Caracas – Fig. 5.41

Blackwell Science, Oxford – Figs. 2.15, 3.2, 3.12b, 3.13, 3.18, 3.19*a*, 3.28*a*, 3.32, 3.42, 4.2, 4.15, 4.16, 5.11, 5.27, 5.29, 5.50

Cambridge University Press – Figs. 5.32, 5.37, 5.38

Ecological Society of America, Washington DC – Figs. 4.20, 5.36

Editions ORSTOM, Paris – Figs. 2.1, 2.9, 2.34, 3.13*b*, 3.14, 3.16, 3.23, 3.35, 3.36*a*, 3.40, 3.43, 4.6*b*, 5.2, 5.3, 5.21*a*, 5.28, 5.53

Editor, *Aquatic Living Resources*, Nancy – Figs. 5.40, 5.56

Elsevier Science, Amsterdam – Figs. 2.3, 3.29

Prof. G. Fryer, Windermere – Fig. 3.33

Japanese Society of Limnology, Tokyo – Fig. 3.20

Kluwer Academic Publishers, Dordrecht – Figs. 2.24, 2.25, 2.28, 2.29, 2.33, 3.5, 3.8, 4.1, 5.4, 5.9, 5.12, 5.14, 5.15, 5.30

Météo-France, Paris – Figs. 4.11*a*, 4.12

Natural Resources Institute, University of Greenwich, UK – Figs. 2.19, 2.37, 3.45, 4.18, 5.33

Oxford University Press – Figs. 2.22, 2.36, 3.25

Publisher and author, *Physiology & Ecology (Japan)*, Kyoto – Fig. 3.17

Royal Society, London – Figs. 2.18, 3.3, 3.12, 3.34

E. Schweizerbart'sche Verlagsbuchhandlung, Stuttgart – Figs. 2.6, 2.10, 2.13, 2.16, 2.35, 3.3, 3.13*a*, 3.22, 4.4, 4.7, 4.8, 4.9, 5.1, 5.39*b*, 5.42, 5.55

SCOPE Publications, Paris – Fig. 2.26

SPB Academic Publishing, Amsterdam (now Backhuys Publishers, Leiden) – Figs. 1.2, 2.20, 4.10

Springer-Verlag, Heidelberg – Figs. 3.19*b*, 3.27, 3.31, 3.37, 3.44, 3.48, 5.8, 5.21*b, c*, 5.46

Dr J. van der Heide, Amsterdam – Fig. 5.6

Zoological Society of London – Figs. 5.43, 5.44, 5.45

1

Introduction

1.1 Scope

Tropical inland waters lie within a wide latitudinal belt in which the noon-day sun can be vertically overhead at some time or times of the year. As lakes and reservoirs, rivers and streams, they have an uneven distribution over three main geographical regions – Central and South America, Africa, and Australasia. Novelty and diversity, reinforced by local practical issues, have encouraged individual or regional description of water-bodies and their inhabitants. In their scientific study broad functional considerations have not been to the fore. Even today, generalized issues of ecology are rarely taken up on a comparative, pan-tropical scale.

In this book the comparative and pan-tropical approach is central. We have concentrated upon dynamic aspects of functioning for environment, population, community and ecosystem; and here the basis of selection needs some explanation. 'Dynamic aspects' are mostly founded upon flows or *fluxes*, that control the magnitude and distribution of *stock* quantities, and like them can be physical, chemical or biological in nature. In rivers attention is captured by the flow or flux, in lakes by the stock quantity of water present. Both fluxes and stocks are important in the natural water economy of a region; the same is true for the energy economy, chemical economy and biological economy. Much environmental physics, geochemistry and biogeochemistry, and population-plus production-dynamics, are implicated. The principles involved are outlined in Chapters 2 and 3. They have been developed mainly from studies in the temperate zones but here are directed to tropical examples. However, tropical material with both analytical and comparative scope is scanty, and generalization is often insecure. Another aspect of ecological dynamics is *time-variability*, which includes flux and stock quantities, and

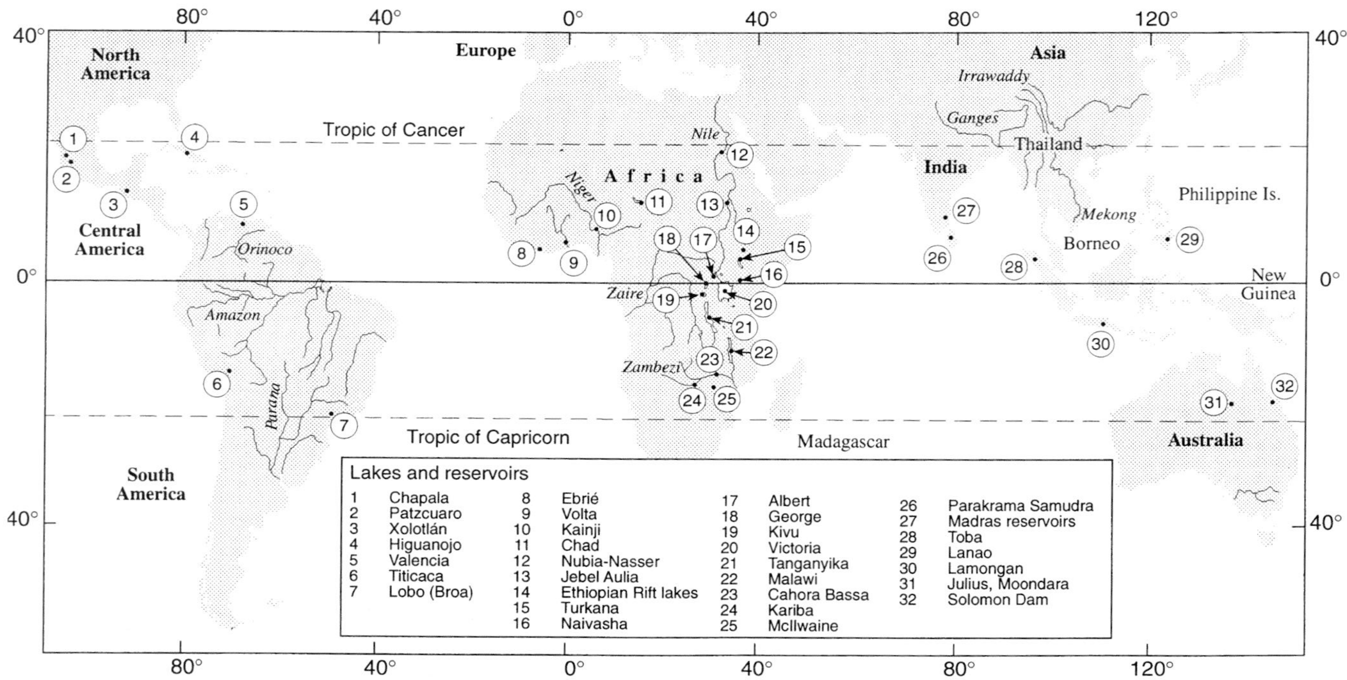

Fig. 1.1. Land-masses of the tropics and subtropics, with major rivers and lakes referred to in the text.

can be illustrated, analysed and compared from a wide variety of tropical studies. It occupies much of this book, in Chapters 4 and 5.

The tropics share with other latitudinal belts the prevalence of three environmental cycles of fixed period – the diel, lunar and annual. Past discussion of tropical distinctiveness has centred on the annual (seasonal) cycle, although long ago Buschkiel (1936) – from the then Dutch East Indies – also emphasized the diel one. Within the tropical belt there is a potential escape from much environmental seasonality inevitable at high latitudes, most notably the thermal correlates of a marked radiation minimum. If this potential is realized, there are profound implications for population ecology. If other modes of seasonality intervene or – by default – cycles of other frequency, what mechanisms and biological response-patterns are operative? If tropical annual cycles of low amplitude are commonplace, do they tend to be more irregular in their phases? Where they are markedly so, the term 'seasonal' becomes unsuitable for intra-annual events. Given two or more interactive cycles of varying phase, are there general consequences of a match-mismatch flexibility as established for some temperate aquatic systems?

The sequence of topics is broadly from simpler to more complex, from the elements of physical–chemical–biological functioning to their expression in time-variability. The latter ends with responses in compound systems, which often integrate characteristics already discussed for individual components. Applied issues are not separated out as such, but man's influences are used to help the understanding of ecological systems. Possibilities arise, for example, with the creation of man-made lakes, with catch-mortality in fish populations, and with species introductions. These include undesirable or controversial impacts.

Our scope is latitudinally defined, referring to the region between the two Tropics at 23°27′ N and 23°27′ S (Fig 1.1). However, latitudinal gradients are not always dominant. East ⇆ West gradients of climate are numerous, often associated with a transverse mountain range (e.g., Andes, Ghats of India) and a transition from oceanic to continental conditions. Altitudinal gradients are also important, with correlations of temperature and rainfall.

The extensive range of the subject matter and its geographical derivation makes a comprehensive treatment impracticable. Nevertheless, we hope that there is sufficient breadth and detail for reference and contemplation. Also, that there is some encouragement from evidence of unity among diversity and from the comparative approach. The bibliography is intended to make a scattered tropical literature more accessible. Our approach will, we hope, complement existing regional and descriptive accounts, and help the emergence of a more generalized and functional tropical science.

1.2 Historical background

The scientific exploration of tropical inland waters (tropical limnology) is largely a product of the twentieth century, especially the second half. Aspects of its early history have been surveyed by Talling (1995*a*) and recent developments by Melack (1996). Progress has tended to lag behind that of temperate limnology, which by 1900 was emerging as a focused science from the efforts of pioneers such as Forel, Apstein, Birge and Whipple. Most early work in the tropics, before 1925, was based upon short-term expeditions with largely taxonomic, faunistic and floristic aims that did not greatly advance the present topic. Notable exceptions were the systematic seasonal collections of plankton from Lake Nyasa (Malawi) in 1899 by the Fülleborn expedition (Fülleborn 1900), and from lakes in Ceylon (Sri Lanka) in 1904–5 by Bogert, that enabled Schmidle (1902) and Apstein (1907, 1910), respectively, to make the first significant contributions to the seasonality of tropical freshwater plankton.

However, another 20 years had to pass before physical, chemical and biological studies were combined and interrelated, most notably in expeditions to lakes in East Africa (Worthington 1930; Beadle 1932*b*; Jenkin 1936) and Indonesia (Ruttner 1931*b*, 1952). These and earlier expeditions are shown chronologically in Fig. 1.2. The recording and analysis of dynamic change advanced even more slowly; the main exceptions were in the hydrology – the science of water fluxes – of important river systems such as the Nile (e.g., Hurst & Philips 1938) and in intensive studies of day–night or diel change that began in 1927 on Lake Victoria (Worthington 1931). Rate-measurements of physiological quantities were represented in the late 1920s and early 1930s, as in the experimental exposures made *in situ* by Jenkin (1936) and Beadle (1932*b*) on the photosynthesis of phytoplankton and aquatic macrophytes in East African lakes. Their interest mainly lay in the delimitation of the photosynthetic zone, and application of the method to assess production rates per unit area had to wait until the early 1950s in Central America (Deevey 1955) and Africa (Prowse & Talling 1958). The 1960s saw the beginnings of intensive rate-measurements in studies of animal production other than of fishes, such as those of the zooplankton of the African lakes Chad (Gras & Saint-Jean 1969) and George (Burgis 1971, 1974).

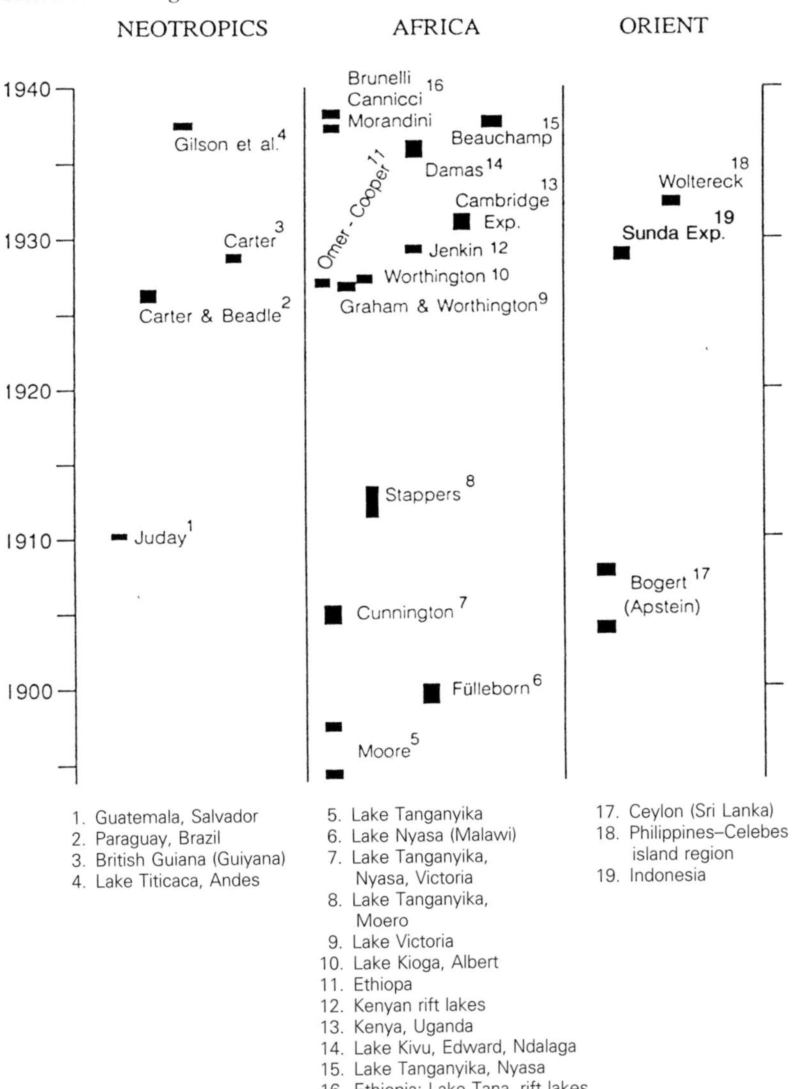

Fig. 1.2. Expeditions of ecological relevance for tropical limnology, arranged chronologically for the period 1894–1940. Numbers identify the locations or lakes involved. From Talling (1995*a*).

In this and later decades several world-wide developments favoured an expansion of work on time-related and dynamic aspects. Some stemmed from development of the subjects themselves, with exploration of the variety of time-sequences, and recognition of the desirability of analysis in terms of individual fluxes and rate-parameters. The expansion (mainly at higher latitudes) of the subjects of population dynamics, production biology and biogeochemistry had obvious influence on tropical as well as general limnology. The same is true of advances in analytical chemistry; spectrophotometry, for example, was first applied widely for nutrient estimation in tropical waters by Ida Talling in 1960–61. Influence of a quite different sort came from the increasing number of bases for research on tropical freshwaters, in universities and institutes. Long-continued recording in time thereby became easier to carry out, as did more focused projects once a background of environmental and biological information was available. 'Expeditions' of individuals or groups could often be assimilated within indigenous institutions, to the potential benefit of both and with the encouragement of time-related studies. A distinctive form of expedition was possible on the Amazon river system, up which a well-equipped oceanographic research vessel – the *Alpha Helix* – first penetrated for thousands of kilometres on several cruises in the 1970s. Earlier, in the 1950s, long research cruises up the Nile in a smaller vessel were mounted from the University of Khartoum. These led to the first monograph on the ecology of a large tropical river (Rzóska 1976).

With many subjects represented, progress was inevitably uneven. Work on the biology and ecology of fish populations has long been maintained and indirectly has supported other research efforts, national and international. Here projects on lakes Victoria, Kariba, Malawi, Turkana and Tanganyika are examples. Tropical swamps have attracted attention and in recent decades research on floodplain systems has intensified. Between 1965 and 1975 there was added stimulus from the International Biological Programme. A practical challenge, and research stimulus, followed the creation since 1960 of very large man-made lakes. Although the ecology of small streams and their invertebrate faunas has been relatively neglected, that of large rivers has not – a reversal of the situation in many temperate regions. Further, the chemical fluxes in large tropical rivers are of considerable global and theoretical interest.

Since 1980 the comparative and generalized aspects of tropical limnology have been developed considerably, as in the works of Serruya & Pollingher (1983), Payne (1986) and Lewis (1987, 1995). This development requires a sufficient background of site-specific or region-specific studies in depth, but also the interplay of reasoning from general to specific with that from specific to general. Here formulations made *a priori*, but not *ex cathedra*, have their place.

1.3 Environmental conditions influenced by tropical latitude

A tropical location makes its impact upon environmental conditions in freshwaters along three main pathways.

Solar radiation input is a primary influence, and has important consequences for the temperature regime (Chapters 2, 4). Its distinctiveness is linked to a high solar elevation at noon, with the twin consequences of normal (perpendicular) or near-normal ray incidence to the earth's surface (the 'geometric factor' of Monteith 1972) and reduced optical air mass (= 1 at normal incidence). The natural photoperiod, daylength, is precisely predictable (tabulations in List 1951). Its seasonal range is very limited: minimum and maximum values, at the northern and southern boundaries of the tropics, are 10.6 and 13.6 hours.

Geostrophic influence from the earth's rotation, as expressed in the concept of the Coriolis force, is minimal. This affects the motion of large moving masses of fluids whether in the atmosphere, oceans or lakes. For the last, a principal predictable effect is for its reduction to allow greater effectiveness of wind-induced vertical mixing and so enhance the depth of an upper mixed layer (Lewis 1987, 1995).

Air-mass circulation is influenced by latitudinal belts of pressure differentials, that include higher pressure regions in the high-insolation areas of the subtropics and a seasonally migrating equatorial trough and inter-tropical convergence zone (ITCZ) of lower pressure and upwelling. There results not only a seasonality in tropical *wind patterns* but also a broad tropical belt of generally elevated and often very seasonal *rainfall*. This leads to much hydrological control of tropical seasonality (Chapter 4.3), with correlated factors of water level, depth and discharge.

In addition, there are more loosely associated features connected with the *chemical denudation* and water-leaching of tropical land-masses. These can be accentuated by a long history of past denudation, due to a predominance of Pre-Cambrian formations embodied in the ancient proto-continent of Gondwana that by disintegration and migration gave rise – outside South East Asia – to most of the tropical land-masses of today. There is also some influence of the modern tropical environment on chemical pathways (e.g., silicate weathering, soil laterization, denitrification) and their chemical species (e.g., silicic acid, nitrate) that in varying concentration enter freshwaters.

These large-scale associations with latitude underlie many details of environmental transfers in tropical waters, now to be surveyed.

2

Environmental transfers in space and time

We will now outline fluxes and exchanges that determine energy balance, water balance and movement, and chemical balance in tropical inland water environments. For each of these topics one must take into account multiple *forms* of transfer, *routes* of transfer, controlling *factors* and the sensitivity of an aquatic *stock quantity* to change. Although basic principles are of general application, features prominent in tropical regions are emphasized and examples drawn from better known tropical situations.

The ultimate control of the processes involved can be visualized as falling into three categories – climatic, geological and biological. These appear in diagrammatic form, with interactions and overlaps, in Fig. 2.1.

2.1 Energy balance

Setting aside the kinetic energy associated with internal water movement (Section 2.3), a water-body acquires and loses energy convertible to heat (= molecular motion) by various forms and pathways. The main route is through the atmosphere–water interface. This pathway has climate-sensitive contributions that comprise: (i) several fluxes of radiant energy, (ii) energy loss by the evaporative water flux; and (iii) exchange of heat that can be directly sensed ('sensible heat') by convection–conduction down temperature gradients.

All these components can be assessed by the energy flux density per unit area, Q (units, e.g., J m^{-2} s^{-1}), at the water surface. They are shown diagrammatically in Fig. 2.2. The largest are usually the radiation fluxes, whose sum is Q_r. Downwards there is a short-wave (largely 0.3–2 μm) and in part visible flux of solar radiation (Q_s) and a long-wave (~10 μm) non-visible flux emitted by the atmosphere (Q_{li}). Upwards there is another flux of long-wave radiation (Q_{lo}) emitted from the water surface.

The downwelling fluxes Q_s and Q_{li} suffer some fractional surface loss, as *albedo* fractions a_s and a_l due to reflection and back-scattering, on entering the water-mass. Their penetrating flux densities $Q_s{}'$ and $Q_{li}{}'$ are hence $(1 - a_s)Q_s$ and $(1 - a_l)Q_{li}$, respectively. The evaporative energy flux density Q_e is governed by rate of evaporation (Section 2.2) and the large (and slightly temperature-dependent) latent heat of vaporization for water, ~2440 J g^{-1}. An evaporation rate of 1 mm day^{-1} has an equivalence of ~2.44 MJ m^{-2} day^{-1}. The convective–conductive energy flux density Q_c can involve flow from atmosphere to water (positive flux) or water to atmosphere (negative flux) according to the temperature gradient.

A second route is by the heat content of water horizontally transferred (i.e. *advected*) by inflow and outflow, to which loss of the heat content of evaporated water – a generally insignificant term – must formally be added. A third route, quantitatively of minor importance, is by heat transfer from water to sediments and its reverse. The sediment to water heat flux includes persistent geothermal flux from the hot interior of the earth, of particular interest in very deep rift lakes. In lakes Tanganyika and Malawi this estimated flux is very small (von Herzen & Vacquier 1967), although in Tanganyika hot vents are now known to exist (Tiercelin *et al.* 1993). In Lake Kivu (within a volcanic region) it is relatively large and accentuated by submerged hot springs, with a total flux calculated to be of the order of 1 J m^{-2} s^{-1} (Newman 1976).

Clearly the first pathway at the water surface is the primary one of comparative importance, and if other terms can be neglected the relationship between its fluxes and changes of heat content ΔH_w stored in the water-body in time interval Δt is:

$$\Delta H_w/\Delta t = Q_s' + Q_{li}' - Q_{lo} - Q_e - Q_c = Q_r - Q_e - Q \qquad (2.1)$$

Here all terms are referred to unit area. This energy budget provides a means of understanding and analysing the changes with time of heat content, and so temperature, of tropical water-bodies.

The water-emissive flux density Q_{lo} is readily estimated from surface water absolute temperature T in degrees Kelvin (K = °C + 273) by the Stefan–Boltzmann relationship:

$$Q_{lo} = E\,\sigma T^4 \qquad (2.2)$$

where the water emissivity $E \simeq 0.97$ (for a perfect 'black body' it is 1) and the Stefan–Boltzmann constant $\sigma = 5.67 \times 10^{-8}$ J m^{-2} s^{-1} K^{-4}.

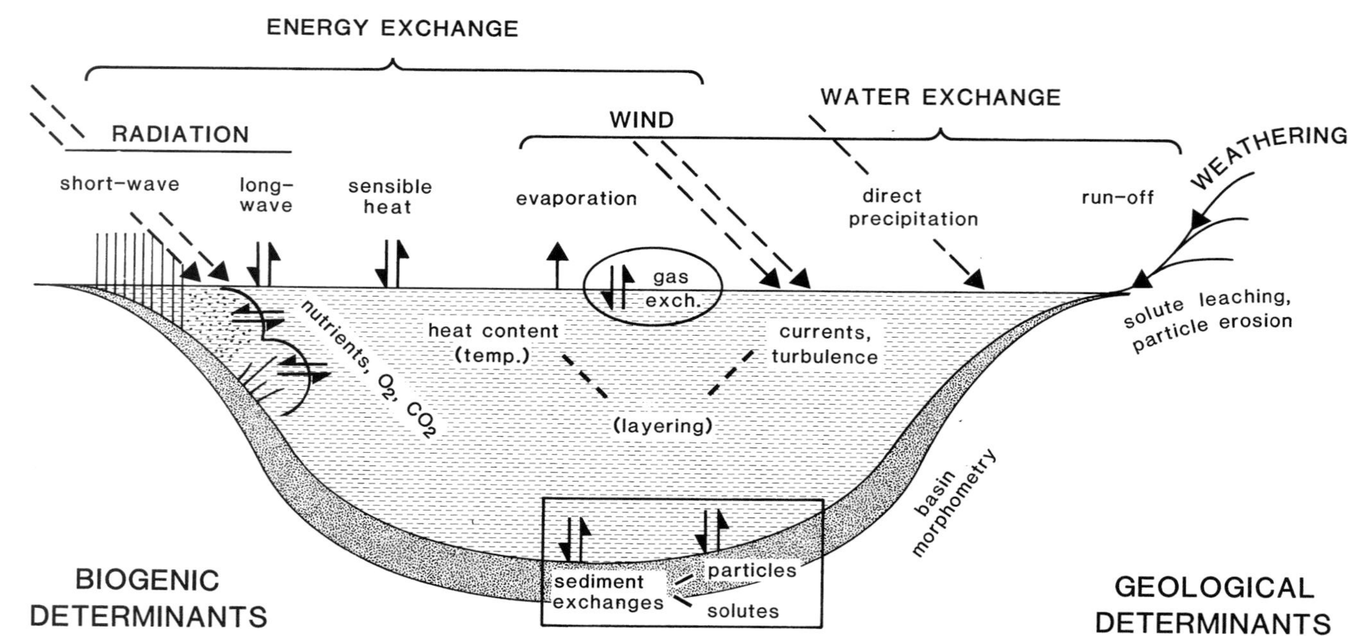

Fig. 2.1 Diagrammatic representation of components involved in the control of a lake environment by climatic–atmospheric, geological and biogenic factors. From Talling (1992).

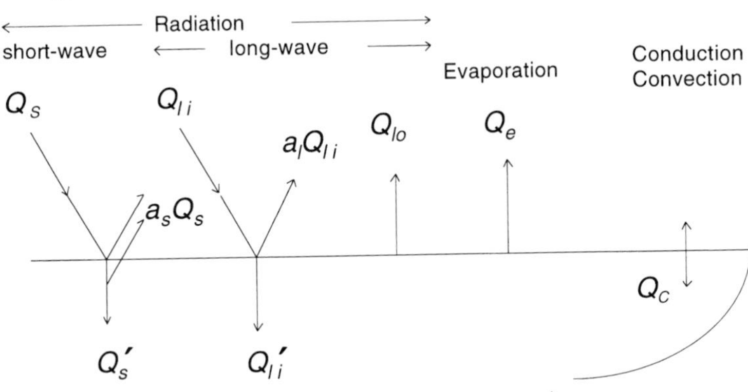

Fig. 2.2. Diagrammatic summary of energy fluxes operating at the air–water interface. For explanation of symbols see text.

This radiant flux is an important though invisible component of the diel energy balance. For example, at a water surface temperature of 15 °C, $Q_{lo} = 32.7$ MJ m^{-2} day^{-1}, whereas at 25 °C it has risen to 37.5 MJ m^{-2} day^{-1}. The difference, of ~5 MJ m^{-2} day $^{-1}$, could exist between typical warm-temperate and tropical lakes and its magnitude is about one-quarter of the incoming solar energy Q_s. In practice the two terms for long-wave radiation are often combined as a resultant quantity *net back radiation* ($Q_{lb} = Q_{lo} - Q_{li}$) that is estimated from an empirical meteorological relationship. This includes cloud cover, which enhances the incoming long-wave radiation Q_{li} and hence reduces net back radiation. Conversely, net back radiation will be considerable at night under a clear sky in arid climates – as on the Jebel Aulia reservoir near Khartoum (Talling 1990). Here the large daytime input of solar radiation Q_s, and output flux of long-wave radiation Q_{lo} both day and night, favoured a strong diel variation of near-surface heat storage and hence of surface water temperature. The term for convective–conductive exchange of sensible heat (Q_c) is usually estimated from the ratio with latent heat transfer, the Bowen ratio Q_c/Q_e, that can be expressed using the meteorological variables of vapour pressure deficit and temperature difference between water and air. These variables, respectively, increase Q_e and Q_c, as does wind velocity. Influence from the atmosphere also appears in the effects of cloud cover and atmospheric water vapour content for reducing the fluxes of incoming short-wave solar radiation and – as already noted – net back radiation.

Surface water temperature is a controlling variable with multiple effects, in its positive relationships with outgoing long-wave radiation

(Q_{lo}), net back radiation (Q_{lb}), conductive–convective loss of sensible heat (Q_c), and evaporative heat flux (Q_e). It is also correlated with heat storage, such that changes in storage (ΔH_w) broadly compensate for divergence between the net radiative flux Q_r ($Q_s' + Q_{li}' - Q_{lo}$, or $Q_s' - Q_{lb}$) and the summed non-radiative output fluxes ($Q_e + Q_c$). Further modification will, of course, arise if the net heat transfer in water inflow minus outflow is appreciable.

Probably the oldest measure of 'heat budget' is the change of heat storage per unit lake area between the annual minimum and maximum values (example in Fig. 4.16), the annual heat income. It is determined by the annual amplitude of surface temperature, the depth of a surface mixed layer, and the depth-distribution of water volume. Of these the first is much reduced in tropical lakes, whereas the second tends to be increased. The net effect, however, is a lower magnitude of annual heat income in deep tropical compared to deep temperate lakes. Typical ranges are 5–10 kcal cm^{-2} in the former, as for lakes Victoria (Talling 1966, 1990), Lanao (Lewis 1973) and Valencia (Lewis 1983*a*), and >25 kcal cm^{-2} in the latter. Although this quantity is formally a net flux over the period (generally ~6 months) concerned, for analysis it is less useful than component energy fluxes resolved for short periods within the diel (24 h) and annual time-scales. Measurements or indirect estimates of these, assembled within an overall energy budget, are not numerous for tropical water-bodies.

The diel energy budget offers an approach by instantaneous rather than time-averaged fluxes. Outlines for two waters (Lake Chad, Jebel Aulia reservoir) in the African Sahel region are given by Talling (1990) and for the shallow reservoir of Parakrama Samudra in Sri Lanka by Dobesch (1983). More rigorous is the study by Pouyaud (1986, 1987*b*), based mainly upon the Bam reservoir in Burkina Faso (Upper Volta) and that of Sene *et al.* (1991) on Lake Toba in Indonesia (see Fig. 2.3). However, both of these were centred upon evaporation–radiation relationships and did not directly evaluate heat storage. This was done by Talling (1990) for successive diel periods at the Jebel Aulia reservoir, where its magnitude was shown to be compatible with estimates of other energy flux densities.

The most variable and hence most evocative component in the diel energy budget is the solar radiant flux density, but Pouyaud (1987*b*) showed the diel variation of the evaporative term could also be considerable (>100 J cm^{-2} h^{-1}) in a dry climate. Balance of the budget in cycles without longer-term carry over implies an element of negative feedback

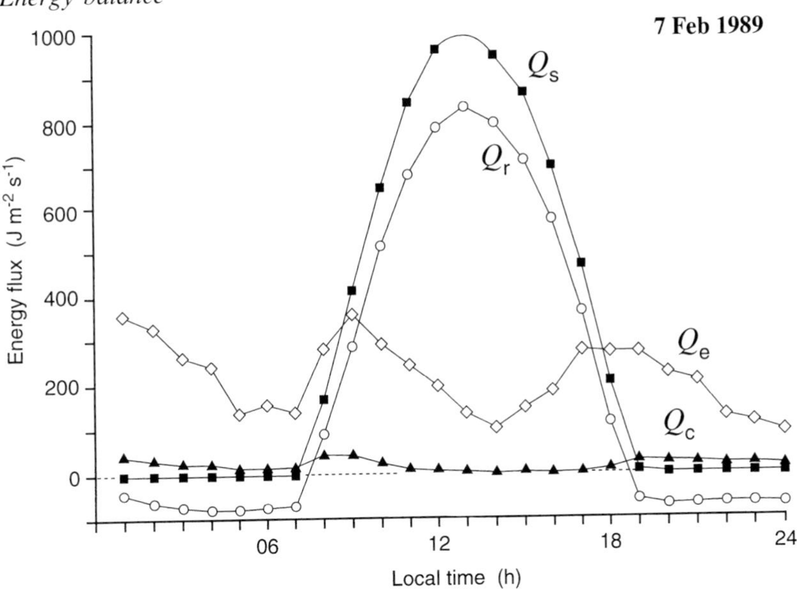

Fig. 2.3. Diel variation of component flux densities in the energy balance of Lake Toba, Indonesia. Components are: Q_s, global short-wave solar radiation; Q_r net downwelling radiation; Q_e, latent heat of evaporation; Q_c, sensible heat loss by conduction/convection. From Sene *et al.* (1991).

that can only be exerted by a flux density sensitive to surface water temperature. The most obvious candidate, 'sensible' or conductive heat transfer between water and atmosphere, would generally seem to be insufficient; thus the other temperature-sensitive fluxes of evaporation (via water vapour pressure deficit) and upward long-wave radiation (proportional to the fourth power of absolute surface water temperature) are also implicated. Examples discussed by Talling (1990) suggest that, excluding horizontal transfers (advection), the *amplitude* of the diel change of energy content stored as heat below unit area of the water-column is 0.5 to 1.0 times the total daily input of solar radiation. Thus a representative value of 20 MJ m^{-2} (2 kJ cm^{-2} or 480 cal cm^{-2}) for the latter might be expected to lead to a transient diel energy storage equivalent to a rise in temperature of 1.2 to 2.4 °C over a depth range of 0–2 m. An actual example of storage of this magnitude was described for the reservoir Parakrama Samudra in Sri Lanka by Bauer (1983), who also evaluated the associated storage (as induced buoyancy) of potential energy supposedly available for later conversion to kinetic energy in water currents. However, a lower amplitude of the diel temperature wave can be expected under windier conditions with enhanced vertical mixing, and also under conditions of clouding (lower solar radiation flux density, lower flux density of net long-wave back radiation) and high atmospheric humidity (low amplitude of evaporative flux density), conditions more prevalent in the humid tropics or in rainy seasons. Examples from a reservoir in Kenya appear in Fig. 2.4.

Representation of the within-year or annual variability of energy fluxes is usually based upon day-average values. It is a valuable means of analysing the annual variability of heat content, and hence of associated temperature and stability of thermal stratification. The study of Pouyaud (1986, 1987*a, b*) on a small West African water-body was

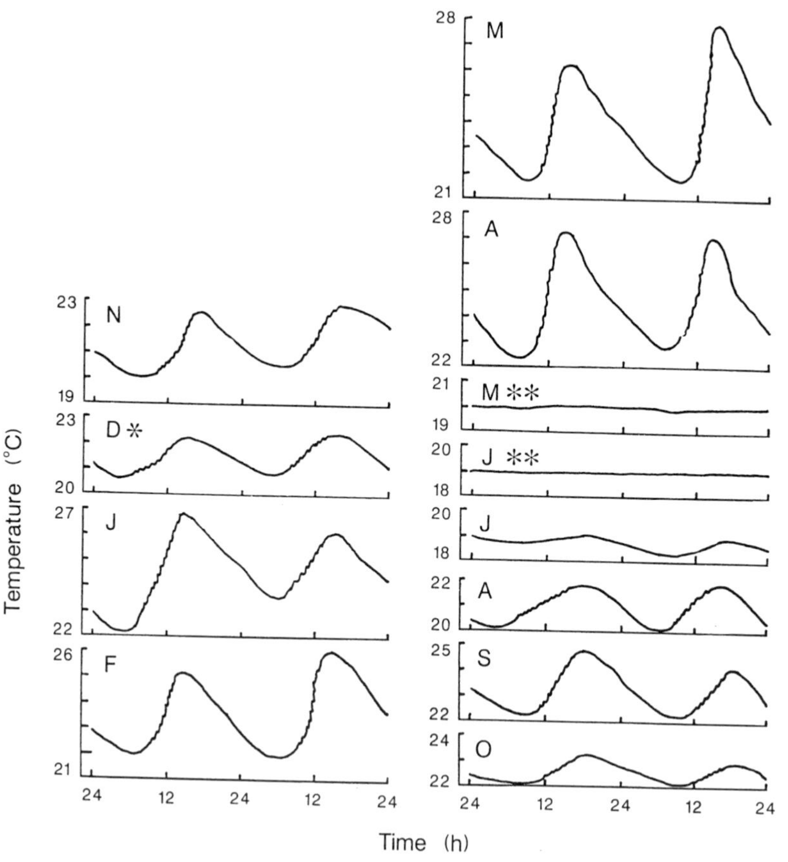

Fig. 2.4. Seasonal shifts in diel cycles of temperature, measured as continuous records at 0.6 m depth in a reservoir (Wellcome Dam) near Nairobi, in or near the last two days of the months indicated during 1971–2 * = short rains, ** = long rains. Modified from Young (1975).

again outstanding for technical instrumentation and direct flux measurement. Here conditions in the dry and wet seasons were compared, incorporating diel differences. The cloudy and more humid wet season yielded lower values for incoming solar short-wave radiation, net outgoing long-wave radiation and latent heat transfer. For this water-body there was the great asset of an independent measure of evaporation rate derived from inflow–outflow–storage relationships. Its annual variation was bimodal, with depressions in the rainy season (especially August–September) and cool dry season (especially December–February), corresponding to the location at 13.5° N.

For continuous information over several years on large tropical lakes, but based on partly indirect estimates of flux densities, there is notable work on Lake Pawlo, Ethiopia (Wood *et al.* 1976), Lake Valencia (Lewis 1983*a*), Lake Titicaca (Kittel & Richerson 1978; Carmouze, Aquize *et al.* 1983; Taylor & Aquize 1984; Carmouze 1992) and two reservoirs in northern Australia (Townsend *et al.* 1997). At least the first three lakes cool and mix deeply in the hemispheric 'winter' but the temperature of the high altitude Lake Titicaca is relatively low. Estimates of the main flux densities for Lake Pawlo and Lake Valencia (Fig. 2.5) suggest that variation in evaporative loss is the largest single contribution to the sharp annual cooling, which chiefly occurs outside the main rainy season and with low humidity plus high vapour pressure deficit. These same conditions, and little cloud cover, also promote increased net long-wave radiative transfer (back radiation). In all examples one obvious link between water and atmosphere – the exchange of sensible heat by conduction–convection – is not a major factor.

At the high altitude Lake Titicaca some special features occur, related to conditions of temperature and atmospheric pressure. According to Carmouze (1992), because of cold winds from adjacent mountains the surface water is warmer than the air above by a mean daily difference of 3.5–5 °C. This difference raises the evaporative flux, also enhanced by low humidity, but its magnitude appears unexceptional for tropical lakes at 4.2–5.3 mm day^{-1}. The low atmospheric pressure reduces the Bowen ratio of conductive–convective to evaporative loss, Q_c/Q_e. The temperature conditions tend to increase the net long-wave radiative transfer or back radiation, for although the outgoing flux Q_{lo} is reduced by low water temperature, the incoming flux from the atmosphere Q_{li} is still further reduced. A combination of increased back radiation and reduction of solar radiation around the winter solstice in June is estimated (Kittel & Richerson 1978; Carmouze 1992) to account for most of the

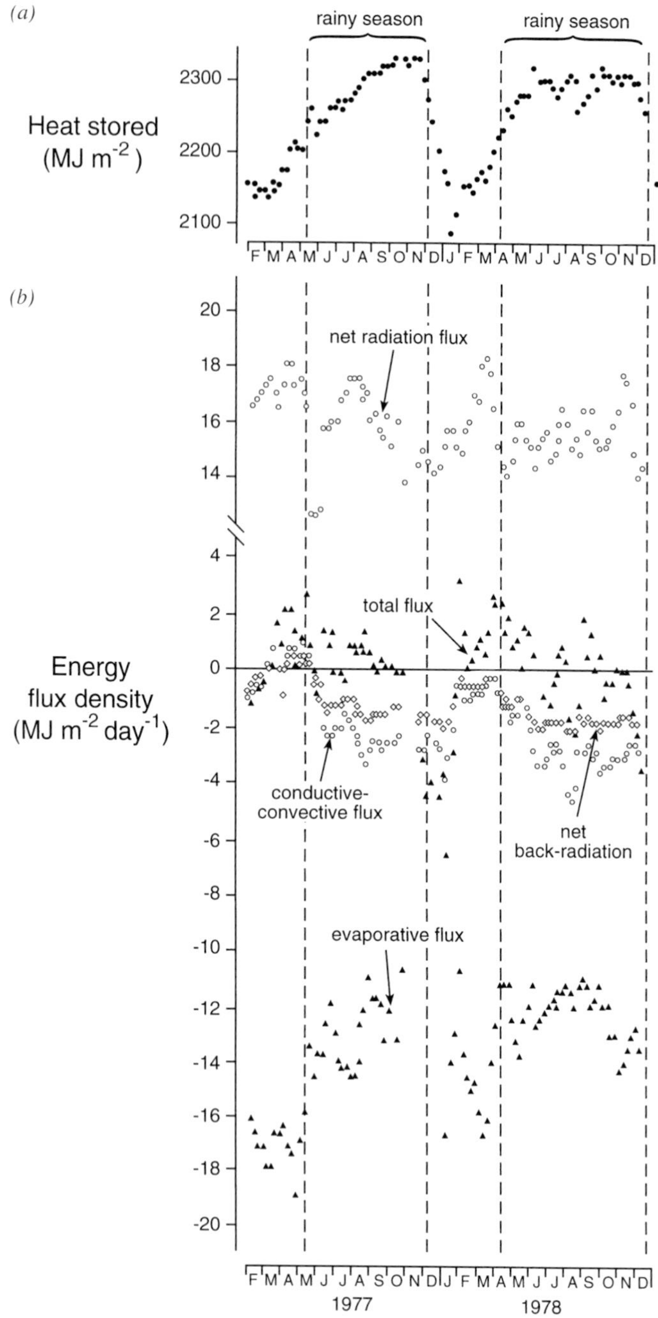

Fig. 2.5. Lake Valencia, Venezuela. Annual variation in (*a*) the mean heat content (above 0 °C) below unit surface area, in relation to (*b*) the estimated major flux densities at the water surface. Modified from Lewis (1983*a*).

increased seasonal loss of heat content in the lake. Long-term variability of some estimated flux components of the energy balance is shown in Fig. 2.6; rate of change of stored heat has been described by Kittel & Richerson (1978).

More generally, the solar radiation factor is likely to be dominant for seasonal cooling towards the limits of the tropics, but at lower latitudes the sensitivity of the energy fluxes associated with evaporation and back radiation can dominate in climates with wet and dry seasons.

2.2 Water balance

Like energy balance, the water balance of a landscape is determined by multiple fluxes that have a variety of pathways. Fluxes in the vapour state involve transport in atmospheric turbulence and circulation; those in the

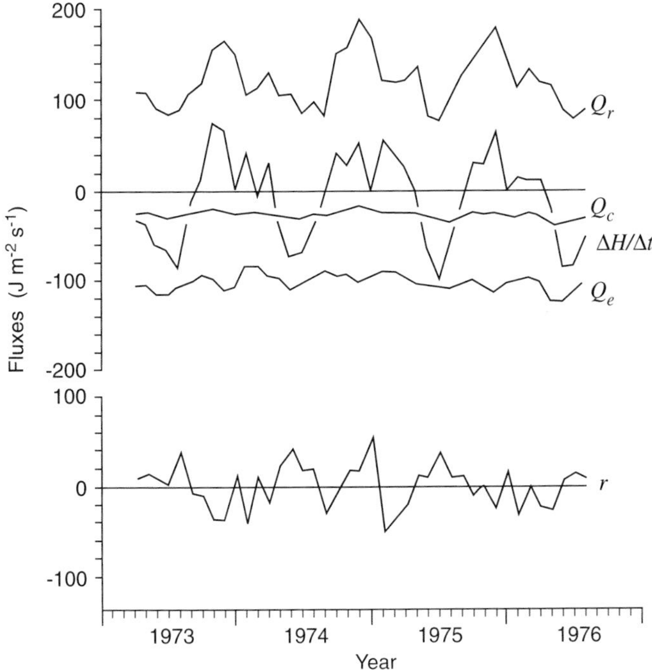

Fig. 2.6. Lake Titicaca, Andes. Seasonal and longer-term variation in four main components of the energy balance as monthly mean flux densities: Q_r net radiation; Q_c sensible heat by conduction/convection; $\Delta H/\Delta t$, stored heat; Q_c latent heat of evaporation. Separately r, the residual quantity required for energy balance (encompassing errors in the above), is also shown. From Taylor & Aquize (1984).

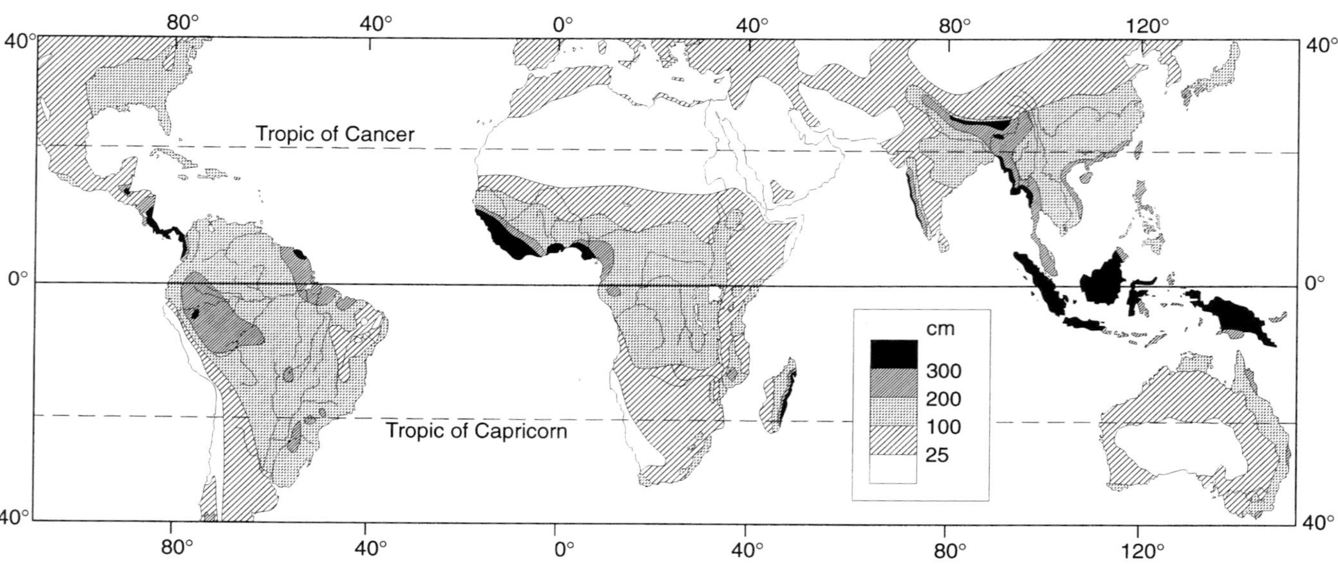

Fig. 2.7. Distribution of mean annual rainfall (cm) in land-masses of the tropics and subtropics.

liquid state are dominated by gravitational fall, in part as surface-bound running waters. In either case height is a crucial factor. As water level is an index of stock, flows of water are often expressed as height or level differences over the surface area under consideration, usually of the catchment or lake. Thus a flux of 1000 m^3 (= 10^9 g) km^{-2} day^{-1} is equivalent to 1 mm day^{-1}.

Because of the proximity of the equatorial trough of low pressure and inter-tropical convergence zone (ITCZ) (see Chapter 4.3b), the equatorial belt is largely a region of high annual rainfall (Fig. 2.7). There are principal subcentres in Amazonia, West Africa and Indonesia that are marked by a stronger prevalence of rising air-masses (Dhonneur 1985). At surface-water level, relatively high tropical temperature raises the operative saturation vapour pressure (e_s) and hence the maximum possible vapour pressure deficit, from saturation, in dry air. The actual deficit, $e_s - e$, also determined by the prevailing partial pressure of atmospheric water vapour e, is a principal factor controlling evaporation from open water surfaces. Estimation of the flux is possible given other meteorological information, including net radiation flux and wind velocity. An equation derived by Penman, and variants summarized by Monteith (1973) and Sene *et al.* (1991), are often used. Applications to tropical lakes include Lake Volta (Penman 1956), Lake Pawlo in Ethiopia (Wood *et al.* 1976) and Lake Toba in Indonesia (Sene *et al.* 1991). Clearly, mean open-water evaporation rates are typically low in the humid tropics, and high – often >8 mm day^{-1} – in the arid tropics and subtropics.

On a catchment scale, evapotranspiration causes losses that reduce the specific discharge of a river system (i.e., discharge per unit drainage area: tabulation for tropical rivers in Serruya & Pollingher 1983) below the corresponding value for rainfall. The fractional reduction, or run-off factor, is <0.2 in many arid tropical catchments. For the humid and forested catchment of the Amazon it is estimated as ~0.6 (from data in Lewis *et al.* 1995). In the short term it is strongly influenced by variable storage in soils, and by alternative modes of surface and channel flow linked to the intensity of rainfall. Thus 'storm-flow' and more persistent 'base-flow' can be distinguished, often with different solute contents that are of ecological significance. This has been well demonstrated by Lesack (1993*a, b*) for a small sub-catchment of Lake Calado in the Amazon floodplain. There, in a particularly wet year, rainfall amounted to 2870 mm, stream-flow export to 1650 mm that included 88 mm storm-flow, subsurface seepage-out 42 mm and a year-to-year difference in soil water storage (soil recharge) of 57 mm. A residual of 1120 mm was ascribed to evapotranspiration.

For further expositions of catchment budget and river hydrology the reader is referred to texts of general hydrology and, for the tropics, to Balek (1977, 1983), Fritsch (1992) and Bonell *et al.* (1993). The following account is mainly directed to lakes and reservoirs; it considers water stocks with flux components and their interrelations chiefly at the annual scale. Time-variability on shorter and longer scales is taken up in Chapter 4.3 and 4.5, respectively.

The hydrology of lakes and reservoirs may be described by two main features: the *water budget* and the *residence time*. The relative importance of the components of the water budget mostly determines the nature of the processes involved in the chemical or ecological regulation of the water-body. In lakes and reservoirs the magnitude and impact of the resulting environmental variations is directly linked with the residence time, which may be expressed as the quotient of water storage/outflow flux (the inverse of the flushing rate) or, alternatively, as the quotient of water storage/total input flux (the inverse of the water renewal rate). The annual water budget of a water-body may be written as:

$$P + R_i + G_i = E + R_o + G_o + \Delta V \qquad (2.3)$$

where P = direct on-lake precipitation, E = evaporation, R_i and R_o = river input and output, G_i and G_o = seepage input and output, and ΔV = change in volume.

For coastal lagoons, two other variables should be taken into account: seawater input and output. All these quantities can be expressed as volumes of water or as the equivalent changes of lake surface level (usually in mm).

Illustrations of evaluated flux components appear in Fig. 2.8 where mean monthly values are represented for the African lakes Victoria and Chad. In Lake Victoria, direct rainfall is the main contribution to the inputs, with a marked seasonality. River inflow and outflow are of comparatively small volume and are relatively constant throughout the year. Evaporation accounts for most of the water losses, with a moderate seasonality directly related to the mean air humidity and wind regime. In Lake Chad, direct rainfall is restricted to three months duration and an annual total of about 300 mm. River inflow, of which 93% is contributed by the River Chari, is the most significant input. There is no surface outlet, but a small net annual seepage outflow. Most of the output results

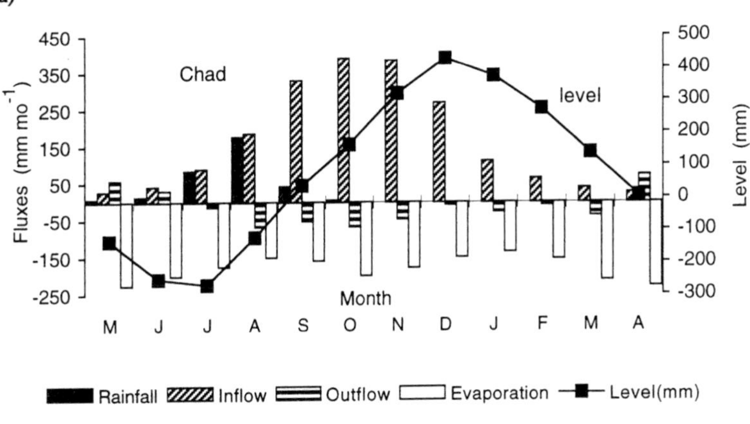

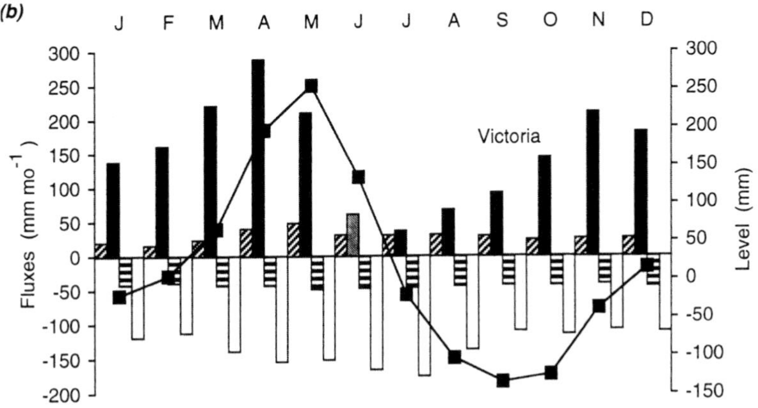

Fig. 2.8. Monthly flux components of the water budget and resulting water level in (*a*) Lake Chad (R_i and E predominating) and (*b*) Lake Victoria (P and E predominating). 'Outflow' is R_o for Lake Victoria and net seepage ($G_o - G_i$) for Lake Chad. Based on Sutcliffe (1988) and Olivry *et al.* (1996).

from evaporation, with some seasonality due to the rainy season (August) and the low air temperature during part of the dry season (January).

Lake Victoria is thus characterized by an atmospheric control, where direct on-lake precipitation P and evaporation E are the main contributors to the water budget. In the endorheic Lake Chad, the surface inflow R_i and E are the most influential components, with surface outflow $R_o = 0$ and a small net value of seepage-out G_o.

During the periods for which these budgets have been determined, both lakes have had a fairly constant mean level, with small seasonal variations resulting from imbalances in the monthly water budgets. Total annual input corresponded, respectively, to 2150 mm yr^{-1} over 68 800 km^2 for Lake Victoria and 2175 mm yr^{-1} over 20 000 km^2 for Lake Chad (Sutcliffe 1987; Olivry *et al.* 1996).

Although these figures may seem similar they involve important differences. Horizontally, a localized river input to Lake Chad results in a south–north movement of water-masses and a corresponding chemical gradient, whereas Lake Victoria appears more homogeneous as both main inputs and losses affect the whole lake area. Vertically, the inputs have to be compared with the mean depth (or volume) of the lake: 39 m in Lake Victoria and 3.5 m in Lake Chad. There results residence times as lake volume divided by water throughput ($V/[R_o + G_o]$) of, respectively, 70 and 20 years.

It is thus clear that the relative magnitude of the different components of the annual water budget has to be taken into account and the importance of the total annual fluxes relative to the lake volume. A normalized budget may be written where all the components are expressed as percentage fractions of the lake volume V:

$$(100/V)(P + R_i + G_i) = (100/V)(E + R_o + G_o + \Delta V) \qquad (2.4)$$

Using this presentation, several types of behaviour are described below (see Table 2.1).

In **large deep basins**, the residence time of water is typically several tens of years. Although in these lakes the level may vary seasonally or in the long term by several metres, they are strongly buffered by the large water volume and most of their functioning is independent of seasonal or other within-year fluctuations in the components of the water budget. An exception lies in the marginal communities, both above and below the surface, sensitive to level variations.

The water budget is often dominated by direct precipitation P and evaporation E (atmospheric-control lakes) but river input R_i and output R_o may also be of some relative importance. Normalized as percentages of lake volume, the annual surface input ($P + R_i$) and output ($E + R_o$) are $< 10\%$. Examples are lakes Victoria, Malawi and Tanganyika in Africa, Lake Titicaca in South America, and Lake Toba in Indonesia (see Chapter 4.3).

Compared to that of deep lakes, evaporation increases in relative importance in **shallow lakes**. This is the case for the shallow and wind-mixed Lake Chapala in Mexico, with a surface area of 1110 km^2 and a mean depth of 7.2 m (Table 2.1). The water level decreased by about 3 m from 1977 to 1983 as a result of the diversion of water from inflow rivers

Table 2.1. *Annual flux quantities in the water budgets of selected tropical and subtropical lakes, reservoirs and floodplains, normalized to storage volume and expressed as % of that volume. For symbols and sources of data, see text of Section 2.2 and Chapter 4.3 (parentheses indicate some irrigation off-take in R_o or pumped input in R_i)*

Water-body	Volume (km^3)	P	R_i	G_i	E	R_o	G_o	ΔV
Open Lakes								
Titicaca	900	0.83	0.95	0	1.54	0.15	0	0.09
Victoria	2700	4.6	0.9	0	4.1	1.4	0	0
Chapala 1953–74	8.0	28	72	0	49	51	0	0
Chapala 1975–84	4.7	38	62	0	58	42	0	0
George	0.8	31	250	0	59	218	0	0
Kyoga	7.6	72	374	0	91	353	0	0
Reservoirs								
Tucurui 1985–87	45	7.5	429	0	4.4	411	0	21
Floodplains								
Yaéré 1968	3.5	243	91	0	301	33	0	–
Okavango Delta	4	125	275	0	385	7.5	7.5	–
Closed lakes								
Turkana	245	<0.6	6.5	0	7.1	0	–	0
Valencia 1977–80	7	4.3	(4.5)	1.5	10.3	0	0	0
Sibaya 1977	0.98	8.1	1.3	2.1	11.2	0	0.4	0
Naivasha 1974	0.72	15.8	28.5	5.9	40.2	(2.0)	5.1	+3
Chad 1964–68	70	9.5	55.9	0	62.4	0	5.1	−2.1
Chilwa 1961–71	0.85	89	41	0	130	0	0	0
Lac de Guiers	0.28	21	174	4	157	(38)	4	0

for agriculture, although modifications on the outflowing Rio Santiago reduced annual output from the lake (Limón *et al.* 1989).

A large river outflow R_o strongly reduces the consequences for change in lake level from within-year (and between-year) variations in the other elements of the budget. This is the case for Lake George (Uganda) where the average flushing rate of 2.8 times per year is seasonally modulated by a well defined bimodal river inflow (maxima in April and October), but where the seasonal level variation is low (± 0.2 m) (Viner & Smith 1973). The transit Lake Kyoga on the Victoria Nile has a similar hydraulic regime (Burgis *et al.* 1987).

Reservoirs used for hydroelectric power are also often characterized by a large outflow to volume ratio. An average residence time of 45 days has been computed for the large Tucurui Reservoir in northern Brazil, with a mean volume of 45 km^3 and a length of 14.3 km. The corresponding large throughput is, however, not sufficient to compensate for oxygen (O_2) depletion in the deep layers as a result of decomposition of the original vegetation (Pereira, 1994). The amplitude of seasonal variation in level in the reservoir, up to 14 m, does not greatly differ from the 10 to 12 m of the natural river (Odinetz-Collart 1987).

Floodplains and temporary water-bodies

As a result of seasonal variations of river discharge, floodplains are highly dynamic transition zones (ecotones) through which aquatic and terrestrial systems exchange nutrients and organic material. Most of the great floodplains nowadays occur in the tropics, mainly as a result of human pressure which has converted those of temperate regions to agricultural land. Welcomme (1979) provides a general descriptive account of most large floodplains of the world. Large tropical examples occur on the river courses of the Amazon, Orinoco, Nile, Niger and Mekong. Associated lakes or lagoons may be numerous (Fig. 2.9).

Their hydraulic functioning may best be approached through their seasonal variations in area and volume (see Chapter 4.3b). However, their annual budget can give complementary information as shown by the comparison between two African floodplains, the Yaéré, a floodplain of the River Logone (Northern Cameroon), and the Okavango inner delta (Botswana) which receives the Kavango River. For both systems, the rain period occurs before the river flood and the first phase of the inundation is largely a result of direct rainfall. But the relative importance of the two inputs may be different. Published data for the Yaéré (Gac 1980) and the Okavango Delta (Mepham 1987) indicate a predominance of direct rain in the Yaéré and of river input in the Okavango, major losses through evaporation in both, and a very short residence time. The mean volume of the water-bodies has been calculated here as the annual mean of monthly values so that these figures are comparable to those of permanent lakes (Table 2.1).

The floodplain of the central Amazon is another region in which seasonal water storage is contributed both by local water run-off and river overflow. One floodplain lake, Lake Calado, was estimated by Lesack & Melack (1995) to export to the river about three times the volume of water it received from that source.

The relationship between annual river inflow (R_i) and outflow (R_o) will obviously depend on the degree and area of flooding. This has been

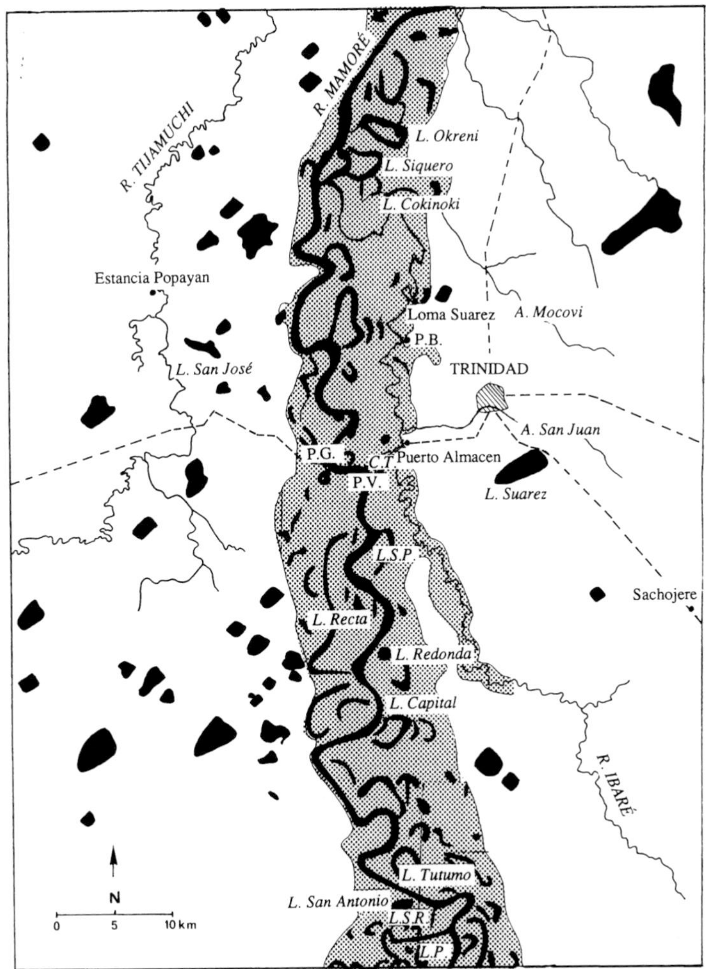

Fig. 2.9. Permanent water-bodies (black) near the Rio Mamoré, Bolivia, in relation to the floodplain (stippled). From Loubens *et al.* (1992).

treated comparatively by Sutcliffe & Parks (1989) for four African floodplains, including the Nile Sudd region for which the net transmission factor was <0.5 at higher annual flows (see Fig. 4.14). Here the transmission loss is an adverse factor for water economy in the more arid regions downstream (Hurst 1952), although the local seasonal inundation has some beneficial consequences for a cattle-tending population (Rzóska 1974).

Lakes without surface outflow

Lakes without surface outlet (natural $R_o = 0$) are usually associated with dry climates. Their budget is characterized by a relatively high evaporation component E. This, with subsurface seepage-out, balances water input in the long term. Due to human use of the water resource, especially for irrigation, some abstraction must be integrated in the budget. This occurs for Lake Naivasha, Kenya (Gaudet & Melack 1981: see Table 2.1), where the budget includes some seepage into the lake, an abstraction of 2% of the lake volume (here noted as a form of surface outflow R_o) and some variation of the level.

A slightly different type of budget applies to the shallower Lac de Guiers (Senegal), a side lake to the River Senegal (data from Gac *et al.* 1987) which is a man-regulated closed lake heavily used for rice and sugarcane irrigation.

Lake Chilwa (Malawi) is a shallow natural closed lake with a highly variable area as a result of climatic irregularities, although in a region with higher rainfall than the two preceding examples (890 mm yr^{-1}). Mean values for the period 1961–71, assuming a mean depth of 1 m (data from Lancaster 1979 and Mepham 1987), lead to the components shown in Table 2.1 with a strong dominance of atmospheric control.

A rather different budget may be calculated for the subtropical (27° S) Lake Sibaya (South Africa), in a region receiving about 1000 mm yr^{-1} annual rainfall. This deeper lake (mean depth ~10 m) has also relatively small river inputs. Its water level reflects climatic variations, with oscillations of over 4 m in a period of a few years (Mepham 1987). Based on mean climatic data, and for a relatively high level as observed in 1977, the normalized budget indicates a limited annual input with a predominance of direct rain. River input dominates in preliminary estimates (Yuretich & Cerling 1983: see Table 2.1) for the large deep endorheic Lake Turkana (mean depth 33 m), that has a history of salinization reflected in Fig. 2.32. In Lake Valencia, Venezuela, the dry-season inflow is sustained at a steady level by known quantities of pumped water. This allowed for a precise estimation of a significant annual groundwater flow, G_i (Lewis 1983*c*).

An extreme case of a closed system is found with Lake Magadi (Kenya), where the dominant inflow is from groundwater, G_i, which is balanced by evaporation E. Most of the lake volume is made of sodium carbonate-rich evaporites. Here the fate of the solute input in a closed lake is clear; this is not always the case for some other endorheic basins (see Section 2.4). In Ethiopia, a wide variation in lake hydrology – with successive lakes in areas of closed drainage – has led to an extensive salinity series (Wood & Talling 1988). Figure 2.33 shows the hydrological background.

2.3 Water movements

Water movements result from three main influences: height difference of surface level, with gravitational flow; density difference with buoyancy or sinking; and surface wind stress with transfer of momentum (mass × velocity) as well as energy.

Gravitational flow is most obviously expressed in river channels. Situations of high rainfall in large drainage basins account for several tropical rivers – including the Amazon, Zaïre (Congo), Ganges and Orinoco – yielding water fluxes or discharge in excess of 500 $km^3\ yr^{-1}$, with the Amazon far pre-eminent at $\sim 6000\ km^3\ yr^{-1}$ (tabulation in Serruya & Pollingher, 1983). The largely tropical Nile, though with a relatively small discharge of $< 100\ km^3\ yr^{-1}$, is the world's longest river at 6695 km. Velocities of water flow relate to altitudinal gradients, sometimes only a few metres fall per 1000 km, and in the absence of tributaries will inversely correlate with cross-sectional area as discharge equals this area times mean velocity. They are reduced under conditions of high hydraulic resistance, as occur in heavily vegetated or meandering channels. The upper limits are reached in waterfalls and rapids that have some spectacular tropical examples. Some, such as the Murchison Falls on the upper Nile and the Victoria Falls on the Zambezi, are notable as zoogeographical barriers and probably as plankton-destructors.

Velocity of flow partly determines the *time of travel* between two points of changes in water levels, that has received much attention from its practical importance in long rivers such as the Nile (Hurst 1952). It, and more especially local backwaters or 'dead zones' for water movement, are factors favouring the development of river plankton (e.g., Rzóska 1976). Higher velocity of flow promotes the suspension and transport of silt, that is often abundant in floodwater, and in deposition can slowly alter channels and landscapes. For African river basins the fluxes of sediment transfer have been estimated by Walling (1984) as extending over the wide range of about 1–4000 $t\ km^{-2}\ yr^{-1}$, although with only limited areas of $> 100\ t\ km^{-2}\ yr^{-1}$. Sediment fluxes are typically greater than the accompanying solute fluxes ($< 1\ t\ km^{-2}\ yr^{-1}$), as is illustrated by estimates of both the Niger and other West African rivers

(Enikeff, 1939; Grove, 1972). However, the reverse situation can be found in forested catchments (e.g., Zaïre). Sediment types can be specific to distant origins, as in those contributed from the Ethiopian highlands to sediments along the Blue Nile and Main Nile below. This system also illustrates how heavy loads are carried during a short seasonal phase of floodwater and, in deposition, have provided the framework for the shallow Delta lakes of Egypt, and the recent massive accumulations of sediment in the reservoirs of southern Lake Nubia (Entz 1976, 1978) and Sennar in the Sudan plain.

In so-called 'standing waters' of lakes and reservoirs, the input–output water flux rarely dominates internal water movements. Some of these are generated, and all are potentially constrained, by vertical differences of density. In most cases the vertical density differences are the consequences of temperature differences, with a non-linear relationship that causes density differences to be accentuated in the higher tropical range of temperature. A contribution to density from dissolved solutes becomes appreciable in waters of higher ionic content; the quantitative basis of this is examined by MacIntyre & Melack (1982) in relation to long-term changes of salinity and stratification in a small Kenyan lake. Some notion of the magnitudes involved is given by the following differences in density expressed in mg dm^{-3} (= mg l^{-1} = g m^{-3}):

(i) For pure water, differences between 5 and 6°, 15 and 16°, and 25 and 26 °C are, respectively, 24, 156 and 261.

(ii) For water of an ionic composition common in East Africa, the difference between waters with electrical conductivity at 20 °C of 100 and 500 μS cm^{-1} is 368.

The first quantitative analyses of stratification in tropical lakes were by Ruttner (1931*a, b,* 1938), who applied to Indonesian lakes the formulation of *stability* after Schmidt. Essentially a static quantity, this is a measure of the work (as force × distance) required to transform a density-stratified to a density-unstratified state without change of heat content. A modern discussion of its formulation and calculation is provided by Idso (1973). Ruttner showed that, because of the non-linear temperature–density relationship, values of stability (which he expressed in kg-m per m^2, or kg m^{-2}; energy units, J m^{-2}, are now preferred) could be as high or higher in tropical than in temperate lakes. However, tropical values are usually lower, as the effect of lessened vertical temperature difference generally predominates over that of the accentuated density–temperature relationship. Later notable applications to tropical lakes have been made by Lewis (1984) in an assessment of annual changes of stratification for Lake Valencia (Fig. 2.10), and Kling (1988) in a comparative survey of stratification and susceptibility to mixing for lakes of Cameroon (Fig. 2.11); also by Townsend 1998 for two reservoirs in northern Australia. Stability increases when a given thermal gradient, or thermocline, is depressed deeper. In general there is a strong relationship between stability and maximum lake depth (see Fig. 2.11), and between thermocline depth and the distance of wind travel ('fetch') over the water surface (Fig. 2.12).

Partly because change in temperature is the most prevalent cause of change in density, partly because temperature can function at depth as a persistent or 'conservative' property, time-series of vertical temperature

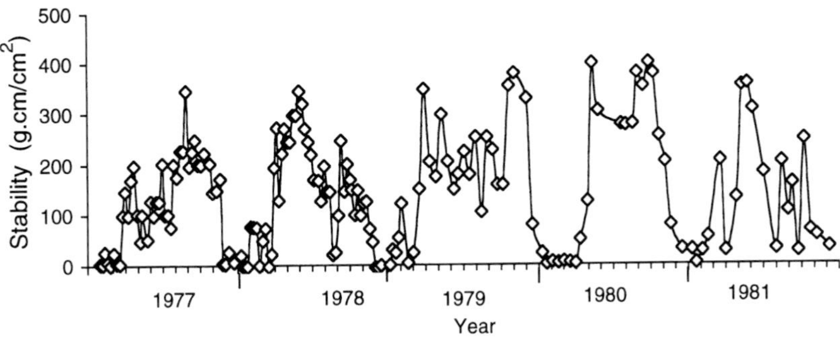

Fig. 2.10. Lake Valencia, Venezuela. Within-year and between-year variation of the stability of stratification, 1977–81. Modified from Lewis (1984).

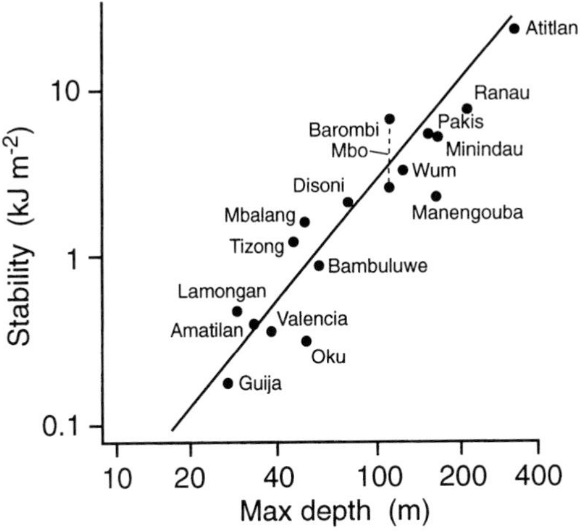

Fig. 2.11. Double-logarithmic plot of the relationship between stability of stratification and maximum depth for tropical lakes. From Kling (1988).

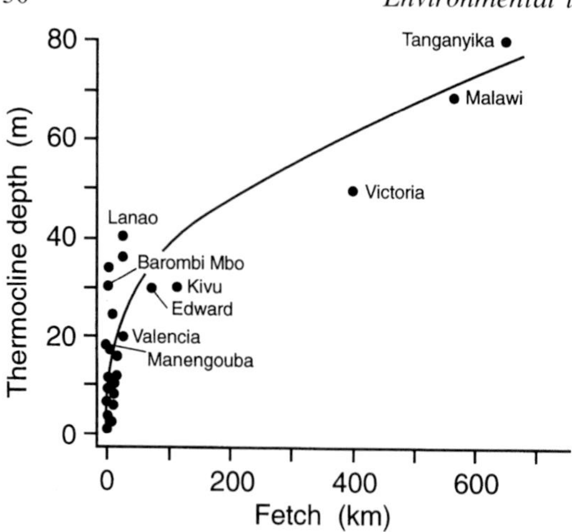

Fig. 2.12. Relationship of thermocline depth and maximum lake fetch for various tropical lakes, including a series of small crater lakes in Cameroon. From Kling (1988).

differences in a lake can give indirect indications of vertical water transport and exchange. Evaluation of the transport of heat downwards by water turbulence in a lake with time has been a classic approach, by defining a coefficient of vertical diffusivity K_z linking the vertical flux F per unit area at depth z (here F and z are deemed positive downwards) and the gradient $d\theta/dz$ of heat concentration (i.e., temperature, θ) over which it operates:

$$F = K_z\left(\frac{-d\theta}{dz}\right) \tag{2.5}$$

The coefficient of vertical (eddy) diffusivity (units $cm^2\ s^{-1}$ or $m^2\ s^{-1}$) so derived can be applied to estimates accompanying chemical fluxes along known concentration gradients, provided that the transfers of heat and chemical quantities are predominantly by water-eddy diffusion and not molecular diffusion (in which they differ). The magnitude of vertical diffusivity in lakes generally increases with lake surface area and wind fetch, without any clear bias for tropical as distinct from temperate lakes. This matter is discussed comparatively by Lewis (1982) from the background of seasonal studies on Lake Valencia (Lewis 1983*a*). Another application of vertical diffusivity coefficients, to assess reduced diffusivity during diel cycles of stratification in Lake Titicaca, was made by Powell *et al.* (1984).

Examples of temperature-time series in tropical lakes appear as depth-profiles in Fig. 2.13, and as depth-time 'contour' diagrams in Figs. 2.15,

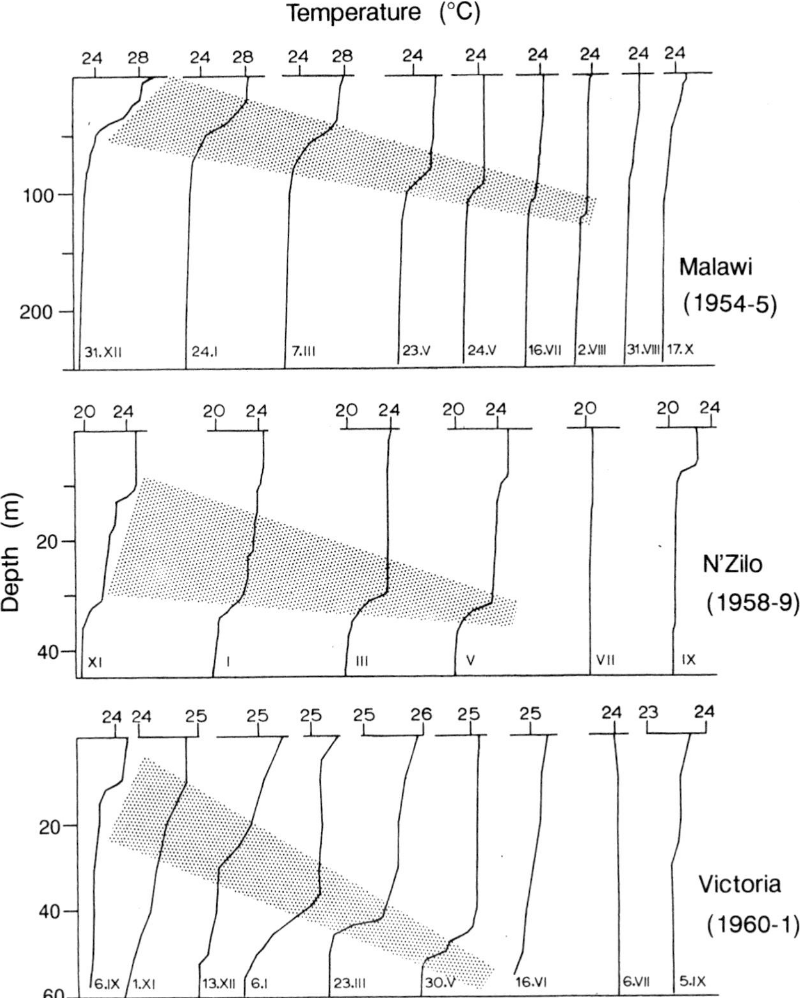

Fig. 2.13. Temperature-depth-profiles illustrating the annual cycle of thermal stratification in lakes Malawi, N'Zilo (Katanga) and Victoria. Stippling indicates the descent of the main thermal discontinuities. From Talling (1969).

4.7, 4.15 and 5.7 where rapid change of temperature in depth or time is marked by the bunching of lines of equal temperature (isotherms). The sequences shown comparatively in Fig. 2.13 demonstrate how, during phases of cooling, the extent of a more mixed surface layer deepens and so *entrains* water that was previously separated below a thermal discontinuity or *thermocline*. Such water is typically richer in plant nutrients than that of the surface layer. In consequence partial entrainment as well as complete vertical mixing is potentially a major influence in the biological functioning of tropical lakes. If entrainment alternates with periods of surface heating, multiple thermoclines of varying duration may emerge, as described for Lake Lanao by Lewis (1973) (see Fig. 2.14), and Lake Pawlo by Wood *et al.* (1976, 1984). Frequent repetition of entrainment, and its vertically deep operation, have been emphasized by Lewis (1987, 1995) as a characteristic of tropical lakes that would promote nutrient recycling or 'turnover'. Although well developed in some tropical climates, as at Lake Lanao in the Philippines, the supposed contrast with temperate lakes is less marked for lakes near the limits of the tropics (e.g., Lake Kariba) or with temperate lakes in relatively unstable oceanic climates.

Vertical water transfers, and mixing, tend for several reasons to be promoted following a surface loss of heat. One mechanism is *penetrative convection*. This is frequently marked by unstable inverse near-surface differences of temperature with cooler and denser water uppermost, which are most evident near the end of night-time cooling. It is relatively more important in the absence of wind stress and consequent *forced convection*, as in sheltered forest-lakes of the Amazon floodplain (MacIntyre & Melack 1988, 1995). Much day-to-day change in the depth of a thermocline can then follow (Fig. 4.8). It is also probably often responsible for a seasonal correlation between vertical mixing in shallow water-bodies and the marked depression of daily minimum – but not maximum – air temperature. Examples of this relation include Opi Lake in Nigeria under the harmattan wind regime (Fig. 2.15). An

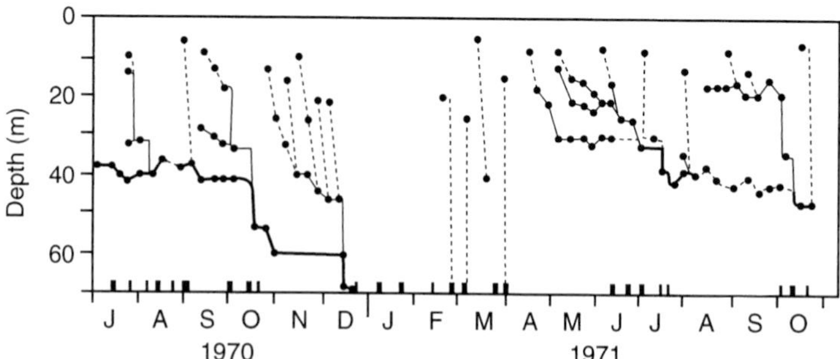

Fig. 2.14. Lake Lanao, Philippines. Within-year changes of thermal stratification, as indicated by the depth locations of thermal discontinuities or thermoclines of varying magnitude and persistence and their attributed origins from breezes (•-----•), squalls (•—•) and storms (•━•). Bars indicate storm periods. Based on Lewis (1973).

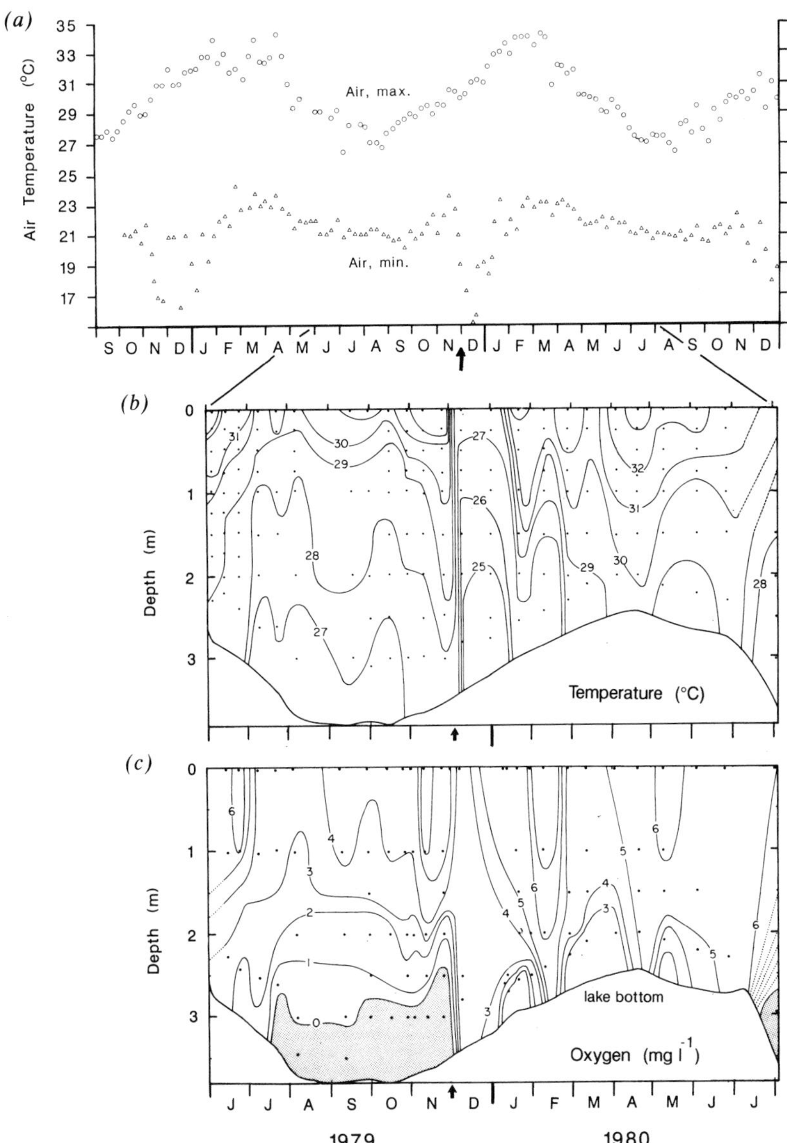

Fig. 2.15. Opi Lake, Nigeria. (*a*) The annual incidence of low minimum air temperature associated with the onset (arrowed) of the harmattan wind regime, and associated effects on the seasonal depth-time distribution of (*b*) temperature and (*c*) oxygen concentration. From Hare & Carter (1984).

importance of night-time surface cooling with resulting convection for the penetration of dissolved oxygen under tropical swamp cover was early postulated by Carter & Beadle (1930), Beadle (1932*a*) and Carter (1934).

Another possible, though subordinate, mechanism is *profile-bound density currents*. These can result from a cooler inflow moving at depth, possibly along the bottom or at its own density level. Examples are recorded from a number of tropical reservoirs, including Gatún Lake of Panama (Gliwicz 1976*a*), Lake Kariba (Begg 1970) and the Guma Dam of Sierra Leone (Mtada 1986). More conjectural is a descent of water, locally cooled (e.g., at night) in a shallow region of a lake, along the bottom to deeper regions. This would constitute profile-bound density currents and was suggested by Talling (1963, 1969) to originate stratification in one African rift lake, Lake Albert (see Fig. 2.16). Further evidence has been obtained from a bay and channel of Lake Victoria (MacIntyre & Melack 1995). It may also contribute to other examples of sloping isotherms at the southern ends of lakes Malawi (Nyasa) and Tanganyika that appear about June–August, near or after the winter solstice. In this season the higher-latitude southern ends have cooler surface water than is present elsewhere along these long lakes that each span 5° of latitude. This feature is well shown by

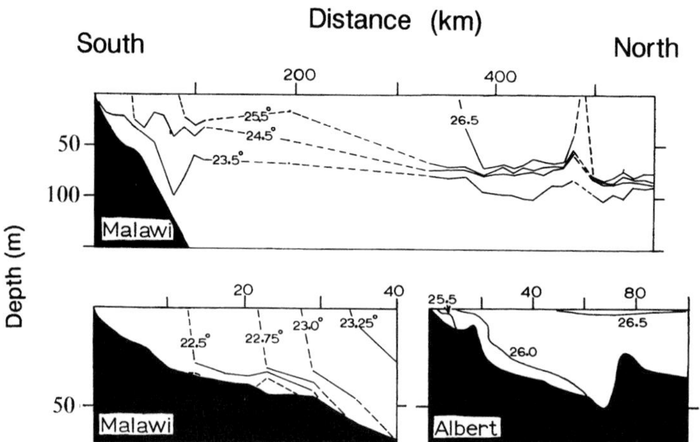

Fig. 2.16. Longitudinal sections of the thermal structure along two large African rift lakes, showing upward tilting of isotherms near the southern ends in May–August associated with wind-induced displacements and/or profile-bound density currents. From Talling (1969).

remote sensing (Wooster *et al.* 1994; Patterson & Kachinjika 1995; Patterson *et al.* 1998: see cover). Another influence at work is *upwelling* of deep cool water, induced by southerly winds tilting the thermocline along the lake-axis. This is most fully documented for Lake Tanganyika (Coulter & Spigel 1991: see Fig. 2.22); it is a basis for upward transport of nutrients and local increase in plankton production and fisheries.

Wind stress is the most powerful agent for generating water movements in lakes. It is roughly proportional to the square of wind velocity, so that stronger winds have disproportionately more influence. Distance of over-water travel, or fetch, is another augmenting factor. When large, an appreciable tilting of the water surface may develop; in Lake Chad this changed sign with the seasonal change in wind direction from northeast (November–May) to southwest (June–October) (see Fig. 2.17). There, the diel wind effects on the level observed in 1969 were eliminated in 1975 by the growth of macrophytes (Carmouze, Chantraine & Lemoalle 1983).

Below the water surface induced motion can take many forms, which Imberger (1985) discusses and summarizes diagramatically. Examples from tropical lakes are considered below. There are very few direct measurements of current velocity and direction. During 1972 Smith deployed free-drifting drogues in Lake George, Uganda, and followed directions and rates of movement at various depths (Viner & Smith

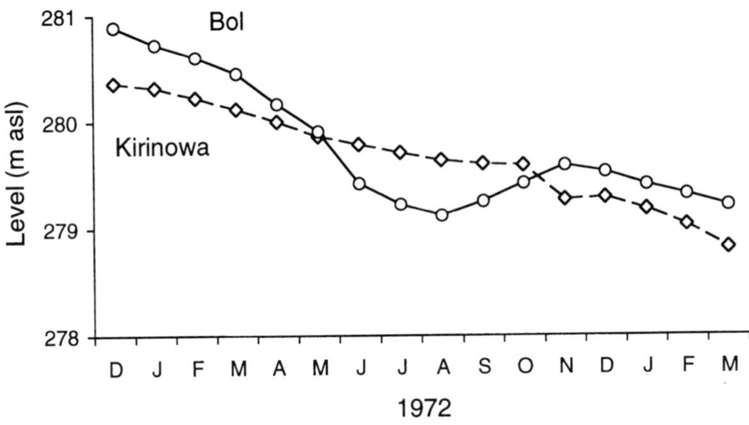

Fig. 2.17. Lake Chad, southern basin. Time-variations of lake level at a southwest station (Kirinowa) and a northeast station (Bol), 1971–73, under the influence of the North East Trade Wind (November–May) and a southwest wind (June–October). Original.

1973). Rates decreased steeply with depth, and in this shallow lake motion at an angle to wind direction appeared to set up anti-clockwise patterns of horizontal circulation (Fig. 2.18). These were likely to at least partly control other examples of concentric distributions found in the lake (Burgis *et al.* 1973), although further shore-related factors were also involved. On a much finer scale of resolution, a hot-bead thermistor flowmeter has been applied by MacIntyre (1984) in a sheltered Kenyan water (Mennell's lagoon, Lake Naivasha) and in the nearby soda lakes of Sonachi and Nakuru (MacIntyre 1981; MacIntyre & Melack 1995). Current velocities, related to wind stress, decreased rapidly with depth within the uppermost 0 to 0.5 m layer. The classical flowmeter of the river-hydrologist with rotating cups is too insensitive for most situations in lakes, but on the Nile (Rzóska 1976) and doubtless elsewhere has enabled measurement of the slowing down of flow velocity in the transition from river to reservoir.

Once set in motion, the water-mass of a lake is a setting for conflict between two tendencies. One is of buoyancy-cum-stability, favoured by larger values of the vertical density gradient $\mathrm{d}\rho/\mathrm{d}z$; the other is of turbulent mixing that occurs between adjacent layers of flow subject to shear in a vertical velocity gradient $\mathrm{d}u/\mathrm{d}z$. The density gradient ($\mathrm{d}\rho/\mathrm{d}z$) in the thermocline region can be used as a measure of stratification either directly or in a simple derivative that equals the square of the Brunt-Väsälä frequency (N^2), where N is an upper frequency limit for energy-rich oscillations:

$$N^2 = \frac{g}{\rho_o} \cdot \frac{\mathrm{d}\rho}{\mathrm{d}z} \tag{2.6}$$

where ρ_o is a mean reference density of water, ~1000 kg m^{-3}, and g the acceleration due to gravity (9.81 m s^{-2}). Remarkably, there appears to be few applications of N^2 to tropical lakes. A seasonal sequence for Lake Tanganyika has been estimated by Coulter & Spigel (1991). They found a minimum value of $N^2 < 1 \times 10^{-4}$ s^{-2} during September at a northeast station when the epilimnion thickness h was deepening most rapidly under the influence of strong southerly winds. MacIntyre (1981) and MacIntyre & Melack (1995) have estimated and used it comparatively with the Richardson Number (R_i: see below) for the Kenyan lakes Sonachi and Nakuru. There has also been limited application of three other parameters that express the interaction between the two opposing tendencies mentioned above.

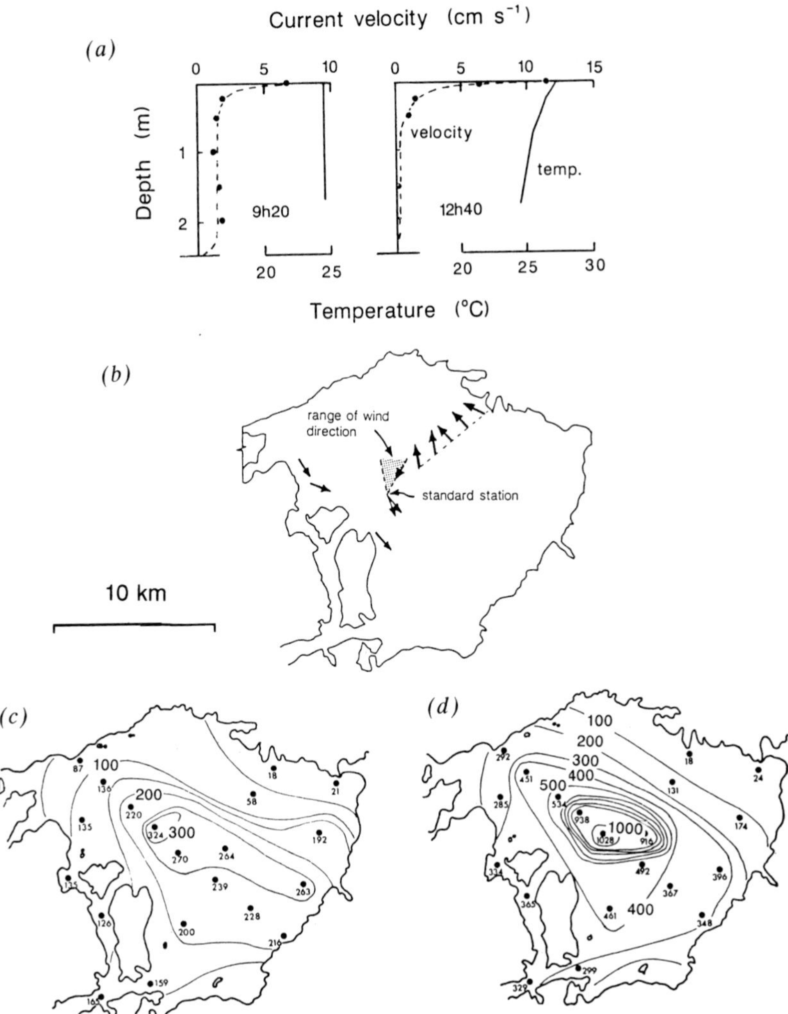

Fig. 2.18. Lake George, Uganda. Examples of (*a*) depth-distribution of current velocity and temperature, and (*b*) the horizontal distribution of current direction, on days when light winds were blowing, with possible consequences in concentric distribution patterns of (*c*) phytoplankton density as mg chlorophyll *a* m^{-3} and (*d*) density of zooplankton Crustacea as µg dry mass l^{-1}. Modified from Viner & Smith (1973) and Burgis *et al.* (1973), after Talling (1992).

The oldest and simplest is the Richardson Number R_i that expresses the ratio between vertical density gradient ($d\rho/dz$) and horizontal velocity shear (du/dz):

$$R_i = \frac{g}{\rho_o} \cdot \frac{\left(\frac{d\rho}{dz}\right)}{\left(\frac{du}{dz}\right)^2} = \frac{N^2}{\left(\frac{du}{dz}\right)^2} \tag{2.7}$$

The approximation of a two-layered model, with upper mixed layer of thickness h and density ρ_1 above a hypolimnion of density ρ_2, and water shear velocity at the water surface of u_*, gives:

$$R_i = \frac{g}{\rho_o}(\rho_2 - \rho_1)\frac{h}{u_*^2} \tag{2.8}$$

R_i is non-dimensional, and values below ~0.25 indicate marked instability. The velocity term is not easily measured, but can be roughly deduced from wind speed. Ganf (1974*d*) used data from Lake George to evaluate it and obtained a correlation with the effects of changing stratification on vertical movements of phytoplankton. Here the diel warming could generate temperature gradients of 2.5 °C over 25 cm depth that led to estimates of $R_i > 1$. However, another such comparison, from the higher and cooler Kenyan soda lakes of Sonachi and Nakuru, yielded no significant correlation (MacIntyre 1981; MacIntyre & Melack 1995). A more indirect assessment of R_i with application to possible sedimentation of phytoplankton in Lake Lanao was given by Lewis (1978*a*).

The two further parameters are also dimensionless and take lake dimensions – length, depth of mixed or surface layer, total depth – into account. They are the Wedderburn Number W and Lake Number L_N, with derivation and significance described by Imberger & Patterson (1990) and – with reference to tropical and subtropical water-bodies – by Allanson (1990). Like R_i, they express the balance between buoyancy and destabilizing agencies.

The definition of the Wedderburn Number (W) recognizes that the prediction of stability or mixing from the Richardson Number can be improved if the ratio between two lake dimensions – depth of the mixed layer (h) and length of the basin at depth h along the direction of the wind (l) – are also taken into account, as:

$$W = R_i \cdot h/l = \frac{g}{\rho_o}(\rho_2 - \rho_1) \cdot \frac{1}{l} \cdot \frac{h^2}{u_*^2} \tag{2.9}$$

Values of W below 1 are then associated with instability, and those above 1 with stability. Figure 2.19 shows an example-application by Patterson & Kachinjika (1995) to Lake Malawi, in which the water shear velocity term u_*^2 is estimated from that of mean daily wind velocity at 2 m height (U^2) multiplied by a factor (1.68×10^{-6}) that is the product of the density ratio of air to water ($\rho_{air}/\rho_{water} = 0.0012$) and a coefficient of drag of 0.0014. Here seasonal mixing within the upper 150 m is marked by a minimum around October of the Wedderburn Number.

In the concept of Lake Number (L_N), account is taken of the classic stability quantity (S), the water velocity-depth gradient or shear (du/dz), the maximum depth of the lake basin (z_m) and its area A_o, and the distances above this depth of the centre of the metalimnion (z_t) and of the centre of lake volume (z_g):

$$L_N = g\ S(1 - z_t/z_m)/\left[\left(\frac{du}{dz}\right)^2 .\ A_o^{1/2}\ (1 - z_g/z_m)\right] \quad (2.10)$$

As with the definitions of R_i and W, uniform units (e.g., metres for length) must be used to set values of the quantities g, S, z, u and A_o.

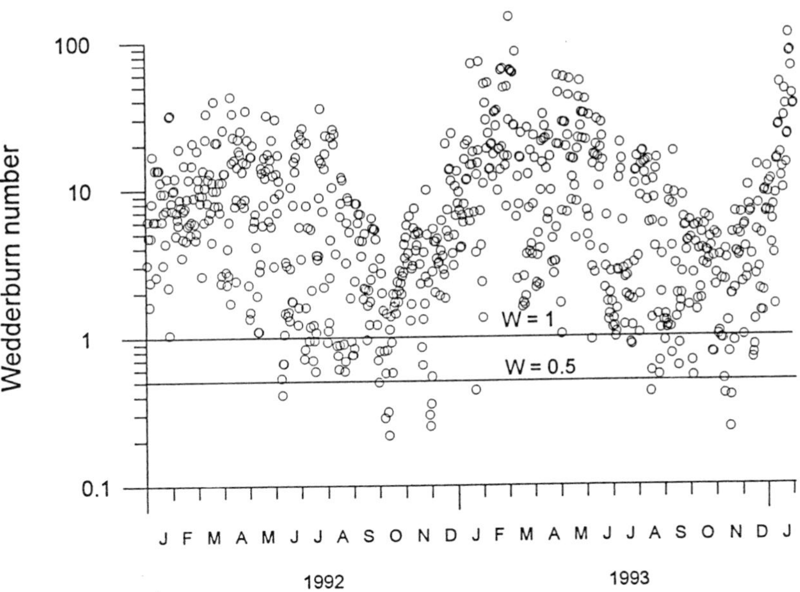

Fig. 2.19. Lake Malawi. Within-year and between-year variation in the Wedderburn Number, showing seasonal minima after the annual cooling. From Patterson & Kachinjika (1995).

Low values of L_N, like those of the Wedderburn Number W, are indicative of susceptibility to vertical instability and mixing.

With these different formulations there is, however, scope for the summarizing Numbers W and L_N to vary independently – especially when applied to basins of different shape. Their combinations can characterize liability to various grades of stability, mixing and entrainment, and persisting internal seiches (see below). Applications to the tropical African reservoirs of Kariba and McIlwaine, and the lakes of Tanganyika and Barombi Mbo (Cameroon), are briefly discussed by Allanson (1990) and Allanson *et al.* (1990). For Barombi Mbo, Allanson (1990) has estimated the decline of L_N with increasing wind velocity U; values of 1 and 0.5 appeared at velocities of >5 and >8 m s^{-1}, respectively. For Lake Tanganyika the values of W and L_N suggest compatibility with end-basin upwelling and large-amplitude internal waves or seiches, as are actually found (Spigel & Coulter 1996). Seasonal change in Lake Number has been estimated for two reservoirs in tropical north Australia and correlated with accompanying changes in dissolved O_2 (Boland & Imberger 1993) and the phytoplankton (Boland & Griffiths 1995). For example, in the relatively deep Lake Julius a predominance of cyanophytes in near-surface water is linked to conditions of stable stratification and high L_N (Fig. 2.20).

In all lakes, but especially in large lakes, much dynamic behaviour cannot be represented by vertical, 1-dimensional characteristics at a single 'representative' station. For example, strong wind stress on a stratified lake is likely to set up a tilted thermocline that descends down-wind. Up-wind, deeper water then rises towards the surface and such *upwelling* can be influential chemically and biologically. It is illustrated for Lake Tanganyika in Fig. 2.22. Should the wind then abate, the thermocline tends to return to a horizontal position. In the process there may be an initial travelling *surge* of returning surface water, and – more prolonged – a periodic seesaw-like oscillation as a standing wave or *internal seiche.* The fundamental mode or uninodal seiche is usually predominant, with a period T_i that is determined for a 2-layered lake by the basin length l, the thickness of upper and lower layers h_1 and h_2, and the density difference between upper and lower layers $\rho_2 - \rho_1$:

$$T_i = \frac{\text{distance traversed}}{\text{velocity of internal wave}} = \frac{2l}{[gh_1h_2(\rho_2 - \rho_1)/\rho_2(h_1 + h_2)]^{1/2}} \qquad (2.11)$$

At any one station, distant from the node, the internal seiche is manifested by a periodic rise and fall of isotherms constructed on a depth-time

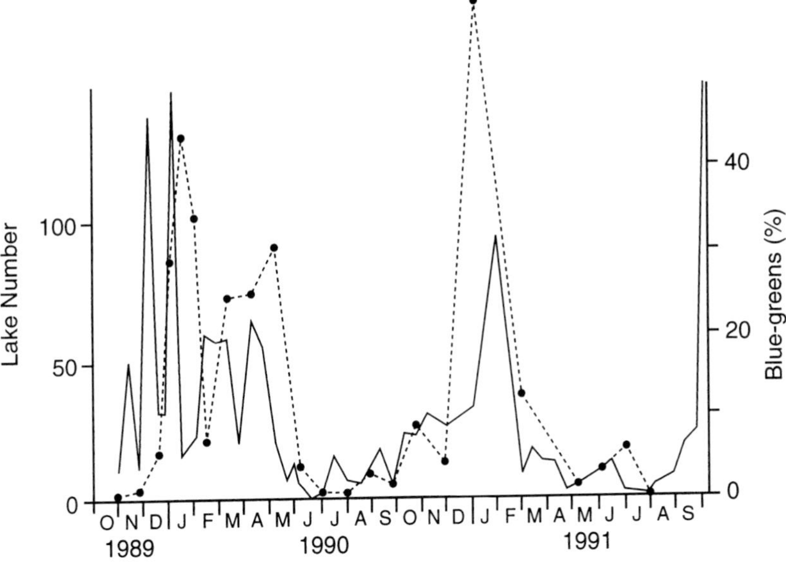

Fig. 2.20. Lake Julius, North East Australia. Within-year variation in Lake Number (–) in relation to that in the % contribution of blue-greens to the phytoplankton biomass in 0.5 m depth samples (---•---). From Boland & Griffiths (1995).

diagram such as appears in Fig. 2.21. In large lakes of temperate latitudes the effect of earth rotation (Coriolis force) often modifies the internal wave motion to that of travelling waves – *Kelvin waves*. In these the vertical displacement of isotherms is greatest towards the shoreline; complementary high and low regions travel around the lake periphery, anticlockwise in the northern hemisphere and clockwise in the southern, with a rotational period that is close to the inertial period determined by latitude (being longer at lower latitude) for the effect of the Coriolis force.

Information on these forms of internal water movements requires long-maintained and frequent observations of thermal structure, preferably at several stations, that are available for few tropical lakes. The best early tropical time-series is that obtained on Lake Victoria in 1951–3 by Fish (1957), from which data are also replotted in a more accessible form by Talling (1957*b*, 1966). Oscillations in deep-water temperature, and derived isotherms, were found at a routine offshore station and along inshore channels. These were interpreted by Fish as evidence of a prolonged internal seiche offshore, whose pump-like effects were transmitted along the stratified channels. From this and later work by Newell (1960)

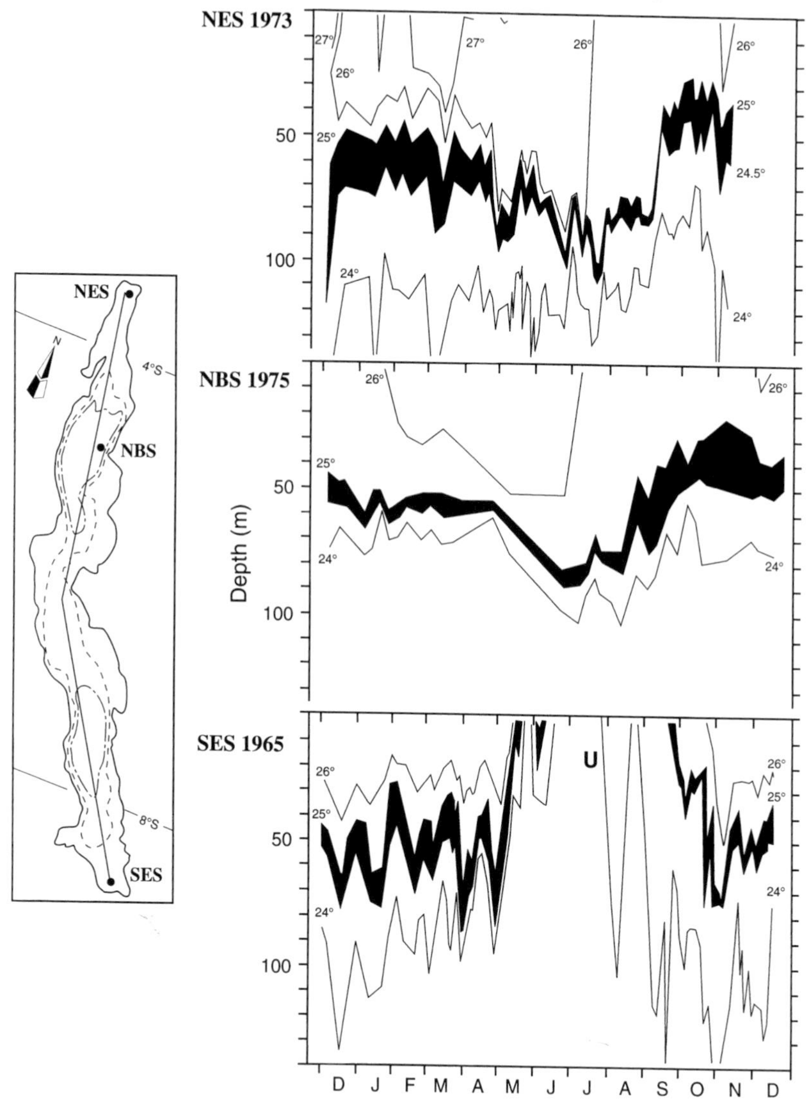

Fig. 2.21. Lake Tanganyika. Within-year variation of thermal stratification, including internal waves and local upwelling (U) observed in different years at three stations along the lake (inset map) and depicted as isotherm displacements on depth-time diagrams. One interval of temperature is shown in black. Modified from Coulter & Spigel (1991).

and Talling (1966), there can be no doubt that major wind-induced tilting of thermal discontinuities is frequent in the lake. In particular, southerly winds induce downtilting from south to north that can lead to periods of temporary isothermy at the standard northern station (Fig. 5.17). The existence of long-continued oscillations of an internal seiche, based on small density gradients, is more controversial: it was denied by Newell (1960), but his own interpretation of a persistent 3-layered structure is also open to criticism (Talling 1966).

At an appreciably higher latitude, Lake Malawi exhibits a stronger seasonal thermocline that is undoubtedly subject to wind-induced tilting and probably to subsequent internal seiches. Internal seiches were first postulated by Beauchamp (1953*a*) from observations in 1939–40, that disclosed some large vertical displacements of isotherms. There is later evidence from 1954–55 (Harding 1963), 1960–61 (Eccles 1962, 1974), 1990–91 (Bootsma 1993*b*) and 1993–94 (Patterson & Kachinjika 1995).

The most advanced analysis of internal water movements is from the neighbouring African rift lake, Tanganyika; it is summarized by Coulter (1988), Coulter & Spigel (1991) and Spigel & Coulter (1996). In the dry season around June–August, strong southerly winds induce a south–north thermocline tilt that allows deeper water to upwell near the southern extremity (Fig. 2.22). Afterwards a return surge is likely and, after warming, later in the wet season (January–June) there are records of pronounced oscillation of isotherms (Fig 2.21) indicative of an internal seiche. Oscillation in the water-mass also appears to be manifest in periodic chemical variation, including pH, O_2 and nutrients. Observations and calculation concur in a remarkably long period of 25–30 days; the entire lake, length 650 km, is involved. A rotational, shore-bound Kelvin wave of similar period does not seem to be appreciably developed, as was earlier supposed (Coulter 1968), because of the low Coriolis force and the relatively small width of the lake.

From near the southern end there is some evidence from daily observations in March 1966 (Coulter 1968) of offshore inertial waves of the Poincaré type, involving a presumed grid of circular patches many kilometres in radius, with anti-clockwise currents and undulating crests lasting about three days. This record is unique for a tropical lake.

In these very deep lakes of Tanganyika and Malawi, surface-derived mixing only extends over a small part – the mixolimnion – of the total depth. A long relative isolation of the surface and bottom water, especially in Lake Tanganyika, is indicated by differences in the concentrations of several isotopes found by Craig *et al.* and Gonfiantini *et*

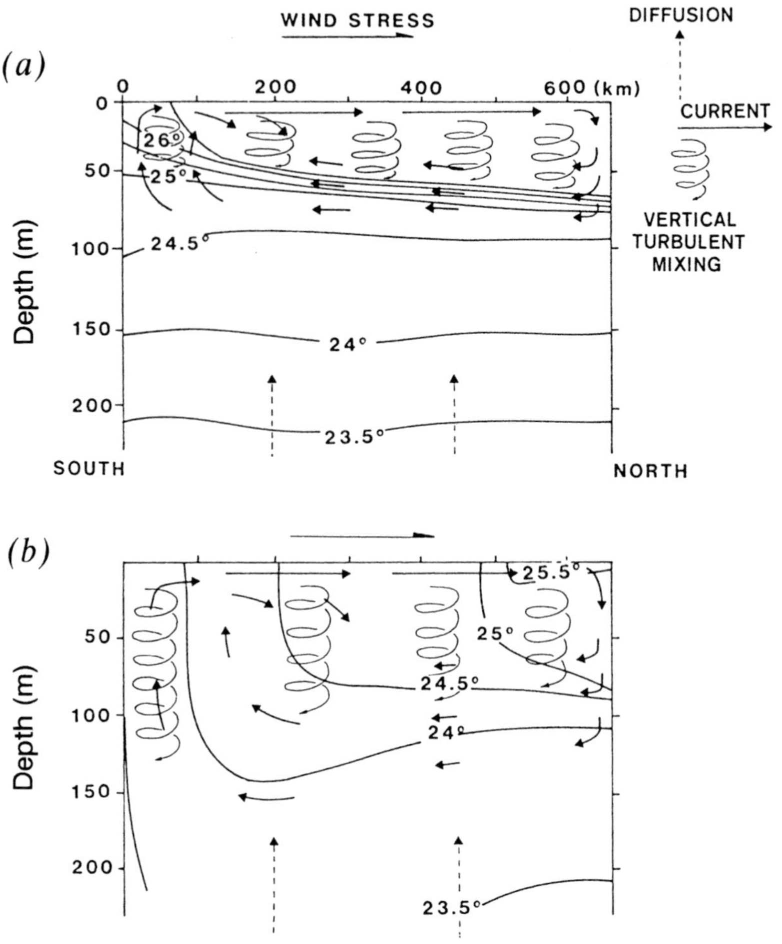

Fig. 2.22. Lake Tanganyika. An interpretation of circulation patterns with south-end upwelling in (*a*) May and (*b*) August during the south wind season. From Coulter & Spigel (1991).

al. (1979) (see Coulter & Spigel 1991), namely tritium, deuterium and O^{18} (Fig. 2.23). Differences in total ionic concentration, and hence its contribution to density difference, are present but small. The situation is particularly interesting in Lake Malawi around July–August, when the upper mixed layer (*mixolimnion*) reaches its thickest and coolest state, and when its temperature difference from the deeper persistent *monimolimnion* can fall to less than 0.5 °C. Halfman (1993) measured a temperature difference across the main region of chemical gradient

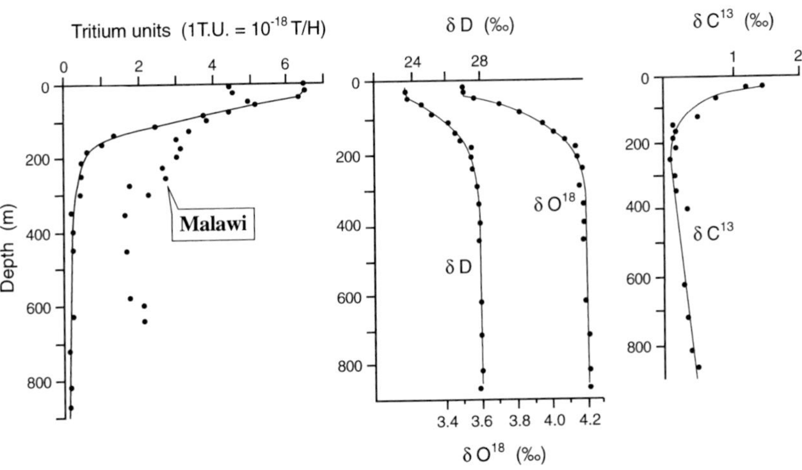

Fig. 2.23. Lake Tanganyika. Depth-concentration profiles of several stable isotopes – tritium, deuterium (δD), oxygen-18 (δO^{18}) and carbon-13 (δC^{13}) sampled at a northern station in February 1973. For comparison, the less sharp stratification of tritium in Lake Malawi in June 1976 is also shown. Based on Craig *et al.* and Gonfiantini *et al.*, modified from Coulter & Spigel (1991).

(*chemocline*) between 150 and 250 m of 0.39 °C, and from conductivity estimated that the expected density difference of 91 mg dm^{-3} would be raised to 103 mg dm^{-3} by the solute contribution. Subsequently the calculations were extended (Wüest *et al.* 1996), and a non-ionic solute – silicic acid – was shown to have a considerable influence on the deep density-stratification. In numerous other long-stratified or *meromictic* tropical lakes the chemical contribution is clearly all-important. In these lakes a strong mid-water inversion of temperature, with cooler water uppermost, can coexist with a stable density stratification (e.g., Lake Simbi: Ochumba & Kibaara 1988) (see Fig. 2.24). Elsewhere, a deep saline layer overlain by more dilute and clear water may acquire relatively high temperature as a result of absorbing solar radiation. Examples occur world-wide; a tropical example near the Panama Canal is described by Bozniak *et al.* (1969). Higher temperature at depth can also result from geothermal heat flux from the earth's interior, in combination with an entry of saline water. In Lake Kivu the heating from below is considerable and has led to a sharply stepwise form in the vertical temperature gradients and associated salinity gradients (Newman 1976).

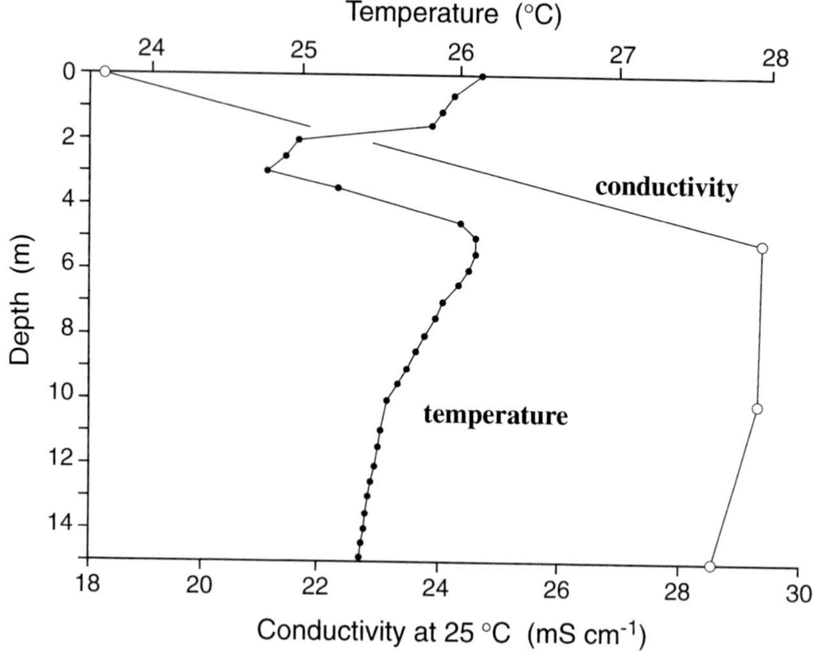

Fig. 2.24. Temperature-depth profile in Lake Simbi, a saline Kenyan crater lake, showing a temperature inversion with minimum in the 0–5 m layer where a conductivity-depth profile indicates a salinity gradient. From Ochumba & Kibaara (1988).

2.4 Chemical balance

(a) Input–output: pathways and fluxes

In most freshwaters the main *chemical input* is linked to the surface flow plus seepage components of the water budget, and – omitting human pollutants – is ultimately mainly derived from (i) the weathering of rocks in the drainage basin, accompanied by (ii) organic matter and its breakdown products derived from life in that basin. Additional fluxes are contributed by (iii) atmospheric precipitation, as wet and dry deposition, that includes 'cyclic salts' carried via spray from the oceanic reservoir; by (iv) exchange of gases across the atmosphere–water interface, with internal consumption of carbon dioxide (CO_2), O_2 and nitrogen (N_2) (by N-fixation) potentially driving important inputs; and by (v) chemical exchange at the water–sediment interface. Likewise, *chemical output* is generally dominated by (vi) surface flow plus seepage-out, supplemented by outward components of exchanges at (vii) the upper air–water and

(viii) the lower sediment–water interfaces. At the lower interface, outputs from the water-mass often involve storage in sediments, as accumulations of inorganic and organic particulates, precipitations from solution after evaporative concentration, 'reverse weathering' with formation of new sediment minerals, and solute incorporation by burial. Within the water-mass also, inequalities between inputs and outputs appear as changes of *chemical storage*.

The *boundary fluxes* numbered (i) to (viii) above will now be considered in the tropical context.

(i) Weathering

This process includes both physical erosion and chemical denudation. It is well known that tropical soil formation can involve some distinctive chemical pathways and products, as in the process of forming laterite soils (laterization) that is widespread in the tropics. The chemical transformations of cation alumino-silicates tend to proceed further to simpler products, hydroxides and oxides including aluminium-(Al-) and iron-(Fe-)sesquioxides, than in temperate soils. In this sequence there is a liberation of cations, especially sodium (Na^+), bicarbonate and silicate, which contributes to their often considerable concentrations found in tropical river and lake waters. Enhanced solubility at higher temperature is another influential factor.

However, as elsewhere, regional chemical characteristics vary with local geological history, affecting the proportions of igneous and sedimentary rocks or deposits, and their element-composition. Individual mineral types are of variable stability; Table 2.2 gives a graded series of those frequent in tropical soils. Most present day land-masses of the tropics derive from the fragmentation and migration of the ancient Palaeozoic proto-continent of Gondwana, with a legacy of long-eroded land surfaces that are particularly evident in Africa and southern India. Later some massive chemical sources have resulted from marine transgressions (e.g., limestone, marine evaporites), tectonic rifting and volcanic activity. Calcareous formations like limestone are not well represented over much of the tropics, which partly accounts for the common preponderance of Na^+ over calcium (Ca^{2+}) in tropical freshwaters. Volcanic lavas can be a ready source of dissolved silicate, and some volcanic regions yield distinctive ionic supplies – as of potassium (K^+) and magnesium (Mg^{2+}) from the Virunga-Bufumbira volcanic field in the African rift valley (Talling & Talling 1965; Viner 1975*a*). In the same area, a remarkable natural output of phosphate enriches lakes Edward

Table 2.2. *Relative levels of mineral stability in tropical soils*

MOST STABLE

Quartz $\gg$
K-Feldspar, Micas $\gg$
Na-Feldspar >
Ca-Feldspar, Amphiboles >
Pyroxenes, Chlorite >
Dolomite >
Calcite >
Gypsum, Anhydrite $\gg$
Halite

LEAST STABLE

Source: From Stallard 1985

and Albert, and injects into the upper White Nile (Talling 1957*c*, 1976); its entry flux there, estimated from concentration ($\sim$170 mg P m^{-3}) and water discharge ($\sim 4 \times 10^9$ m^3 yr^{-1}), is $\sim$700 t P y^{-1}. From the same volcanic area, but in the opposite direction, another considerable flux of phosphorus (P) appears to be carried by the Ruzizi River (reported PO_4-P concentration 167 mg m^{-3}, water discharge $\sim 2 \times 10^9$ m^3 yr^1) towards Lake Tanganyika (Dubois 1958; Hecky 1991). In another continent, and on a smaller scale, local volcanic sources of phosphorus clearly account for much variation of P-content in streams draining the Caribbean slopes of Costa Rica (Pringle *et al.* 1990).

Contrasting with such sites of active chemical denudation are catchments where hard granitic rocks are prevalent. Stream waters of very low ionic content, and hence electrical conductivity, can result. One example, examined in detail by Lewis (1986*b*) and Lewis *et al.* (1987), is the Caura River drainage of the Orinoco river system. If silicate export was used as an index of chemical weathering, a mean physical rate of 1.8 cm per 1000 yr was obtained. This value is not inconsiderable, and is influenced by a warm wet climate; from it and information on mean rock composition the weathering-derived flux of other elements could be estimated (Table 2.3). For P this was appreciably larger than was exported in river discharge from the catchment, corrected for atmospheric P deposition, implying some overall removal (largely abiotic?) *en route*. Chemical denudation in the Orinoco system as a whole has also been examined by Edmond *et al.* (1995). They computed that weathering of the primary basement or shield, lacking limestones or evaporites, was about 1.0 cm per 1000 yr; also that weathering proceeded to the minerals kaolinite and gibbsite as

Table 2.3. *Components of mass-balance for the Caura River basin, Venezuela, expressed in kg ha^{-1} yr^{-1} as the sum of dissolved and particulate fractions*

The equivalents of Si weathering (4) are estimated from the Si flux and the element-composition of the predominant rock type (3). Deviations from discharge output minus atmospheric deposition (5 = 4 – [1 – 2]) are interpreted as net retention if positive, and as selectively faster weathering than Si if negative.

	Na	K	Ca	Mg	P	Si
1. basin discharge	41.4	19.5	18.2	7.52	0.47	152
2. atmospheric deposition	8.21	1.03	1.32	0.29	0.14	0
3. fractional rock composition	0.026	0.027	0.027	0.012	0.0011	0.31
4. Si-weathering equivalent	12.6	13.35	13.15	5.84	0.54	(152)
5. retention	−20.6	−5.12	−3.73	−1.39	+0.21	–

Source: From Lewis *et al.* (1987)

products, with active laterization. The mineral products of weathering, and so the export of dissolved silicate, are influenced by climate and especially the mean annual water runoff. From a study of river basins in Kenya, Dunne (1978) associated low runoff with a predominance of kaolinite, and high runoff (to 2000 mm yr^{-1}) with that of montmorillonite and a greater proportion of silicate in the solutes exported.

Large tropical river catchments typically include subregions with very different geology and rock-types, with differences in chemical denudation that are both quantitative and qualitative. A strong chemical divergence is seen (Talling 1976) in the two main limbs of the Upper Nile, with cationic Ca^{2+} dominance in the Blue Nile and Na^{+} dominance in the White Nile. Differences in the Amazon system are reflected in tributaries of the types 'black-water' (acidic, low ionic content) and 'white-water' (higher ionic content) with relationships to geology that are examined broadly by Sioli (1984) and quantitatively by Stallard & Edmond (1983, 1987). Headwaters in the Andes are in a region that bears carbonates and evaporites, and is susceptible to relatively rapid chemical denudation – unlike the lower shield areas of old siliceous rocks (Fig. 2.25). Carbonate erosion has constraints from the solubility products linked to calcite ($CaCO_3$) and possibly dolomite ($MgCO_3$); it accounts for a correlation between concentrations of Ca + Mg and alkalinity in river water. Weathering of siliceous rocks yields soluble silicon (Si) correlated non-

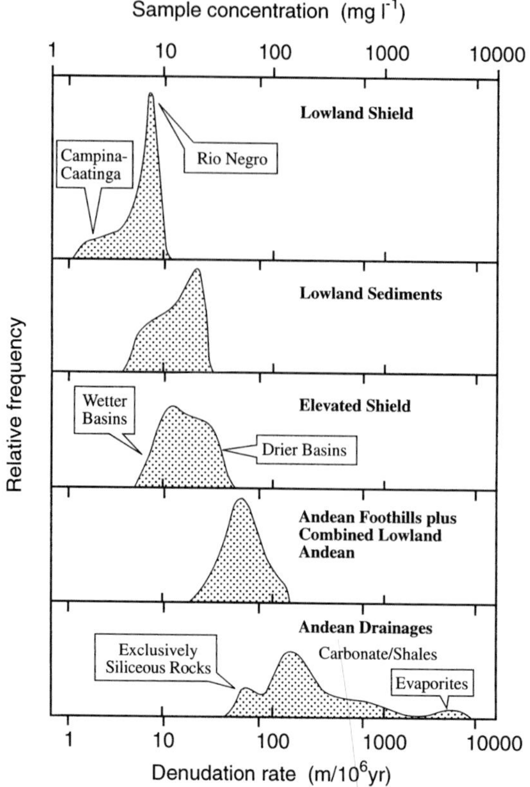

Fig. 2.25. Relative frequency distributions of water samples from five geographical-drainage regions of Amazonia in relation to concentration of dissolved solutes and estimated magnitude of denudation rate. From Stallard (1985).

linearly with Na + K concentrations (corrected for salt of marine origin), with the possibility of stabilization of Si-bearing kaolinite and quartz by soluble Si already produced. Also correlated with altitude are kinetic factors, so that weathering limitation on steep slopes is succeeded by *transport* limitation on shallow slopes with thick soils (Stallard 1985). A study by Lesack (1993*a*) of solute export from a small forested sub-catchment off the middle Amazon demonstrated chemical divergence between several transport pathways – in base-flow, storm-flow and sub-surface soil-flow, storm-flow here being (unusually) the richest in solutes.

(ii) Organic inputs

Inland waters receive from their drainage basins greater or lesser amounts of organic matter, of terrestrial origin, in dissolved (DOM) or particulate form. Thus litter-fall on an Amazonian forest stream is likely to be in the region of 77 g dry weight m^{-2} yr^{-1} (McClain & Richey 1996). Additional quantities originate internally. In large water-bodies the DOM (expressed as carbon) is typically in the concentration range of 2–10 mg C l^{-1}. Much higher concentrations, to over 100 mg C l^{-1}, can exist in detritus-rich pools such as some analysed from the African Okavango system (Cronberg *et al.* 1996). Features of time-variability are discussed in Chapter 5.1e. Further computations of fluxes have been made, especially for the river systems of the Amazon (e.g., Hedges *et al.* 1986; Richey *et al.* 1990) and Orinoco (e.g., Lewis & Saunders 1989). From these systems the main-river flux of DOM per unit area of drainage basin was in the range 2–12 g C m^{-2} yr^{-1}. It was the greater part of the total organic (dissolved plus particulate) flux.

Persistence of compounds resistant to chemical breakdown is naturally favoured, yielding the broad classes of humic and fulvic acids. More labile compounds, such as purines and vitamins, are likely to have a considerable biological importance that is as yet almost unexplored in tropical waters. It is conceivable that their influence lies behind a supposed qualitative tendency towards 'pond-plankton' in tropical lakes, on which several pioneer planktologists (e.g., Schmidle 1902) commented.

The most obvious physical property linked to dissolved organic matter is the capacity for the strong absorption of short-wave visible and ultra-violet radiation (Chapter 3.1a). Its effects are especially marked in tropical rivers of the 'black-water' type, first distinguished in Amazonia by Alfred Russel Wallace, such as the rivers Negro, Orinoco and Zaïre. These are of low base and total ionic content other than hydrogen ions (H^+), acidic in reaction, with only moderately high concentrations of dissolved organic carbon, and brown in colour. The capacity for ultra-violet absorption is also an agent of self-destruction (Chapter 3.4), and appreciable rates of photochemical mineralization to CO_2 have been detected in the Amazon system (Amon & Benner 1996).

Other physico-chemical properties include the capacity to combine (chelate) with metals such as Fe and maintain them in solution; also to contribute components, including H^+, active in the pH-buffer system. The last effect is likely to be particularly significant in the HCO_3-poor black-water rivers. Neither has been adequately studied from tropical waters.

(iii) Atmospheric precipitation

Besides water, rain provides a 'wet' chemical flux from the atmosphere that is augmented by the dry deposition of solids. These two fractions constitute 'bulk precipitation'. Their combined quantitative significance is poorly known for tropical freshwaters, although there are important pioneer studies for several African lakes, Lake Valencia in Venezuela, and the Amazon region (Table 2.4). None of these show significant persistant acidification of surface waters by 'acid rain', probably since large-scale industrial or urban combustions are few in the tropics – although four regions likely to experience future acidification have been distinguished (Rodhe & Herrera 1988: see Fig. 2.26). However, an organic acid component of plant origin has been reported from forested regions, in West Africa (Lacaux *et al.* 1987, 1992) and Amazonia (Lesack & Melack 1991), and occasional very acid samples (pH <4.5) appear elsewhere – as early in the rainy season at Lake Valencia (Lewis 1981; Lewis & Weibezahn 1981*a*). Wider interest attaches to the contribution of major ions (Na^+, K^+, Mg^{2+}, Ca^{2+}, HCO_3^-, SO_4^{2-}, Cl^-), and of N and P as critical plant nutrients.

Sea-salt or 'cyclic salt' – detached in spray and borne in rain – contributes to major ions, especially Na^+, Cl^- and Mg^{2+}, in concentrations that are likely to decline sharply inland. This is directly demonstrated by the study of Stallard & Edmond (1981) on the Amazon region. Chloride is often used as an index concentration from which other sea-derived ions can be estimated (e.g., Gaudet & Melack 1981; Stallard & Edmond 1981; Lewis *et al.* 1987; Wood & Talling 1988), using the ion-ratios characteristic of seawater. Other chemical sources are exposed soil surfaces and, a feature common in many tropical areas, burning vegetation (Crutzen & Andreae 1990). Soils are particularly liable to contribute inorganic particulates, that with particulates from burning vegetation are favoured by dry-season conditions. Intermittent rainfall typically has a scrubbing-action, with higher concentrations found early in the rainy period and lower ones later by prior elimination and dilution. Examples are provided by studies at Lake Chad (Lemoalle 1973*b*), Lake Valencia (Lewis 1981; Lewis & Weibezahn 1981*a*) and Lake Malawi (Bootsma *et al.* 1996).

The important nutrient elements N and P are contributed in both inorganic and organic forms. Again burning vegetation may be a major source in the tropics. The ratio of inorganic to organic forms is apparently very variable, being high at Lake Malawi (Bootsma *et al.* 1996) and low at Lake Valencia (Lewis 1981). The tropical analyses suggest that of

Table 2.4. *Mean chemical concentrations and annual fluxes estimated for wet atmospheric deposition, with varying exposure to dry deposition, at five tropical sites applicable to the adjacent lakes indicated*

For Lake Calado, values for predominantly wet and dry seasons are distinguished as W_1 and W_2, respectively.

	Chad	Ebrié	Malawi	Valencia	Calado	
					W_1	W_2
A. concn. ($\mu mol\ l^{-1}$)						
NH_4-N	38.2	15.4	5.4		5.2	13.3
NO_3-N	12.9	19.6	3.2		2.2	9.2
Total dissolved N	–	102	8.4		–	–
Total N	–	–	9.4			
PO_4-P	1.19	3.6	0.18		0.04	0.16
Total dissolved P	–	–	0.10		–	–
Total P	–	–	0.14			
Na	–	–	2.9		1.9	5.1
K	–	–	0.6			
Ca	–	–	5		0.95	2.9
Mg	–	–	1.1		0.65	1.05
Cl	–	–	1.5		4.0	8.2
SO_4			2.5		1.75	4.45
B. flux ($mmol\ m^{-2}\ yr^{-1}$)						
NH_4-N	24.7	33	4.1	17.3		
NO_3-N	8.7	42	2.4	9.1		
Total N	–	220	7.8	53.1		
PO_4-P	0.77	7.4	–	1.0		
Total P	–	–	0.16	5.4		
Na	–	–	2.2	72.1		
K	–	–	0.5	10.9		
Ca	–	–	15	21.5		
Mg	–	–	0.8	22.5		
Cl	–	–	1.1	54.0		
SO_4	–	–	1.8	16.9		

Source: Lake Chad: Lemoalle (1973*b*)
Lake Ebrié: Lemasson & Pagès (1982)
Lake Malawi: Bootsma, Bootsma & Hecky (1996)
Lake Valencia: Lewis (1981)
Lake Calado, Amazonia: Lesack & Melack (1991)

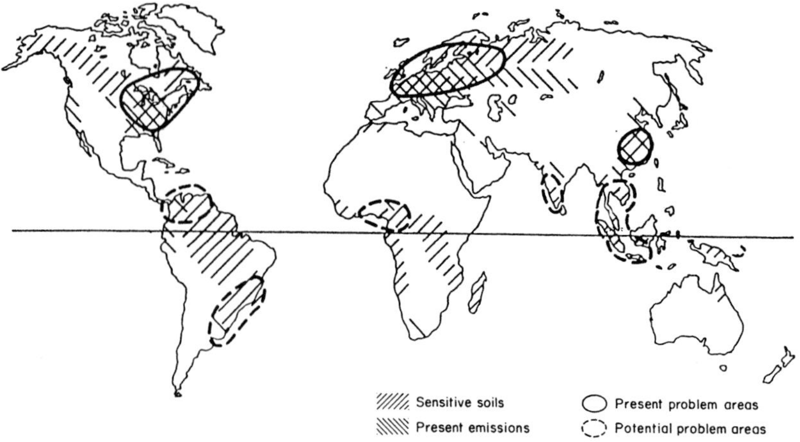

Fig. 2.26. Schematic map showing world regions (heavy rings) where present problems of acidification arise from the combination of sensitive soils and acid emissions, and others (broken rings) where such problems might become severe in the future. From Rodhe, in Rodhe & Herrera (1988).

the two main forms of inorganic N, NO_3-N is less often predominant over NH_4-N compared with temperate regions. The N:P ratio also may be lower (Lewis 1981; Bootsma *et al.* 1996). Examples of concentrations and estimated annual loading fluxes are given in Table 2.4. Some curiously high values for P (>3 μmol l^{-1}) from Uganda are summarized by Livingstone & Melack (1984) and Hecky (1991, 1993). However, the sequential analyses of rainwater at Lake Chad tabulated by Lemoalle (1973*b*) show that much variability between samples (e.g., <0.1–6.0 μmol PO_4-P l^{-1} can exist. Another source of variability is the varying incorporation in the analytical estimates of dry deposition, a flux that extends beyond a rainy season and which can sometimes surpass wet deposition – as with estimates of annual P flux for lakes Valencia (Lewis 1981) and Malawi (Bootsma & Hecky 1993; Bootsma *et al.* 1996). Dry deposition, as dust carried by the local harmattan wind, is suspected to be a major source of several ions (especially Na^+) to Lake Bosumtwi in West Africa (Turner *et al.* 1996).

(iv) Gaseous exchange

Net input of a gas at the air–water interface is proportional to its partial pressure deficit between surface water and atmosphere. Flux densities are also slightly influenced by the increase in specific diffusivity with temperature, and more strongly by turbulence in the uppermost water layer. The last factor is widely modelled in terms of a supposed stagnant surface film as a rate-limiting diffusive path, whose thickness (generally <1 mm) is an inverse correlate of wind velocity. If the partial pressure deficit changes sign, a net output will occur under the same controls. Examples for Amazonian waters are given by Melack & Fisher (1983) and Richey *et al.* (1988); see (vi) below.

For computations the surface film model can be expressed as:

$$F = (D/z).(c_w - c_s) \qquad (2.12)$$

where F = the gaseous flux per unit area from water to air, D = diffusivity of the gas at surface water temperature, z = surface film thickness, c_w = concentration of the gas in surface water, and c_s = corresponding concentration at air-equilibrium.

For compatibility of units, if F is measured in mol m^{-2} s^{-1}, D is expressed in $m^2\ s^{-1}$, z in m, and c_w and c_s in mol m^{-3} (mmol l^{-1}). The component D/z is also known as the piston velocity (units, e.g., m s^{-1}), and is sometimes estimated empirically without postulating a stagnant surface film or boundary layer. A film-replacement model may indeed be physically more realistic. In any case, in flowing water the wind influence is likely to be replaced by turbulent shear of internal origin.

Three gases – O_2, CO_2 and N_2 – are well defined constituents of the air and are also liable to biological consumption in freshwaters. Surface-input (and -output) fluxes are therefore driven by deviations from the corresponding air-equilibrium concentrations. For CO_2 these deviations are indicated by the associated deviations in pH, which are illustrated by Talling & Talling (1965) for several African waters in a series of increasing alkalinity. Thus the surface water of offshore Lake Victoria could then reach pH 8.7, and now still higher, due to the photosynthetic consumption of CO_2, although the air-equilibrium pH is near 8.1 at prevailing temperature and pressure. Air-equilibrium concentrations of the gases decrease with rising temperature, and are much lower for CO_2 than for N_2 and O_2 because of its relatively low molar fraction and hence partial pressure (pCO_2) in the air. Consequently the CO_2 input flux density is fundamentally limited by the low atmospheric pCO_2, although a chemical enhancement becomes appreciable at pH >9.

In productive and oxygenated waters there are often large, inverse, and roughly equimolar gains and losses of O_2 and CO_2 by the processes of photosynthesis and respiration. Magnitudes can reach $\sim$1 mol m^{-2} day^{-1}, and transient storages ensue on both the diel and annual time-scales. On the diel scale, relations between these storages and the generating fluxes have been examined for a few tropical waters, including a Nile reservoir (Talling 1957*a*) and some African soda lakes (Talling *et al.* 1973; Melack & Kilham 1974). On the annual scale, changes in deep accumulations of free CO_2 per unit area closely matched the simultaneous depletions of O_2 in the water-column of Lake Victoria during 1960–61 (Talling 1966). Such matching would not be expected in alkaline waters above pH 8.5, as then the ionic reserves of HCO_3^- and CO_3^{2-} become significantly involved as CO_2-sources or acceptors.

The input of atmospheric N_2 has been little studied from air-equilibrium considerations, as its prime interest is through the process of biological N_2-fixation. Rates of light-dependent fixation have been estimated from exposures of phytoplankton in submerged vessels in a variety of tropical freshwaters, including Lake George (Horne & Viner 1971; Ganf & Horne 1975), Lake Valencia (Levine & Lewis 1984, 1986), two Amazonian floodplain lakes (Doyle & Fisher 1994; Kern & Darwich 1997; Kern *et al.* 1998) and from laboratory tests for Lake Titicaca (Wurtsbaugh *et al.* 1985). Diel patterns in the first two lakes are illustrated in Chapter 5.2. Activity largely results from the presence of heterocystous blue-green algae or cyanophytes. It undoubtedly does much to compensate for the low availability of inorganic combined N, as deduced from concentrations, in many tropical waters. In an Amazonian floodplain lake, Lake Calado, much N-fixation was associated with the periphyton of floating macrophytes rather than the phytoplankton (Melack & Fisher 1988; Doyle & Fisher 1994). Fixation rates were depressed by even low concentrations of nitrate-N, $<0.5\ \mu mol\ l^{-1}$, common in river floodwater. For the high-altitude Lake Titicaca, it seems likely that reduced atmospheric pressure (only 0.64 atmosphere) also has an indirect significance for N-availability. According to Vincent *et al.* (1985), the low atmospheric pO_2 leads to reduced oxygenation, and so more ready seasonal deoxygenation of deep water – conditions under which the formation and accumulation of nitrate as N-source are impaired by denitrification. In this lake there is experimental evidence that seasonal N-fixation modifies the timing of the prevalent N-limitation to phytoplankton (Wurtsbaugh *et al.*, 1985).

(v) Chemical input from exchange at the sediment–water interface

Inputs from this interface can be regarded as largely redistributions of earlier inputs to the water-mass, but there is also an element of 'new' input from pre-existing material. This is important, for example, in the early nutrient enrichment of newly formed reservoirs (Chapter 5.1), such as Mitchell (1973) describes for Lake Kariba. Very rarely submerged hot springs (hydrothermal vents) exist, as in Lake Tanganyika where their content of sulphide appears to sustain local mats of a prokaryote, possibly *Beggiatoa* (Tiercelin *et al.* 1993). Sediment to water transfers are at least partly responsible for the deep-water accumulation in stratified lakes of several chemical constituents, including Fe^{2+}, Mn^{2+}, CO_2, HCO_3^-, PO_4-P and NH_4-N. Some of these are soluble reduced species, other products of organic decomposition.

There are very few studies from tropical waters of the dynamics and fluxes involved. The African Lake George is a shallow and very productive lake where the rather fluid sediments could be expected to contribute, or recycle, nutrients as products of decomposition or 'mineralization'. The forms of vertical profiles of water-removable NH_4-N and PO_4-P in the sediment indicate losses from the 0–10 cm layer by periodic wind-induced disturbance. The vertical extent of organic decomposition depends upon the depth of penetration of O_2 (Viner 1975*b*), whose consumption has been followed experimentally in relation to nutrients generated (Golterman 1971; Ganf 1974*a*). Although release of NH_4-N and PO_4-P from the undisturbed surface of isolated sediment cores has been followed (Viner 1975*c*), Viner (1975*c*, 1977*b*) believed that most recycling of these nutrients occurred in the plankton-rich water-column itself. Both here and in the productive Lake Nakuru, a Kenyan soda lake, aerobic conditions at the sediment–water interface were associated with phosphate uptake (Viner 1975*d* ; Melack & MacIntyre 1992) that was probably largely abiotic (Viner 1975*c*, 1975*d*). Uptake was also marked in studies of sediment–water exchange at Lake Kariba (Lindmark 1997), but liberation could be induced under anoxic conditions.

(vi) Water-borne chemical output

In most lakes, and all rivers, this is the largest chemical output. Dependent upon water throughput, its composition reflects water characteristics in the outflow region. Where throughput is strong, and retention time short, the chemical flux in the surface outflow can approximate that of the inflow and be borne at similar concentrations. In floodplain lakes this can apply to the high level period, within which most boundary flux occurs, although a later isolation can develop. Such a system is Lake Tineo on the Orinoco River floodplain, for which estimates of chemical inputs, outputs and retention were derived by Hamilton & Lewis (1987). Some net depletion occurs with P, N and K, of which large amounts were incorporated during the growth phases of such grasses as *Paspalum repens* (cf. Fig. 2.34), and some by phytoplankton after the filling phase. Net release occurred during periods of grass decomposition. The output from some floodplains is much reduced in particulate material. Thus the Yaéré floodplain in northern Cameroon was estimated by Gac (1980) to have a mean annual input of 897 000 t, mostly inorganic, but an output of only 27 000 t.

In arid regions the water output can be much smaller than the input; hence, if other chemical outputs are inconsiderable, the water-body will necessarily be more saline than average input water. Saline inland waters are widespread in the drier tropics. A particularly extensive and graded series in Ethiopia was surveyed by Wood & Talling (1988) and Kebede *et al.* (1994); the former attempted to estimate the factors of evaporative concentration involved. That for the large Lake Turkana, once connected to the Nile, is influenced by the loss of surface water throughput, like that of the more recently isolated Lake Valencia in Venezuela (Lewis & Weibezahn 1981*b*). However, modes of solute loss to sediments (see (vii) below), and by seepage, can determine relatively low levels of salinity in basins without surface outflow. Lakes Naivasha, Awasa and Chad are African examples (Beadle 1981).

Water throughput in elongate water-bodies – lakes, reservoirs and river systems – often involves chemical modifications during passage. These can be illustrated from a few large-scale longitudinal surveys along tropical river systems. The White Nile was the first so investigated. Here sulphate is largely removed in a very large swamp region, the 'Sudd' (Talling 1957*c*, 1976) and the nutrients phosphate and nitrate are seasonally depleted by phytoplankton development in the Jebel Aulia reservoir downstream (Prowse & Talling 1958: see Fig. 3.21). In the nearby Blue Nile, the seasonal nutrient depletion is now spread over two consecutive reservoirs and a downstream stretch, as a cascade sequence (Hammerton 1972, 1976). Along the main channel of the Amazon below Manaus heterotrophic processes predominate, maintaining pCO_2 above the atmospheric level (Richey 1981; Wissmar *et al.* 1981; Stallard & Edmond 1987: see Fig. 2.27).

(vii) Outputs from the air–water interface

Outputs from the air–water interface are gaseous. The only large and widespread fluxes (other than of H_2O!) are of O_2 and CO_2, but under some conditions hydrogen sulphide (H_2S), methane (CH_4) and ammonia (NH_3) could be significant. The quantitative regulation of transfer has already been outlined in (iii) above. The driving difference of partial

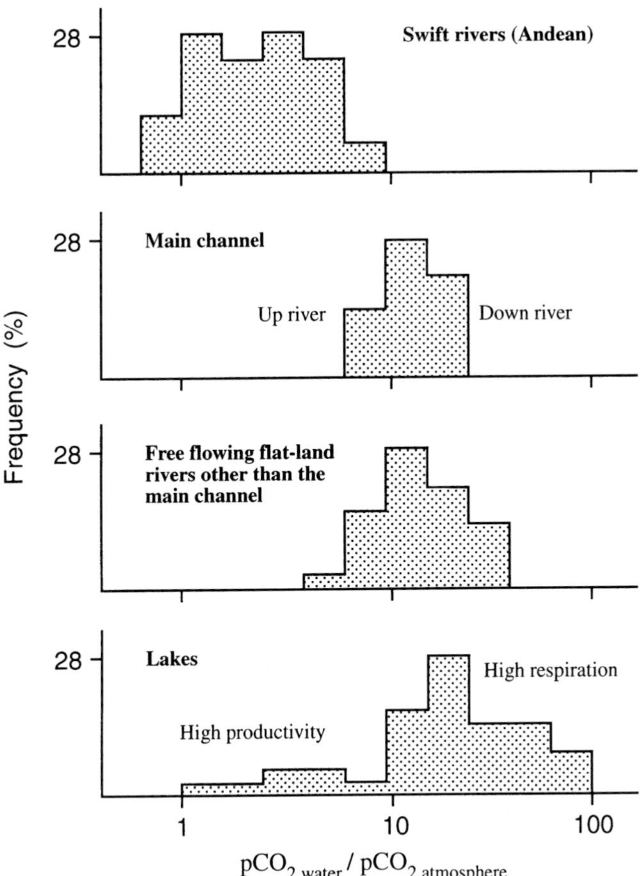

Fig. 2.27 Relative frequency distributions of samples from hydrologically distinctive water types of Amazonia in relation to the ratio of CO_2 partial pressure (pCO_2) in water and atmosphere. From Stallard & Edmond (1987).

pressure can arise in small measure from the change in temperature and hence saturation concentrations, but more considerably from biological activity in productive water for O_2 and CO_2 and under anoxic conditions for the reduced species of H_2S and CH_4. For in-water CH_4 loss, bacterial oxidation is also important if there is access to O_2; fluxes of 2.2–8.8 mmol m^{-2} yr^{-1} have been estimated for the deep African rift lakes of Kivu and Tanganyika (Jannasch 1975; Rudd 1980).

Both concentrations and output fluxes often have a periodicity that is diel (Chapter 5.1) or annual. Rare episodes can have large local impacts, as in the sudden liberation of stored CO_2 of magmatic origin from two crater lakes of Cameroon (Kling *et al* 1991, and Chapters 4.5, 5.1).

Concentrations of dissolved gaseous ('free') CO_2 tend to build up above air-equilibrium levels in waters with considerable organic input, that include most rivers (Fig. 2.28). Most river waters, and to a lesser degree lake waters (African examples in Cole *et al.* 1994), tend to have raised pCO_2 levels and a net surface loss of CO_2. In the Amazon this is promoted by large imports of organic material, and has been estimated by Wissmar *et al.* (1981) and Richey *et al.* (1988). The latter's estimate is an average of 2080 g CO_2 m^{-2} yr^{-1} for the main-stem of the river, with a considerable associated loss of CH_4 here and in the flooded *várzea* region. More generally, with a net CO_2-enrichment in surface water, the exit or evasion flux of CO_2 is potentially large. It may or may not roughly balance an invasion flux of O_2, driven into sub-saturated surface water. Evasion of CO_2 has been studied intensively for the Amazon (Devol *et al.* 1987; Richey *et al.* 1988) in relation to alternative carbon transport in this long river (Richey 1981, Richey *et al.* 1980, 1990; Hedges *et al.* 1986). Two independent methods – using radon transfer and O_2 balance – suggested equivalence to a surface film thickness of about 50 μm (Devol *et al.* 1987). Mean rates per unit area of CO_2 evasion, O_2 invasion and oxidation *in situ* were of similar magnitude, around 3–8 μmol m^{-2} s^{-1} (Richey *et al.* 1990). However, the oxidation rates appeared to vary over the annual cycle, being greatest during rising water level when there is an input of more labile organic material to the river water (Richey *et al.* 1980). The long river main-stem is a transport channel for carbon in several forms, inorganic and organic, dissolved and particulate. The gaseous CO_2 component is important in that its oxidative (respiratory) origin accounts for most of the flux of dissolved inorganic carbon (DIC), with mineral weathering subordinate (Richey 1981). This comparison is supported by evidence from isotopic composition, as $\delta^{13}C$. Further, CO_2 evasion from the river constitutes a large loss flux, that Devol *et al.* (1987) estimated to be equal to approximately half of the input DIC over a long river stretch.

Since high inputs of particulate organic carbon (POC) in swampy terrain are associated with O_2 depletion and CH_4 formation, an alternative gaseous pathway for carbon loss as CH_4 is often significant. This may occur by detached bubbles (ebullition) as well as surface diffusion. Loss by bubbles appeared to predominate in the Amazon floodplain Lake Calado, where a detailed study over two months at falling water level

(*a*)

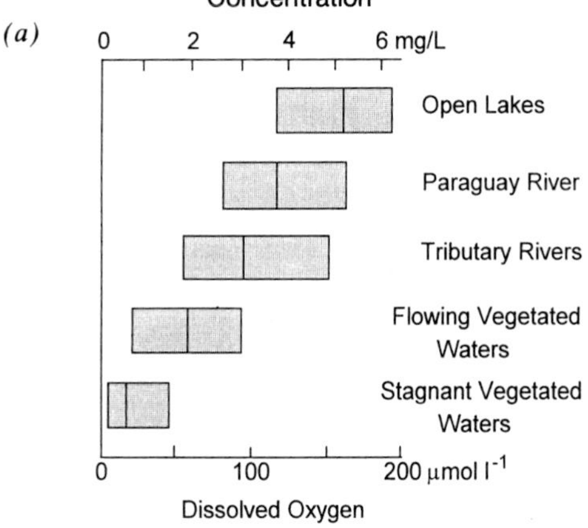

(*b*)

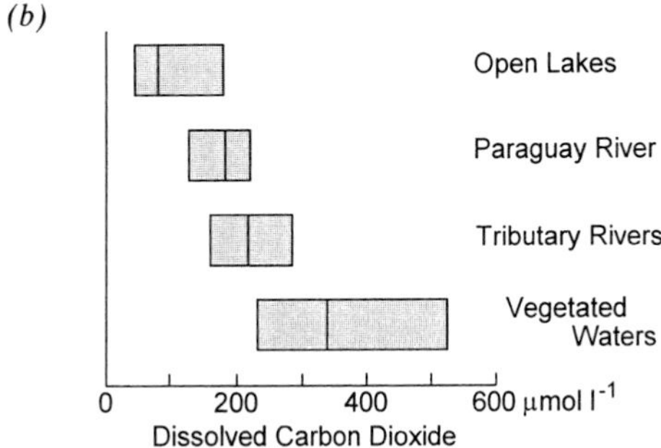

(*c*)

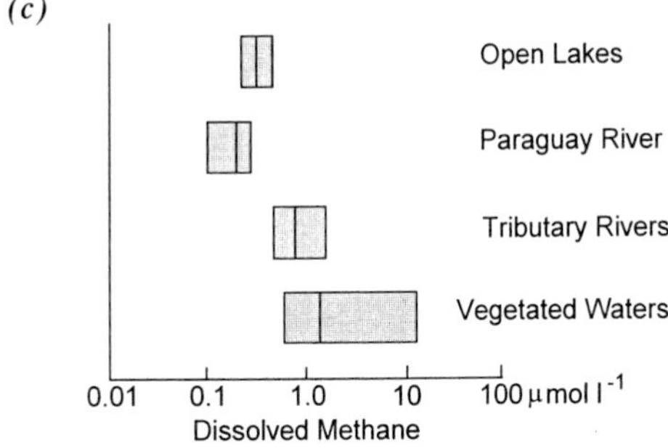

Fig. 2.28. Concentrations of (*a*) oxygen (*b*) carbon dioxide (*c*) methane dissolved in near-surface water along a series of water-bodies in the Pantanal region of Brazil. Stippled boxes enclose median values and the 25 to 75 percentile range. From Hamilton *et al.* (1995).

indicated an average total surface loss of 27 mg CH_4 m^{-2} day^{-1}, a progressive accumulation rate in the water-column of ~100 mg m^{-2} day^{-1}, and a diffusive escape from sediments of ~85 to 118 mg m^{-2} day^{-1} (Crill *et al.* 1988). For Amazon floodplain waters in general Richey *et al.* (1988) estimated the mean decomposition flux of organic carbon as ~1600 g C m^{-2} yr^{-1}, with ~300 g C m^{-2} yr^{-1} as CH_4. Comparative estimates of gaseous carbon loss as CO_2 and CH_4 have been made by Hamilton *et al.* (1995) in a range of waters within the Pantanal wetland of Brazil. Heavily vegetated and stagnant waters yielded the highest concentrations of these gases, and the highest emission fluxes (Figs. 2.28, 2.29). Bacterial methanogenesis accounted for about 20% of the gaseous carbon efflux as CO_2 + CH_4, with about 10% as CH_4 (Fig. 2.29).

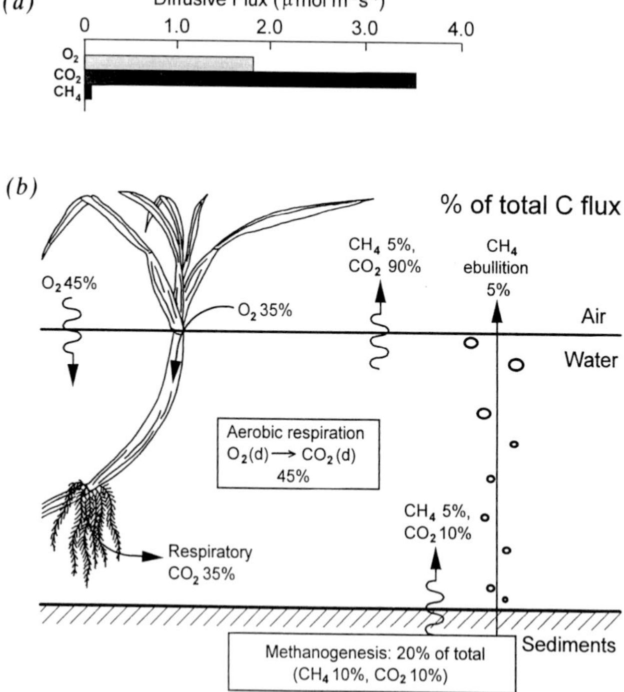

Fig. 2.29. Conditions affecting gaseous concentrations and fluxes (O_2, CO_2, CH_4) in vegetated waters of the Pantanal wetlands, Brazil. (*a*) Estimated diffusive fluxes at the water surface of one vegetated site. (*b*) Flux routes and magnitudes expressed as equivalent % of total C flux. From Hamilton *et al.* (1995).

(viii) Outputs through the water-sediment interface

Losses from the water-column to sediments are mainly by sedimentation of particulates. They are not outputs from the lake basin and may or may not be long-term outputs from the water-mass. Organic sedimentation often relates to local biological production; it includes living and dead plankton organisms, and other non-living detritus including sludge as remains from floating vegetation such as *Cyperus papyrus*. Some chemical features of residues from papyrus and other swamps are described by Gaudet (1979*b*) and Gaudet & Muthuri (1981*a, b*). They are major local incorporations of nutrient elements (N, P, S), but most mineral cations are lost at an early stage. Inorganic sediment-building is partly from products of distant erosion, partly by the mineral fraction of organisms such as diatoms (diatomite, SiO_2) and molluscs ($CaCO_3$).

Generalizations for tropical waters are impracticable or at least insecure. A comparison by McLachlan (1974) indicated a generally lower organic content of sediments from tropical than from temperate lakes and reservoirs. However, highly organic tropical lake sediments are known – as from Lake George (Viner 1977*a*), Lake Victoria (Hesse 1958*a*; Beauchamp 1958; Hecky 1993) and several lakes of the Rio Doce valley in Brazil (Saijo *et al.* 1991). These, and the probable occurrence of lacustrine petroleum deposits (Fleet *et al.* 1988), qualify the old expectation (Ruttner 1931*b*, 1952) of necessarily higher rates of organic decomposition under the much warmer deep water conditions of most tropical lakes. Refractory organics and exposure to O_2 are other variable factors of importance.

Dry tropical climates induce varying degrees of evaporative concentration, the greater being in closed drainage basins. The major ionic composition of a salinity series (Fig. 2.30), from lakes in East and Central Africa, indicates the loss by insoluble combination of Ca^{2+} and Mg^{2+} from the resulting alkaline and saline waters. At the highest salinities there is crystallization of white deposits of trona ($Na_2CO_3.NaHCO_3.2H_2O$), and other minerals including silicate (Maglione & Maglione 1972), as in many waters where Na^+, HCO_3^- and CO_3^{2-} are principal ions. The solubility relations involved have been examined in detail for some African soda lakes, including the commercial site of Lake Magadi (e.g., Monnin & Schott 1984). An example of trona deposits at Lake Nakuru in Kenya is shown in Fig. 2.31. The chemical basis for the long-term evolution of such carbonate-rich brines is further discussed by Eugster & Jones (1979), Gac (1980), Rippey & Wood (1985) and

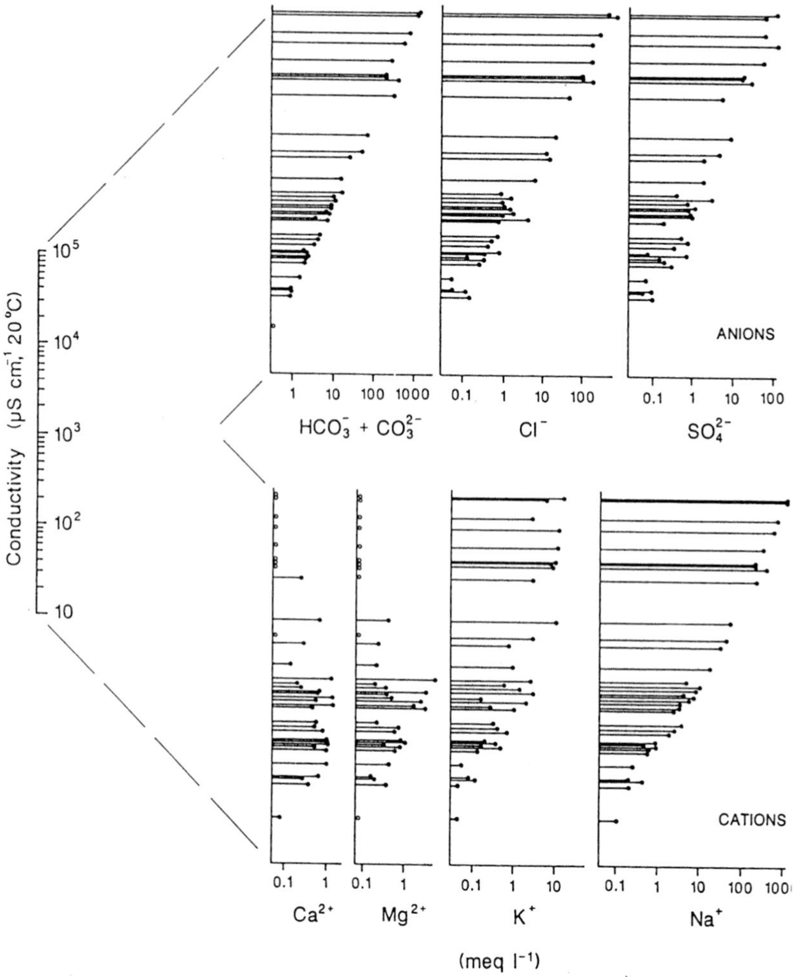

Fig. 2.30. Concentrations of major anions and major cations in relation to conductivity, as an index of total solute concentration, in a series of East and Central African lake waters. Modified from Fryer & Talling (1986).

Kilham (1990*b*); besides the elimination of Ca^{2+} and Mg^{2+}, the proportion of Cl^- among anions tends to rise. The quantities of eliminated solutes have been assessed by Carmouze (1983) for Lake Chad in relation to its changing water storage and chemical budget.

In Lake Chad, and probably many other alkaline tropical lakes rich in dissolved silicate, there appears to be a large-scale chemically selective ('incongruent') elimination of solutes from water-mass to sediments. This

Fig. 2.31. Lake Nakuru. An African soda lake at low level in 1961, with precipitated deposits of trona (some blown by wind) and flamingoes.

transfer is non-biological: sometimes called 'reverse weathering', it involves transformations of sediment minerals as new formation of clay smectites, with consumption of silicate, cations and HCO_3^- and with production of CO_2 (Carmouze *et al.* 1976: Carmouze 1983; Darragi & Tardy 1987). There is thus a net conversion of solutes to particulate material. It has also been invoked, mainly from considerations of mass balance, in Ethiopian rift lakes by von Damm & Edmond (1984), although the evidence here is probably less conclusive (Wood & Talling 1988). In Lake Turkana an appreciable long-term loss of solutes to the sediments has probably occurred by ion-exchange and burial (Yuretich & Cerling 1983), with a form of 'reverse weathering' acting to remove Mg^{2+}.

(b) Chemical budgets (mass balances)

These interrelate input, output and storage fluxes over some given period, help an overall understanding, may provide a quantitative check, and indicate the relative magnitudes of component fluxes. Among the few tropical examples, annual budgets for major ionic components have been estimated for the shallow African lakes of Chad (Carmouze 1983) and Naivasha (Gaudet & Melack 1981), the deeper Lake Turkana (Yuretich & Cerling 1983), and Lake Titicaca in the Andes (Carmouze *et al.* 1982). The first three lakes are exceptional in lacking surface outflows and none is so saline as might be expected from a history of unmitigated evaporative concentration. The water and solute budgets (Tables 2.1, 2.5) indicate that seepage-out and/or net sediment-uptake mainly account for the relatively low salinity. These loss terms have been estimated in different ways between the four lakes, as from water budget quantities (Lake Naivasha), ion exchange of sediment samples plus burial in sediment columns (Lake Turkana), or by difference as residual quantities in an ion budget, that may use presumed conservative (unreactive) chemical species – Na^+ or Cl^- – as a quantitative guide to seepage-out (Lake Chad, Lake Titicaca).

A mean annual solute budget for Lake Chad (1954–72) was described by Carmouze (1983); it is summarized, with related hydrological quantities, in Table 2.5. Chloride and sulphate are not included but were in very low concentration. Although the estimated inputs and outputs balance, this is generally not an independent check on the budget validity as some components are estimated by difference. Sedimentation fluxes are calculated from ionic ratios and the assumption that the sedimentation of Na is negligible. Considerable between-year variation exists behind many of the mean values cited (e.g., of solute concentrations).

The budget indicates how a lake of relatively low salinity can exist without surface outlet in a tropical region with high open-water evaporation. Although solute concentration in the river input (predominantly the Chari River) is low, this would not prevent a progressive rise of salinity in the absence of losses other than evaporation. The other important net losses are by seepage and sedimentation. The first is unselective with respect to solute-components, but occurs chiefly from the deeper northern basin whose input is already with ionic content increased and qualitatively modified from that of the main southern inflow river. Sedimentation is partly by biological agents, with $CaCO_3$ deposition by molluscs, Si and K incorporation by macrophytes (Carmouze *et al.* 1978), and Si incorporation by diatoms (Lemoalle 1978; Carmouze 1983) all quantitatively important. For example, Lévêque (cited by Carmouze 1983) has estimated that annual production by the rich molluscan benthos in 1970 would remove 7×10^5 t of Ca, equal to four times the

Table 2.5. *Estimated components of the water balance and major ionic balance of lakes Chad and Titicaca, as mean annual values for the periods 1954–72 and 1964–78 respectively*

Lake	Quantity	Water ($10^9 m^3$)	Component							
			Na	K	Ca	Mg	HCO_3 (10^9 mol)	SO_4	Cl	$Si(OH)_4$
Chad	lake stock	72								
	annual flux:									
	river inflow	41.5	5.6	2.05	4.35	3.30	22.4	–	–	16.1
	rainfall	6.3					negligible			
	surface outflow	0					0			
	seepage-out	3.8^{x}	5.6	1.65	2.40	1.95	15.6	–	–	3.0
	evaporation	44					0			
	sedimentation^{+}	–	0*	0.40	1.95	1.35	6.8	–	–	13.1
Titicaca	lake stock	900								
	annual flux:									
	river inflow	8.51	13.9	0.90	8.23	2.84	11.04	6.43	12.69	1.37
	rainfall	7.47					negligible			
	surface outflow	0.22	1.85	0.095	0.33	0.325	0.42	0.58	1.73	0.004
	seepage-out	1.36	10.6	0.54	2.22	1.96	2.98	3.59	9.68	0.041
	evaporation	13.8					0			
	sedimentation^{+}	–	0.04	0.195	5.38	0.295	7.25	1.79	0*	1.32

Note: × by difference from the Na^{+} balance
\+ by difference
* assumed

Source: From Carmouze (1983), Carmouze *et al.* (1981, 1982) and Roche *et al.* (1992)

annual river input or half the dissolved stock in the lake. Dissolution and recycling of Ca are therefore important. Non-biological transfer to sediments occurs on a large scale by transformations of sediment minerals (= neoformation of clay smectites, 'reverse weathering') favoured by an alkaline medium rich in soluble silicate. The latter is consumed, together with quantities of Ca^{2+} and Mg^{2+}, and some HCO_3^- transformed to CO_2. Finally, some precipitation of calcite, $CaCO_3$, occurs especially in the northern basin.

A solute budget for another lake without surface outflow, Lake Naivasha, has been estimated by Gaudet & Melack (1981). Here also some components (water seepage-out, sediment exchange of solutes) were estimated by difference; all major ions, plus fluoride (F^-), were included. The relatively low salinity of the lake was ascribed to the major dilute inputs of river water and direct rainfall, an appreciable unselective loss by seepage (+ irrigation off-take), and a selective net accumulation of solutes in sediments. Estimated by difference, in which constituent-errors may be compounded, the absolute magnitude of the last and especially its resolution into (? overestimated) input and output components must be uncertain. There is good evidence for uptake and sedimentation of the Si stock by diatoms, but appreciable effects from sedimentary neoformation ('reverse weathering') were considered unlikely. The study was further notable for a direct use of seepage meters and a Cl^--based assessment of the proportion of cyclic sea-salt in the river solute input, and so by difference the proportion of the total input (~30%) attributable to chemical denudation. The input of some ions (especially K^+, Cl^- and SO_4^{2-}) appeared to be chiefly from the content in rain, whereas surface weathering was dominant for Ca^{2+} and HCO_3^-.

For Lake Turkana, Yuretich & Cerling (1983) estimated a considerable input of solutes to sediments. This occurred mainly by the processes of ion-exchange and solute burial; the former was quantified for three sediment minerals, and the latter estimated from the water content and deposition rate (here high) of sediment. The ion exchange was dominated by Na^+ uptake and Ca^{2+} release with subsequent precipitation of Ca^{2+} as calcite that formed about 5% of deposited sediment. The smaller, but also near-complete, uptake of Mg^{2+} probably involved new sediment mineral (reverse weathering) other than a carbonate species. It was calculated that >40% of the present day lake input of Na^+ could be taken up by ion-exchange, and that ~40% of Cl^- be removed by solute burial. For both these ions the apparent accumulation time required for current concentrations in the lake water was thus raised to approximately 2500 years (Table 2.6), the greater part of the estimated closed basin duration of approximately 3500 years. Later this treatment was further modelled and used to predict and compare the long-term increase of Cl^- concentration in lakes Turkana and Baringo (Barton *et al.* 1987: see Fig. 2.32).

The main inflow to Lake Turkana, the Omo River, is one of three large drainage systems of Ethiopia that end in closed basins. These systems include intermediate 'transit' lakes of lower salinity and one or more terminal lakes – veritable sumps for solutes – of higher salinity. Water fluxes and solute accumulations in the three drainage systems were surveyed by von Damm & Edmond (1984) and Wood & Talling (1988); some are illustrated in Fig. 2.33. Chloride was assumed to be an unreactive or conservative chemical species that could be used to assess levels of evaporative concentration, as between inflow and lake water, and also the relative contributions of cyclic sea-salt and chemical denudation.

Lake Titicaca is notable for the active chemical denudation in its Andean catchment, resulting in inflow concentrations that are generally between 5 and 10 mmol l^{-1}. Losses by the small surface outflow are insufficient to account for the moderate salinity of the lake, *c.* 11 mmol l^{-1}.

Table 2.6. *Mass balance of Lake Turkana and the effect of cation exchange on cation response times*

	Annual Fluxes					Mass	Response Time		
	1.	2.	3.	4.	5.	6.	7.	8.	9.
		10^9 moles per year				10^{12} moles		years	
Na	9.11	0.07	3.71	2.46	3.01	7.99	880	1460	2650
K	0.94	0.07	0.25	0.04	0.72	0.13	138	171	181
Ca	4.72	1.51	0.17	0.01	6.06	0.028	5.9	4.6	4.6
Mg	2.44	0.68	0.10	0.01	3.02	0.023	9.4	7.6	7.6
Cl	2.17			0.94	1.23	3.08	1420		2500
T_{CO2}	22.4			1.32	20.1	4.29	192		203
SiO_2	5.3			0.02	5.3	0.081	15.1		15

Note:
Column:
1. river input (dissolved load)
2. load carried by clays in rivers assuming 100 meq/100 grams exchanging mineral (smectite)
3. annual load removed from lake water by cation exchange
4. loss due to burial of water by sedimentation
5. net input after exchange and burial of interstitial water (1 + 2 – 3 – 4)
6. mass present in lake
7. response time neglecting cation exchange or water burial (6 ÷ 1)
8. response time after cation exchange 6 ÷ (1 + 2 – 3)
9. response time after cation exchange and burial of interstitial water (6 ÷ 5)

Source: From Yuretich & Cerling (1983)

Carmouze *et al.* (1982) deduced, assuming an unreactive conservative Cl^- component, a larger and chemically non-selective loss by seepage and, by difference, varying losses by biogeochemical sedimentation (see Table 2.5). In the latter silicate, Ca^{2+} and HCO_3^-/CO_3^{2-} were prominent. Thus the apparent accumulation or retention time for silicate was estimated as only 20 years for the lake as a whole, as opposed to 479 years for Na^+ and Cl^-.

Another type of mass balance can be applied to a river, equating the chemical flux at a single cross-section to the ultimate sources and transmission factors in the catchment upstream. Table 2.7 gives estimates from three tropical rivers. In one comprehensive study, of the Caura River in Venezuela, Lewis *et al.* (1987) estimated that chemical weathering took

Table 2.7. *Estimated characteristics for the chemical mass balance of the catchments of three tropical rivers*

Gambia River, inland continental basin, 1980–81 (area 42000 km^2, mean rainfall 0.94 m yr^{-1}, discharge 4.6×10^9 m^3 yr^{-1}, runoff factor 0.12); Malewa River, Kenya, 1973–74 (area 1730 km^2, mean rainfall 0.773 m yr^{-1}, discharge 0.14×10^9 m^3 yr^{-1}, runoff factor 0.11); Caura River, Venezuela, 1982–84 (area 47500 km^2, mean rainfall 4.50 m yr^{-1}, discharge 115×10^9 m^3 yr^{-1}, runoff factor 0.54).

	major cations				major anions			major nutrients		total		units
	Na	K	Ca	Mg	HCO_3	SO_4	Cl	P	Si	solutes	partics	
1. areal discharge												
Gambia	3.9	1.4	4.0	2.0	20.3	0.4	0.6	0.048	7.0	48	49	kg ha^{-1} yr^{-1}
Malewa	6.8	2.1	5.7	1.9	55	4.2	2.5	–	5.6	91	–	
Caura	27.6	14.6	15.5	6.0	124	10.0	10.3	0.24	91.5	593	274	
2. atmospheric deposition												
Malewa	4.6	2.6	1.6	1.8	0.82	5.7	3.0	–	–	–	–	
Caura: collection site	10.6	1.33	1.7	0.37	–	11.0	15.1	0.18	0	–	15	kg ha^{-1} yr^{-1}
Caura: from Cl^- flux[a]	8.2	1.03	1.3	0.29	–	8.5	11.7	0.14	0	–	–	
3. surface chemical weathering[b] (as net export)												
Malewa	2.2	−0.48	4.2	0.13	54	−1.6	−0.52	–	–	–	–	kg ha^{-1} yr^{-1}
Caura	19.4	13.6	14.2	5.7	124	1.5	−1.4	0.10	91.5	281	–	
4. areal solute discharge/ rainfall												
Gambia	4	2	4	2	22	0.4	0.6	0.05	7	51	–	kg ha^{-1}yr^{-1} /m yr^{-1}
Malewa	9	3	7	3	71	5	3	–	7	118	–	
Caura	6	3	3	1	28	2	2	0.05	20	132	–	= mg l^{-1} × 10

Table 2.7. *Continued*

	major cations				major anions			major nutrients		total		units
	Na	K	Ca	Mg	HCO_3	SO_4	Cl	P	Si	solutes	partics	
5. atmospheric fraction of river solute discharge												
Malewa	0.68	1.23	0.27	0.93	0.02	1.37	1.21	–	–	–	–	–
Caura	0.20	0.05	0.07	0.04	–	0.82	1.12	0.30	0	–	–	
6. marine fraction of atmospheric deposition												
Malewa[c]	0.52	0.03	0.04	0.22	–	0.10	1.00	–	–	–	–	–
Caura	0.79	0.23	0.19	1.00	–	0.19	1.00	0	0	–	–	
7. river output mean concentration[d]												
Gambia	3.4	1.17	3.4	1.7	17.3	0.37	0.49	0.066	6.0	41	42	
Malewa	9.0	4.3	8.0	3.0	70	6.2	4.3	–	8.0	122	–	$mg\ l^{-1} = g\ m^{-3}$
Caura	1.14	0.60	0.64	0.25	5.1	0.41	0.42	0.003	3.8	24.5	11.3	

Note:
[a] as Cl areal discharge minus rock weathering estimate, adjusted, times observed quotients to Cl in precipitation
[b] as areal discharge minus atmospheric deposition
[c] estimates from East African data: Wood & Talling (1988)
[d] discharge-weighted values for Gambia and Caura rivers

Source: From Lesack *et al.* (1984), Gaudet & Melack (1981) and Lewis *et al.* (1987).

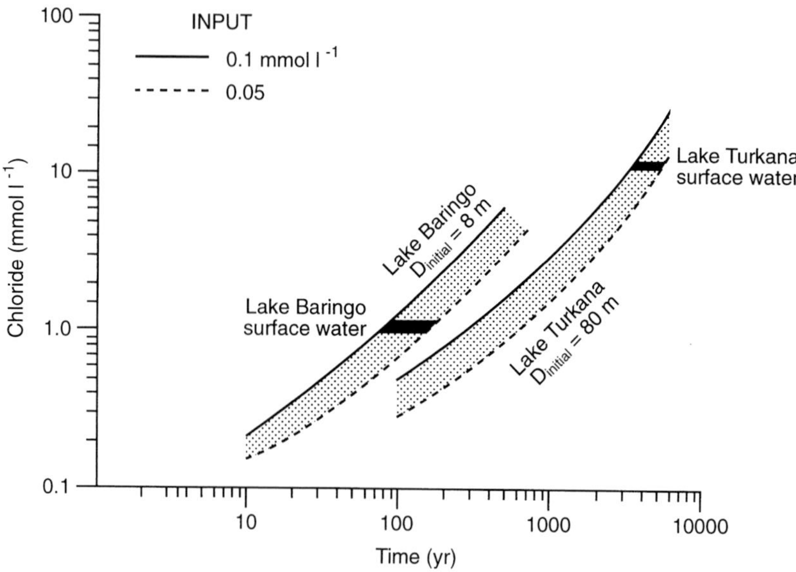

Fig. 2.32. Graphical model of the long-term increase, after basin closure, of Cl concentration in two Kenyan closed-basin lakes with differing initial mean depths ($D_{initial}$) and inflow concentrations set at two levels. Black bars denote the range of concentrations in lake water now observed; the initial, pre-closure, concentrations are taken as those of the river inflows. From Barton *et al.* (1987).

precedence over atmospheric deposition for fluxes of most but not all component major ions, that there was a large N-fixation, and that a possibly abiotic interception of phosphorus occurred. The marine-derived component of atmospheric deposition was here considerable, especially for Mg^{2+} and S; lower contributions of this origin were estimated for more inland sectors of the Amazon (Stallard & Edmond 1981). For the small Malewa River in Kenya, local geology also supported the assumption of an atmospheric origin for the content and flux of Cl^-, and for a deduced major contribution to several other ions – especially K^+, Mg^{2+} and SO_4^{2-}. Gaudet & Melack (1981) estimated the specific fluxes of major ions per unit area of drainage basin, as kg ha^{-1} yr^{-1}. If expressed per unit annual rainfall, these were generally of magnitudes similar to those for the tropical Caura and Gambia rivers (see Table 2.7) and for some North American streams.

Although the Caura catchment is largely on hard siliceous rocks, the geomorphology and high mean rainfall of 4.5 m yr^{-1} lead to considerable annual net discharges (solutes + particulates) of various elements. That of Si, as a major component of both rocks and the river's chemical discharge,

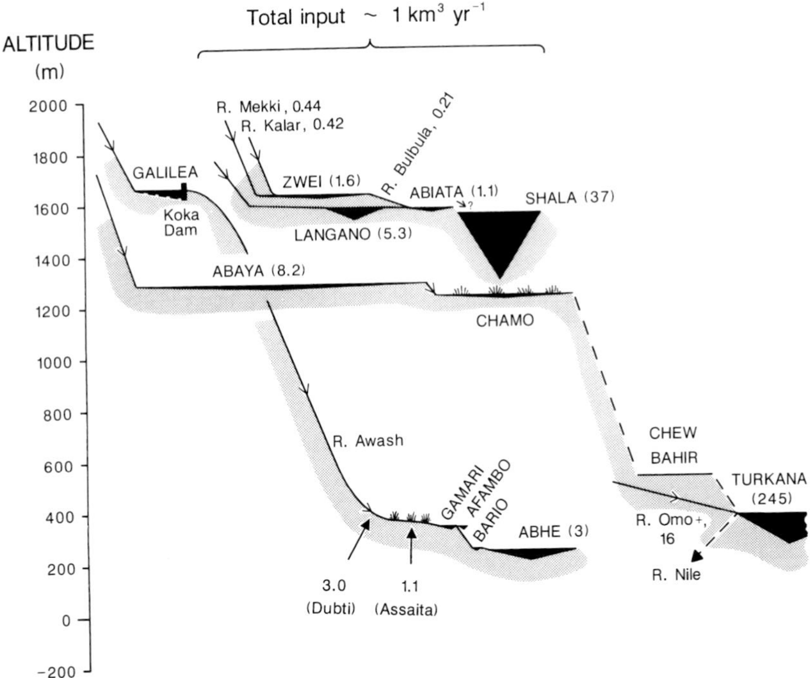

Fig. 2.33. Diagrammatic hydrological sections along three Ethiopian lake–river cascade systems of closed drainage, ending in the salt-accumulant lakes of Shala, Abhé and Turkana. Shared scales are used for altitude, lake area and maximum lake depth. Numbers in brackets indicate lake volume in km^3, and other numbers estimated water-flows in $km^3\ yr^{-1}$. Broken lines indicate past or intermittent connections. From Wood & Talling (1988).

was used to estimate a mean weathering rate of 1.8 mm thickness per 1000 years. Those of other elements were interpreted (see Table 2.3) in terms of weathering fluxes to be expected from the Si-weathering flux and the mean chemical composition of the catchment rock-type, plus a catchment 'retention flux' that was positive for P and negative – indicating preferential weathering – for the major cation elements. However, the soluble component of Si-discharge is in another respect a special case, its concentration being known to show an atypically low responsiveness to water discharge for reasons not well understood. Examples of this are described and discussed for the other two tropical rivers (Malewa, Gambia) by Gaudet & Melack (1981) and Lesack *et al.* (1984). Thus at high flows a 'dilution effect' is not well marked, indicative of large mineral reserves.

Some tropical rivers traverse sectors in which they lose much of their discharge, with chemical consequences. African examples appear in the

'Sudd' swamp region of the Upper Nile (Talling 1957*c*, 1976) and the 'internal deltas' of the Niger and Okavango. For the Okavango system, comparisons of input and output indicate a water loss of ~95% but only a roughly four-fold increase of solute concentration. Here the behaviour of Cl^- (viewed as a conservative quantity) indicated some water loss by infiltration. Ratios with other chemical quantities suggested major losses of other solutes, especially Mg^{2+}, Ca^{2+} and Si, by conversion into particulates (Cronberg *et al.* 1996). In South America, contact of the Paraguay River with the Pantanal wetland could lead to deep periodic anoxia (with fish-kill) in the river, that possibly derived in part from the bacterial oxidation of CH_4. Most major ions, however, behaved conservatively in mixing and distribution (Hamilton *et al.* 1997).

Budgets that summarize movements of the biologically incorporated elements N, P and (for diatoms) Si have special ecological interest. However, on a whole-lake scale, well-established examples appear to be almost lacking for tropical lakes. The outstanding exception is Lake Calado on the Amazon floodplain, for which components of input, regeneration and output for N and P as nutrients are estimated by Melack & Fisher (1990) and Fisher *et al.* (1991) (see Tables 2.8 and 3.3). Here within-lake regeneration is the predominant source of nutrients at low water level. In another floodplain lake, Lake Camaleão, a mass balance for N included large contributions from N-fixation and denitrification (Kern & Darwich 1997; Kern *et al.* 1998). Incomplete annual estimates for Lake George, that do not include some possibly important fluxes such as denitrification and sedimentary P deposition, have been brought together by Livingstone & Melack (1984); they are shown in Table 2.9. These figures, and budgets involving more extensive data for N-fixation in other tropical and some subtropical lakes (e.g., Ashton 1979, 1985*a*), suggest that this process can be a major input of combined N. For Lake Valencia it was estimated to provide ~23% of the N-input, not enough to eliminate a prevailing N-deficiency (Levine & Lewis 1986). For Lake Calado the corresponding estimate for 1989–90 was only ~8% (Doyle & Fisher 1994).

There have been a few attempts to construct Si budgets for tropical lakes in relation to consumption by diatoms, and occasionally also macrophytes, often evident as much reduced surface concentrations in deep lakes. Explicit budgets (mass balances) exist for the shallow lakes of Chad (Carmouze 1983) and Naivasha (Gaudet & Melack 1981), and for the deep lakes Titicaca (Carmouze 1992) and Malawi (Hecky *et al.* 1996) where the estimated net fractional retentions of input Si exceed 0.95.

Table 2.8. *Lake Calado, Amazonia floodplain. Estimates of input, sedimentation, regeneration and output of N and P, expressed as average fluxes per unit area with component percentages*

An average lake depth of 6 m is assumed; regeneration in the water column assumes a 3-m deep epilimnion and little activity below in anoxic water.

	flux (mmol m^{-2} day^{-1})		percentage	
Source/component	N	P	N	P
1. *inputs – total*	1.44	0.045		
direct rainfall			8	6
surface runoff			42	11
groundwater			8	19
adjacent lakes			8	13
Amazon River			34	51
2. *particulate sedimentation*	6.0	0.53		
3. *regeneration – total*	21.3	4.36		
macro-zooplankton			5	2
other water-column regn.			65	78
sediment-to-water transfer			30	20
4. *outputs – total*	2.25	0.11		
groundwater			4	5
burial in sediment			28	31
lake outflow			68	64

Source: From Fisher *et al.* (1991)

Table 2.9. *Phosphorus and nitrogen budgets for Lake George, Uganda in t yr^{-1}; 1000 t yr^{-1} = 4 g m^{-2} (lake area) yr^{-1}*

	N	P
Inputs		
River discharge	1153	182
Rain	277	119
Hippopotamus excretion, etc.	99	15–26
Fixation in lake	1280	–
Total	2809	316–327
Outputs		
Effluent	3180	220
Sediment	655	
Export of fish	75	
Total	3910	
Balance	–1101	

Source: From Livingstone & Melack (1984)

Possibly more biologically relevant budgets for some nutrients, involving chemical stocks and fluxes, have been estimated for plant-productive sectors of a lake. One is floating papyrus swamp vegetation. This has been studied quantitatively in the form of floating stands of papyrus (*Cyperus papyrus*), especially on lakes Victoria (Gaudet 1976) and Naivasha (Gaudet 1977*b*, 1979*b*). Dense stands, of ~5 kg dry weight m^{-2}, incorporate large stocks of N, P and K (Gaudet 1977*b*; Chale 1987) that are shown comparatively with other tropical stands in Fig. 2.34. Export of these elements from the papyrus stands is predominantly to the layer of decomposing 'sludge' deposited below. Import of N from rain and

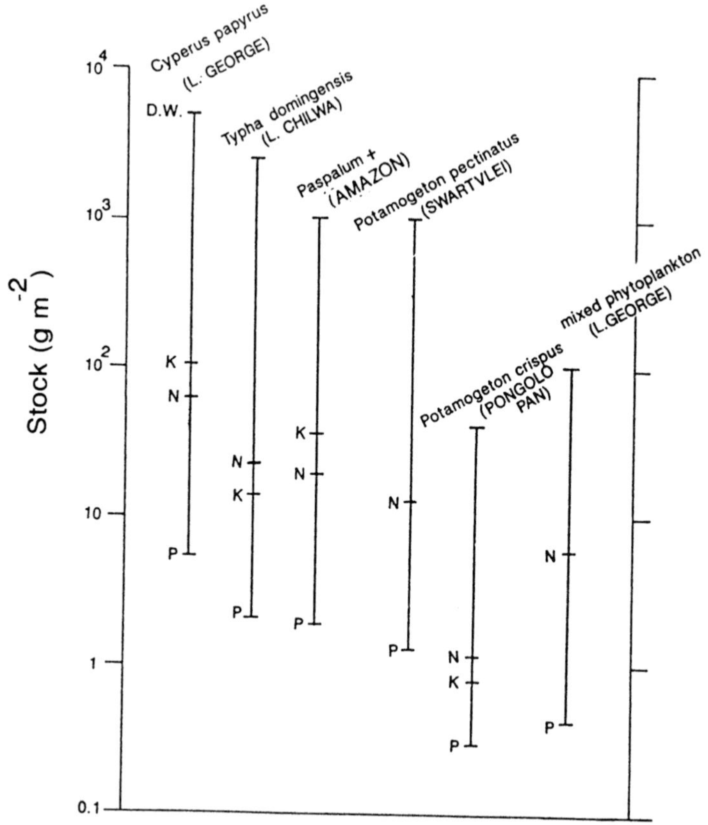

Fig. 2.34. Stocks per unit area of the elements K, N and P in some dense tropical and (Swartvlei, Pongolo Pan) subtropical stands of reedswamp, floating and submerged macrophytes, and phytoplankton of varied dry weight (D.W.). From various authors, after Talling (1992).

surface run-off seemed too small to maintain the larger quantity exported, as estimated from sediment traps, and a major contribution from N_2-fixation has been proposed (Gaudet 1979*b*). Other indications of N-fixation (Calder 1959; Viner 1982*b*) involved very much smaller fluxes, and the status of the (large) sediment trap-based estimates as net export has been questioned (Viner 1982*b*). However, there is now some evidence for considerable nitrogenase activity, with N-fixation, in the root system of papyrus (Mwaura & Widdowson 1992). Floating 'meadows' of another highly productive plant, the grass *Echinochloa polystachya*, are abundant on the Amazon floodplain where nutrient uptake and release are governed by the annual cycle of rising and falling water level. At one site this vegetation was estimated to have an annual consumption and temporary accumulation of 37.7 g N m^{-2}, 5.1 g P m^{-2} and 114 g K m^{-2} (Piedade *et al.* 1997). The floating-mat situation differs from the strongly soil-rooted condition of tropical reedswamps, for which some estimates have been made of nutrient input and output from lakes Chilwa (Howard-Williams & Lenten 1975; Howard-Williams & Howard-Williams 1978) and Naivasha (Gaudet & Muthuri 1981*a, b*: see Fig. 2.35). The subject is reviewed comparatively by Howard-Williams & Gaudet (1985). Rooted reedswamps can be said to act as 'nutrient pumps', in that nutrient uptake from subsurface soils eventually leads to biomass decay with liberation to shallow water (Fig. 2.35). Such circulation is probably also important for the floating meadows of grasses in Amazonian floodplain lakes (Junk 1997).

Another productive sector, on a larger scale, is the mixing-prone upper layer or *mixolimnion* of deep and indefinitely stratified (meromictic) lakes. Two outstanding tropical examples of such lakes are Lake Tanganyika (Fig. 2.36) and Lake Malawi (Fig. 2.37), for which Hecky (1991), Bootsma & Hecky (1993) and Hecky *et al.* (1996) have made speculative estimates of nutrient income to the respective mixolimnia (Table 2.10). Roughly the upper 200 m is included here, within which a seasonal thermocline induces further but temporary compartmentation. The estimated inputs of P and Si are predominantly by mixing from below, whereas for N in Tanganyika this source is held to be largely eliminated by denitrification near the base of the mixolimnion. In this region notably low concentrations of NH_4-N and NO_3-N have been measured (Fig. 2.36), although possible variation with horizontal location and time is not known. If the P-output from the mixolimnion is supposed to be mainly by biotic incorporation and sedimentation, the accompanying N-output – assessed from the expected N:P in biomass – is large compared to esti-

3

Resource utilization and biological production

We now pass from mainly environmental transfers to the movement of energy and materials into assemblages of replicating living organisms. Beyond the initial interception, the energy diverted is a very small fraction of the main source-flux of solar radiation. In fact consequences other than dissipation as heat are usually neglected in environmental energy budgets. However, chemical uptake can consume large parts of the chemical source-fluxes. Here we consider, in the tropical context, fluxes of organic production in relation to these inputs and to the active biomasses responsible. Significant here is the relatively maintained level of solar radiation on the annual scale and of elevated temperature as a near-universal rate-promoting factor. Sequences of consumption operate from primary producer to top-predator and involve the general composition of tropical aquatic communities.

3.1 Primary utilization: energy

The flux of photosynthetically available solar radiation (PAR) that penetrates the water surface sustains primary or photosynthetic production by underwater communities of aquatic plants, both small (microphytes) and large (macrophytes). The kinetics of this utilization are most readily studied for the dispersed phytoplankton, which provides most of the tropical information discussed here. There is also interception, above the water surface, by emergent aquatic macrophytes (e.g., *Cyperus papyrus*: Fig. 3.19a) that fringe water-bodies or dominate large tropical swamps. For these the direct information on photosynthetic fluxes is scanty but the general principles of energy input, interception and utilization by tropical macro-vegetation (Monteith 1972) are applicable.

General characteristics of the incoming solar radiation – including its variation with latitude, season, diel period and atmospheric interception – are outlined in Chapter 4. There fluxes are given in terms of their total energy content, of which photosynthetically active radiation (PAR) is one spectral component or wavelength band (roughly 400–700 nm, fraction 0.46) and which is also often expressed in terms of photon flux, as μmol (= 'μEinstein') m^{-2} s^{-1}. Here 1 μmol equals 6.02×10^{17} photons. Approximate equivalences for daylight, dependent on its spectral composition, with absolute magnitude corresponding to a solar elevation of 45° in a clear sky, are 700 J m^{-2} s^{-1} or 1.0 cal cm^{-2} min^{-1} as total energy-flux density; alternatively, as PAR, ~320 J m^{-2} s^{-1} or ~1600 μmol photons m^{-2} s^{-1}. Here 1 J m^{-2} s^{-1} ≈ 5 μmol photons m^{-2} s^{-1}. With a vertically overhead sun, the maximum flux density on a horizontal surface is about 1.5 cal cm^{-2} min^{-1} (total solar radiation) ≈ 480 J m^{-2} s^{-1} (PAR) ≈ 2400 μmol m^{-2} s^{-1} (PAR).

(a) Underwater light penetration and interception

Penetration below a water surface is first subject to a 'surface loss' or albedo, mean order of magnitude about 10%, by reflection and by back-scattering with upwelling light. The reflected component is increased at low solar elevation and calm conditions; the back-scattered component relates to the particulate content of the water. The total loss has apparently never been studied critically in tropical inland waters, although the back-scattered component has occasionally been used to provide information (e.g., on turbidity) by remote sensing. Examples exist for Lake Kainji (Abiodun & Adeniji 1978) and Lake Chad (Lemoalle 1979*b*). For short-term assessments with high solar elevation near midday, a loss factor of 5% has been assumed (e.g., Talling 1965*a*). In further penetration, the diminution $-\mathrm{d}I$ of a flux density I over a vertical depth interval $\mathrm{d}z$ can be expressed in terms of a vertical attenuation (extinction) coefficient K as:

$$-\mathrm{d}I = K.I.\mathrm{d}z \qquad (3.1)$$

or in integrated form between two depths z_1 and z_2 with flux densities I_1 and I_2,

$$K = \frac{1}{(z_2 - z_1)}(\ln I_1 - \ln I_2) \qquad (3.2)$$

If the water column is optically homogeneous, the coefficient K (units m^{-1}) is near-constant so that the flux density I decreases exponentially and its logarithm decreases linearly with depth for any spectral (colour) component. However K will vary for different regions of the spectrum, with a minimum value K_{min} at some intermediate wavelength or colour region.

Spectral relationships were first demonstrated in tropical freshwaters of Guiana by Carter (1934) using photoelectric measurements with colour filters. Variation with time was followed by Beauchamp (1953*a* and unpublished) during 1939–40 in Lake Malawi. More extensive applications to the analysis of underwater photosynthesis were made in the 1950s (e.g., Levring & Fish 1956; Talling 1957*a*), 1960s (e.g., Talling 1965*a*; Ganf 1974*c*) and 1970s (e.g., Lemoalle 1973*a*, 1979*a*; Lewis 1974; Robarts 1979; Melack 1979*a*, 1981). Examples that include the African lakes Victoria and George are shown in Fig. 3.1. These lakes differ in that attenuation is much higher in Lake George, where the spectral 'window' with minimum attenuation K_{min} is displaced to the red region, as compared with the green region for Lake Victoria and the blue region for pure water. A corresponding 'red shift' occurs in the spectral distribution of radiant energy at depth (Fig. 3.2*b*).

An important link to biology is the lower limit of the *euphotic zone*, the depth zone within which appreciable photosynthetic production is possible. This is by convention taken as the depth z_{eu} where the flux I of PAR is reduced to 1% of the surface-penetrating value I_0'. A simple and rough but useful guide to z_{eu} can be obtained from *transparency* measured visually, as image-loss, by Secchi disc. A multiplying factor between two and three is generally applicable; it will vary especially with conditions of light-scattering. Lemoalle (1979*a*) has defined values specifically applicable to different water types in Lake Chad. If light reduction is supposed to conform, approximately, to a single 'average' or 'effective' vertical attenuation coefficient K_e, then $z_{eu} = 4.6/K_e$. Strictly coefficients cannot be so averaged, and a summation of spectral components gives a slightly curvilinear relation between ln I and z (Fig. 3.2c). An alternative is to define the euphotic zone depth z_{eu} in terms of the coefficient K_{min} applicable to the most penetrating spectral component. Empirical tests on various African lakes have indicated as best relation $z_{eu} \approx 3.7/K_{min}$ (Talling 1965*a*; Ganf 1974*c*; Belay & Wood 1984). This relation has also been applied in several other studies of photosynthesis by phytoplankton in African lakes (Talling *et al.* 1973; Robarts, 1979). Further, $K_e \approx 1.15$ to $1.3\ K_{min}$ (Talling, 1965*a*; Ganf, 1974*c*; Erikson *et al.* 1991*b*).

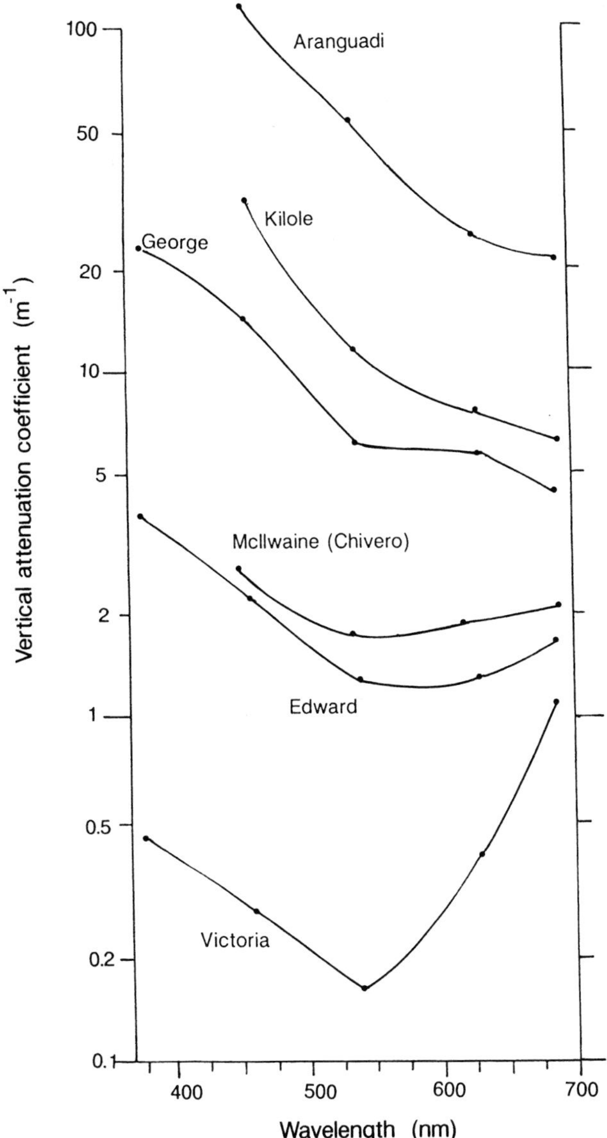

Fig. 3.1. Variation, over the visible spectrum, of the vertical attenuation coefficient *K* measured with selenium cell and colour filters in a range of tropical African lakes. Based on Talling (1965*a*), Talling *et al.* (1973) and Robarts (1979).

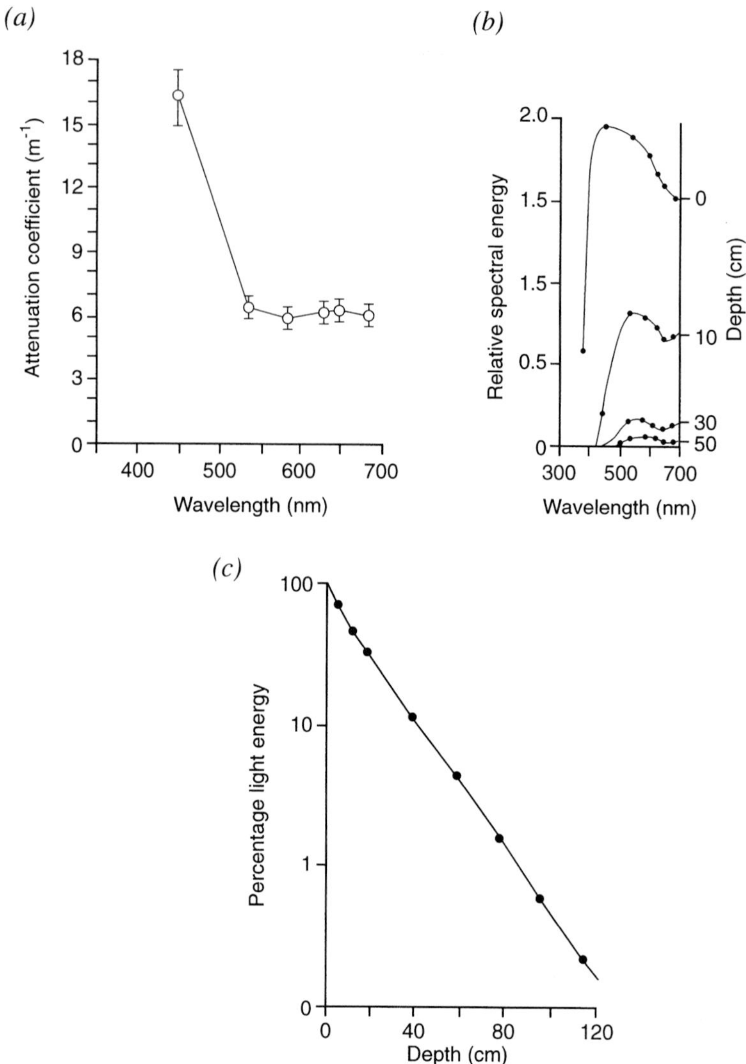

Fig. 3.2. Light penetration in Lake George, Uganda, as defined by (*a*) the spectral variation of mean vertical attenuation coefficients, (*b*) the derived spectral distributions of radiant energy at four depths, and (*c*) the further derived semi-logarithmic plot of % total energy flux (PAR) with depth, showing slight curvilinear character. From Ganf (1974*c*).

Since about 1980 instruments to measure underwater photon flux density, as PAR, have been widely available. They bypass the time-consuming measurements with multiple colour filters followed by indirect summation, but with loss of spectral information.

Especially in rivers it is not uncommon to meet conditions of high mineral turbidity with transparency below 0.1 m, very high values of attenuation coefficients (e.g., White Nile: Talling 1957*c*) and little opportunity for photosynthesis – that is confined to a zone a few centimetres deep. Similar effects of river sediment load may be transmitted to a receiving impoundment, such as Lake Dalrymple in North East Australia (Griffiths & Faithful 1996).

Of prime biological interest is the component of the total attenuation coefficient due to light interception by photosynthetic phytoplankton. If biomass concentration B is assessed by chlorophyll a concentration [chl-a], and an index coefficient of light attenuation – K_e or K_{min} – plotted against this, a linear relationship is typically obtained if [chl-a] is more variable than other major determinants of attenuation such as silt content. Figure 3.3 shows example plots of K_{min} versus [chl-a] from the

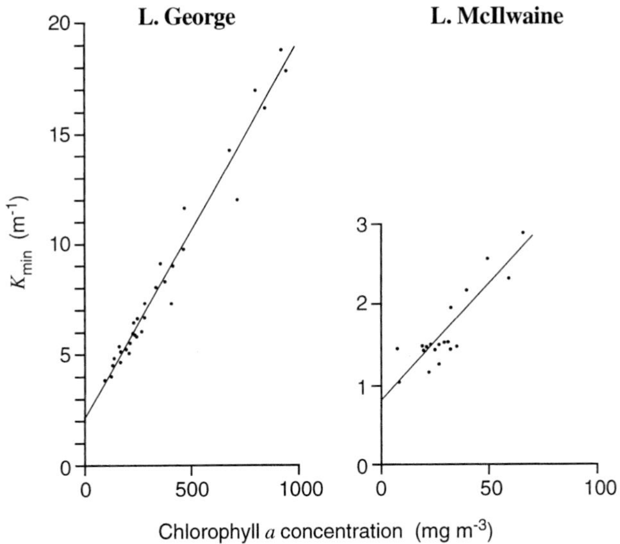

Fig. 3.3. Linear regressions relating the spectral-minimum vertical attenuation coefficient (K_{min}) to the chlorophyll a concentration of phytoplankton in two African lakes. Based on Ganf & Viner (1973) and Robarts (1979), after Talling (1992).

African lakes George and McIlwaine (=Chivero). The gradient of each plot (k_s) is a measure of light interception (ln units m^{-1}) per unit biomass concentration (mg chl-*a* m^{-3}), yielding ln units of attenuance per mg chl-*a* m^{-2}. Values for these lakes are, respectively, 0.016 and 0.0207 (mg chl-*a* $m^{-2})^{-1}$; they would be slightly increased (factor approximately 1.2 to 1.3) as k_s' if the coefficient K_e fitted to mean PAR attenuation were used (as by Hawkins & Griffiths 1986) instead of K_{min}. The tropical values, including those of Lemoalle (1983) for Lake Chad, Erikson *et al.* (1991*b*) for Lake Xolotlán, and Chacón-Torres (1993*a*) for k_s' in Lake Patzcuaro, are consistent with world-wide experience of a limited variability of k_s, at ~0.015 ± 0.010 (mg chl-*a* $m^{-2})^{-1}$. In this variability, mean particle size and spectral effects contribute significantly. The expectation of low k_s values with large particles (the 'package effect') is apparently realized with the blue-green *Spirulina fusiformis* (*S. platensis* auct.) (Melack 1979*b*).

The absorption spectra of plant pigments are varied and distinctive. In phytoplankton-rich waters they influence the spectral distribution of light underwater; however, for tropical waters there is a dearth of measurements *in situ* with spectroradiometers of suitably high resolution. Townsend *et al.* (1996) describe two examples from reservoirs in northern Australia. Laboratory measurements (Talling 1970) on plankton-rich water from Lake George, and derived filtrate, have shown the varying spectral attenuation of the plankton component, plus the enhanced blue absorption of the filtrate with organic 'yellow-substance' (also known as Gelbstoff, gilvin). Material of the latter type is of obvious influence in tropical 'black-water' rivers such as the Rio Negro, Orinoco and Zaïre, as well as in many highly productive tropical lakes (e.g., Lake Chilwa: Moss & Moss 1969; Lake Kilole: Talling *et al.* 1973).

If the euphotic zone is the significant site of photosynthetic production, the biomass-content of that zone per unit area ($\Sigma_{eu}B$, e.g., as mg chl-*a* m^{-2}) is a key quantity for production dynamics. Actual contents are equal to the product of mean concentration B and depth z_{eu}, and will have an *upper limit* $[\Sigma_{eu}B]_{max}$ when the biomass-associated attenuance is equal to that defining the euphotic zone. Formulating in terms of K_{min}:

$$z_{eu} \approx \frac{3.7}{K_{min}}, \text{ and } K_{min} = B.k_s \quad (3.3)$$

$$\text{so } [\Sigma_{eu}B]_{max} = z_{eu}B \approx \frac{3.7}{K_{min}}\, B \approx \frac{3.7}{Bk_s}\, B \approx \frac{3.7}{k_s} \quad (3.4)$$

Thus representative values of k_s, the attenuance per unit biomass content, of 0.01 and 0.02 (mg $m^{-2})^{-1}$ then correspond to maximum euphotic contents of approximately 370 and 185 mg chl-*a* m^{-2}, respectively. Values of this magnitude are known from several productive tropical lakes (Talling *et al.* 1973; Ganf 1974*c*; Lemoalle 1981*a*; Melack 1981; Erikson *et al.* 1991*b*). Lower values of k_s for *Spirulina fusiformis* apparently led to much higher euphotic content, up to 665 mg chl-*a* m^{-2}, in a Kenyan crater lake, Lake Simbi (Melack 1979*b*).

(b) Depth-profiles of photosynthesis

The great variation of light penetration into tropical inland waters implies a corresponding variation in the depth-range of photosynthetic activity. Early tests of this variability were made in 1929 by Jenkin (1936) on a Kenyan lake, in 1931 by Beadle (1932*b*) and in 1953 by Levring & Fish (1956) on various East African lakes. These studies measured oxygen production by standardized samples (often net-phytoplankton) exposed in light and dark bottles suspended over a range of depths. In principle they were mainly bioassays of light (PAR) penetration rather than estimates of production, although Jenkin (1936) derived some estimations of specific rates per unit biomass of the blue-green *Spirulina fusiformis* (then referred to *Arthrospira platensis*).

Work intended to assess rates of photosynthetic production *in situ* began in the early 1950s with experimental exposures in two Central American lakes (Deevey 1955) and in the upper Nile system (Talling 1957*a*; Prowse & Talling 1958). During the 1960s and 1970s they were extended, including some longer time-sequences and measurements in incubators, to many other tropical water-bodies. Examples, listed in Table 3.1, can be drawn from Africa and South East Asia. Before 1980 various pioneer studies had also been made in Central and South America, including Amazonia, Venezuela, Panama, Cuba and Lake Titicaca; also small reservoirs and a forest-lake (Carioca) in southeast and northeast Brazil. Since 1980 there have been notable studies on Lake Titicaca, on the Central American lakes of Xolotlán and Chapala, as well as work on numerous smaller lakes and reservoirs of Brazil, Ecuador, Venezuela and Cuba. Also from 1980, work in Africa has included Lake Malawi, Lake Victoria, several Ethiopian rift lakes and a Nigerian river-lake. Overall, studies have become fewer than in previous decades. In South East Asia recent studies are also few in number, but include work on largely man-made water-bodies in India, Sri Lanka, Bangladesh and Papua New Guinea.

Table 3.1. *Measurements of photosynthetic production by phytoplankton*

Period	Neotropics	Africa	Australasia
pre-1960	*Central America* Deevey 1955	*Nile System* Talling 1957*a*[1] Prowse & Talling 1958	
1960–80	*Amazonia* Hammer 1965 Schmidt 1973*a*, *b*[2], 1976[2], 1982 Fisher 1979[1] Melack & Fisher 1983 *Venezuela* Gessner & Hammer 1967 Lewis & Weibezahn 1976 *Central America* Gliwicz 1976*b*[2] Pérez-Eiriz *et al.* 1976, 1980 Romanenko *et al.* 1979[1] *Titicaca* Richerson *et al.* 1977[2], 1986, 1992[2] Lazzaro 1981[2] *Brazilian Lakes* Tundisi *et al.* 1978 Barbosa & Tundisi 1980 Hartman *et al.* 1981	*East Africa, Sudan, Ethiopia* Talling 1965*a*[2] Ganf 1972, 1975[1,2] Talling *et al.* 1973 Melack & Kilham 1974[1] Ganf & Horne 1975[1] Melack 1979*a*[2], *b*, *c*, 1980, 1981, 1982 Vareschi 1982[1,2] Harbott 1982 Belay & Wood 1984 *West Africa, Zaïre, Chad* Lemoalle 1969, 1973*a*, 1975, 1979*a*[1], 1981*a*, 1983[1,2] Freson 1972[2] Thomas & Radcliffe 1973 Karlman 1973[2], 1982[2] Pagès *et al.* 1981 Dufour 1982 Dufour & Durand 1982 *Southern and Central Africa* Robarts 1979[2] Hecky & Fee 1981 Degnbol & Mapila 1985 Cronberg 1997	*India* Sreenivasan 1965 Hussainy 1967 Ganapati & Sreenivasan 1970 Michael & Anselm 1979 Kannan & Job 1980c[1] *Malaysia* Prowse 1964, 1972 Richardson & Jin 1975 *Philippines* Lewis 1974[1,2]
1980+	*Central America* Erikson *et al.* 1991*a*[1], *b* Lind *et al.* 1992[2] *Venezuela* Gonzales *et al.* 1991 *Ecuador* Miller *et al.* 1984 *Titicaca* Vincent *et al.* 1984, 1986[2] Richerson 1992 *Brazil* Reynolds *et al.* 1983 Gianesella-Galvão 1985 Barbosa *et al.* 1989[1] Forsberg *et al.* 1991 Tundisi *et al.* 1997[1,2]	*East Africa* Mugidde 1993[2] Mukankomeje *et al.* 1993 Patterson & Wilson 1995[1] *Ethiopia* Belay & Wood 1984 Kifle & Belay 1990[2] Gebre-Mariam & Taylor 1989*a* Lemma 1994 *West Africa* Nwadiaro & Oji, 1986 *Malawi* Degnbol & Mapila 1985 Bootsma 1993*a*	*India* Saha & Pandit 1987[2] Durve & Rao 1987 Kundu & Jana 1994 *Sri Lanka* Dokulil *et al.* 1983[1] Silva & Davies 1986[1], 1987[2] *Bangladesh* Khondker & Parveen 1993[2] Khondker & Kabir 1995 *Papua New Guinea* Osborne 1991

Note: [1] with diel series
[2] with annual series

Figure 3.4 shows a diversity of depth-profiles of photosynthetic activity from East African lakes. The fluxes shown are based on O_2 production during short exposures of 1–3 h, are expressed per unit water volume. As differences from dark exposures they are interpreted as gross photosynthesis. The vertical extent of the profiles corresponds closely to that of light in the most penetrating spectral band, and hence to the reciprocal $1/K_{min}$ and the euphotic zone depth z_{eu} (Talling 1965*a*). The horizontal extent is mainly determined by biomass concentration and its inverse relation to

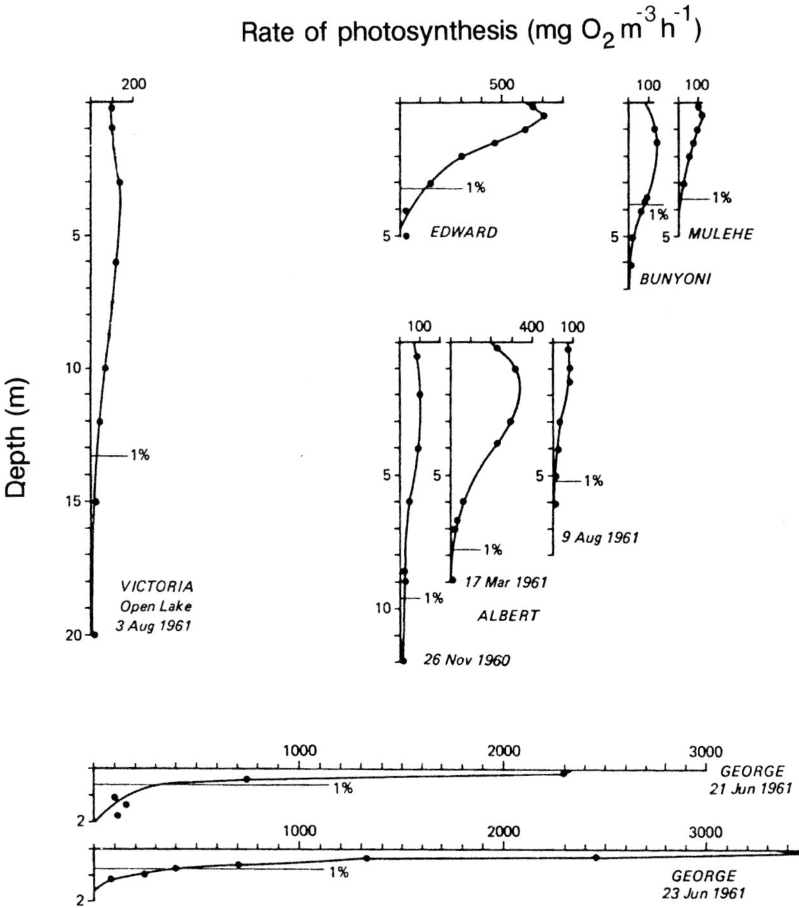

Fig. 3.4. Photosynthesis-depth-profiles in East African lakes, showing diversity of absolute rates and euphotic zone depths (1%). Modified from Talling (1965*a*).

vertical extent is partly due to the significance of biomass-associated attenuance ($k_s B$) in the total attenuance (K_{min}).

Excluding these questions of vertical and horizontal scaling, there is a similar intrinsic shape to the profiles that includes three main regions – near-surface light-inhibition, light-saturation and a lower light-limitation. The 'intrinsic shape' can be isolated by a plot of profiles on double logarithmic scales. It will clearly depend on the relationship between specific photosynthetic rate ϕ and radiation (or photon) flux density ('light intensity') I, which can be modelled quantitatively and is susceptible to variation with such factors as temperature, previous light-exposure and type of algae. The magnitude of the surface-penetration flux I_0' will determine the operative extent of the $\phi - I$ relationship; in dull daylight, for example, the inhibition region is likely to be absent and the entire depth-displacement of the profile reduced.

Finally, the shape of the profile may be influenced by differences of biomass with depth, quantitative or qualitative. Such differentiation is small or absent in the examples of Fig. 3.4, and similar results were obtained with homogeneous surface-derived material as with depth-related samples. Elsewhere, in certain inshore bays of Lake Victoria, the biomass was sometimes strongly depth-stratified and this modified the profiles of activity per unit water volume (Talling 1965*a*: see Fig. 3.5). Sedimentation of diatom biomass from near the water surface led to a false simulation of photoinhibition.

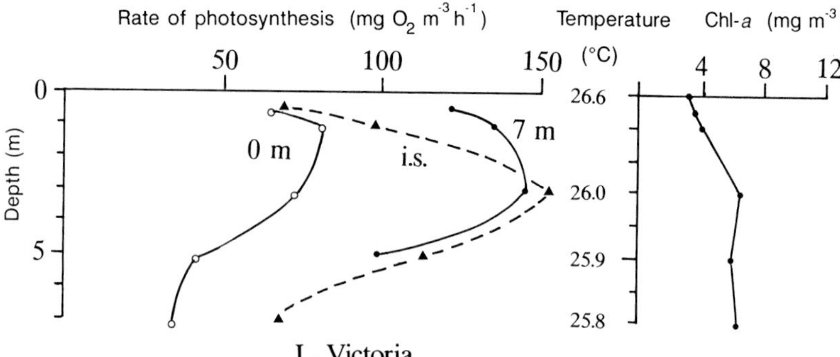

Fig. 3.5. Depth-profiles of photosynthesis per unit water volume (A), temperature and chlorophyll *a* concentration in a weakly stratified bay of Lake Victoria, showing true near-surface inhibition of photosynthesis for 0 m and 7 m collected material, and additional depression due to diatom sedimentation for material exposed *in situ* at the depths of collection (i.s.). Based on Talling (1965*a*, after Talling 1995*b*).

Examples of depth-profiles from other tropical regions appear in Figs. 3.6, 3.7 and 5.49. Most are based on short exposures (< 6 h) of phytoplankton samples within bottles *in situ*, with fluxes estimated either by O_2 evolution or by ^{14}C fixation. These two methods were run in parallel, with broadly similar results, by Lewis (1974) on Lake Lanao in the

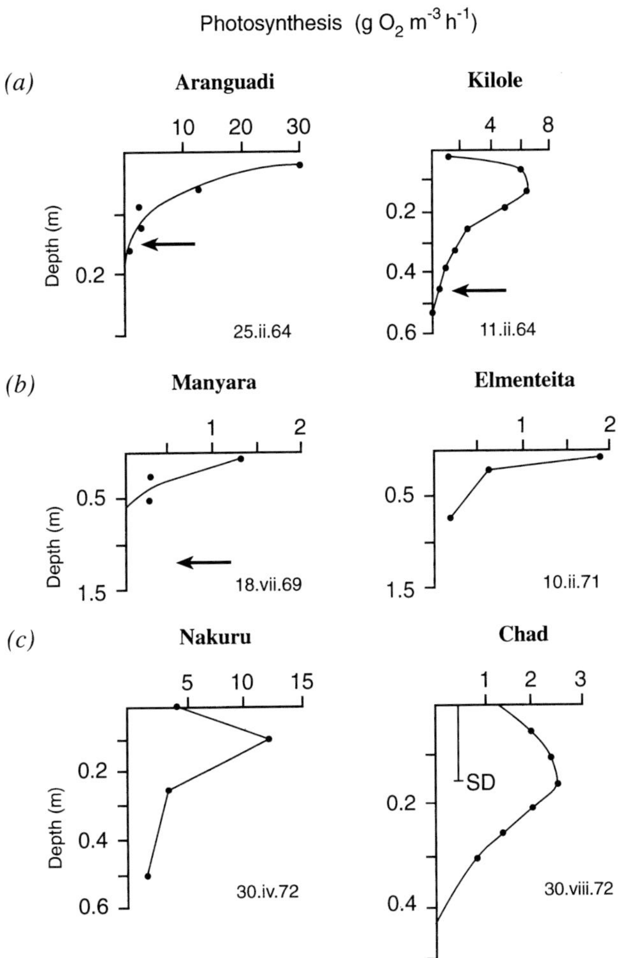

Fig. 3.6. Vertically condensed depth-profiles of photosynthesis by dense phytoplankton in highly absorptive lake waters on various dates: (*a*) two Ethiopian crater lakes, (*b*) two East African soda lakes, (*c*) Lake Nakuru, Kenya, and Lake Chad. Euphotic depths are indicated by arrows in (*a*), and Secchi depth by a bar in (*c*). Modified from Talling *et al.* (1973), Melack & Kilham (1974), Vareschi (1982) and Lemoalle (1983).

Philippines and by Ganf & Horne (1975) on the highly productive waters of Lake George in Uganda. In Lake George, and many soda lakes of East Africa with abundant phytoplankton (Talling *et al.* 1973; Melack & Kilham 1974; Melack 1979*b*, 1981, 1982; Vareschi 1982), and in highly absorptive river waters such as the White Nile (Prowse & Talling 1958) and parts of the Orinoco system (Lewis 1988), the euphotic zone is less than 1 m deep and photosynthesis correspondingly condensed (Figs. 3.6 and 3.7).

The situation of a dense phytoplankton with a condensed photosynthetic zone can have implications of intense CO_2-demand by photosynthesis per unit water volume. Such demand leads to considerable CO_2-depletion, sometimes HCO_3-depletion (Wood, Kannan & Saunders, 1984), and diurnally raised pH (to >10) in waters of relatively low alkalinity (<3 meq l^{-1}) such as Lake George, where bicarbonate is probably used directly as a CO_2-source (Ganf 1972). The same use is likely, but unproven, in soda lakes of high alkalinity and pH. There the strong buffering reduces diurnal pH excursions to values that are small (Talling *et al.* 1973) or imperceptible (Vareschi 1982), although a continuing CO_2-depletion can exist relative to the concentration expected if air and water were in gaseous equilibrium (Talling & Talling 1965).

(c) Photosynthetic characteristics

Finer analysis of photosynthesis depth-profiles depends upon characterization of the rate of photosynthesis (A or ϕ) versus light flux (I) relationship. Omitting high-light inhibition, this can be described by parameters representing absolute and biomass-specific rates at light-saturation (A_{max}, ϕ_{max}), the onset of light saturation (I_k), and the initial gradient at low light of the ϕ–I relationship ($\alpha = \phi_{max}/I_k$). These are illustrated in Fig. 3.7. The parameters of absolute rate per unit water volume (A_{max}) or biomass-specific rate (ϕ_{max}) are available from many locations. The others have been estimated from profiles at a few tropical sites, including a Nile reservoir (Talling 1957*a*), several East African lakes (Talling 1965*a*; Ganf 1975; Ganf & Horne 1975), Lake McIlwaine (= Chivero) in Zimbabwe (Robarts, 1979), and Lake Chad (Lemoalle 1979*a*, 1983). An alternative approach has been to expose samples within a constant temperature bath in which irradiance is either constant (Tundisi *et al.* 1978; Lemoalle 1979*a*; Barbosa *et al.* 1989), set at various fractions of incident solar radiation by calibrated filters (Fisher 1979), or held at a series of constant values under artificial illumination (Hecky & Fee 1981;

Neale & Richerson 1987; Lewis 1988; Forsberg *et al.* 1991; Bootsma 1993*a*).

The magnitude of parameters so derived are ultimately crucial for ecosystem function. That of the light-saturation characteristic I_k is independent of biomass, but shares the temperature dependence of rates at light-saturation and is of the order of one-tenth of full sunlight. Its ratio with changing values of surface-penetrating radiation (I_o) determines the depth-displacement of the profiles and their incorporation of light-saturation behaviour – lessened on dull days or with low solar elevation. The initial gradient α of the ϕ–I relationship is geometrically equal to ϕ_{max}/I_k. It is determined by light absorption per unit biomass (related to k_s) and maximum conversion efficiency. It is largely independent of temperature and possible tropical correlates, and although influenced to some extent by spectral region (within PAR) – reflecting algal pigmentation and photon (quantum) yield – has a typical magnitude of 2–4 mmol O_2 or CO_2 (mg chl-*a*)$^{-1}$ (J m^{-2})$^{-1}$ or, in terms of photon flux density PAR, 0.5–1.0 mmol O_2 or CO_2 (mg chl-*a*)$^{-1}$ (mol photons m^{-2}s^{-1}). A slightly higher value of ~1.8 mmol O_2 (mg chl-*a*)$^{-1}$ (mol photons m (s^{-2})$^{-1}$) was estimated by Reynolds *et al.* (1983) for a deep population of *Lyngbya* in a Brazilian lake. The expected magnitude was used by Lewis (1988) to estimate very roughly values of chl-*a* concentration in the Orinoco River from the initial gradient of unnormalized A–I curves. Conversely, given the surface-incident light flux, the attenuation coefficient K_{min} of most penetrating light and the chl-*a* concentration, the deep light-limited 'tail' of a photosynthesis (A) – depth-profile can be roughly estimated.

Although the great range of the saturation rate parameter per unit water volume, A_{max}, is chiefly due to variable biomass concentration B, the specific rate per unit biomass ϕ_{max} (A_{max}/B) – or photosynthetic capacity – presents a challenging range of variation. Contributing here are factors of algal type, including cell size, of 'healthiness', nutrient supply, CO_2 supply and temperature. Figure 3.7 illustrates an example from Lake George of progressive diurnal decline, accompanied by a fall in the initial gradient parameter α. World-wide, examples of an inverse correlation between photosynthetic capacity and population density are often encountered. This trend is also represented in the tropics, as at Lake George (Ganf 1972) and Lake Ebrié (Dufour 1982). A notable example from seasonal variability is described for Lake McIlwaine (Chivero) by Robarts (1979). However, and possibly uniquely, tropical soda lakes can show a combination of very high biomass concentration and above-average values of photosynthetic capacity. Examples are described and

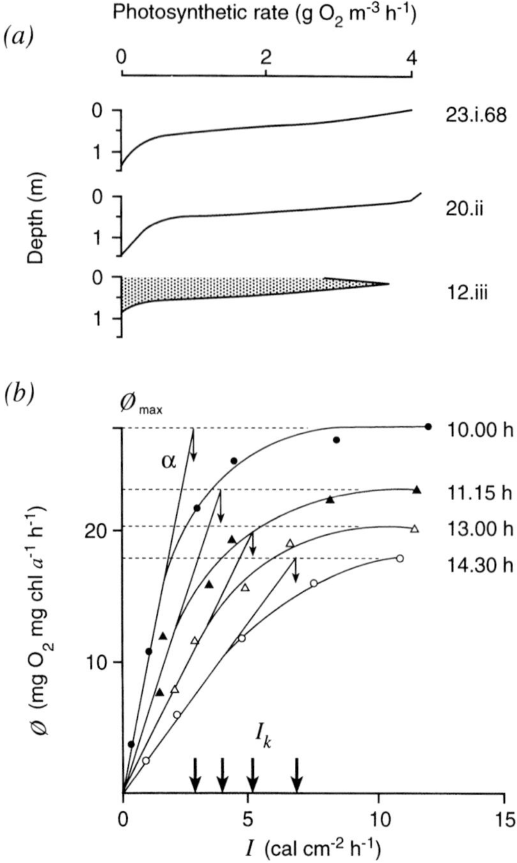

Fig. 3.7. A series of photosynthesis (*A*)-depth-profiles measured in Lake George, Uganda, with (below) a diurnal sequence from one date of derived specific rate ϕ versus irradiance I characteristics that show associated shifts in the parameters for photosynthetic capacity ϕ_{max}, light-saturation onset I_k and initial gradient α. Modified from Ganf (1975).

discussed by Talling *et al.* (1973), Melack (1979*b*) and Lemoalle (1981*a*, 1983). One favourable factor in such waters is the large reserve of CO_2 for localized photosynthetic activity in condensed euphotic zones (Talling *et al.* 1973).

Various authors have commented on relatively high values of the photosynthetic capacity ϕ_{max} obtained from tropical waters, whether biomass is assessed in terms of cell volume (e.g., Talling 1957*a*) or – more usually – chlorophyll *a*. Lemoalle (1981*a*) has compared mean rates of gross photosynthetic production in the euphotic zones of temperate and tropical lakes (Fig. 3.14), and concluded that the generally higher tropical rates stem mainly from higher values of photosynthetic capacity. The explanation is probably to be sought in usually much higher tropical temperature. Most algae show a positive relationship between capacity and temperature in the critical range of 20–30 °C (example in Fig. 3.9), although the temperature coefficient (Q_{10}) is there usually much less than in lower ranges – and for some algae becomes negative above 25 °C (e.g., Mariazzi *et al.* 1983). Although seasonal comparisons of the rates are liable to have other confounding factors, a depression occurred during the markedly cooler season of lakes McIlwaine (Robarts 1979) and Chad (Lemoalle 1979*a*, 1983).

(d) Losses: surface inhibition and respiration

The near-surface inhibition or depression of rates is a common feature of tropical, as of temperate, profiles. It is linked to high irradiance and sometimes also an appreciable component of ultra-violet radiation. From Lake Kariba, with relatively clear water, there is evidence that ultra-violet radiation depresses bacterial production – measured by uptake of labelled leucine – in the 0–3 m layer (Lindell & Edling 1996).

The significance of surface-inhibition of photosynthesis for column production has been controversial, because of dependence upon previous light-history and hence possible over-representation in experimental exposures maintained at fixed depths. However, an outstanding kinetic study, on the high-altitude Lake Titicaca, has shown that an inhibition-associated change in fluorescence characteristics is developed diurnally in the lake itself (Fig. 3.8). There it is promoted by the daily regime of temperature–density stratification which restricts the vertical circulation of phytoplankton (Vincent, Neale & Richerson 1984; Neale & Richerson 1987; Vincent 1992). During darkness the effect on fluorescence decays within a diel period, with fast or slow recovery of photosynthetic capacity dependent on the prior light-exposure. The changing response of photosynthetic rate was modelled in terms of a sensitivity factor and the excess of photon flux density (of PAR) above a threshold value I_T (Neale & Richerson 1987). The threshold value was lower for deep than for near-surface phytoplankton. If exceeded during a high-I exposure, a reduced capacity ϕ_{max} follows even at lower flux densities.

Respiratory fluxes accompany those of photosynthesis. Although an essential part of growth, they constitute a loss term in a carbon budget.

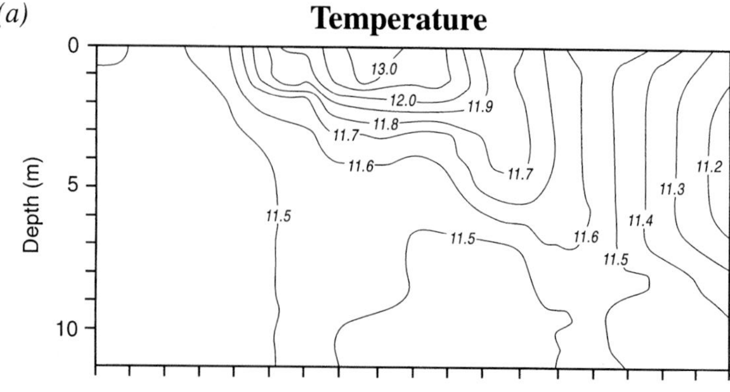

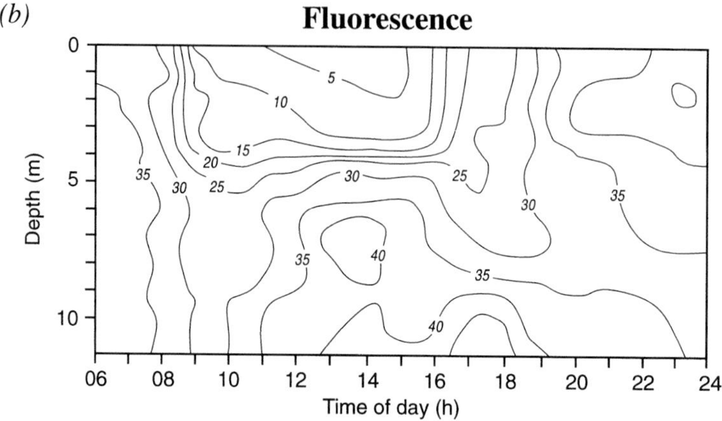

Fig. 3.8. Lake Titicaca, Andes. Depth-time representation of (*a*) the diel cycle of temperature stratification, with isotherms in °C, and (*b*) the associated diel cycle of the depression of an index of phytoplankton fluorescence, indicative of photosynthetic capacity, with contours in relative units. From Vincent (1992).

Usually measured by O_2 consumption in dark exposures, their magnitude relative to photosynthesis is very variable. Besides a contribution from zooplankters, which has been studied specifically in Lake George by Ganf & Blažka (1974), there can be a large component due to heterotrophic bacteria in waters with considerable inputs of organic matter, as in Amazonia (e.g., Wissmar *et al.* 1981; Melack & Fisher 1983). However, in situations dominated by phototrophs, either naturally or by experimental segregation, the fractional value of respiration to light-saturated photosynthesis is typically only about 0.05. A corresponding absolute specific rate is ~1 mg O_2 (mg chl-*a*)$^{-1}$h^{-1} (e.g., Talling 1965*a*; Ganf 1974*a*; Lemoalle 1983). This rate is positively temperature-dependent (e.g., Ganf 1972: see Fig 3.9), and – in the absence of temperature-compensation behaviour – is likely to be raised by temperature levels of the tropics. Notably high values were often found by Dokulil *et al.* (1983) in a Sri Lankan reservoir, but may have been influenced by accompanying heterotrophs.

Rates of respiration may be influenced by light in two main ways. The direct mechanism, photorespiration, depends upon a bifunctional enzyme system, ribulose *bis*phosphate carboxylase/oxygenase, and could invalidate the operational assumption that a gross photosynthesis rate can be separated as the difference between O_2 changes in light and dark exposures. However, its significance in field measurements with tropical freshwater phytoplankton is uncharted, and dealing only with measured net rates (photosynthesis minus respiration) would confuse contributions from phototrophic and heterotrophic organisms. Consequently 'gross'

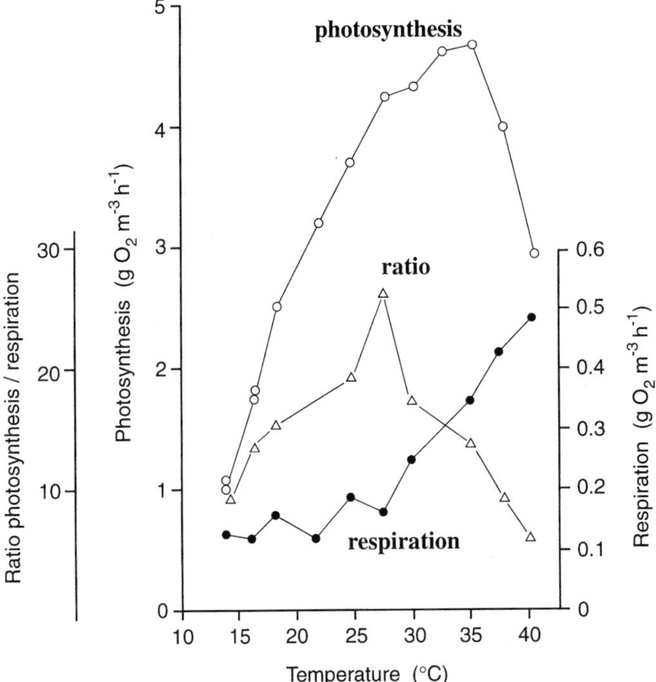

Fig. 3.9. The temperature-dependence of rates per unit water volume for gross photosynthesis, respiration, and their ratio from phytoplankton of Lake George, Uganda. Modified from Ganf (1972).

rates are still often estimated from O_2 fluxes. For other reasons ^{14}C fluxes, used in many measurements, are not easily interpreted – unequivocally – as either gross or net rates of photosynthesis. A second effect on respiration, from *prior* light exposure, is by the stimulation of growth and the production of respirable intermediates. It is illustrated by one outstanding tropical study, by Ganf (1974*a*) on the phytoplankton of Lake George. Here a diurnal stratification segregated the community between upper illuminated and deeper dark regions, and between more buoyant and sedimenting fractions. Specific, chl-*a* based rates of respiration, measured from short exposures, varied in a consistent pattern with depth and time: of a range from ~1.0–4.5 mg O_2 (mg chl-*a*)$^{-1}$ h^{-1}, higher rates were obtained from regions with prior illumination (Figs. 3.10, 3.15) and related to cumulative prior photosynthesis. A similar finding was obtained by Dokulil *et al.* (1983) from the Parakrama Samudra reservoir in Sri Lanka.

(e) Rates per unit area

In the tropics, as elsewhere, rates of photosynthesis per unit water volume have often been integrated over depth to obtain corresponding estimates, gross or net, of rates per unit surface area. It is convenient to distinguish, as ΣA, short-period fluxes (e.g., in units of g O_2 or g C m^{-2} h^{-1}) and, as $\Sigma\Sigma A$, daily fluxes. The latter are widely used as a descriptive character-

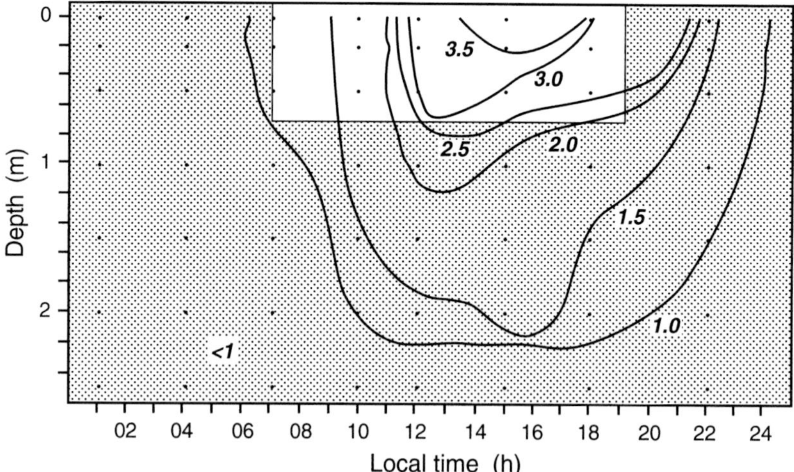

Fig. 3.10. The prevalence of the dark sector (shaded) over depth and diel time in Lake George, Uganda. Contours indicate the distribution of variable specific rates of respiratory oxygen uptake referred to chlorophyll *a* (mg O_2 mg chl-a^{-1} h^{-1}) and measured during short dark exposures. Modified from Ganf & Viner (1973).

istic of a water-body and with more uncertainty as a basis for estimating or interpreting other fluxes of production.

Less frequently, area-based rates have been assessed without a computed depth-integration. In productive, absorptive and vertically compressed euphotic zones it is practically possible to encompass the zone as a 'core' in a closed transparent cylinder of known depth. Then the production-derived change of mean concentration of the mixed contents can be converted into a change per unit area. Tropical applications have been made to an Ethiopian soda lake (Kilotes = Kilole) by Talling *et al.* (1973), and to the Kenyan soda lakes of Simbi (Melack 1979*b*), Nakuru (Vareschi 1982) and Elmenteita (Melack 1988). Grobbelaar (1985) described a more detailed application to a small subtropical reservoir, Wuras Dam, in South Africa.

Another, non-experimental, method is based on the analysis of day–night (diel) changes of concentration *in situ* of a gas – generally O_2 – induced by photosynthesis. It was first used in the tropics by Talling (1957*a*), for a variety of water-bodies on the upper Nile and Lake Victoria. Later applications were made to productive soda lakes in East Africa by Talling *et al.* (1973), Melack & Kilham (1974) and Melack (1981, 1982), to the newly created Lake Brokopondo in Suriname by van der Heide (1982), and to floodplain lakes of the Amazon by Melack & Fisher (1983). Example-sequences are illustrated in Figs. 3.11 and 3.12. Contents of dissolved O_2 are assessed per unit area, obtaining rates of change in daytime that are then corrected for the consequences of accompanying air–water exchange and respiratory consumption. A combined correction based upon mean night-time rates is a simple but rough solution (Talling 1957*a*; Talling *et al.* 1973), or alternatively the air–water exchange can be separately and indirectly estimated from known diffusion coefficients and wind stress (Melack 1982; Melack & Fisher 1983). This exchange involved a large and variable background flux in the markedly sub-saturated Amazonian lake.

Most estimates of rates of photosynthetic production per unit area of water surface are based on the depth-integration of rates from exposed samples, an integral equivalent to the area enclosed by the photosynthesis (A)-depth (z) profile. Reference to Fig. 3.4 will show that similar integrals, and hence areal rates, can result from profiles with very different combinations of activity and depth range. These two components tend to be inversely related, partly because waters of higher background attenuation tend to support denser populations, partly because those populations themselves induce a higher attenuation and reduced photosynthetic

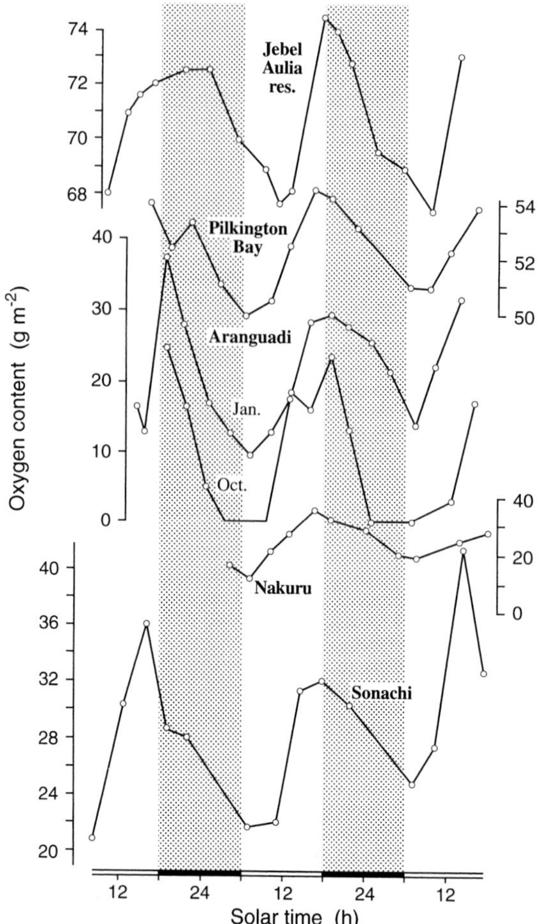

Fig. 3.11. Examples of diel variation in oxygen content below unit surface area in productive African lakes: Jebel Aulia reservoir, White Nile; Pilkington Bay, Lake Victoria; crater lake Aranguadi, Ethiopia: Lake Nakuru, Kenya; Lake Sonachi, Kenya. Modified from Talling (1957*a*), Talling *et al.* (1973), Melack & Kilham (1974) and Melack (1982).

zone. A few very productive waters, as seen in Lake George, some soda lakes and Lake Xolotlán, yield gross hourly rates (ΣA) in excess of 2 g O_2 or ~0.7 g C $m^{-2}h^{-1}$ near midday (Fig. 3.13).

Daily yields ($\Sigma\Sigma A$) exceed near-midday hourly rates by a factor of about 9 (± 1) in a number of tropical waters investigated by sequential diurnal exposures. Examples are described by Talling (1957*a*), Ganf & Horne (1975), Lemoalle (1979*a*, 1983), Lazzaro (1981), Vareschi (1982),

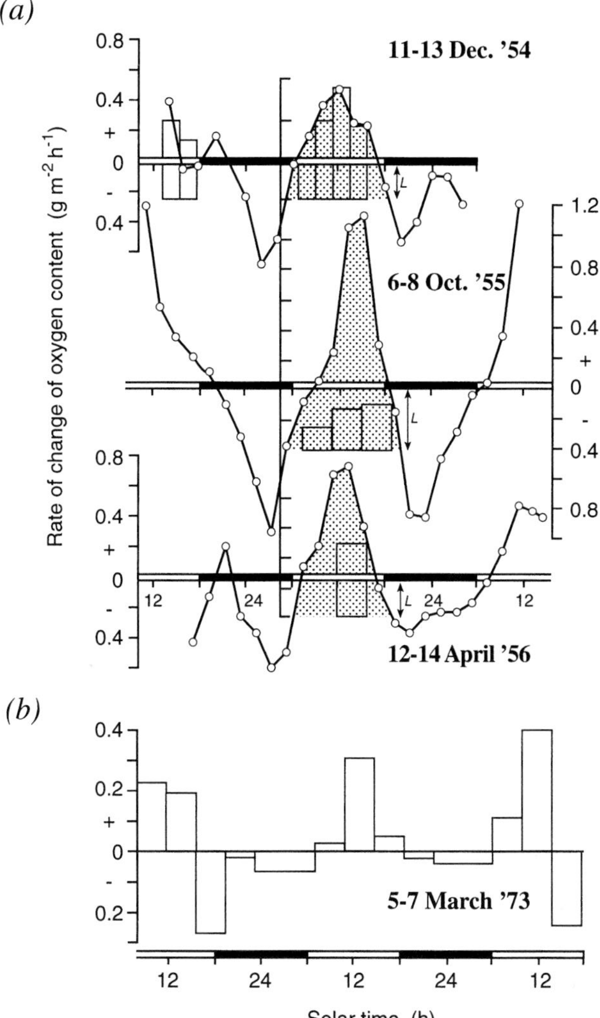

Fig. 3.12. Diel variation in the rates of change of column oxygen content (g m^{-2} h^{-1}) as a measure of gross or net photosynthetic activity, from (*a*) the Jebel Aulia reservoir, White Nile (December, October) and Pilkington Bay, Lake Victoria (April), (*b*) Lake Sonachi, Kenya. In (*a*) accompanying experimental gross rates are superimposed as histograms, constructed above the mean rate of nocturnal depletion *L*. Shaded areas indicate time-integrals of daily areal photosynthesis deduced from changes *in situ*. Modified from Talling (1957*a*) and Melack (1982).

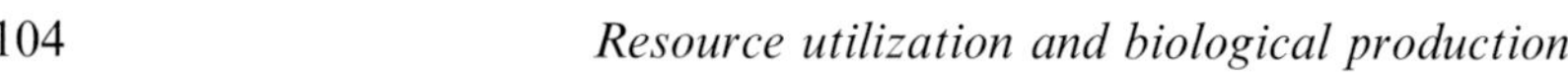

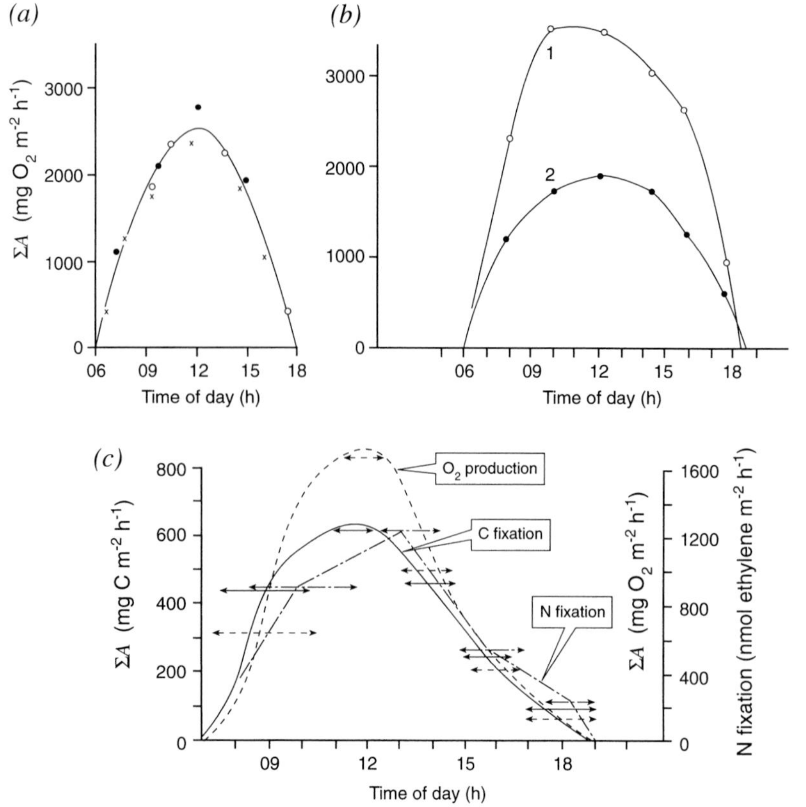

Fig. 3.13. Examples, obtained from bottle exposures, of symmetric and asymmetric diurnal variation in rates of photosynthetic production per unit surface area (ΣA): (*a*) Lake Xolotlán, Nicaragua, three diurnal series, (*b*) Lake Chad, two series from different regions, (*c*) Lake George, Uganda, with comparative data from ^{14}C assimilation, gross O_2 production and N-fixation assessed by acetylene reduction. Modified from Erikson *et al.* (1991*a*), Lemoalle (1983) and Ganf & Horne (1975).

Dokulil *et al.* (1983), and Erikson *et al.* (1991*a, b*). Variability of this factor is reduced by the limited variation of daylength but increased if site-examples both include (e.g., Lake George, Lake Titicaca, Lake Tanganyika) or not include (e.g., Lake Xolotlán) an afternoon change – usually depression – of specific activity (see Fig. 3.13). If combined with the near-upper limit of hourly rate cited above, one obtains near-maximal daily rates of *c.* 18 g O_2 or $\sim$6.3 g C m^{-2} day^{-1}. There are but few markedly higher estimates based upon exposures of enclosed samples (e.g., Ganapati & Sreenivasan 1970), but possibly stronger evidence from analysis of diel variation in water-column content (e.g., Talling *et al.*, 1973; Melack, 1982). As discussed by Melack (1982), it seems likely that the enclosure of dense phytoplankton in vessels held at fixed depths results in some underestimation of the actual rates *in situ.*

Two parameters have had some success as proportionate predictors of the integral of area-based activity ΣA. One is the quotient of maximum activity in unit water volume to vertical light attenuation coefficient, namely A_{max}/K_{min}, or A_{max}/K_e, reflecting obvious direct relations of profile area with A_{max} and euphotic depth z_{eu}. Close linear dependence of ΣA on A_{max}/K_{min} or A_{max}/K_e has been demonstrated from East African lakes (Talling 1965*a*; Mukankomeje *et al.* 1993), Gatún and Madden lakes, Panama (Gliwicz 1976*b*), Lake McIlwaine (Robarts 1979), Lake Chad (Lemoalle 1979*a*, 1981*b*), Lake Xolotlán (Erikson *et al.* 1991*a*) and Dhanmondi Lake, Bangladesh (Khondker & Parveen 1993). The dimensionless gradient of proportionality lay between 2.0 and 2.7 (for A_{max}/K_{min}) or 2.7 and 3.6 (for A_{max}/K_e) in these examples from exposures near midday. It chiefly expresses the relative vertical displacement of the profile in the euphotic zone, which is a function of the ratio between surface-penetrating radiant flux density (I'_o) and light-saturation characteristic (I_k). This ratio, and hence the function f (I'_o/I_k), is not dissimilar in magnitude to values obtained at higher latitudes, as both components I'_o and temperature-dependent I_k are typically somewhat higher in tropical waters.

The second correlated parameter is the euphotic biomass content, $\Sigma_{eu}B$. As already seen, this incorporates the ratio of mean biomass concentration $\overline{B}$ to the attenuation coefficients K_{min} or K_e. Unlike the parameter A_{max}/K_{min}, that equals $B.\phi_{max}/K_{min}$, it does not incorporate a measure of photosynthetic capacity (ϕ_{max}) and so can be expected to be a poor predictor of areal production if such capacity is very variable. This situation was found for Lake McIlwaine (Robarts 1979), but not for East African lakes (Talling 1965*a*), Lake Chad (Lemoalle 1979*a*, 1981*a*, 1983: see Fig. 3.14) or Lake Xolotlán (Erikson *et al.* 1991*a, b*). However, Lemoalle (1981*a*) showed that a given euphotic biomass content was correlated with markedly higher values of areal production in tropical inland waters than in temperate ones (Fig. 3.14). The governing factor was the generally higher values of photosynthetic capacity in the former.

Rates of respiration by phytoplankton can also be integrated over depth to obtain areal values. However reliable estimates are few: as normally used the ^{14}C method does not yield measures of respiration, and the O_2 method does not readily distinguish between uptake due to phyto-

day^{-1}. These correspond to mean generation times of less than one day, and are of especial interest in relation to other estimates of high planktivore production in this lake (Section 3.5c). Their validity depends on the accuracy of indirect population census as C by microscopy and on the adequate incorporation of respiratory loss. Later estimates of specific growth rate g from Lake Malawi, by Bootsma (1993*a*), yielded similarly high values.

(f) Mathematical models

Factor interactions and their implications can be assessed by means of generalized mathematical models. Only empirical or semi-empirical models (e.g., Dufour 1982) have been developed purely for tropical waters, but two general models that describe photosynthetic production have been applied to them. Using the previous symbolism, these models have as basic construction:

$$\Sigma A = \frac{A_{max}}{K_e} f(I) \qquad (3.5)$$

and so relate photosynthetic production per unit area (ΣA) to photosynthetic activity per unit volume (A_{max}), vertical light attenuation (K_e) and light flux density (I). The version of Talling (1957*d*, 1965*a*) was intended for vertically homogeneous populations, and neglects light-inhibition behaviour. It evaluated the effective attenuation coefficient K_e as $1.33K_{min}$, and the function of irradiance f(I) as ln $(I_o'/0.5I_k)$. As already noted, $A_{max} = B.\phi_{max}$, and the initial gradient $\alpha = \phi_{max}/I_k$. The model has yielded values of ΣA that are in general close agreement with measured (planimetric) values when tested on several tropical water-bodies. These include a Nile reservoir (Talling 1957*a*), Lake Victoria (Talling 1965*a*), Lake George (Ganf 1975), Lake McIlwaine (Robarts 1979), Lake Chad (Lemoalle 1979*a*, 1983), and Lake Nakuru (Vareschi 1982). It has also yielded values of the factor relating hourly and daily assimilation ($\Sigma A, \Sigma\Sigma A$: Talling 1965*a*) that are in accord with empirical experience (e.g., Lemoalle 1979*a*; Erikson *et al.* 1991*a, b*). Further, it can be used to derive a parameter that evaluates conditions for equality between photosynthetic gain and respiratory loss in a water-column of specified depth over a 24-h period. Applications to Lake George (Ganf & Viner 1973; Ganf 1974*c*; Ganf & Horne 1975) have shown that in this lake the 'column compensation point' is quite readily approached under normal conditions. The same conclusion can be reached from the measurements of Vareschi (1982) on Lake Nakuru. In these lakes, and elsewhere, it would be misleading to directly interpret – as often done – estimates of area daily production ($\Sigma\Sigma A$) from ^{14}C exposures as net C-fixation potentially available for supporting phytoplankton growth and transfer to secondary production. This reservation holds regardless of whether short-term ^{14}C-fixation is considered to approximate net photosynthesis (Lewis 1974) or gross photosynthesis (Ganf & Horne 1975), as its estimation excludes respiratory losses in unsampled dark depth-time sectors.

The model version of Fee (1969, 1973*a*, *b*) has been applied to Lake Tanganyika by Hecky & Fee (1981), and to Lake Malawi by Bootsma (1993*a*) and Patterson & Kachinjika (1995). It utilizes an experimentally determined relationship between rate of photosynthesis (A) and light-flux density (I), and performs the integration of rate over depth by computer summation rather than by a formal (analytical) integral. Accommodation of varying biomass and activity in stratified populations is possible. The method is primarily designed to calculate values of areal production from quantities measurable from exposures in an illuminated incubator combined with environmental information on solar radiation I_o and underwater light attenuation K_e. The sensitivity of the output values to variation of input factors can thereby be assessed.

(g) Efficiencies

A rate of photosynthetic production per unit surface area, gross or net, can readily be expressed as a dimensionless conversion efficiency from the incoming flux density of solar radiation if both input and output are framed in terms of energy flux. Photosynthetic production of biomass as carbohydrate implies equivalences of 1 mol $O_2 \equiv$ 1 mol C $\equiv$ 469 kJ or 112 kcal. In reality, the average composition of biomass is more reduced. As rounded figures, often used, a more realistic equivalence between O_2 and C is 3 g $O_2 \equiv$ 1 g C, corresponding to a photosynthetic quotient ($\Delta O_2/\Delta CO_2$, by moles) of 1.125, and between C and energy content 10 kcal $\equiv$ 1 g C or 502 kJ $\equiv$ 1 mol C. As mean totals of daily solar radiation income are not very variable within the tropics (a standard tropical value of 400 cal cm^{-2} day^{-1} has been proposed by Lewis 1974 for comparisons), these efficiencies generally reflect the absolute output flux densities. Mechanistically of greater interest are their relationships to upper limits of production, set by inherent limits to component factors.

This general approach, treating the overall efficiency as the product of a multiplicative chain of efficiency factors, is well set out by Monteith

(1972) in relation to tropical terrestrial vegetation. A fundamental component is the maximal photochemical efficiency (ε_{max}) in photosynthesis, set by maximum quantum yield over the PAR spectral band. In the aquatic context, other determining factors that intervene between surface-incident radiant flux density I_o (solar constant $\times$ geometric factor $\times$ atmospheric transmission factor: see Monteith 1972) and net production yield per unit area (ΣA_n or $\Sigma\Sigma A_n$) are the fractional penetration ε_p at the water surface, the spectral fraction ε_s of PAR to total solar radiation, the fractional interception ε_i of PAR underwater by plant pigments, the fractional area ε_a or depth-profiles of gross photosynthesis to the area predicted in the absence of light-saturation and -inhibition behaviour, and the fraction ε_r that the net photosynthetic (primary) yield bears to the gross yield after deducting the respiratory cost. Thus for short exposures:

$$\Sigma A_n = I_o.\varepsilon_p.\varepsilon_s.\varepsilon_i.\varepsilon_a.\varepsilon_{max}.\varepsilon_r \qquad (3.6)$$

and for day-periods:

$$\Sigma\Sigma A_n = \Sigma I_o.\varepsilon_p.\varepsilon_s.\varepsilon_i.\varepsilon_a.\varepsilon_{max}.\varepsilon_r \qquad (3.7)$$

Of the six efficiency factors ε_{max}, ε_p ε_s and ε_i will vary little between short and long periods, whereas ε_a and ε_r are susceptible because of the varying representation of conditioning I_o values and dark depth-time sectors. Values of ε_p and ε_s are close to 0.9 and 0.46, respectively, the former being likely to be raised slightly at high solar elevations. The interception factor ε_i is very variable and so influential. This influence has been quantified by Lemoalle (1983) for Lake Chad and other African lakes (see Fig. 3.16), partly in relation to photosynthetic production per unit area and partly to the total efficiency of energy conversion. The interception factor can be evaluated as the ratio of cellular attenuation (equal to the product Bk_s, or Bk_s') to the total attenuation (K_{min} or K_e). Wood *et al.* (1979) and Megard *et al.* (1979) give examples, including the Ethiopian rift lakes of Zwai, Langano and Abijata, and Lake George. It is roughly proportional to the euphotic content of biomass $\Sigma_{eu}B$, and is approximated by the fraction which that content constitutes of the maximum euphotic content – the latter set by the k_s characteristic of cellular attenuation and typically in the range 200–500 mg chl-*a* m^{-2}.

The light-saturation factor ε_a is mainly set by the ratio of surface-penetrating radiant flux density to the saturation characteristic, I_o'/I_k, and for vertically uniform biomass can be calculated (Talling 1957*d*) as:

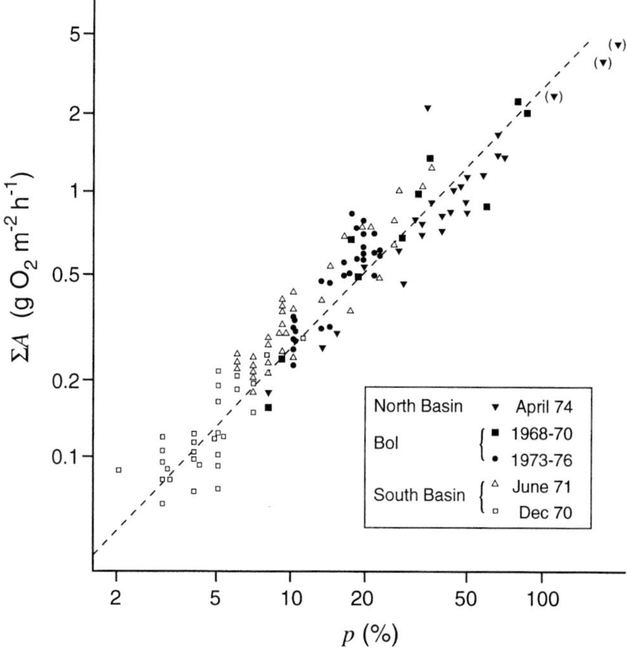

Fig. 3.16. Hourly rates of photosynthesis per unit area (ΣA) in Lake Chad in relation to estimates of percentage light interception by phytoplankton (p). From Lemoalle (1979*a*, 1983).

$$\varepsilon_a = \frac{I_k}{I'_o} ln \frac{I'_o}{0.5 I_k} \tag{3.8}$$

with some additional contribution from time-dependent light inhibition. Lower values of ε_a are therefore found at higher values of the surface-incident radiant energy flux (I_o). Neither saturation nor inhibition appear to be tropically distinctive, judging from the general form of recorded depth-profiles. A tropical trend to higher values of the surface fluxes I_o and I'_o is largely balanced by another to higher values of light-saturation onset I_k, which – as ϕ_{max}/α – results from a general rise of capacity ϕ_{max} but relative constancy of α that represents the basic photochemical conversion. This conversion determines the maximum conversion efficiency ε_{max} of absorbed radiant energy (PAR), at a magnitude of ~0.2 that is determined by the ratio α/k'_s.

The ε_r factor that reflects respiratory cost is too poorly known for secure comparisons. However, the probability of low values (<0.5) in some productive tropical lakes, and the general temperature-dependence of respiration rate, are suggestive. If the ε_r factor is omitted by considering gross photosynthetic production, the corresponding efficiency of total radiant energy conversion reaches values of 1–2% (e.g., Lemoalle 1983).

With some assumptions, a similar analysis of overall photosynthetic efficiency can be applied to observed yields *per unit volume* of strata within photosynthesis depth-profiles. If radiant power consumption per unit volume (m^3) is used as the reference quantity, a dimensionless true efficiency is obtained. In a depth-element δz, power consumption is equal to the product of irradiance I_z, attenuation coefficient K and δz, less a small and usually neglected correction for back-scattering. If the cellular component of attenuation can be estimated, by the product $B.k_s$, the corresponding cellular component of power consumption can be approximated. This in turn can be used, with the $I - z$ relationship, to estimate conversion efficiencies in strata throughout the profile. Two contrasting tropical examples, from lakes George and Victoria, are given by Talling (1982). They include logarithmic profiles of three types of power consumption, the last photosynthetic, with magnitudes conditioned by efficiency factors. The depth-profiles show that higher efficiency is achieved at deeper levels below the zone of light-saturation. This would be anticipated on theoretical grounds and is further illustrated by profiles of efficiency calculated by Vareschi (1982) for Lake Nakuru.

Although a true efficiency is ideally dimensionless, and is obtained from ratios of energy fluxes, photochemical fundamentals lead to the expression of photosynthetic yield as mol O_2 (or linked C) per mol photons (6.02×10^{23} quanta). Areal yields on six East African lakes have been measured as in the range 0.8–12.4 mmol O_2 (mol photons, PAR)$^{-1}$, respectively (Melack 1979*b*, 1981; Kifle & Belay 1990). These values are far below the maximum of ~50–100 mmol (mol photons)$^{-1}$ potentially possible, the difference being probably largely due to the product of the efficiency factors ϵ_i and ϵ_a that are concerned with light-interception and light-saturation.

(h) Non-planktonic systems: microphytobenthos and macrophytes

Other forms of photosynthetic cover are represented by surface-attached communities of micro-algae (microphytobenthos or periphyton) and by submerged or emergent aquatic macrophytes. Both micro- and macro-forms of cover often achieve near-complete light (PAR) absorption, but their respective densities of total biomass per unit area typically differ by several orders of magnitude. Their photosynthetic production in tropical inland waters has not been widely investigated.

In three large African rift lakes – Turkana, Tanganyika and Malawi – microphytobenthos on sediment (epipelic) or rock (epilithic) is well developed over some shallow margins and is an important food resource for some browsing fishes. For each lake there have been preliminary studies of photosynthetic production within enclosures *in situ* by O_2 or ^{14}C based methods (Harbott 1982; Takamura 1988; Bootsma 1993*a*). The rates obtained per unit area of surface were considerable, in the range 0.2–0.7 g C m^{-2} day^{-1}, comparable with those estimated for the phytoplankton of the two less productive lakes, Tanganyika and Malawi. In these, sequential exposures provided information on the diurnal pattern of areal production (Fig. 3.17), with some evidence for within-day depression of rates.

The distinction between phytoplankton and phytobenthos is blurred in the shallow, productive Lake George, where a partly living layer of phytoplankton material joins the superficial sediments after sedimentation. Ganf (1974*c*) has shown that this retains appreciable photosynthetic capacity (see Fig. 3.18) and by resuspension can augment the density of the dispersed phytoplankton.

Work on the microphytobenthos of the flowing River Limon in Venezuela (Lewis & Weibezahn 1976) has shown the presence of two communities with very different rates of areal photosynthetic production. Mats that included filamentous algae could show high rates, estimated as >1 g C m^{-2} day^{-1}. In contrast, the thin crusts of the red alga *Hildenbrandtia* yielded only around 1% of these values – probably linked with slow but persistent growth.

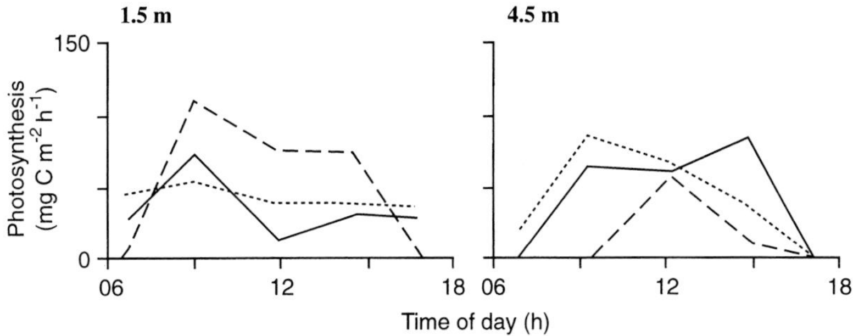

Fig. 3.17. Photosynthesis per unit area of benthic algae, as diurnal patterns by cover on rocks of Lake Tanganyika, for two depths at three stations. From Takamura (1988).

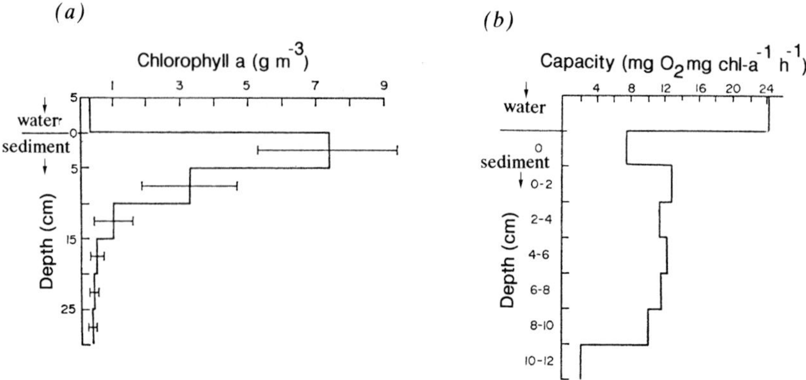

Fig. 3.18. Sedimented phytoplankton of Lake George, Uganda, showing (*a*) concentrations as chl-*a* incorporated in the uppermost layers of sediment, with (*b*) the corresponding vertical distribution of photosynthetic capacity at light-saturation and 27 °C. From Ganf (1974*c*).

The areal production rates of emergent macrophytes are generally assessed from changes in biomass as dry weight, rather than from measurements of photosynthesis. Examples in dense tropical stands (reviewed by Jones 1986) range widely; they include values of > 10 g dry weight m^{-2} day^{-1} for *Cyperus papyrus* (Thompson, cited by Westlake 1975) and ~5 g dry weight m^{-2} day^{-1} for the sedge *Lepironia articulata* in the Malaysian swamp Tasek Bera (Ikusima and co-workers, in Furtado & Mori 1982). Examples of light interception in stands of these two species, and of *Typha domingensis*, are described by Jones & Muthuri (1985), Jones (1986, 1988) and Ikusima (1978). Relationships between rate of photosynthesis and irradiance or illuminance here indicate fairly high levels of saturation (Ikusima 1978; Jones 1986, 1987, 1988), although that for *C. papyrus* (see Fig. 3.19) is low for a supposed 'C_4' plant (Jones & Milburn 1978). Such plants include many tropical grasses; their photosynthetic pathway includes a four-C metabolite, with an efficient CO_2-concentrating mechanism and typically high rates of photosynthesis at light-saturation.

The high potential production of a C_4 plant appears to be realized by the floating grass *Echinochloa polystachya* on the Amazon floodplain. Here much production is invested in the stem internodes required to cope with the seasonal rise of 8 m in water level. Measured rates of photosynthesis as CO_2 uptake per unit leaf area were high, with variations related to the diel and annual (floodpulse) cycles (Piedade *et al.*

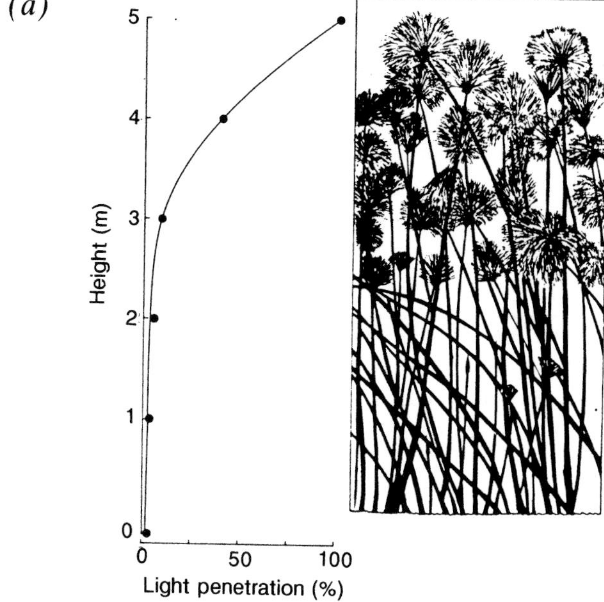

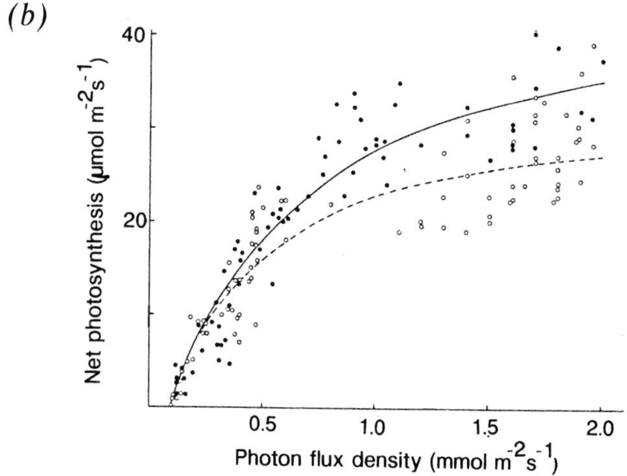

Fig. 3.19. Papyrus (*Cyperus papyrus*) swamp. (*a*) Vertical penetration of light (PAR) measured downwards through a stand, with biomass as culms and umbels of photosynthetic tissue illustrated, (*b*) relationship between rate of net photosynthesis per unit area of bracteoles and photon flux density, PAR, with measurements on two days distinguished by • and ○. From Jones & Muthuri (1985) and Jones (1987).

1994). The annual net production of biomass was estimated in one stand by Piedade *et al.* (1991) as 9.9 kg dry weight m^{-2} yr^{-1}, a very high value approximately equivalent to an average maintained rate of 12 g C m^{-2} day^{-1}, or 1 g C per MJ of intercepted solar radiation. Further high estimates for other grasses in this region included 5 kg dry weight m^{-2} yr^{-1} for the successive growth of three annual species and 7 kg m^{-2} (8 $mo)^{-1}$ for the perennial *Paspalum fasciculatum* (Junk & Piedade 1993).

Production of tropical submerged macrophytes has been followed only infrequently by repeated census of biomass. An example from southern India (Haniffa & Pandian 1978) involved dense stands of *Chara fragilis* and *Hydrilla verticillata* in a seasonally contracting pond, that were estimated to have a net annual yield equivalent to 0.7% of incident solar energy. By another approach, enclosing portions of stands in vertical cylinders of glass or plastic, several investigations have followed O_2 changes influenced by photosynthesis and respiration as well as air–water exchange. Examples are described by Ikusima (1978, 1982) for the Tasek Bera swamp in Malaysia, by Harbott (1982) for a gulf of Lake Turkana and by Ikusima *et al.* (1983) for the Broa (Lobo) Reservoir in Brazil. Day–night alternations of activity (Fig. 3.20) could be followed in the first and last of these shallow-water systems, dependent on a photosynthesis P–light flux I relationship that involved light-saturation at relatively low radiant flux densities.

A dense cover of submerged macrophytes is present in some shallow areas of Lake Titicaca (Andes), with a depth-zonation of major species. The net production of several over the warmer season has been studied by O_2 changes with enclosed samples *in situ* and by recolonization of cleared areas (Collot *et al.* 1983). Despite the high altitude and relatively low temperature, specific rates of production per unit dry weight were considerable; estimated photosynthetic production per unit area was greatest for the abundant *Chara* spp.

Production of abundant submerged macrophytes in the littoral (0–5 m) of Lake Kariba has been studied (Machena *et al.* 1990) by a combination of biomass survey, growth increments on individual plants and diel records of O_2 changes in cylinder-enclosures. The derived net rates of biomass increase were not especially high; an average value of 3.5 mg C (g dry $weight)^{-1}$ day^{-1} was obtained from growth at 4 m depth and less from net O_2 change at 1 m depth. However, in combination with surveys of biomass distribution, a considerable mean littoral (0–5 m) production rate of 244 g C m^{-2} yr^{-1} was estimated.

(a)

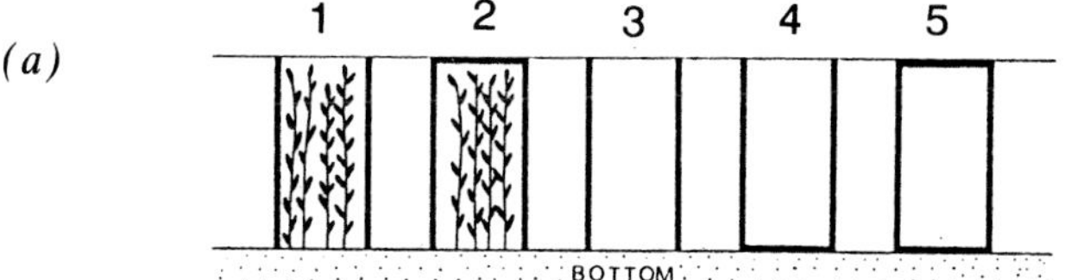

(b)

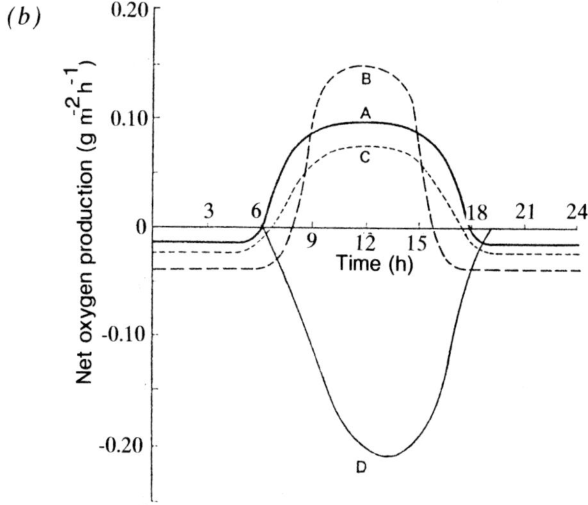

Fig. 3.20. Measurements of oxygen exchanges by a submerged plant stand dominated by *Mayaca sellowiana* in Broa (= Lobo) reservoir, southern Brazil: (*a*) arrangements of five enclosing transparent plastic cylinders, differing in openness to the atmosphere or sediment and in the presence or absence of plant material, (*b*) derived schematic patterns of diel variation in net rates of oxygen change. A, oxygen production by macrophytes and their epiphytes; B, production by algae on the sediment surface; C, production by phytoplankton; D, loss by diffusion from water to atmosphere. Modified from Ikusima *et al.* (1983).

3.2 Primary utilization: nutrients

The interception and conversion of solar radiation by autotrophic communities is accompanied by uptake and further assimilation of nutrients, originally present largely in inorganic forms. In many aquatic environments, especially those with active microbial biomass, the corresponding chemical fluxes are closely enmeshed with fluxes linked to heterotrophs. Operationally, therefore, many studies emphasize the transition from dissolved to particulate phases and sometimes the size-fractions, as well as between inorganic and organic components.

Nutrient pathways in both deep and shallow tropical waters, and some attempted mass-balances, will be discussed in Section 3.4. Few of the elements now known to be essential macro- or micro-nutrients have been demonstrated in a role of controlling or limiting factors for production. Attention has concentrated on nitrogen and phosphorus as universal macro-nutrients, present in inorganic and organic forms of varying availability, and to a lesser degree upon silicon as a major though less widespread requirement for diatoms. Inorganic carbon is a major and universal requirement for photosynthesis that here is treated as a nutrient. Its utilization has rarely been studied in tropical freshwaters for its intrinsic interest, but rather as a quantitative index of photosynthetic production. The uptake of potassium and of sulphur involves quantities that are, respectively, of similar magnitude to those of nitrogen and phosphorus. However, these elements have generally not attracted attention as potentially limiting nutrients, because the main environmental sources – K^+ and SO_4^{2-} – are (with Na^+, Ca^{2+}, Mg^{2+}, HCO_3^- and Cl^-) major ions in most natural waters and are usually supposed to be present much in excess of requirements.

At another quantitative extreme are nutrients usually present in solution in extremely low concentrations and required in very small amounts relative to biomass. These include trace metals, such as Fe, Mn and Mo, and vitamins such as B_1 (thiamin) and B_{12} (cobalamin) required by some autotrophic algae. Knowledge of their uptake or possible limiting role in tropical inland waters is lacking or rudimentary.

Most 'major ions' are plant nutrients, but they also determine qualities of the general ionic environment that may be favourable or unfavourable to specific organisms. These qualities are those of salinity, pH and ionic balance. Correlations with the distribution of tropical aquatic organisms are most systematically developed for diatoms, notably in East Africa (e.g., Hecky & Kilham 1973; Gasse *et al.* 1983). Other examples of ecological importance involve bloom-forming cyanophytes such as *Spirulina fusiformis* in soda lakes (e.g., Iltis 1968; Vareschi 1982; Melack 1988; Kebede 1997). Dynamic consequences principally concern time-sequences such as salinization, or its reverse; the latter apparently led, in Ethiopia, to the loss of *S. fusiformis* from Lake Besaka (Metahara) (Kebede *et al.* 1994) and of *Chroococcus minutus* from Lake (Hora) Kilole (Lemma 1994).

(a) Nutrient uptake and environmental availability

As biomass is a token of past nutrient uptake, circumstances will exist under which biomass concentration is inversely correlated with ambient nutrient concentration. This relationship can be expressed in horizontal, vertical and temporal variability. Horizontal manifestation is well seen in flowing water systems where a common nutrient background is locally depleted in a sector where hydrological conditions like impoundment favour biomass accumulation. Thus the inverse correlation appeared in longitudinal sections of a Nile reservoir (Prowse & Talling 1958: Fig. 3.21), and in a broad survey over main-stems and lateral waters of the

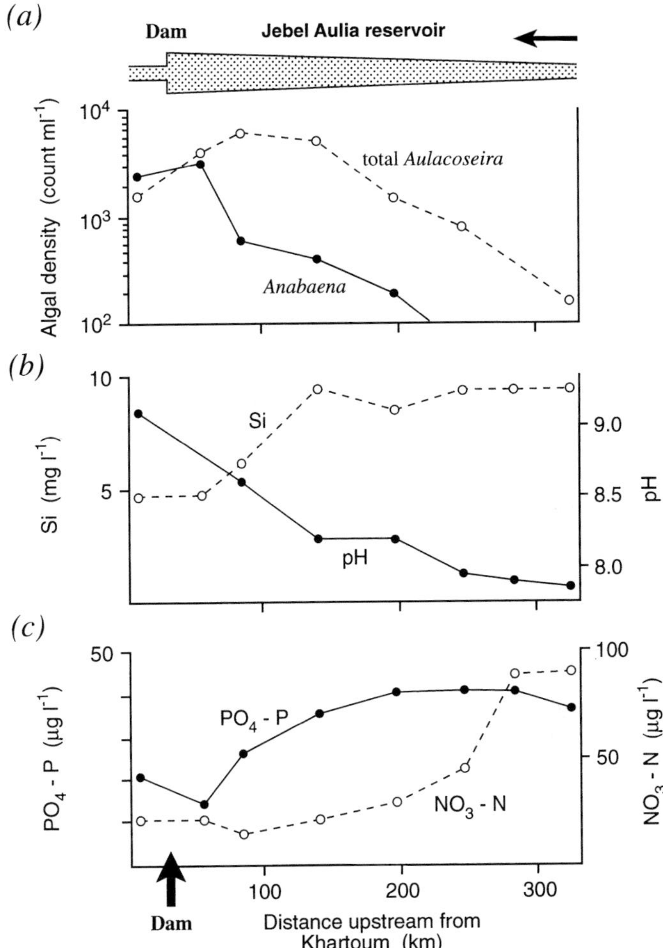

Fig. 3.21. Longitudinal section along a reservoir on the White Nile above Khartoum, October 1954, showing (*a*) the increase of two major components of phytoplankton towards the dam, with correlated (*b*) elevation of pH and (*c*) depletion of nutrients. Modified from Prowse & Talling (1958).

Amazon system (Wissmar *et al.* 1981), where other multiple correlations involving nutrient concentrations have been distinguished (Forsberg *et al.* 1988). The vertical inverse correlation is ubiquitous in deeper stratified lakes, where phytoplankton production is light-restricted to an upper layer that has become nutrient-depleted by uptake and by sedimentation of particulates. This situation is illustrated by nutrient distribution in the indefinitely stratified (meromictic) lakes Tanganyika (Fig. 2.36) and Malawi (Fig. 2.37).

The inverse correlation over time is seen in most studies of phytoplankton time-variability accompanied by nutrient analyses. World-wide, it has been a classic if unspecific approach to the analysis of phytoplankton periodicity. Its prospects are seen best in natural analogues of batch-culture growth based on an initial nutrient stock and least in analogues of continuous-culture growth supported by later nutrient fluxes, as from recycling. Natural situations in the tropics range between these extremes. Perhaps the most intensely studied batch-culture-like analogue is that of the reservoir on the White Nile (Prowse & Talling 1958), refilled annually and with a succession of population maxima (Fig. 5.19); a continuous-culture analogue is the continuously productive Lake George, Uganda (Ganf & Viner 1973; Ganf 1974*b*).

Lake George exemplifies the situation of a dense, photosynthetically active biomass of phytoplankton in a medium depleted of inorganic combined nitrogen and phosphorus. Viner (1973, 1977*c*; Ganf & Viner 1973) showed that experimental additions of these nutrients (N as NO_2^- or NH_4^+) were taken up rapidly, especially if the phytoplankton cells involved had a history of illumination and photosynthetic C assimilation. Limitation of uptake by the natural low concentrations can be inferred, although here information from uptake rate–concentration relationships that can be expressed as half-saturation constants is lacking. Such information is very sparse for tropical freshwater phytoplankton. It is contributed by the work on nitrogen uptake in Lake Titicaca by Vincent, Wurtsbaugh *et al.* (1984), in Amazonian waters by Fisher *et al.* (1988) and in the Rio Doce Valley lakes of southern Brazil (see Fig. 3.22) by Mitamura *et al.* (1995); also that of Lehman & Branstrator (1994) on P uptake by phytoplankton of offshore Lake Victoria.

Nutrient availability deduced from concentrations of free-water solutes will be modified if solid particulate phases are closely involved in nutrient supply. Such a supply of P has been demonstrated from the particulate loads carried by the Purari River, Papua New Guinea (Viner 1982*a*) and by the Amazon (Grobbelaar 1983; Engle & Sarnelle 1990). In Lake

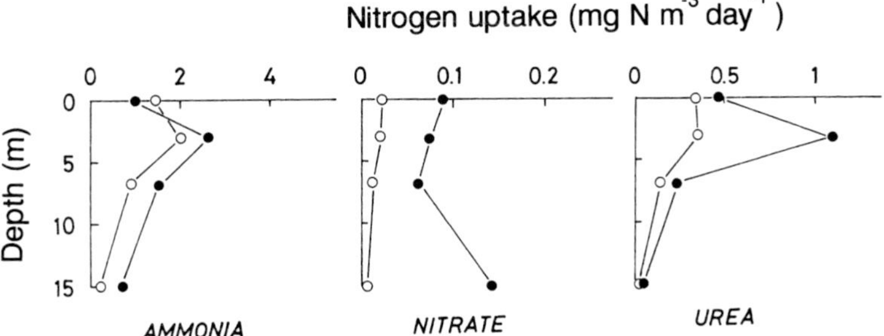

Fig. 3.22. Depth-profiles of light-influenced N uptake by phytoplankton, as rates of uptake of three N-sources deduced from ^{15}N-labelled additions in Lake Dom Helvécio, Brazil, during rainy (•) and dry (○) seasons. From Mitamura *et al.* (1995).

George some sedimented phytoplankton is periodically resuspended from the bottom sediments, which may consequently act as an extended source of nutrients presented at higher concentration. Soils and sediments are often the primary source of nutrients (other than CO_2) for rooted macrophytic vegetation, from which by subsequent decay an overall 'nutrient pump' from solid substratum to water can operate. This feature is illustrated and discussed in relation to macrophyte stands of Africa by Howard-Williams & Lenten (1975) and in Denny (1985), and to those of the Lobo (Broa) Reservoir, Brazil, by Barbieri & Esteves (1991).

(b) Element composition of biomass

The composition of biomass is useful for assessing absolute and relative nutrient demand, and subsequent release, and for indications of specific nutrient deficiency (see Section 3.2.d). For these reasons there have been some analyses of tropical phytoplankton (separated by net or, as seston, by finer filters) and of aquatic, mostly floating or emergent, macrophytes. For phytoplankton the earliest analyses applied to the ecology of seasonal growth were those of Prowse & Talling (1958) for a Nile reservoir. Here phosphorus and silicon concentrations estimated for peak populations were of similar magnitude to those (as inorganic forms) in water entering the reservoir. However, the nitrogen present in peak blue-green algal populations was considerably higher, suggestive of N-fixation.

Probably the most sustained study of the chemical composition of tropical phytoplankton is that of Viner (1977*c*) on Lake George.

Variation over a year was limited, with the sestonic (particulate) *mass* ratio C:N:P:chl-*a* averaging 100:13:0.9:1. These proportions were used by Talling (1992), together with a standard *molar* ratio proposed by Redfield for C:N:P of 106:16:1, roughly to estimate the concentrations of the elements likely to be incorporated in various maximal densities of phytoplankton recorded in African inland waters. The nomogram so derived is reproduced in Fig. 3.23.

In later years the 'Redfield ratio' has had widespread use as a standard for sestonic comparison, relative nutrient availability and nutrient defi-

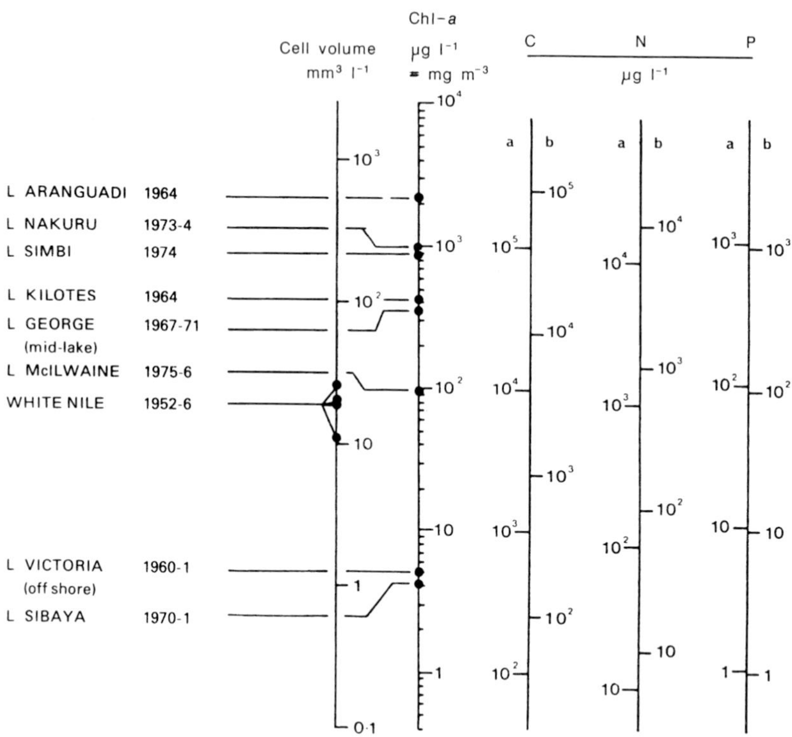

Fig. 3.23. Approximate interrelations, read horizontally across vertical logarithmic scales over four orders of magnitude, of concentrations of two indices of phytoplankton biomass (cell volume, chlorophyll *a*) as maxima observed in nine African waters, and estimates of associated quantities of cellular C, N and P. The interrelations are based on a chl-*a* content per unit cell volume of 4 µg mm^{-3}, C/chl-*a*, and C/N and C/P ratios that are either (a) the mean values for Lake George (Uganda) recorded by Viner (1977*c*), or (b) a C/chl-*a* mass ratio of 40, and the generalized Redfield values for C:N:P. From Talling (1992).

ciency symptoms. In Lake Ebrié, West Africa, variability of the C:N:P proportions of suspended particulates (seston) has been studied in time and space (Dufour *et al.* 1981*a*, *b*). Some results appear in Fig. 3.29. Hecky *et al.* (1993) describe an extensive comparative survey that includes the African lakes of Malawi, Kivu, Victoria, Albert, Kyoga and George, as well as various temperate lakes. Deviations from the Redfield composition were common in these tropical lakes (Table 3.2), suggestive of nutrient deficiencies that varied from lake to lake (see Section 3.2.d). Some additional results from Lake Malawi, partly available as vertical depth-profiles, are further described and discussed by Kilham (1990), Bootsma (1993*a*) and Guildford *et al.* (in press). Low ratios of particulate N and P to C (a biomass measure) were especially frequent in surface water and in the thermocline region.

Analyses of floating tropical macrophytes have particular relevance in relation to water-borne nutrient supply and uptake. Those of the invasive water-fern *Salvinia molesta* have been made in a series of freshwaters from Papua New Guinea to Northern Australia (Room & Thomas 1986 *a*, *b*). Low nitrogen content was often linked to reduced growth rates, expressed seasonally or latitudinally (Fig. 5.47). Stocks of nutrients in biomass per unit area can be particularly large in floating papyrus mats, floating grasses and in rooted reed-swamps; examples are shown in Fig. 2.34. The chemical composition of such tropical vegetation has received considerable study in four regions: East Africa (Gaudet 1975, 1977*b*; Gaudet & Muthuri 1981*b*), Central southern Africa (Howard-Williams 1972, 1979*a*; Howard-Williams & Lenton 1975; Iltis & Lemoalle 1983), Amazonia (Junk 1970; Howard-Williams & Junk 1976; Junk & Howard-Williams 1984; Junk & Furch 1991; Furch & Junk 1992; Piedade *et al.* 1997), and southern Brazil (Esteves & Barbieri 1983; Barbieri *et al.* 1984; Barbieri & Esteves 1991). This attention has partly been in relation to issues of nutrient interception and release (e.g., for Lake Naivasha, Kenya: Gaudet 1979*b*; Lobo (Broa) Reservoir, Brazil: Esteves & Barbieri 1983; Barbieri *et al.* 1984; Barbieri & Esteves 1991; Amazonia: Piedade *et al.* 1997). It is perhaps not generally appreciated that these plant stocks contain potassium in similar quantities (often about 1–4% dry weight) to nitrogen, and that a not inconsiderable silica content can lead to an appreciable consumption relative to the inflow Si in a well-vegetated water-mass such as Lake Chad at low level (Carmouze *et al.* 1978; Carmouze 1983).

Table 3.2. *Element-composition of suspended particulate matter (seston) in samples (n, number) from 6 African lakes, as mean absolute concentrations in μmol l^{-1} and atomic ratios*

Region	Lake	*n*	C	N	P	C:N	C:P	N:P
Deep	Malawi (offshore)	29	16.1	1.3	0.07	12.2	257	21
	Kivu	7	38.1	3.5	0.07	10.9	531	48
	Victoria	16	68.3	8.2	0.64	8.5	110	13
	Albert	11	99.7	6.8	0.76	15.4	175	11
	Mean deep lakes		55.6	5.0	0.4	11.8	268	23
Shallow	Kyoga	4	607.8	58.4	1.43	9.6	399	39
	George		4160.0	210.0	7.75	19.8	537	27

Source: From Hecky *et al.* (1993)

(c) Fluxes of uptake and regeneration

With the exception of one element, fluxes of nutrients between external medium and biomass have rarely been studied for tropical situations with relatively fine experimental resolution. The exception is inorganic carbon, with photosynthetic assimilation measured generally by the ^{14}C tracer or by the correlate of O_2 evolution. Some useful physiological information has resulted, especially on carbon flux–light relationships (Section 3.1). However, interest has centred on use of the photosynthetic fluxes as an accessible rate-meter of primary, autotrophic production, rather than for chemical exchange *per se*.

Carbon uptake fluxes involve several problems that are significant for tropical inland waters. It appears that the highest uptake fluxes in dense phytoplankton, greater than *c.* 0.3 mmol l^{-1} h^{-1}, are only likely to be sustained in waters of high inorganic C content (e.g., soda lakes: Talling *et al.* 1973). Rate-ceilings may also be set by rates of CO_2 hydration – dehydration kinetics, as has been proposed for Lake George (Ganf 1972). If the net flux of CO_2 entry at the atmospheric–water interface is exceeded by the net photosynthetic uptake per unit area, and other carbon inputs are minor, CO_2-depletion (and raised pH) will develop – as is actually found in many productive tropical lakes. The issue of bicarbonate use as a C-source then becomes important. This question has not been taken up experimentally in any tropical water, with the possible exception of Lake George (Ganf & Milburn 1971; Ganf 1972). In most instances CO_2 depletion has been deduced from pH-alkalinity relationships, but there are a few applications of further acidometric titration (Sreenivasan 1964; Dunn 1967) and total CO_2 measurement (Ganf & Milburn 1971; Wood, Kannan & Saunders 1984), some of which indicated the conversion of bicarbonate- to hydroxide-alkalinity as CO_2 was assimilated.

Phosphorus uptake can be studied through removal or cellular accumulation from relatively high concentrations of added phosphate, or from movement of an added radioisotope (^{32}P or ^{33}P) that does not appreciably alter the natural initial concentration. These approaches have been applied to uptake by phytoplankton in a few tropical lake waters.

The very dense and persistent phytoplankton of Lake George, Uganda, incorporates about 0.2–0.4 mg P l^{-1}, whereas the concentrations of ambient soluble reactive P (SRP) are lower by a factor of $\sim 10^{-3}$ (Viner 1977*c*). If the system is enriched to PO_4-P concentrations maintained at 0.5–1.5 mg l^{-1}, the initial net uptake flux is large at $\sim$0.5 mg P l^{-1} day^{-1} (Viner 1973). However, this appears to have the character of 'luxury consumption', as the consequent increase of cellular carbon is much less than the original proportion of cellular C to P would suggest.

The capacity for short-term (2 h) luxury uptake of P, after enrichment, to alter the cellular C:P ratio was tested over an annual cycle in Lake Titicaca by Vincent, Wurtsbaugh *et al.* (1984). This indication of luxury uptake potential varied with season and the offshore or bay character of the station, but in general was less marked than the C:N shift and strong N uptake after N-enrichment.

The isotope-tracer method has been widely applied in temperate lakes, with which direct comparisons are possible (e.g. Kalff 1983). The estimated quantity is the relative decrease rate of added soluble reactive P, corrected for back-flux from particulates, and usually expressed as its reciprocal or 'turnover time' (= pool/flux). The 'turnover times' so obtained can be too short owing to unreliability of the back-flux correction under experimental conditions, with some cell-leakage (Fisher & Lean 1992), and possible adsorption onto inorganic particles. However, their comparative value can probably be accepted and in general correlates with the original concentration of soluble reactive P. Thus, in the original tropical work in East Africa, Peters & MacIntyre (1976) measured slow relative uptake and estimated long 'turnover times' in the phosphate-richer Lake Nakuru and the converse conditions in the phosphate (SRP)-poorer Lake Sonachi and Lake Elmenteita. For the neighbouring Lake Naivasha, Kalff (1983) found a pronounced seasonal variation of relative uptake rate. Fairly slow relative uptake rates have been estimated for the moderately P-rich offshore waters of lakes Victoria (Lehman & Branstrator 1994) and Titicaca (Vincent, Wurtsbaugh *et al.* 1984). In both these lakes, however, much faster rates were obtained from shallow inshore areas (Fig. 3.24) that were richer in phytoplankton but with lower concentrations of soluble reactive phosphorus. The results from offshore Lake Victoria water with graded P additions suggested a half-saturation constant of about 1.2 μmol P l^{-1}, a concentration exceeded in the initial lake water.

Work on ^{32}P uptake in the surface water of an Amazonian floodplain lake, Lake Calado, is summarized by Fisher, Doyle & Peele (1988) and Melack & Fisher (1990). It illustrates two features encountered elsewhere:

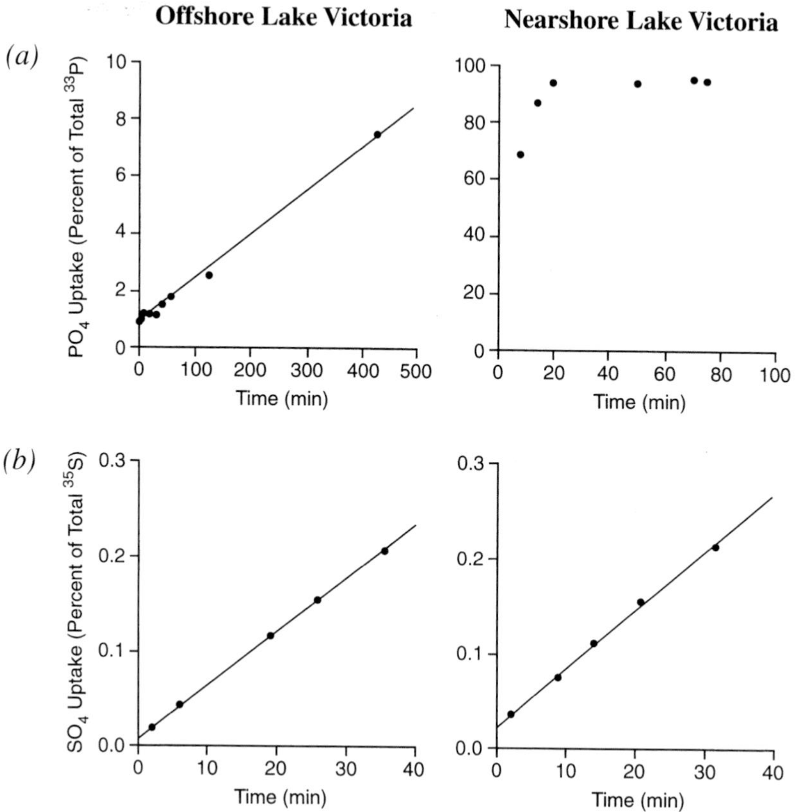

Fig. 3.24. Comparison of the time-course of uptake by particulates of added (*a*) ^{33}P-PO_4 and (*b*) ^{35}S-SO_4 from offshore and nearshore surface water of Lake Victoria. Modified from Lehman & Branstrator (1994).

that very small organisms < 3 μm, including heterotrophic bacteria, can dominate both uptake and regeneration, and that the absolute movement (as specific rate constant × ambient P-pool size) is much greater than the flux of net P uptake obtained from chemical analysis of P-enriched samples. The last difference can be indicative of rapid recycling and turnover, but could also have some contribution from an experimental artefact involving overestimated values of the specific rate constant (Fisher & Lean 1992). Features relevant to P-uptake, but discussed later as tests of nutrient limitation (Section 3.2d), are phosphatase activity on organic P-substrates and luxury accumulation of P deduced from release by heat-treatment.

Nitrogen uptake, like phosphorus uptake, can be followed by chemical analyses after addition of an N-source, or by isotopic means. The possibilities are extended by the several N-sources potentially available. These include ammonium-, nitrate-, nitrite and organic-N (e.g., urea); also, for nitrogen fixers, dinitrogen (N_2) itself. There is evidence for the occurrence of hydroxylamine in at least one Ethiopian lake (Baxter *et al.* 1973). In unpolluted surface waters nitrate is the only form likely to accumulate in quantity, above 100 μmol l^{-1}; such accumulation is however, uncommon in warm tropical waters. Ammonium-nitrogen is generally taken up preferentially in relation to nitrate-nitrogen. This feature, plus its importance in the regeneration of inorganic nitrogen and frequent deep accumulation in stratified waters, has led to its use in many studies of N-enrichment and uptake. Older work, before about 1960, was handicapped by the lack of an effective analytical method for ammonium-N at low concentrations.

Enrichments of NO_3-N, NO_2-N and especially NH_4-N were applied by Viner (1973, 1977*c*) at relatively high concentration (0.5–1.5 mg N l^{-1}) to follow net uptake by the dense phytoplankton (~3 mg particulate N l^{-1}) of Lake George. In this lake the concentrations of these N-forms were generally below limits of detection over much of the day. Enrichment induced high rates of N-uptake over several days (~0.5 mg N l^{-1} day^{-1}) that were promoted by light in both laboratory and lake exposures, and with accompanying increase of cellular C. Some enhancement by added phosphate was also obtained. The larger and smaller algal forms of the phytoplankton showed considerable difference in behaviour, a fraction of smaller forms being much more active in N-uptake.

Radioisotope technique can be applied to measure the gross rate of N-uptake by use of an ammonium analogue, ^{14}C-methyl-ammonium. Rates were so measured by Vincent *et al.* (1984) over an annual cycle in Lake Titicaca. Expressed per unit chlorophyll *a* they were very high compared to values measured elsewhere, suggesting a prior N-limitation. However, rates fell after winter mixing led to replenishment of inorganic nitrogen in surface water. Tests with graded concentrations of the methyl-ammonium indicated a half-saturation value for uptake of about 1 μmol N l^{-1}.

From application of another isotopic tracer, ^{15}N, the same half-saturation value (corrected for isotopic dilution) of ~1 μmol l^{-1} was estimated for $^{15}NH_4$-uptake at light-saturation in Lake Calado, an Amazonian floodplain lake (Fisher *et al.* 1988; Fisher, Doyle & Peele 1988; Melack & Fisher 1990: see Fig. 3.25). The actual NH_4-N concentrations of epilimnetic water, after the river inflow was lost, were generally less than this – indicating N-limitation of uptake rate. Higher rates were possible under N-enrichment, and when expressed per unit chl-*a* the N-saturated rate was ~0.01–0.04 μmol N mg chl-a^{-1} h^{-1}. The rates were enhanced by

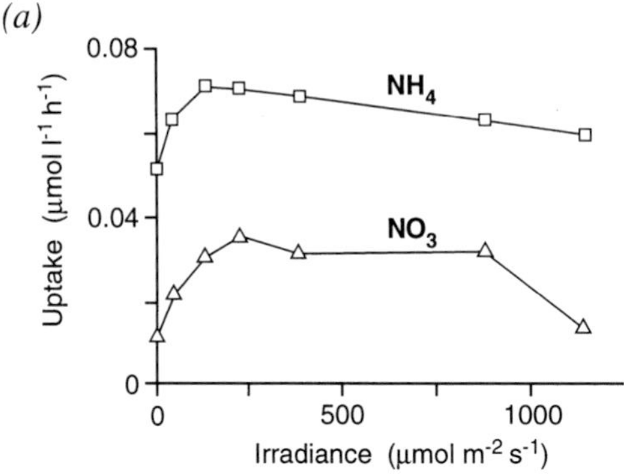

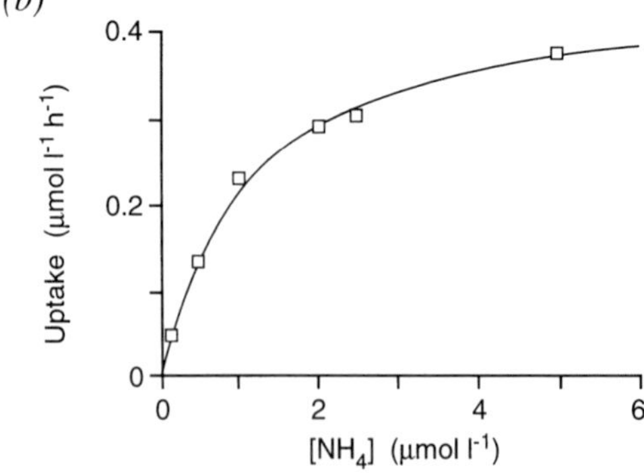

Fig. 3.25. Rates of uptake of NH_4-N and NO_3-N by particulates in samples of water from Lake Calado, Amazonia: (*a*) in relation to solar irradiance as photon flux density, (*b*) in relation to ambient concentration. From Fisher *et al.* (1988).

light, although less markedly so than rates of NO_3-N uptake (Fig. 3.25), and so in nature most N-assimilation occurred during daylight hours. However, the predominant limitation of NH_4-N uptake was by low ambient concentrations. Uptake of NO_3-N, also ^{15}N labelled, was comparatively minor. Size-fractionation of the plankton showed that much uptake and regeneration of N, like those of P, were mediated by small cells (< 3 μm) of bacterial size (Fisher, Doyle & Peele 1988) – a possible reflection of the mainly heterotrophic metabolism of this forest-lake.

At a higher latitude in southern Brazil, rates of uptake from several N-sources were studied in the Rio Doce Valley lakes with ^{15}N tracer by Mitamura *et al.* (1995). Ammonium- and urea-N were sources preferred to NO_3-N; all had rates subject to a light-influenced depth dependence (Fig. 3.22), with a considerable parallelism to photosynthetic C uptake.

N_2-fixation is a capacity widespread among blue-green algae (cyanophytes, cyanoprokaryotes) and is likely to make a considerable contribution to the N-utilization of communities in which heterocystous forms of this group are abundant. Confirmation by direct measurements, using ^{15}N or the acetylene reduction assay, is available for a few tropical lakes that include Lake George (Horne & Viner 1971; Ganf & Horne 1975), Lake Valencia (Lewis & Levine 1984; Levine & Lewis 1986), Lake Titicaca (Wurtsbaugh *et al.* 1985) and Lake Kariba (Moyo 1991, 1997). The light-dependence of fixation leads to a diel cycle of rates in nature (Chapters 2.4a, 5.2), as found in Lake George (Ganf & Horne 1975) and Lake Valencia (Fig. 3.26). As well as in the plankton, the process can be important in littoral attached communities, such as the periphyton of Lake Calado (Melack & Fisher 1988; Doyle & Fisher 1994) and Lake Camaleão (Kern & Darwich 1997).

Silicon uptake, predominantly by diatoms, can induce conditions of marked Si-depletion in some tropical waters, expressed in space (e.g., along a Nile reservoir: Fig. 3.21) or in time (Chapter 5.1). The kinetics of uptake have not been studied for tropical situations, but from other work it appears that half-saturation constants are likely to range considerably between species or genera common in tropical phytoplankon, and specifically in that of African lakes and rivers. It has been conjectured that the ambient ratio of Si:P is influential for their distribution (Kilham *et al.* 1987). A direct culture study has been made by Kilham of the relationship between silicon concentration and the specific growth rate of a clonal isolate of a widespread species, *Aulacoseira (Melosira) granulata*, from Lake Mulehe in Uganda. The result (Kilham 1990*a*: see Fig. 3.27) indicated a half-saturation level of about 5 μmol Si l^{-1}. This is

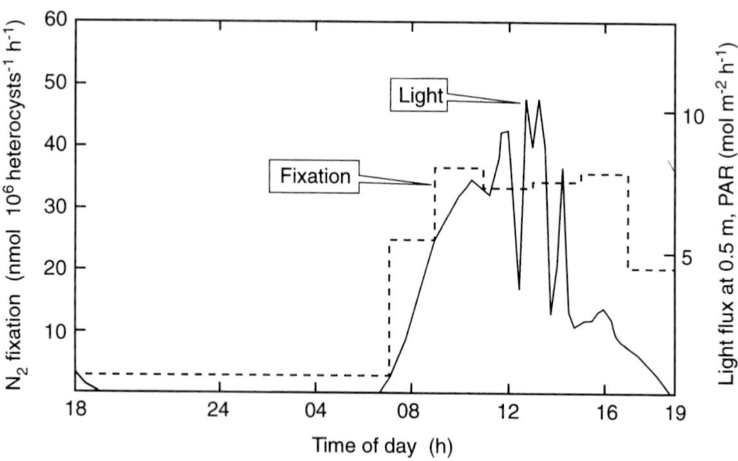

Fig. 3.26. Lake Valencia, Venezuela. Diel variation in rates of nitrogen fixation by planktonic cyanophytes (blue-green algae), estimated as specific rates of fixation per unit number of heterocysts (heterocytes) in relation to light. From Levine & Lewis (1984).

much higher than that of some other species, such as *Stephanodiscus minutus*, but low relative to the concentrations of soluble reactive Si prevalent in most tropical waters.

Sulphate-S uptake has received a single study by Lehman & Branstrator (1993, 1994), stimulated by the unusually low concentrations of 3–4 μmol l^{-1} (0.15–0.22 mg SO_4 l^{-1}) now established for Lake Victoria and the earlier belief that sulphate deficiency limited phytoplankton production in this lake (Beauchamp 1953*b*; Fish 1956). Measurements with tracer-labelled sulphate, $^{35}SO_4$, yielded uptake rates of 11.5 ± 2 nmol l^{-1} h^{-1} and relatively long turnover times in excess of ten days. Enrichments with 5–50 μmol SO_4 l^{-1} did not significantly increase the uptake rate, suggesting that even the low original concentrations were well above the half-saturation constant for sulphate uptake.

Fluxes of nutrient regeneration coexist and in part support those of nutrient uptake in any water-mass. Studies on a fine scale, other than by mass budgets (Chapter 2.4) or seasonal time courses (Chapter 5.1), are few for tropical waters. In the productive Lake George, decomposition or mineralization of the organic particulate material (largely phytoplankton) has been followed over periods of 1–3 days by loss of organic C and N, with correlation to O_2 consumption (Golterman 1971; Ganf 1974*a*). Fractional loss rates were of the order of 0.05 to

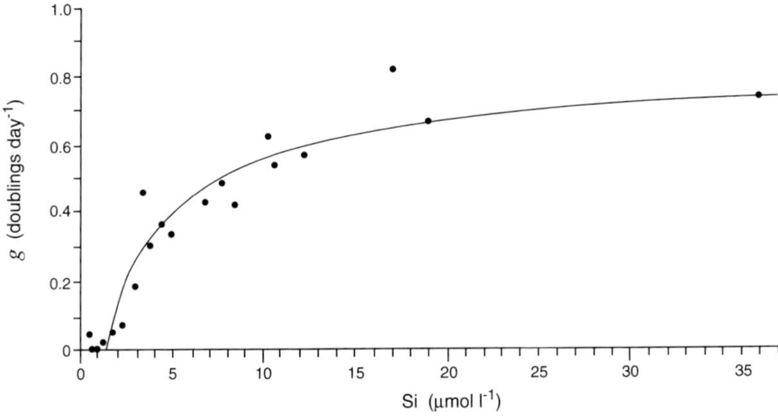

Fig. 3.27. Relationship to Si concentration of the specific growth rate (g) of an African clone isolate of the diatom *Aulacoseira (Melosira) granulata* from Lake Mulehe, Uganda, grown at 20 °C in batch cultures. From data of S.S. Kilham, after Kilham (1990*a*).

0.1 per day, but showed considerable variation. Further variability appeared in the loss of non-particulate organic C and N ascribable to bacterial activity.

One component of the regenerative flux is that from nutrient excretion by animals. This has been estimated with laboratory measurements on isolated samples of the larger zooplankton from at least three tropical lakes – George (Ganf & Blažka 1974), Titicaca (Pawley & Alfaro 1984), and Calado (Lenz *et al.* 1986). Diel variation of rates per unit water volume could arise from variation in specific rates per animal biomass (Lake George) or from variation in animal abundance allowing for migrations (Lake Calado). In lakes Titicaca and Calado the estimated contribution from this macrozooplankton source was small relative to estimated demand from the phytoplankton (Table 3.3). For Lake George it was probably appreciable: from the short-term data of Ganf & Blažka (1974), Ganf & Viner (1973) estimated mean excretion rates of 36 mg NH_4-N m^{-2} day^{-1} and 7 mg PO_4-P m^{-2} day^{-1}, rather greater than the mean annual-based rates of N and P entering the lake from inflows. Other more indirect estimates based on feeding rates and element-composition of the phytoplankton (Viner 1977*c*) are of the same order.

Nutrient excretion by benthic animals can be illustrated by a single study, that of Kiibus & Kautsky (1996) on the littoral mussels of Lake Kariba. Measurements of NH_3-N and PO_4-P release per unit biomass and hour were combined with those of mollusc stocks in the lake littoral (0–12 m), allowing for size structure. The excretion appeared to be an important agent of nutrient regeneration in the lake; it was estimated to be equivalent to the mineralization of ~ one-quarter of the external load to the lake of P and ~8 times that of N.

Measurements of N-regeneration in relatively undisturbed water samples from Lake Calado were made by Morrissey & Fisher (1988) using the technique of isotope dilution. In this $^{15}NH_4$-N was added, followed by incubation in 10-l containers exposed for up to 12 h in the epilimnion. The relative isotopic composition and size of the NH_4-pool was assessed at intervals, after precipitation by an Hg-reagent; an isotopic dilution took place due to the addition of unlabelled NH_4-N as regeneration from particulate-N. From the product of rate of change of isotopic composition and pool size, the rate of regeneration was calculated. Its average magnitude in epilimnetic water, 0.86 ± 0.15 μmol NH_4-N l^{-1} h^{-1}, was roughly in balance with uptake rate limited by low ambient concentration of NH_4-N; it implied a short NH_4-N turnover time of <1 h. Much regeneration, like uptake, appeared to be predominantly linked to small particles, <3 μm, that could be largely micro-heterotrophs such as bacteria and small Protozoa (Fisher *et al.* 1988).

Some parallel measurements of P-regeneration rates were also made in Lake Calado (Fisher *et al.* 1988), using $^{33}PO_4$-P as a tracer and calculations from isotope dilution. The rates obtained for original, unfractionated samples were very variable (0.01–1.1 μmol P l^{-1} h^{-1}) and some probably overestimated; any relationship with particulate size-fractions could not be resolved.

Table 3.3 summarizes estimates from Melack & Fisher (1990) of regeneration fluxes of N and P in Lake Calado, expressed per unit area. These are separated between epilimnion and hypolimnion; also for excretion by macrozooplankton and output from the sediments.

(d) Limiting nutrients

Almost any attention to issues of primary autotrophic production at a particular site tends to raise one question: what is the main limiting nutrient? This question could be difficult or even unrealistic in some circumstances, as when growth rates are overwhelmingly limited by physical factors, when multiple nutrient limitations operate, when the components of communities have different requirements or sensitivity to depletion and when different limitations follow in a time-sequence. All

Table 3.3. *Comparison of direct measurements of N and P regeneration in Lake Calado. Values from various sources re-calculated per unit area*

Source	μmol m^{-2} h^{-1} ammonium production	μmol m^{-2} h^{-1} phosphate production
Epilimnion (whole)	2600 ± 260	720 ± 50
(Macrozooplankton)	26 ± 3	3.3 ± 0.3
Hypolimnion (whole)	380 ± 60	–
Sediments	270 ± 60	37 ± 19
Total	3750 ± 270	

Source: From Melack & Fisher (1990)

these circumstances can be illustrated from tropical waters. However, a number of approaches or experimental tests have been used to detect and identify limitations.

Four main approaches are recognizable. These, respectively, focus on **ambient nutrient concentrations**, in absolute magnitude, mutual ratios and time-correlations with biological events; on the **elemental composition of biomass**, especially concerning C:N:P ratios and the implications for net nutrient uptake of observed population change; on **growth response** induced experimentally by combinations of added nutrients; and on **short-period physiological responses** of the original or nutrient-enriched biomass, such as phosphatase and nitrogenase activity, rates of nutrient uptake, and enhancement of dark uptake of ^{14}C-labelled CO_2. No general consensus exists on their advantages and limitations (see, e.g., Hecky & Kilham 1988), but effective application of multiple tests is well exemplified by the work on the Ebrié lagoon and Lake Titicaca summarized below.

There is a long history of related studies on waters and phytoplankton of lakes and reservoirs in East and Central Africa. Here early analyses of ambient nutrient concentrations in surface water, especially of PO_4-P, NO_3-N and Si, showed the frequent occurrence of relatively high values for PO_4-P and Si but low values for NO_3-N and (less distinctively) NH_4-N. This information was augmented, compiled and summarized by Talling & Talling (1965), who suggested that N-limitation rather than P-limitation might be regionally prevalent. For this there was some construed support from time-correlations of nutrient concentration and algal population dynamics in the White and Blue Niles (Prowse & Talling 1958; Talling & Rzóska 1967) and in offshore Lake Victoria (Talling

1966), that included biomass analysis with low percentage N content. Early analyses, showing sulphate in Lake Victoria water at very low concentration, < 1 mg l^{-1}, were interpreted by Beauchamp (1953*b*) as indicative of sulphate limitation. Although these results were questioned (Talling & Talling 1965; Talling 1966), three decades later the application of an improved analytical method (ion chromatography) also yielded very low concentrations, of 3–4 μmol (0.3–0.4 mg) SO_4 l^{-1} (Lehman & Branstrator 1993, 1994).

Potentially stronger evidence of nutrient limitations is obtainable from growth bioassays, that also have a long history in East and Central Africa. As applied to Lake Victoria, Fish (1956) obtained responses with inoculated test algae that he interpreted as a major sulphate limitation with some additional P- and N-limitation. Evans (1961), working with the original phytoplankton, found positive response to added phosphate by an *Aulacoseira (Melosira)* component. All these results can be criticized as involving unrealistically high concentrations of added nutrients with resulting dense biomass and the possibility of distortion of 'wall effects' (e.g., adsorption) in small glass vessels. The same reservations apply to two series of bioassay tests in Central Africa. The earlier, by Moss (1969), were on nine water-bodies in Malawi, including lakes Malawi, Chilwa and Malombe. Those of Robarts & Southall (1975, 1977) utilized inoculation of a test alga (then known as *Selenastrum capricornutum*) on samples from reservoirs that included lakes Kariba and McIlwaine (Chivero). Much later the same alga was again applied in tests on Lake Kariba by Lindmark (1997). Another test alga, *Monoraphidium minutum*, was used by Liti *et al.* (1991) on water from Lake Turkana, Kenya. These various tests often, but not always, indicated N as the primary limiting nutrient; those of Moss gave very clear-cut and consistent responses, with positive response to added N (as NO_3^-) far predominating (Fig. 3.28c). The same can be said of the more recent bioassays of surface water from offshore Lake Victoria (Lehman & Branstrator 1993, 1994) with added NH_4-N, PO_4-P and SO_4-S (see Fig. 3.28b). Here responses were rapid, with only N stimulation, after two days in large polyethylene bags with enrichments in low concentration. Probably the largest scale of experimentation has involved entire fishponds such as those in Tanzania enriched by Payne (1971) with NH_4-N and PO_4-P. On one occasion response to N, but not to P, led to a considerable increase of chl-*a* concentration that was later associated with increased rate of growth of the fish *Tilapia zillii*.

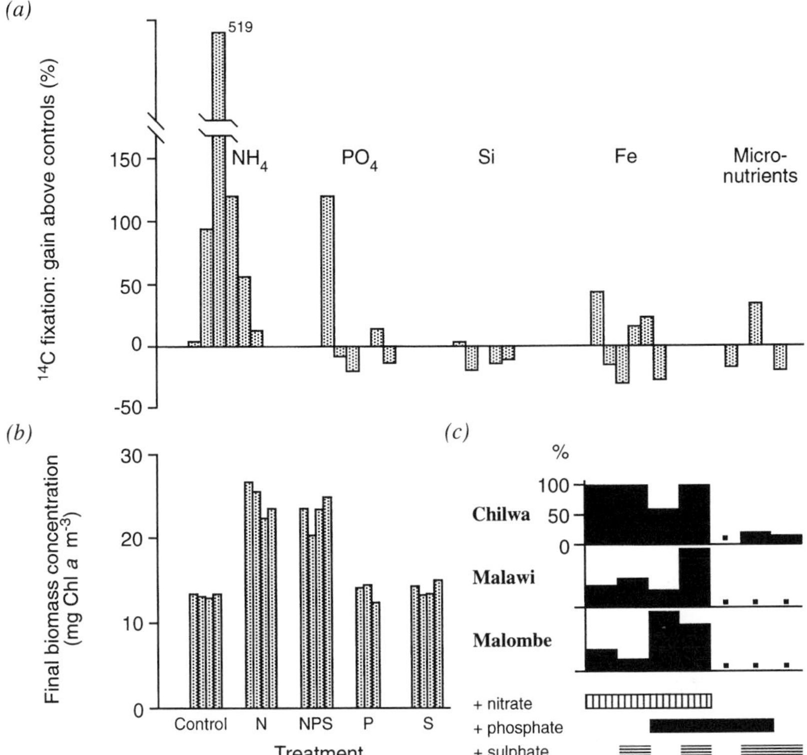

Fig. 3.28. Comparative responses to various added nutrients in bioassays of water from (*a*) Lake Titicaca, (*b*) Lake Victoria, (*c*) lakes Malawi, Chilwa and Malombe. The responses involved were as ^{14}C fixation after separate 5–6 day incubations (*a*), as concentrations of chl-*a* after replicate 48 h incubations (*b*), and as increase after several weeks in chl-*a* as percentage of the maximum such increase in each separate experiment (*c*; ▪, no increase over control). From Wurtsbaugh *et al.* (1985), Lehman & Branstrator (1994) and Moss (1969).

There is evidence for P-limitation in some adjacent tropical lakes in East Africa, although these are relatively cool by virtue of their altitude of around 1800 m. The best supported case is the small Lake Sonachi, notable for water with a considerable concentration of total P (~150 μg l^{-1}) if not soluble reactive P. Here an enrichment experiment, involving large plastic enclosures *in situ*, showed positive stimulation of biomass by PO_4-P in excess of that by NH_4-N addition. There was supporting evidence from changes of C:N:P ratios in particulate matter (Melack *et al.* 1982). Overlapping work by Kalff (1983) also pointed to P-deficiency by fast relative rates of ^{32}P uptake here and in the nearby lakes Oloidien and

Elmenteita; also possibly by seasonally rapid rates in Lake Naivasha. Additional evidence sought by Kalff from release of 'surplus' seston-P by boiling and from the ratio total P:total N in water samples, was, respectively, negative in Lake Sonachi or probably biased by unutilizable organic N.

Elsewhere in Africa, bioassays by multiple nutrient enrichment have been applied to the coastal and slightly saline Ebrié lagoon, Ivory Coast (Dufour *et al.* 1981*a, b*). These, and less direct evidence from C:N:P ratios in particulate matter (seston) and relative concentrations of ambient nutrients, suggested co-limitation by N and P, but predominantly N. For example, increase in the biomass of unenriched phytoplankton occurred mainly above an N/C quotient (by atoms) of 0.1 (Fig. 3.29).

In South East Asia a bioassay study was made by Anton *et al.* (1996) on a large Malaysian reservoir, Pansoon. Additions of NO_3-N, PO_4-P and Si were made to plastic-walled isolation columns *in situ*. All showed positive growth response above controls after two weeks, with decline later. The largest increase was produced by addition of N; the qualitative effect on species representation varied with the N:P ratio.

Other tropical studies have largely been in four regions of South America. That on Lake Valencia, Venezuela, was mainly based on correlative responses of the phytoplankton to concentrations and fluxes of inorganic N and P in nature and in enriched enclosures (Lewis 1983*b*,

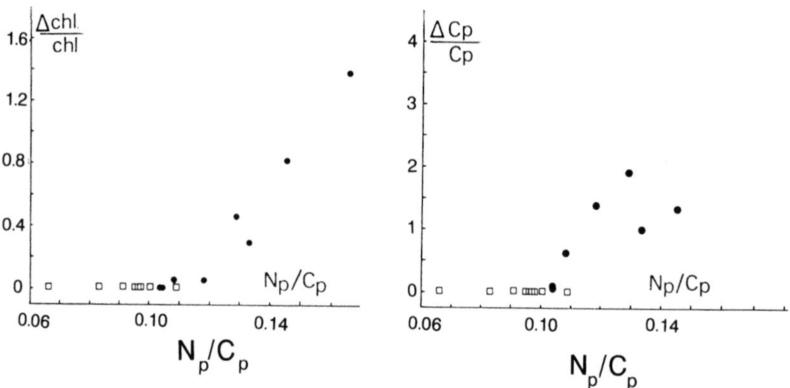

Fig. 3.29. The threshold for relative growth increments during laboratory exposures, assessed by chl-*a* (left) and particulate C (right), in terms of the initial (•) ratio of N to P in particulate matter (N_p/C_p, by atoms) from Lake Ebrié, West Africa. □, ratios after growth ceased. From Dufour, Lemasson & Cremoux (1981).

1986*a*), with additional information on N-fixation (Levine & Lewis 1984); N was deduced to be the principal limiting nutrient. In Amazonia, the floodplain lake Lake Calado has been studied intensively by enrichment growth bioassays (Setaro & Melack 1984; Pinheiro 1985 cited by Melack & Fisher 1990), by various short-term physiological responses (Setaro & Melack 1984), and by comparative tracer measurements of P and especially N fluxes. Seasonal sequences related to water level were all important, without overall limitation by a single nutrient. At a high level neither P nor N appeared limiting, with enhancing additions, in assays; later there was evidence of a transition from mainly P to mainly N limitation at falling and low water levels. This could be connected with the dissimilar main sources of these nutrient elements – river-based flux for P input and lateral run-off for N input; also with the low N:P ratio in products of regeneration (Fisher *et al.* 1991). The low level phase of predominant N-limitation also appeared to have a parallel in another floodplain lake, Lake Jacaretinga, where mainly growth bioassays showed strong positive responses to N- or N + P- enrichment but little or no significant response to P-enrichment alone (Zaret *et al.* 1981; Henry *et al.* 1985*b*), or to that of humic and fulvic acids (Devol *et al.* 1984).

Other work in central and southern Brazil has centred on the Lago Paranoá reservoir at Brasilia (Lindmark 1976), the Lobo (Broa) and Barra Bonita reservoirs near São Paulo (Henry & Tundisi 1982, 1983; Henry *et al.* 1984; Henry *et al.* 1985*a*) and two of the Rio Doce Valley lakes (Henry *et al.* 1997). Growth bioassays, of 14-day duration, with chl-*a* and cell numbers or biovolume as indices, showed positive responses mainly to N- or N + P-enrichment and little to P-enrichment alone in the Lobo Reservoir. In the larger and relatively nitrate-rich Barra Bonita Reservoir and in the sewage-enriched Lago Paranoá, positive response to P- and P + N-enrichment was found. The relative importance of P- and N- responses varied with season and the stratification cycle in the Rio Doce Valley lakes of Dom Helvécio and Carioca. Additions of Mo also elicited a positive response in the Lobo Reservoir (Henry & Tundisi 1982), but not those of the metal-complexing agent EDTA during a single experiment in February 1980 (Henry & Tundisi 1983).

Probably the most thoroughly investigated tropical lake for nutrient limitations is the high-altitude Lake Titicaca. Work there is summarized by Wurtsbaugh *et al.* (1992). The tests applied have included several longer period, growth-sensitive responses to enrichment (Carney 1984; Wurtsbaugh *et al.* 1985: see Fig. 3.28a) and five types of short-period physiological assays (Vincent *et al.* 1984). The latter included alkaline phosphatase activity (low), ^{32}P relative uptake (slow), changes of seston composition after enrichments, uptake rates of ^{14}C-labelled methyl ammonium (high) and NH_4-enhancement of dark ^{14}C fixation (pronounced). All tests concurred in the prevalence of N-limitation, except during the annual cool phase of extended vertical mixing in the main lake (Lago Grande) when light-limitation probably replaces nutrient limitation of the phytoplankton and higher concentrations of nitrate accumulate (see Fig. 3.30). The lake is situated in a region with high rates of chemical denudation, which contribute to considerable levels in the lake of total phosphorus and – outside bays – of soluble reactive phosphorus.

The replacement of nutrient limitation by light-limitation is doubtless frequent in many turbid tropical waters. It is probably a year-round phenomenon in the large but shallow Lake Chapala of Mexico. Here Dávalos *et al.* (1989) have explored potential nutrient limitation by growth bioassays with both the natural phytoplankton and a test alga, *Ankistrodesmus bibraianus*, using exposures in laboratory and lake. Nitrogen limitation was strongly indicated in illuminated laboratory samples enriched with nitrate or phosphate or both, but in the lake itself light-limitation was predominant, as was also found in studies of primary production (Lind *et al.* 1992). Analogous results were obtained in some subtropical waters of the Paraná floodplain, Argentina (Carignan & Planas 1994).

All the enrichment tests so far mentioned have centred on phytoplankton. The response of periphyton has been studied by Pringle *et al.* (1986) for a lowland stream in Costa Rica, using submerged agar plates with a nutrient-agar-sand substratum to introduce combinations of nutrients *in situ*. The results suggested that, in this region of volcanic soils, there was not a primary limitation by N or P but probably one by micro-nutrients.

To summarize, it appears that although various forms of nutrient limitation for phytoplankton are regionally or seasonally possible, N-limitation is particularly widespread in tropical lakes. Less extensive work on floating macrophytes, especially *Salvinia molesta* in Papua New Guinea and northern Australia (Room & Thomas 1986*a*), supports this view. There is also conclusive evidence for N-limitation of *Eichhornia crassipes* in the subtropical Paraná floodplain (Carignan *et al.* 1994). As Vincent, Neale & Richerson (1984) remark, it differs from the picture of predominant P-limitation reached largely from studies of north-temperate glacial lakes. Whether or not the tropical-temperate difference is generally and significantly valid is disputed (Kalff 1983, 1991) and remains to be resolved.

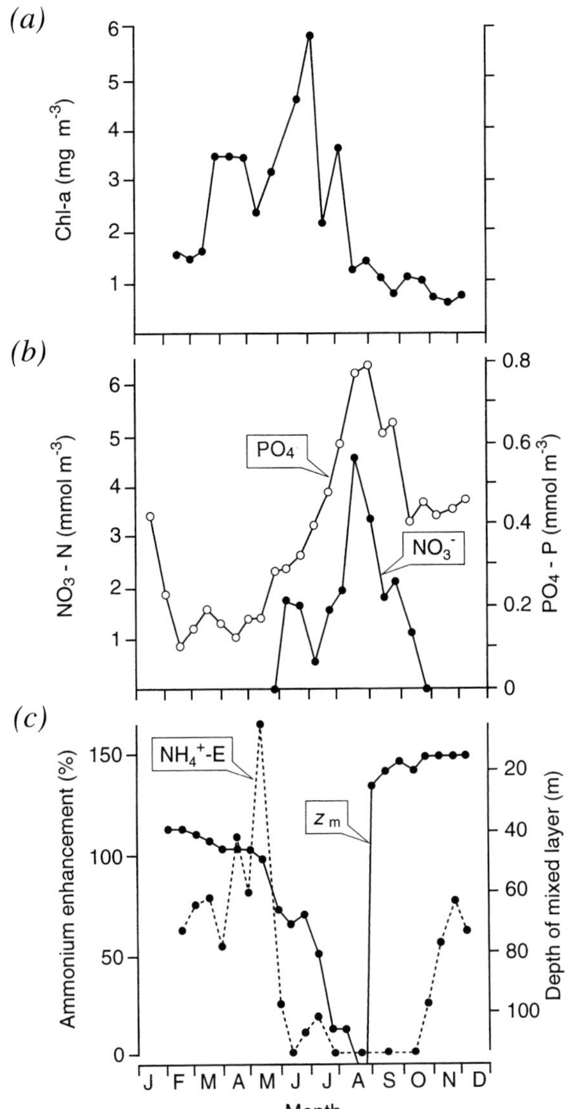

Fig. 3.30. Lake Titicaca, offshore area. Annual variation of concentrations within the euphotic zone of (*a*) chl-*a*, and (*b*) PO_4-P and NO_3-N in relation to (*c*) ammonium enhancement (NH_4^+-E) of dark ^{14}C fixation, as a test for N-limitation. In (*c*) the depth of the surface mixed layer z_m also shown. From Vincent *et al.* (1984).

3.3 Secondary utilization

Organisms dependent on pre-formed organic material for their income of energy – supporting metabolism, growth and reproduction – are of diverse microbial, invertebrate and vertebrate affinities. The microbial sector, overall the most influential in ecological processing, is the least explored in tropical freshwater environments. Some gross features are considered in the following Section 3.4. Here the activities of invertebrate and vertebrate consumers are discussed.

(a) Food acquisition

The mechanisms of food acquisition relate to its character as plant, animal or detrital material (with herbivores, carnivores/predators, detritivores, omnivores); to the absolute and relative size of particles (microphages, macrophages); to the degree of dispersion or compaction (planktivores, shredders, scrapers); and to relative passivity or encounter-activity of the consumer (filter-feeders, substratum ingesters, ambush predators, pursuit predators). These distinctions and specialisms are general rather than tropical in significance. However, there are some indications of tropical bias and some mechanisms have been analysed in depth in relation to the natural economy of tropical sites at which they occur.

Over recent decades studies of temperate running waters have increasingly used modes of food acquisition to subdivide faunas into functional groups. Thus shredders, grazers, collector-gatherers, filter-feeders and predators are distinguished. Although corresponding work on tropical waters (e.g., Yule 1995) is scanty, Dudgeon & Bretschko (1995, 1996) have surveyed and compared such functional representation in tropical South East Asia and temperate Europe. One apparent difference is the fewer vegetable shredders recorded from the tropical region, which may be connected with a greater rate of breakdown or greater prevalence of unpalatable leaf types – a factor discussed by Stout (1989). A connected feature may be the deficiency of tropical amphipods and isopods, compared with their abundance in many temperate streams. Throughout the world, benthic invertebrates detached as a flowing 'drift' (Chapter 5.2g) are a possible food source for fishes. A close relationship in one tropical river, draining Mount Kenya, has been studied by Mathooko (1996). Here the introduced Rainbow Trout, *Oncorhynchus mykiss*, functioned as a natural 'drift sampler'.

In a very different habitat, the lake pelagial, the tropical potential for year-long stocks of plankton can provide a food resource that is less interrupted over time than in most temperate lakes. Probably in part-response, there is an extensive development of planktivores at the vertebrate level in tropical lakes. Among fishes, the cichlids and clupeids (both originally marine groups) provide examples in many African lakes. Instances for which food intake has been studied qualitatively and quantitatively include the cichlids *Oreochromis niloticus* and *Haplochromis nigripinnis* in Lake George (Moriarty *et al.* 1973) and, in less detail, *Sarotherodon galilaeus* in Lake Chad (Lauzanne 1978, 1983).

The Lesser Flamingo *Phoeniconaias minor*, abundant in some East African soda lakes, is a unique bird planktivore. It wades in shallow water, feeding by a rapid pumping and filtration mechanism (Jenkin 1957; Vareschi 1978). The tongue moves through a beak (lower mandible) space of about 0.5–0.8 ml volume at around 20 cycles s^{-1}, propelling lake water past filter-processes that catch a high proportion of the larger phytoplankters (Fig. 3.31). For Lake Nakuru when the large phytoplankter *Spirulina fusiformis* (*S. 'platensis'*) was dominant, Vareschi (1978) estimated a filtration efficiency of *c.* 64–86%. This was reduced to low levels when the large *Spirulina* was replaced by much smaller and less filterable forms. Food intake per bird fell and the flamingo population moved to other lakes. As a warm-blooded homeotherm, each bird has a considerable energy requirement, indirectly estimated by Pennycuik & Bartholomew (1973) as 5.8 J s^{-1} or 500 kJ day^{-1} for resting metabolism and 15 J s^{-1} or 1300 kJ day^{-1} for normally active life plus $\sim$1 J s^{-1} for pumping energetics. These estimates were compared with the energy equivalent of practicable food assimilation (A, in J s^{-1}), determined by the food concentration (c) and energy content (e), pump-space volume (v), pumping frequency (n), filtering efficiency (ε_f) and food energy utilization efficiency (ε_l), and maximum fractional time spent feeding (f) as:

$$A = (c.e)(v.n.f)\ \varepsilon_f.\varepsilon_l \qquad (3.9)$$

Adopting estimates of $e = 20 \times 10^3$ J (g dry weight)$^{-1}$, $v = 0.5 \times 10^{-6}\text{m}^3$, $n = 20\ \text{s}^{-1}, f = 0.8$, $\varepsilon_l = 0.8$ and implicitly assuming $\varepsilon_f = 1$, it was shown that a food concentration of c of 125 g dry weight m^{-3} ($\simeq$125 mg l^{-1}) was required to maintain adult birds in normal activity ($\sim$1400 kJ day^{-1}), with some further increase at times of breeding. These high concentrations only occur in exceptionally productive lakes such as the Kenyan soda lakes.

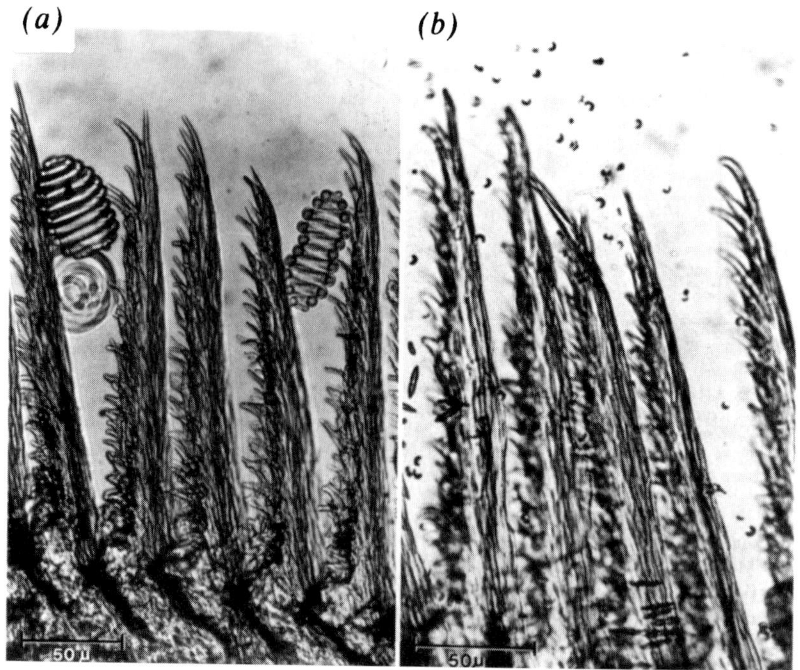

Fig. 3.31. Processes of the filter system of the Lesser Flamingo, *Phoeniconaias minor*, from Lake Nakuru, Kenya in relation to (*a*) the larger phytoplankter *Spirulina fusiformis*, (*b*) various smaller phytoplankters. From Vareschi (1978).

Vareschi (1978) re-evaluated the parameters above, made additional direct measurements of flamingo feeding rates by plankton removal, and greatly extended the ecological relevance by application to flamingo populations of known size and food resource in Lake Nakuru during 1972–77. The general requirement for dense phytoplankton food, >100 g dry weight m^{-3}, was confirmed. In 1972–73, when the lake was occupied by a flamingo population averaging 915 000 birds or 0.023 birds m^{-2}, the estimate of mean consumption rate was 0.69 g dry weight m^{-3} day^{-1}. This is approximately equivalent to 12.6 kJ m^{-3} day^{-1} or 0.31 g C m^{-3} day^{-1}. Corresponding area-based rates are 1.6 g dry weight m^{-2} day^{-1}, 29 kJ m^{-2} day^{-1} and 0.70 g C m^{-2} day^{-1}. These are very high rates of herbivore consumption.

Within the same lake and period, the value of 12.6 kJ m^{-3} day^{-1} can be compared with parallel estimates of mean consumption by a small planktivorous cichlid fish, *Oreochromis alcalicus grahami* – 3.4 kJ m^{-3} day^{-1}; by the principal zooplankter, the calanoid copepod *Lovenula africana*

(*Paradiaptomus africanus*) – 6.5 kJ m^{-3} day^{-1}; also by a rotifer, *Brachionus dimidiatus*, feeding on both microparticulate plankton and detritus – 12 kJ m^{-3} day^{-1}. The high flux of microparticulate consumption by the rotifer is striking considering the relatively low rotifer biomass involved, only about 3% of that of the flamingo. A similar emphasis on microbial consumption rates is indicated by a later (for 1985) rough order-of-magnitude estimate of consumption here by ciliates (Finlay *et al.* 1987) – 3–14 g dry weight m^{-3} day^{-1}, a very large relative quantity. Both these high consumption fluxes depend on the high ratio of consumption to biomass (C/B) in small organisms. The estimate for ciliates adopts a C/B value of 12 day^{-1}. By contrast, values of C/B calculated – from diel variation in stomach contents – for various fishes in the pelagial of Lake Malawi were only 0.012–0.066 day^{-1} (Allison *et al.* 1996).

Much lower rates of consumption than in Lake Nakuru have been estimated for planktivores elsewhere, even in waters with dense phytoplankton. Lake George provides examples: largely phytoplankton consumption by the two main fish consumers, *Oreochromis niloticus* and *Haplochromis nigripinnis*, was summed at 0.034 g C m^{-2} day^{-1}, and that by the main zooplankter consumer *Thermocyclops hyalinus* estimated as 0.50 g C m^{-2} day^{-1} (Moriarty *et al.* 1973). These and many other estimates of consumption rates are based on the diel variation of gut contents, which is often generated by visual feeding. Such variation may include one or more peaks per day, and can be modelled by fitted rates of consumption and evacuation. Examples for two planktivorous fishes of Lake Malawi are given in Fig. 3.32.

Limitation by size is clearly an important general factor for food consumption. It may operate at a pre-ingestion stage, as in filter-feeders, other particle collectors and larger predators; also at the ingestion stage, as in mouth gape-limited feeders. Applications to planktivorous fishes are considered in detail by Lazzaro (1987). On a much smaller scale, Haney & Trout (1985) showed from feeding experiments with ^{14}C-labelled food that zooplankton components of Lake Titicaca had varied preferences for particles above and below a size of 10 μm. Diel variability of intake is another widespread factor, illustrated in Fig. 5.54. Although this is ultimately related to the light–dark cycle, additional influences on quantity and quality of food appear when zooplankters migrate diurnally in a stratified water-column. Lewis (1977) described an example for *Chaoborus* in Lake Lanao, where a changing prey selectivity, and vertical distributions of prey and predator, interact.

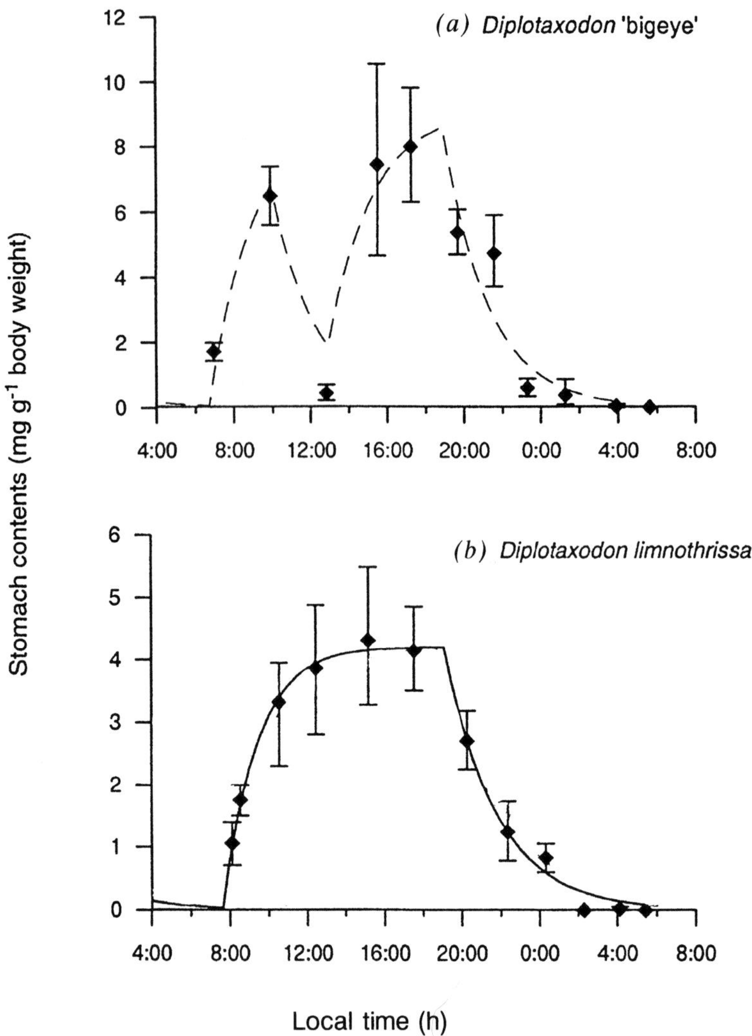

Fig. 3.32. Diel variation of stomach contents (with 95% confidence limits) in two planktivorous fishes of Lake Malawi (*a*) feeding upon pupae and then larvae of the lakefly *Chaoborus edulis*, (*b*) feeding mainly upon planktonic Crustacea. Line fits are in (*a*) as two feeding periods with constant ingestion rate, in (*b*) as one feeding period with constant ingestion rate; constant exponential evacuation rates apply to both. From Allison *et al.* (1996).

Consumption by filter-feeding populations generally can be formulated in terms of the population density, the water volume (v) processed or 'swept clear' by an individual in unit time and the retention efficiency. An example has already been given for the Lesser Flamingo; the large value of v here, about 50 l h^{-1} (Vareschi 1978), can be contrasted with an estimate of 70 μl day^{-1} for the rotifer *Brachionus plicatilis* (Vareschi & Jacobs 1984). Intermediate values (e.g., 5–30 ml day^{-1}) have been attributed to members of the important groups of planktonic Cladocera and Copepoda, although directly obtained estimates of volume filtered or 'swept clear' are rare for tropical waters. One of these is the notably high upper value of 28 ml day^{-1} for the main planktonic herbivore, the calanoid *Tropodiaptomus cunningtoni*, of Lake Malawi (Hart *et al.* 1995). For Lake Kariba a well-developed benthic community of lamellibranch molluscs (Machena & Kautsky 1988) appeared to be capable of processing the volume of the lake per year (Kiibus & Kautsky 1996; Kautsky & Kiibus 1997). For small individuals of *Aspatharia wahlbergi* the filtration rate per unit dry shell-free biomass was ~500 ml g^{-1} h^{-1}.

The quantitative impact of a consumer on its food source will depend on the specific rate of consumption, or volumetric clearance, per unit of consumer biomass and on that biomass concentration. Impact may be tested experimentally, using either natural or enhanced concentrations of the consumer. Thus the feeding of Lake Victoria zooplankton has been studied after pre-concentration by transfer of radioisotope from labelled phytoplankton in 15 minute exposures and at natural densities from chlorophyll *a* change after 48 h exposures. In both cases the grazing rate appeared small (Branstrator *et al.* 1996).

Much debate and speculation has surrounded evolutionary and ecological aspects of predation in tropical rivers and lakes. Issues include its possible role in the speciation of fishes (Fryer 1965), in selection favouring diel migrations of zooplankton (Zaret & Suffern 1976) and in the prevalence of a smaller size in many representatives of tropical zooplankton (reviewed by Fernando 1980*a*, *b*). Large top-predators include animals of partly or mainly non-aquatic habits such as the Nile Crocodile and the African Fish Eagle *Haliaeetus vocifer*. In many African waters, species of the Tigerfish *Hydrocynus* are major piscivores. Effects of these and other fish piscivores (e.g., *Lates*) on fish stocks and fisheries are surveyed in D. Lewis (1988). Over tropical standing waters generally, larvae of the dipteran genus *Chaoborus* are influential planktivores; they are transparent with minimal visibility and apparently an exceptionally low energy cost for maintenance (Cressa & Lewis 1986). Evidence for appreciable, sometimes dominant quantitative impact on their prey populations is provided by Lewis (1975, 1979) for Lake Lanao, Saunders & Lewis (1988*a, b*) for Lake Valencia and Twombly & Lewis (1987, 1989) for a Venezuelan floodplain lake (Fig. 5.25). In Lake Valencia the evidence included a comparison, for various prey herbivores, of the estimated magnitude of specific predation rates and specific growth rates.

For any consuming species, flexibility or its converse regarding types of food intake appear in several ways. Evidence from behaviour and gut contents can indicate greater or lesser selectivity, as measured for example by Ivlev's index of electivity in organisms of lakes George (Moriarty *et al.* 1973) and Chad (Lauzanne 1983). Ontogenetic development is usually correlated with change, notably in animals with a multi-stage life history (e.g., copepods) or with extended development (e.g., crocodile: Cott 1954). Thus shifts between herbivore and carnivore habits, or between plankton and benthos as food, occur quite often. Further, some species are relative omnivores, taking food of diverse character. These, and detritivores, include some of the tropical fishes most successful in aquaculture or reservoir introductions, such as *Oreochromis niloticus* and *O. mossambicus*. In the long term, evolutionary radiation into diverse feeding habits may characterize a particular group such as lacustrine cichlids (Fryer & Iles 1972). Other members of lacustrine fish faunas are generally adapted from river-inhabiting ancestors; the adaptation involves food and feeding habits, as described by Corbet (1961) for the non-cichlid fishes of Lake Victoria.

A component of non-living detritus may be a functional part of the diet especially in bottom-feeders. In detritivores it becomes the dominant and often relatively invariant source of food. This is extensively developed in many benthic animals and in the fish fauna of the Amazon (Bowen 1984), where large components of the external input of decaying plant material are not necessarily much used (see Section 3.5f). The balance between sources of utilized detritus from within and without the water-body can change with time, as in the African lakes Kariba and Chilwa (McLachlan 1977). In Lake Chad, detritivory supports much fish production, alongside herbivory and zooplanktivory (Lauzanne 1983: see Fig. 3.40).

(b) Food assimilation and use

Besides its capture, effective utilization of a food source depends upon chemical processing between ingestion/consumption (C) and assimilation

(*A*). This challenge tends to be greatest for detritivores and herbivores, which typically have long guts compared with those of carnivores and omnivores (Fig. 3.33), and for which food material is often abundant. For example, high densities of phytoplankton in productive tropical lakes are usually derived from blue-greens (cyanophytes) – organisms that have acquired a reputation as unfavourable food material for both zooplankton and fishes, including tropical examples (e.g., Fish 1955).

Nevertheless, work on several tropical lakes has shown that high levels of secondary production can be based on an effective utilization of planktonic blue-greens. The most detailed and now classic studies are those of D.J.W. Moriarty and his co-workers on the two principal herbivorous fishes – *Oreochromis niloticus* and *Haplochromis nigripinnis* – of Lake George, Uganda. These are described by Moriarty (1973) and Moriarty & Moriarty (1973*a, b*), and summarized in Moriarty *et al.* (1973). Digestion depends upon the exposure of ingested food to pH values of below 2.0 in the stomach. This pH sensitivity was demonstrable *in vitro* using ^{14}C-labelled food. In nature both food ingestion and acid secretion follow diel cycles, with feeding over before sunset when the lowest pH

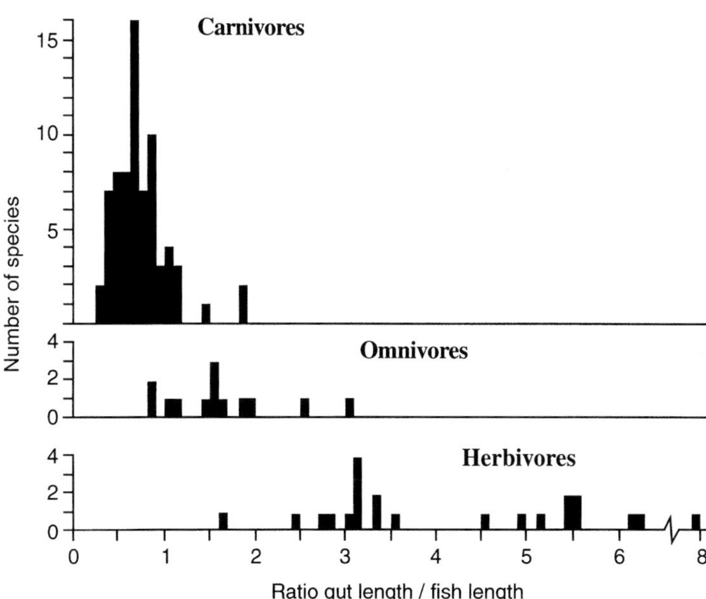

Fig. 3.33. Frequency distribution of three groups of fishes in Lake Tanganyika and its affluents, in relation to the ratio of gut length to fish length. From Fryer & Iles (1972).

values of 1.4 were reached in the fundus of the stomach. This sequence, and a reconstructed time-course of carbon assimilation in the intestine, is shown in Fig. 3.34. The percentage efficiency ($A/C \times 100$) of carbon assimilation from the ingested blue-greens most abundant in the lake, *Microcystis* and *Anabaena* spp., could reach maximal values of 70–80%, but average values over a 24-h day were about 43% for *O. niloticus* and 66% for *H. nigripinnis*.

The same phytoplankton also supported a considerable population of the copepod *Thermocyclops hyalinus* (= *Th. crassus*), a raptorial (grasping) feeder that can eat larger blue-greens such as *Microcystis*. In Lake Chad *Th. neglectus* similarly ingests large *Anabaena* spp. (Gras *et al.* 1971; Iltis & Saint-Jean 1983). Work with ^{14}C-labelled food on *Th. hyalinus* by Tevlin (in Moriarty *et al.* 1973) included measurements for carbon of ingestion, defaecation and efficiency of assimilation. Values for

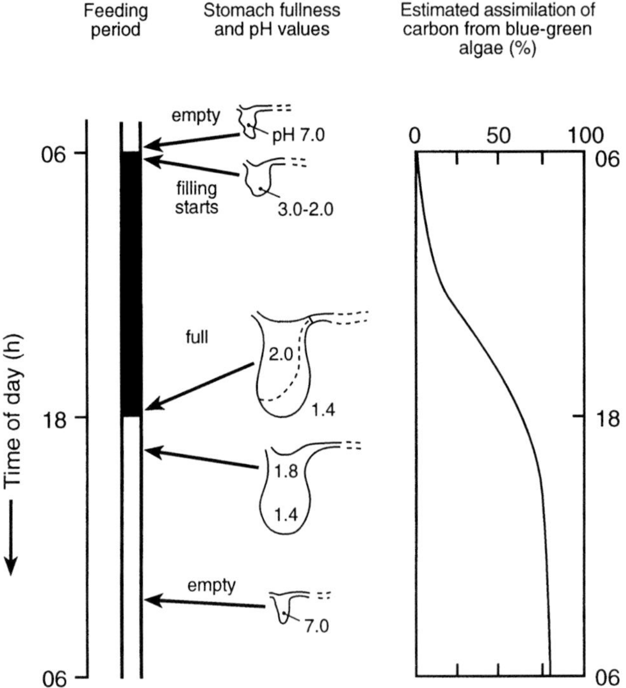

Fig. 3.34. The role of acidity in digestion and assimilation of blue-green algae by the cichlid fish *Oreochromis niloticus*: the distribution of pH in the stomach with time of day and food intake, in relation to the estimated cumulative assimilation efficiency for carbon. From Moriarty *et al.* (1973).

the latter were about 35% for copepodids and adults, and about 58% for nauplii. Finally, information on ingestion rate and population density was combined to obtain estimates of herbivore food intake per unit area. These total 0.54 g C m^{-2} day^{-1}, exclusive of plankton-derived input to the zoobenthos that is partly detrital.

The crucial role in assimilation of exceptionally low values of stomach pH (<2.0) is also indicated by measurements on the planktivorous fish *Oreochromis alcalicus grahami* of Lake Nakuru (Vareschi & Jacobs 1984). Again a dense population can be supported by blue-green phytoplankters. There is no specialized filter mechanism, and elsewhere – at the parent Lake Magadi – the fishes feed on blue-greens that grow on stones (Coe 1966). Remarkably, the acid secretion involved must initially eliminate the high ambient alkalinity of these saline soda lakes.

Non-living detrital material is generally viewed as low-value food, for which a low efficiency of utilization is anticipated. However, it is often combined with living material on the surface of sediments and other substrata. One such combination, a 'periphytic detrital aggregate' on the bottom of Lake Valencia, was studied by Bowen (1979, 1980, 1981) for chemical composition and utilization. It supported the main constituent of the fish biomass, the introduced detritus-feeder *Oreochromis mossambicus* of African origin. Digestion and assimilation of organic matter again showed high efficiency, estimated as *c.* 60–80%, in this species, with stomach pH known to fall below 2.0 (Bowen 1976). In Lake Valencia the detritus content of non-protein amino acids was deemed all-important for growth (Bowen 1980).

Aquatic macrophytes also present a potential food source not often grazed extensively by herbivores. Even when heavily grazed, as observed in diel feeding of the fish *Puntius filamentosus* on *Ceratophyllum*, the energy derived may be too little for daily requirements (Hofer & Schiemer 1983). However, the situation is different in the African cichlid fish *Tilapia zillii*, for which non-cellulosic components have been measured to have assimilation efficiencies of $>50\%$ (Buddington 1979). Another cichlid, *Tilapia rendalli*, can exploit the semi-aquatic grass *Panicum repens* that is commonly found in periodically flooded habitats in southern Africa. Daily ingestion at Lake Kariba has been estimated by Caulton (1977) from the diel cycle of gut content. Its energy content, $\sim$0.34 kJ g^{-1} fish biomass day^{-1}, was compared with that in palatable material on unit area of flooded grassland ($\sim 8.43 \times 10^4$ kJ ha^{-1}). The conclusion was reached that food quantity was not a limiting factor for the population densities of fishes observed.

For all secondary producers, use of the quantity of assimilate (A) can be attributed to growth by biomass increment (ΔB) and to respiratory metabolism (R). Thus, if excreta (faeces + urine) are denoted by E and consumption by C:

$$C = A + E = \Delta B + R + E \qquad (3.10)$$

The quantities involved can be expressed in terms of dry weight, carbon content or energy content. The (dimensionless) efficiency of assimilation, A/C, has already been illustrated. First and second order efficiencies of growth (classically denoted by K_1 and K_2, here by ε_1 and ε_2) can also be defined as $\Delta B/C$ and $\Delta B/A$, respectively, with interrelation as an efficiency chain (cf. primary production):

$$\Delta B/C = \Delta B/A.A/C \qquad (3.11)$$

Empirical estimates of $\Delta B/C$ exist for some tropical populations, although they are largely unaccompanied by independent estimates of A/C. Variation with the character and digestibility of food can be expected. Considering phytoplanktivorous fishes, Lauzanne (1978, 1983) obtained from field data first-order efficiency or $\Delta B/C$ values for *Sarotherodon galilaeus* in Lake Chad of up to 5.5% (dry weight basis) or 19% (energy basis). A corresponding value for *Oreochromis alcalicus grahami* in laboratory experiments at Lake Nakuru, from Vareschi & Jacobs (1984), is 41% (dry weight basis). Lauzanne (1983) has surveyed estimates from a range of cold-water and warm-water fishes and noted a tendency to higher values in the warm-water species. For two tropical predators, the small planktivorous *Alestes baremoze* and the large piscivorous *Lates niloticus*, he obtained values of 45 and 27%, respectively. The work at Lake Nakuru also yields estimates for the dominant zooplankter, *Lovenula africana*, of 15–32% (dry weight basis).

When food consumption (C) and assimilation (A) are largely devoted to supporting respiration so that $R >> \Delta B$, the growth efficiency factors $\Delta B/C$ and $\Delta B/A$ will be low. An extreme instance, approaching 0, is provided at Lake Nakuru by the population of Lesser Flamingoes. Breeding and therefore young do not occur at this lake, but at Lake Natron, and the maintenance of homeothermy involves a relatively high respiratory cost. An opposite extreme, with very low respiratory cost, is apparently shown by the widespread aquatic larvae of the phantom midge *Chaoborus*. Work by Cressa & Lewis (1986) over one year on *C. brasiliensis* at Lake Valencia yielded a very high growth efficiency ($\Delta B/A$, dry weight basis, 0.59–0.76) for the instars II–IV. This suggested a low maintenance cost, in concordance with measured rates of respiration that were low for organisms of this body size. Compatible but more indirect estimates were obtained for *Chaoborus edulis* in Lake Malawi (Allison *et al.* 1995). Near-neutral buoyancy of larvae is conferred by air sacs and low visibility of the 'phantom' enables successful ambush-predation without much active swimming movement. A contrary emphasis on *high* energetic costs at an elevated temperature for growth and respiration of early stages (e.g., instar I) was made by Halat & Lehman (1996), from their indirect approach by modelling.

The energetics of tropical zoobenthos are known mainly from two studies on communities dominated by relatively large sedentary molluscs. The first was from Lake Chad (Lévêque 1973*b*). Measured rates of respiration and biomass increment were combined with previous population studies (Lévêque 1973*a*) and transformed, using bomb calorimetry, into population estimates of production (P), respiration (R) and assimilation ($A = P + R$) in units of kcal m^{-2} day^{-1} or yr^{-1}. One long series of estimates for the species *Cleopatra bulimoides* is shown in Fig. 3.35, without indication of regular seasonality. The relative contribution of assimilated food (A) to production ($P = \Delta B$) is expressed by the production to assimilation quotient P/A, with which there is positive correlation of estimates of annual $P/\overline{B}$ (see Section 3.3c) and to relative (instantaneous) rates of growth, g (Fig. 3.36). The annual mean values of P/A derived for four species of molluscs at various stations ranged from 0.089 to 0.325;

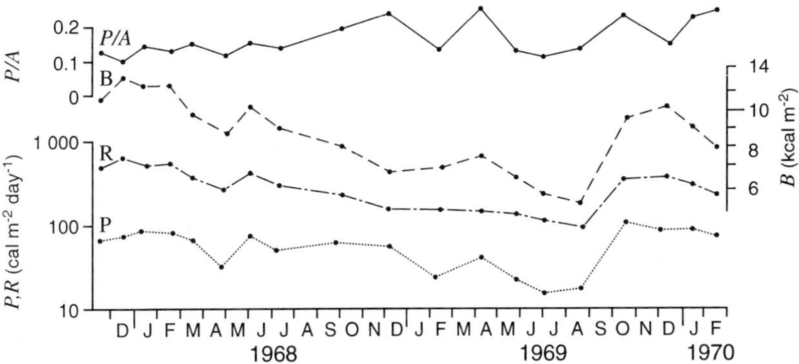

Fig. 3.35. Lake Chad, zoobenthos. A time-series at one station of estimates per unit area for the mollusc *Cleopatra bulimoides*, based upon components of its energy budget, comprising biomass (B), daily rates of production (P) and respiration (R), and the efficiency of production to assimilation (P/A). From Lévêque (1973*b*).

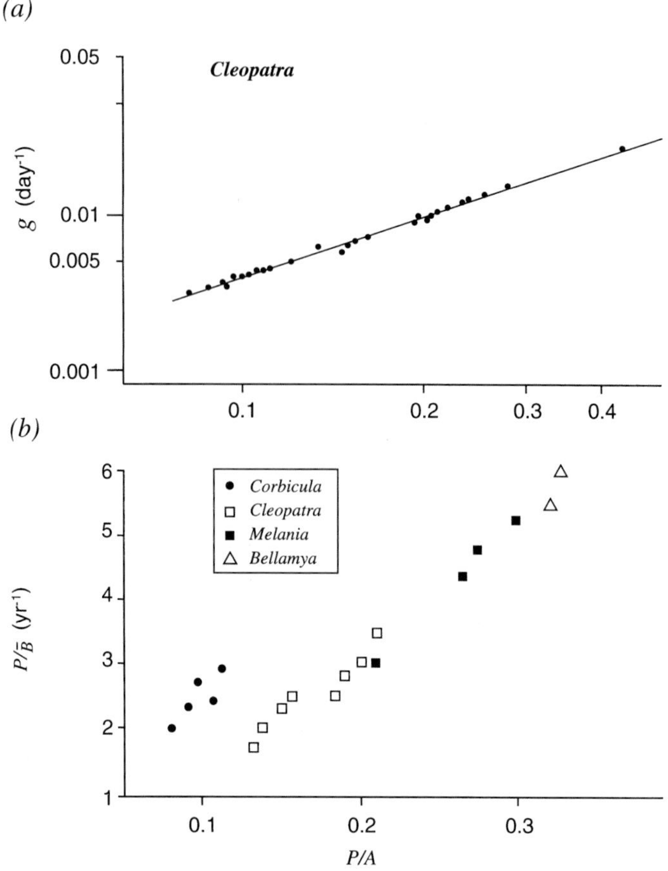

Fig. 3.36. Lake Chad, zoobenthos. Relationships to the estimated ratio of annual production to assimilation (P/A) of (*a*) the instantaneous growth rate g of the mollusc *Cleopatra bulimoides*, logarithmic scales, (*b*) the annual $P/\bar{B}$ ratio for four molluscs. From Lévêque (1973*b*).

thus much the greater part of assimilation was expended in respiration. Respiration was also measured directly in the second study (Kiibus & Kautsky 1996; Kautsky & Kiibus 1997) on the benthos of Lake Kariba. However, the estimates of biomass production there depended heavily upon $P/\overline{B}$ values obtained from the literature.

A study by Hart (1980, 1981) of the energetics and production of a largely tropical shrimp or prawn, the decapod *Caridina nilotica* from the subtropical Lake Sibaya in South Africa, is also of relevance. This linked features of bioenergetics, production and population dynamics at littoral sites. Besides providing estimates of the efficiencies $\Delta B/C$ and $\Delta B/A$, Hart & Allanson (1981) suggested that at temperature values above 24 °C there was evidence for homeostasis or temperature compensation in respiratory expenditure, reducing the energy requirement otherwise anticipated at higher temperature. Later Ignatow *et al.* (1996) combined the experimental rate relationships of Hart with population data from Lake Victoria to model the energetics and production of *C. nilotica* in the plankton of that lake.

(c) Production–biomass relationships

Whether at the level of individual or population, a supporting biomass (B) must determine further production. Considering production as biomass increment $\mathrm{d}B$ in time element $\mathrm{d}t$, the simplest relationship is:

$$\mathrm{d}B = g.B.\mathrm{d}t \tag{3.12}$$

where g is the specific or instantaneous growth rate, with dimension time^{-1}. If maintained unchanged over an extended period ($t_2 - t_1$), without an opposing specific loss or mortality rate m, increase in B from B_1 to B_2 is exponential and defined by the equation:

$$B_2 = B_1 e^{g(t_2 - t_1)} \tag{3.13}$$

or

$$\ln(B_2/B_1) = g(t_2 - t_1) \tag{3.14}$$

If the number N of organisms can be taken to represent biomass, then the specific rate g can be evaluated from an exponential population increase from N_1 to N_2. This is especially useful for microbial organisms with short generation times. With long generation times and a complex (e.g., multi-stage) life history, g varies between stages of development (examples in Fig. 3.37) and a composite of values determines expression – assuming stable age distribution, zero mortality and density-independent growth – in the overall specific rates of population increase, now generally known as intrinsic rates of population increase (r_m).

In the real world the above three conditions are often not met and specific rates of population increase (r) are often lower than r_m. In particular, specific rates of mortality (m) are likely to operate unequally over different stages of the life history. Further, sets of individuals that origi-

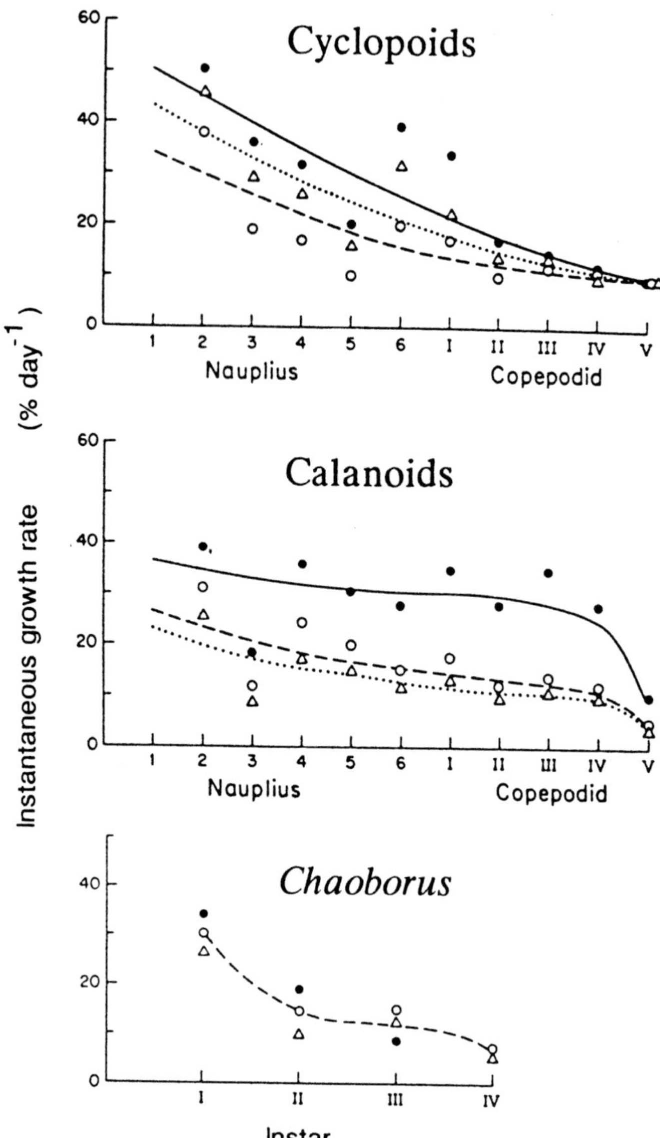

Fig. 3.37. The decline of instantaneous growth rate (*g*) over successive developmental stages of three components of the zooplankton of Lake Lanao, Philippines. Symbols distinguish estimates from different cohorts. From Lewis (1979).

nate at a particular time and constitute an age-class or *cohort*, may or may not overlap in time with other such cohorts. On the annual time-scale, cohorts may be single (univoltine) or multiple (multivoltine); the cohort production interval(s) may leave an unfavourable portion of the year unoccupied, so proportionately reducing potential annual production.

For a population or community, the relation of production to biomass involves integrating both quantities over a chosen time interval, from which the integral production P and mean biomass $\overline{B}$ can be expressed as the $P/\overline{B}$ ratio. Both daily and annual values are often used. In tropical examples, as elsewhere, P has been estimated by at least three mathematical procedures, compatible with various sets of assumptions especially with respect to the incidence of mortality (Rigler & Downing 1984). The 'egg ratio method' for estimating birth rate from egg numbers and development time, and hence deriving demographic turnover, may involve unwarranted assumptions; it is most applicable to the short life histories of parthenogenetically reproducing rotifers and, with adaptation, to parthenogenetic reproduction of Cladocera. African site-examples include Lake George for a copepod (Burgis 1971, 1974; also Rigler & Downing 1984), Lake Chad for Cladocera (Gras & Saint-Jean 1978, 1983; Lévêque & Saint Jean 1983) and Lake Nakuru for rotifers (Vareschi & Jacobs 1984). With larger organisms and multi-stage life histories (e.g., copepods) the mass increments in successive stages of known abundance are best calculated separately by linear or exponential fits and converted into rates by reference to stage-durations. The latter have been estimated from the times for transition from stage to stage observed in natural cohort sequences (e.g., Lewis 1979; Saunders & Lewis 1988*a*) or from experimental samples in the field or laboratory (e.g., Lévêque 1973*a*; Ferguson 1982; Gras & Saint-Jean 1981, 1983; Vareschi & Jacobs 1984; Saunders & Lewis 1988*a*; Mengestou & Fernando 1991*b*; Mavuti !994; Irvine 1995*a*).

For organisms in general, the overall relation of production to biomass is dominantly influenced by individual body size (Peters 1983). This applies whether the production index adopted is the intrinsic rate of population increase (r_m) or the $P/\overline{B}$ ratio, the former setting, for any organism, an upper limit to the latter. Their correlative relationships with body mass have been expressed by Fenchel (1974) and Banse & Mosher (1980), respectively, as illustrated in Fig. 3.38. Against this background values obtained for tropical aquatic animals can be compared, as is shown by the inserted examples from Lake Nakuru. Components of

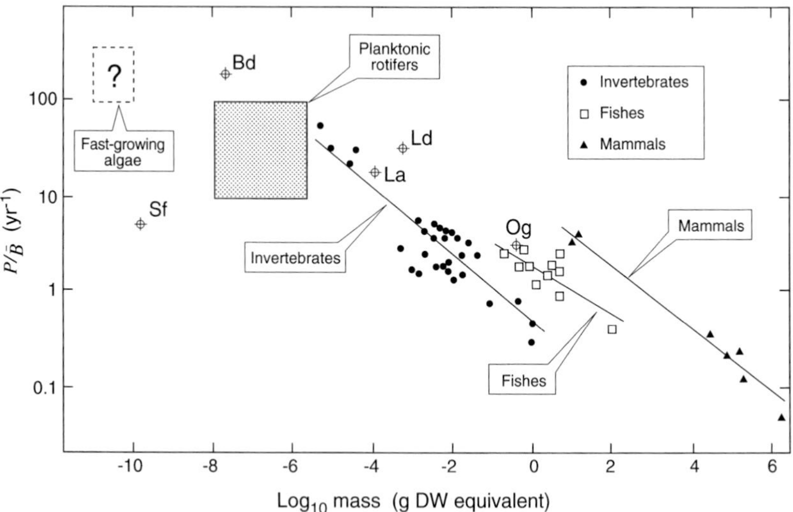

Fig. 3.38. Regression relationships for the annual $P/\bar{B}$ ratio to individual mass at maturity in major groups of organisms. An approximate interconversion of 1 g dry weight $\simeq$ 4.5 kcal is assumed and five data points are added from estimates made by Vareschi and co-workers on Lake Nakuru, 1972–73. Sf, *Spirulina fusiformis* (cyanophyte); Bd, *Brachionus dimidiatus* (rotifer); La, *Lovenula africana* (copepod); Ld, *Leptochironomus deribae* (chironomid); Og, *Oreochromis alcalicus grahami* (cichlid fish). Modified from Banse & Mosher (1980) and Vareschi & Jacobs (1985).

zooplankton, zoobenthos and fish populations are represented, forming a community-series with increasing body mass that broadly accounts for decreasing values of r_m and $P/\overline{B}$.

Of direct tropical relevance is the sensitivity of both these production parameters to temperature. For r_m an excellent example is the study by Pourriot & Rougier (1975) of growth in culture of the rotifer *Brachionus dimidiatus* isolated from a Chad soda lake. Rates increased in the range 20 to 30 °C, and at 30 °C – the optimal temperature – and with good nutrition on *Spirulina* were over 1.0 (ln units) day^{-1}. At this temperature both embryonic and post-embryonic development times were short, and the minimum time between parthenogenetic generations (i.e., egg to egg) was only ~30 h. The decline of stage-duration times, including that of eggs, with temperatures in the ecological range (e.g., 15–30 °C) has also been shown by various other studies of tropical zooplankton Crustacea (Gras & Saint-Jean 1969, 1976, 1978; Burgis 1978; Ferguson 1982, Mengestou & Fernando 1991*b*; Mavuti 1994; Saint-Jean & Bonou 1994) and Rotifera (Duncan 1983). For high rates of relative growth

(g), population increase (r_m) and production ($P/\overline{B}$) attained around 30 °C, the parthogenetic cladoceran *Moina micrura* is outstanding. Respective values of >1 day^{-1} (young stages), 0.9 day^{-1} and ~1 day^{-1} were reported from populations in Lake Chad (Gras & Saint-Jean 1978, 1983) and fish ponds of the Ivory Coast (Saint-Jean & Bonou 1994). In both places the estimated $P/\overline{B}$ values were markedly depressed in the cooler season (~20 or ~26 °C, respectively) – a depression that at Lake Chad extended to the zooplankton as a whole (Fig. 5.53). For this community there have been long-term studies by Gras & Saint-Jean of growth and production, summarized in Gras & Saint-Jean (1983) and Lévêque & Saint-Jean (1983). In Cladocera the life cycle, egg to egg, could be as short as 52–81 h at 30 °C, with values of $P/\overline{B}$ >0.5 day^{-1}. With the Copepoda, representative values of $P/\overline{B}$ were much lower, around 0.2 day^{-1} for cyclopoids and 0.04 day^{-1} for calanoids.

The work on *Brachionus dimidiatus* and *Moina micrura* also provided direct (experimental) and indirect (observational) evidence of the regulation of r and $P/\overline{B}$ by the quantitative and qualitative adequacy of food supply. For values from natural conditions without nutrient limitation, other estimates for herbivores from two shallow, equatorial and highly productive waters in East Africa – lakes George and Nakuru – deserve attention. In Lake George the dominant copepod *Thermocyclops hyalinus*, a raptorial (grasping) feeder when adult, was considered by Burgis (1971) to have an excess of available food as relatively large and digestible blue-green algae. Burgis (1974) estimated the mean value of $P/\overline{B}$ as 0.078 day^{-1} or 28.5 yr^{-1}; another method of calculation for P used later by Rigler & Downing (1984), better in principle but with limited input data, yields 0.14 day^{-1} (51 yr^{-1}) for $P/\overline{B}$. Both sets of $P/\overline{B}$ values are similar to those derived from work on temperate zooplankton (e.g., Morgan *et al.* 1980).

At Lake Nakuru, Vareschi & Jacobs (1984) obtained during 1972–73 estimates of $P/\overline{B}$ for five principal producers, representing zooplankton (the rotifers *Brachionus dimidiatus* and *B. plicatilis*, and calanoid copepod *Lovenula africana*), zoobenthos (larvae of the chironomid *Leptochironomus deribae*) and fishes (*Oreochromis alcalicus grahami*). Of these the copepod and fish fed largely on the large and very abundant blue-green alga *Spirulina fusiformis* (*S. 'platensis'*); the rotifers probably also utilized an indeterminable food proportion as bacteria and the chironomid was mainly a detritus feeder. For all, food was probably available in excess, except possibly for the rotifer whose population maxima may have correlated with increased bacterial abundance. Mean 1972–73 estimates for biomass, production and consumption are given in Table 3.4,

Table 3.4. *Lake Nakuru, Kenya, 1972–73. Separately ranked estimates of biomass, production and consumption by six groups of consumers, expressed in energy units*

Biomass ($kJ\ m^{-3}$)					
Lesser Flamingoes	fishes	copepods	chironomids	pelicans	rotifers
128	51	32	9	5	4
Production ($kJ\ m^{-3}\ day^{-1}$)					
Rotifers	copepods	chironomids	fishes	pelicans	Lesser Flamingoes
1.7	1.5	0.7	0.4	0.03	< 0.01
Consumption ($kJ\ m^{-3}\ day^{-1}$)					
Lesser Flamingoes	rotifers	copepods	chironomids	fishes	pelicans
12.6	12	6.5	3.4	3.4	0.4

Source: From Vareschi & Jacobs (1984)

using units of energy content per unit water volume for the three quantities. The different ranking of producers when based on biomass and on production is striking. The rotifer is again an outstandingly efficient producer, with high $P/\overline{B}$; the $P/\overline{B}$ values for chironomid, copepod and fishes are rather high but not exceptional for the groups involved (cf. Fig. 3.38). The planktivorous Lesser Flamingo has a negligibly low production at the lake because of the absence there of breeding and young birds.

Contrasting with presumed situations of excess food in these productive shallow lakes, work on the deep and relatively unproductive Lake Malawi has yielded an apparent case of anomalously inadequate food availability in the pelagial. Here the food dependency of development times for stages of a dominant copepod, *Tropodiaptomus cunningtoni*, was studied by Hart *et al.* (1995). Controlled doses of phytoplankton food from algal cultures were used. Whereas the duration of naupliar stages was insensitive to food availability, possibly due to utilization of lipid reserves, the metabolic maintenance demand of adult females in relation to measured clearance rates (up to 28 ml $indiv^{-1}\ day^{-1}$) seemed to require higher concentrations of food than were available in near-surface water of the lake. However, the higher concentrations of phytoplankton often present at depth may have been utilized through vertical movements of the copepod. Other experimental studies of growth limitation at low food concentrations are rare for tropical zooplankters. With the calanoid *Phyllodiaptomus annae* isolated in culture from a Sri Lankan reservoir, growth rates declined rapidly at food concentrations below 50 μg C l^{-1} (Piyasiri 1985). Also in Sri Lanka, further estimates have been made of the natural variability of production rates of this and other zooplankters, assessed per unit water volume. When combined with other tropical estimates, production rate P – but not the ratio $P/\overline{B}$ – showed moderate positive correlation with phytoplankton abundance (Amarasinghe *et al.* 1997).

Further widening of scope for production–biomass relationships is obtained from three more studies on African lakes. In Lake Chad, molluscs are abundant and relatively large components of the zoobenthos, whose production biology was studied during 1967–72 by Lévêque (1973*a, b*; Lévêque & Saint-Jean, 1983). The animals were amenable to experimental manipulation with enclosures, from which cohorts could be distinguished, with size progression and survival in time. In the eight main species, instantaneous growth rates fell either more or less rapidly over the life span of 1–3 years, with an inverse relationship to individual mass. The magnitude of $P/\overline{B}$ values was 2–6 yr^{-1}; it showed positive correlation with the ratio of production to assimilation (P/A) (see Fig. 3.36) and a pronounced negative correlation with longevity (Lévêque *et al.* 1977). In the same zoobenthos, larval insect components were of shorter longevity (down to 13 days, egg to adult, for *Chironomus pulcher* in laboratory cultures at 30 °C: Dejoux 1971) and probably a higher $P/\overline{B}$ relationship. The $P/\overline{B}$ magnitude of around 3 yr^{-1} for the molluscan component is also typical of much temperate-zone zoobenthos, but may be influenced in opposite directions by relatively large individual size and by continuous year-round production and recruitment.

A second production study, by Mengestou & Fernando (1991*b*), concerns the crustacean zooplankton of Lake Awasa, a rift lake in Ethiopia. Because of its altitude of 1708 m, the water temperature (21–24 °C) did not reach the high values of ~30 °C at which the exceptionally high values of r and $P/\overline{B}$ were derived from Lake Chad and a pond in West Africa. However, the highest $P/\overline{B}$ value of 221 yr^{-1} (0.61 day^{-1}) was again for a cladoceran, *Diaphanosoma excisum*. The aggregate value for the dominant Crustacea was 56 yr^{-1} (0.15 day^{-1}), rather high, though equalled by some records from temperate lakes.

A third study, exemplifying fish production biology of economic relevance, is that of Coulter (1981) on the pelagic clupeids of Lake Tanganyika. These fishes, and especially *Stolothrissa tanganicae*, are small planktivores with a life span of about one year. Recruitment and production are continuous; the main population check is mortality due to predation by large endemic centropomid fishes, relatives of the Nile

Perch. Cohort $P/\overline{B}$ was estimated, by the Allen curve method, as 3.9, and a value of about 3.5 yr^{-1} is probably a good estimate of annual $P/\overline{B}$. Allowing for the size factor and life span, this is not dissimilar to temperate values.

Tropical equivalents of the numerous temperate studies of P–$\overline{B}$ relationships in stream zoobenthos are almost non-existent. For ten insect species in the small Bovo river on Bougainville Island, Papua New Guinea, Marchant & Yule (1996) have estimated the ratio between cohort production and annual production, and used it as a guide to larval life span or 'cohort production interval'.

These examples support the conclusion that, relative to biomass, production rates of tropical populations tend to be somewhat high but not consistently so. Probably only a few species exploit high temperature (e.g., 30 °C) to achieve outstandingly high values of daily r and $P/\overline{B}$ – a feature also known in microbial ecology, including that of phytoplankton (Eppley 1972). Besides temperature, several other factors could contribute to a trend towards higher tropical values of annual $P/\overline{B}$. These include a potential year-long duration of continuous growth and recruitment with active biomass – although multivoltine life cycles are possibly a correlate rather than a determinant (Rigler & Downing 1984). Another factor is possibly high mortality rates, which tend to increase $P/\overline{B}$ by reducing the representation of older biomass with lower relative growth rate (g). The improbability of simple effective generalization, especially for temperature, is suggested by the survey of invertebrate production in non-tropical streams by Benke (1993). This showed that although some remarkably high values of annual $P/\overline{B}$ were obtained for insect components of subtropical streams that seasonally reached 30 °C, correlative relationships of $P/\overline{B}$ with temperature could be both positive and negative according to taxonomic group. Another recent survey, by Brown (1994) on African snail populations that included the much-investigated schistosome vector *Bulinus globosus*, demonstrated that r can correlate both positively and negatively with temperature. Temperature-compensation behaviour (Bullock 1955), with downward adjustment of rate at higher temperature, may also be widespread.

(d) Rates of production per unit area

Table 3.5 lists absolute rates of secondary production for components of zooplankton, zoobenthos and fish communities in eight tropical lakes. All were estimates obtained mainly by relatively direct methods, based on

Table 3.5. *Estimates of annual rates of secondary production, as g C m^{-2} yr^{-1}, in eight tropical lakes. Conversions from values in original publications assumed 1 g C $\simeq$ 2 g dry weight $\simeq$ 10 kcal $\simeq$ 42 kJ*

	George	Chad	Nakuru	Awasa	Valencia	Lanao	Malawi	Tanganyika
mean depth (m)	2.4	3–4	1.5–2.5	11	19	60	290	570
phytoplankton density	high	high	high	mod.	high	mod.	low	low
ZOOPLANKTON								
(a) major herbivore								
Thermocyclops hyalinus[1]	7.2							
Tropodiaptomus cunningtoni[2]							20.6	
Lovenula africana[3]			29.6					
(b) mixed herbivores[4]					39.1	27.3		
(c) total Crustacea[2,5]		22		25			30.5	
(d) main rotifer (herbivore-bacterivore)								
Brachionus dimidiatus[3]			33.5					
B. plicatilis[3] (1974)			202					
(e) major predator								
Mesocyclops aequatorialis aequatorialis[2,9]							2.3	
Chaoborus edulis[6]							2.1	
ZOOBENTHOS								
(a) part-detritivores molluscs[7]		7.2						
Leptochironomus deribae[3]			15.8					
FISHES								
(a) major herbivore								
Oreochromis alcalicus grahami[3]			7.9					
(b) major planktivores								
Stolothrissa tanganicae[8]								5.6
Engraulicypris sardella[9]							1.0	

Source;
1. Burgis 1974
2. Irvine 1995*a*
3. Vareschi & Jacobs 1984
4. Lewis 1979; Saunders & Lewis 1988*a*
5. Lévêque & Saint-Jean 1983; Mengestou & Fernando 1991*b*
6. Irvine 1995*b*
7. Lévêque & Saint-Jean 1983
8. Coulter 1981
9. Allison *et al.* 1995

variants of the egg ratio/recruitment rate/turnover or growth increment summation procedures, rather than on expected $P/\overline{B}$ values. Table 3.5 also shows that three of the lakes are shallow with mean depth <5 m, and rich in phytoplankton; two are of intermediate depth but also with a considerable phytoplankton concentration; and three are deep lakes with low phytoplankton concentrations per unit volume.

Despite this background diversity, seven of the eight lakes yielded production rates for crustacean zooplankton dominants or assemblages, that were largely or entirely herbivorous, within the range 20–40 g C m^{-2} yr^{-1}. For such areal rates the increased vertical dimension in deep lakes will be a compensating factor. In Lake George the corresponding low estimate may possibly reflect high mortality of nauplii (Burgis 1971); its magnitude was slightly increased by an alternative mode of calculation (Rigler & Downing 1984). For Lake Tanganyika there is no direct estimate, but a tentative indirect one – based on measured zooplankton biomass and expected $P/\overline{B}$ values – was high, at 50–60 g C m^{-2} yr^{-1} (Burgis 1984, 1986).

Production rates that are probably largely sustained by bacterivory or detritivory also appear for lakes Nakuru and Chad. In Lake Nakuru two species of the rotifer genus *Brachionus* yielded exceptionally high production rates during periods of decline and decomposition of dense phytoplankton. Rates for a benthic chironomid were lower but still high compared with temperate zone equivalents. The same applies to the rate estimate for the very different mollusc community of Lake Chad, with larger individuals, higher longevity and a lower $P/\overline{B}$ value of $\sim$3 yr^{-1}. If $P/\overline{B}$ values of similar magnitude apply to the mollusc-dominated littoral biomass of Lake Kariba, a mean estimate of production for the 0–12 m zone is $\sim$11 g shell-free dry weight or $\sim$5 g C m^{-2} yr^{-1} (Kiibus & Kautsky 1996).

For invertebrate planktonic predators, consumption and production have been studied in species of the insect genus *Chaoborus* in several lakes (Lewis 1979; Cressa & Lewis 1986; Saunders & Lewis 1988*a*, *b*; Irvine 1995*b*). At Lake Valencia the mean annual biomass of *Chaoborus* could exceed that of accompanying herbivores, whose abundance it then appeared to control, with a large consumption flux and efficient conversion. At Lake Malawi, where aerial clouds of emerged adult 'lakeflies' are a frequent feature of the landscape, the estimate for mean production rate of the aquatic larval instars and pupae during 1992–93 was 2.1 g C m^{-2} yr^{-1}. Another important invertebrate predator of the Malawi zooplankton was the copepod *Mesocyclops aequatorialis aequatorialis*, with an estimated production of the same magnitude (Irvine & Waya 1993; Irvine 1995*a*; Allison *et al.* 1995).

Table 3.5 also includes three estimates for production rates of planktivorous, in part mainly herbivorous, fishes. The highest is for a small specialized cichlid, *Oreochromis alcalicus grahami*, in the soda lake Nakuru. Here algal food was abundant and probably usually in excess. The dominant examples from the very different deep lakes of Malawi and Tanganyika are chiefly reliant upon zooplankton Crustacea as food, present in lower concentration. Although these production rates are lower, they are very considerable within any global comparison. The higher value of 5.6 g C m^{-2} yr^{-1} from Lake Tanganyika is equivalent, using previous conversion factors, to *c*. 560 kg fresh weight ha^{-1} yr^{-1}. Most estimates for fish production in temperate freshwaters are well below this level; higher production of 1000–2000 kg ha^{-1} yr^{-1}, and economic yield, are realized mainly in fishponds with artificially enhanced food supply. In addition, a single very high and less established estimate of 1224 kg ha^{-1} yr^{-1} from the early phase of Lake Kariba (Mahon & Balon 1977) can be mentioned. Nevertheless, the Tanganyika estimate could be indicative of an unusually efficient food chain (see Section 3.5), as well as representing a fish with small size, early maturation, year-round growth and high mortality, all favouring a relatively high $P/\overline{B}$ ratio.

Using examples already discussed, a broad survey of P–$\overline{B}$ relationships is presented in Fig. 3.39. Although mean biomass per unit area ($\overline{B}$) is the most obvious supporting variable for P, for dynamic range, its influence is often not more significant than that of $P/\overline{B}$, a measure of effective specific activity. High extensions of this activity can be gauged from the maximum instantaneous rates of increase (~1 day^{-1}) of a few small Metazoa (e.g., *Brachionus dimidiatus, Moina micrura*) at 30 °C; further increase will occur in the microbial world of heterotrophic Protozoa and bacteria. According to M. Bouvy (personal communication), recent work on bacterioplankton of small reservoirs of the Ivory Coast, West Africa, has given estimates of mean microbial doubling time of the order of 1–2 h; this compares with estimates from temperate lakes, obtained from the same method based on uptake of labelled thymidine or leucine, of the order of a day or week. Similar work on the productive Lake Xolotlán (Nicaragua), however, indicated a mean doubling time of ~1 week (Bell *et al.* 1991).

3.4 Decomposition and recycling

Continued availability of primary nutrient resources often depends on the decomposition of pre-formed biomass. This process is largely carried out

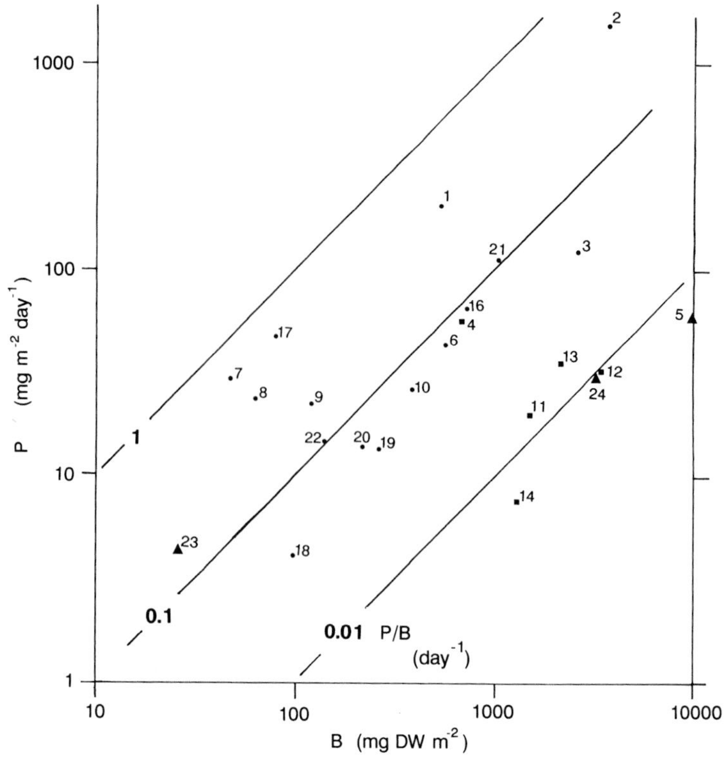

Fig. 3.39. NAKURU 1973 or* 1974: 1. *Brachionus dimidiatus*; 2. *Brachionus plicatilis**; 3. *Lovenula africana*; 4. *Leptochironomus deribae*; 5. *Sarotherodon alcalicus grahami*.
CHAD Eastern Archipelago or *Baga Kawa sta.: 6. *Thermocyclops hyalinus*; 7. *Moina micrura*; 8. *Diaphanosoma excisum*; 9. Cyclopoids; 10. Calanoids; 11. *Melania tuberculata**; 12. *Cleopatra bulimoides**; 13. *Bellamya unicolor**; 14. *Corbicula africana*.
AWASA: 16. *Thermodiaptomus oblongatus*; 17. *Diaphanosoma excisum*; 18. *Thermocyclops consimilis*; 19. *Mesocyclops aequatorialis similis*.
MALAWI: 20. *Mesocyclops aequatorialis aeq.*; 21. *Tropodiaptomus cunningtoni*; 22. *Chaoborus edulis*; 23. *Engraulicypris sardella* (larvae).
TANGANYIKA: 24. *Stolothrissa tanganicae*.

Double logarithmic plot of production (P) – mean biomass ($\bar{B}$) relationships from five tropical African lakes. Symbols distinguish zooplankton (●), zoobenthos (■) and fishes (▲). Based on data in Vareschi & Jacobs (1984), Lévêque & Saint-Jean (1983), Mengestou & Fernando (1991*b*), Irvine (1995*a*, *b*), Allison *et al.* (1995) and Coulter (1981).

by microbial decomposers, although some autolysis and animal digestion plus excretion will also contribute. The mineralized products in part sustain further production by recycling, given a re-contact with active biomass. This re-utilization of old resources is sometimes called 'regenerated production', as contrasted with 'new production' dependent upon entry of new nutrient resources. Intergrades are of course possible.

An entirely different mode of breakdown of dissolved organic matter (DOM) is photodecomposition in sunlight. It is associated mainly with the ultra-violet component and is largely uncharted for tropical inland waters. The only detailed study has been on the Amazon system (Amon & Benner 1996), where for the Rio Negro the rates in illuminated surface water (~4 μmol l^{-1} h^{-1}) exceeded those for bacterial decomposition – although rates for the entire water-column were less. There is also preliminary evidence of a photochemical loss of 'colour' in the clear surface water of Lake Tanganyika (Sarvala & Salonen 1995) and indirect evidence of such loss from water-bodies in northern Australia (Townsend *et al.* 1996).

More generally, rates of decomposition can be assessed from the associated gas exchange ($-O_2$ $+CO_2$), from the decrease in mass of parent stock or from the appearance of products in the medium. Only rarely are these followed simultaneously and cross-referenced. Measurements on tropical waters, which are not numerous, relate to four main types of sub-system.

Open water contains stocks of organic material that are particulate (POM) and dissolved (DOM). Their decomposition (mineralization) under microbial action, and respiration of intact organisms that comprise phytoplankton and zooplankton, are accompanied by oxygen uptake. The aggregate uptake is often measured under standardized but arbitrary conditions of temperature and duration – involving bacterial multiplication – as an index of organic decomposition: biological oxygen demand (BOD). This measure is not easily related to actual fluxes *in situ*. A correlative relationship may however be useful, as in general surveys.

In one productive equatorial lake, Lake George, high concentrations of particulate organic carbon (~18 mg C l^{-1}) and dissolved organic carbon (~10 mg C l^{-1}) are combined with considerable rates of dark uptake of O_2. There is evidence, from chemical inhibitors and size fractionation, that an appreciable part of the uptake is by heterotrophic bacteria (Golterman 1971; Ganf 1974*a*). Loss rates of total (i.e., dissolved + particulate) organic carbon and nitrogen by mineralization, measurable in dark incubations of one or two days, were of a magnitude equivalent to that expected from cellular composition and rates of O_2 consumption (Ganf 1974*a*). As the particulate concentrations, largely phytoplankton, did not vary strongly with time, and as mineralized or inorganic nitrogen did not accumulate, the collective measurements are evidence for an active recycling.

More specific measurements of N regeneration rates have been made (Fisher *et al.* 1988; Morrissey & Fisher 1988) on Lake Calado, an Amazonian floodplain lake. These involved short incubations after the addition of $^{15}N\text{-}NH_4^+$ and the determination of its isotope dilution in total recovered NH_4-N by regenerated ^{14}N. With a mean value of 0.86 μmol NH_4-N l^{-1} h^{-1}, the regeneration activity at 28–32 °C was somewhat higher than most values from temperate waters. The regeneration flux was in near-balance to the uptake flux estimated for natural concentrations of NH_4-N in the lake.

Sediments receive and incorporate organic material whose mineralization depends on many factors. It might be expected to have most ecological influence in shallow waters because of the greater ratio of sediment area to water volume and opportunities for vertical transport. However, shallow waters can also be very productive, with intense metabolism – including mineralization – within the water-mass. For Lake Valencia, Lewis *et al.* (1986) believed, from indirect calculations based on bacterial abundance, that such mineralization would fall short of net phytoplankton production and would leave an organic residue for potential incorporation in the sediments. The example of Lake George has already been mentioned; here Viner (1975*b, c*, 1977*b*) believed that decomposition within the sediments was subordinate to that above them and was controlled mainly by the limited penetration of O_2. Periodic disturbance of the surface sediment by wind action was one factor (Viner 1977*a*). Release of NH_4-N and PO_4-P from relatively undisturbed sediment cores could nevertheless be measured (Viner 1975*c*), although absorption of PO_4-P – probably by non-biological mechanisms – was substantial (Viner 1975*d*).

Much earlier, Beauchamp (1958, 1964) drew attention to the apparent resistance to breakdown of sediment from mainly shallow areas of Lake Victoria, unless pretreated by heating or drying. For this Hesse (1958*b*) described measurements by respirometry. From personal observation, it is possible that particularly refractive algal remains were involved, such as of the genus *Botryococcus*. Bradley (1966) has given examples from elsewhere. For Lake Victoria there was a particular interest in the distribution of SO_4-S, a nutrient in exceptionally low concentration in the lake water. The belief that there was much unmineralized organic-S in the sediment (Hesse 1958*b*) has not been supported (Blomfield *et al.* 1970).

Although also affected by input quantity, the proportion of organic material (or C) in lake sediment will be influenced by rates of mineralization (diagenesis). If these rates are generally higher in tropical than in temperate lakes, a tendency to low levels of organic content could be anticipated. In practice there seems to be a wide range of values, both relatively low (McLachlan 1974) and relatively high (Saijo *et al.* 1991; Hecky 1993), preventing generalization. The topic has economic interest in relation to the possible accumulation of petroleum precursors – kerogens – in tropical lake sediments (Fleet *et al.* 1988).

The variation of organic-C with depth in the sediment is also partly dependent on rates of mineralization. Profiles from two Amazonian floodplain lakes, with cores dated by ^{210}Pb, were used by Devol, Zaret & Forsberg (1984) to model the depth-distribution of rates and hence deduce, by integration, average rates of C oxidation per unit area.

As decomposition in sediments is primarily by bacterial activity, independent measures of such activity are of comparative value. Ahlgren *et al.* (1997) assessed bacterial production by incorporation of added ^{3}H-thymidine, and compared this in two Swedish lakes (at 10 °C) and two Nicaraguan lakes (at 30 °C). Rates in the 0–10 cm layer were of similar magnitude (about 30–80 mg C m^{-2} day^{-1}) in the two sets. Values from the Nicaraguan lakes were much smaller fractions of the corresponding phytoplankton production and of estimated pelagic bacterial production. Despite the higher temperature they yielded lower estimates of mean specific growth rate. It was suggested that the activity was temperature-limited in the temperate lakes and substrate-limited in the tropical lakes. In part these *intensity* and *capacity* aspects must coexist and operate interactively.

Macrophytes introduce organic material with subsequent decomposition that is important in many shallow waters, as the parent biomass density is high. The nature of the material aids its standardization and the subsequent measurements. 'Litter bags' are often used, but are open to several artefacts (Robarts 1987). World-wide comparisons as well as inter-tropical ones are available (e.g., Howard-Williams & Davies 1979; Howard-Williams & Junk 1976; Furtado & Verghese 1981; Gaudet & Muthuri 1981*a, b*; Esteves & Barbieri 1983; Petersen 1984; Polunin 1984; Pearson *et al.* 1989; Junk & Furch 1991; Furch & Junk 1992; Leguizamon *et al.* 1992; Gupta *et al.* 1996). As the decay of biomass is often approximately exponential with time, although frequently diphasic with a rapid initial phase of ionic leaching (Howard-Williams & Howard-Williams 1978), the time taken to reduce the initial biomass (as dry weight) to half has been used as a comparative parameter (Howard-Williams & Davies 1979). Values, with temperatures, include 100, 12 and 10 days for *Salvinia auriculata, Paspalum repens* and *Eichhornia crassipes*, respectively, Central Amazonia, 25–34 °C; 93 days for *Typha domingensis*, Lake Chilwa, 26 °C; and 35 days for *Potamogeton pectinatus*, Swartvlei (temperate South Africa), 15–26 °C. There is clearly an overwhelming influence from the more or less refractory character of the initial material, also reflected in associated respiratory activity (Olah *et al.* 1987), and some influence from coexisting invertebrates (Petersen 1984), but also a trend to shorter times (i.e., higher specific rates of decay) in warmer tropical conditions. Decomposition can be held up by a microbial need for elements other than C, and which – like N – may be immobilized in the microbial biomass itself.

Inputs to freshwaters of terrestrial plant material occur world-wide, especially as leaf-fall. They are particularly significant for forest lakes, and in most streams and rivers. Tropical situations are affected by the altered, and often much reduced, periodicity of leaf-fall. This is illustrated in the comparison of Dudgeon & Bretschko (1995, 1996) between flowing waters of Central Europe and South East Asia. In the latter region there are some estimations of early breakdown products as coarse and fine particulate organic matter (CPOM, FPOM), important as food for some collector-species of the zoobenthos.

Further decomposition can accentuate O_2 depletion in both standing and running waters, and augment primary nutrients. This has occurred on a large scale in the early flooding phase – with drowned vegetation – of tropical man-made lakes that include Kariba in Africa (Fig. 5.32) and Brokopondo in South America (Fig. 5.6). For one Amazonian reservoir, Tucurui, decomposition of the plant material has been modelled with a distinction of three fractions of differing biodegradability (Pereira *et al.* 1994). Fungi, rather than bacteria, may be principal decomposers, especially in waters with low pH. Padgett (1976) substantiated their role in a rainforest stream of Costa Rica, where exposed leaf discs were attacked and often lost half their dry weight in ~12–16 days. There are intergrades between detrital inputs from terrestrial plants and those from swamp vegetation. Examples of the latter are inputs from the areas of *Typha domingensis* that surround and influence Lake Chilwa (Howard-Williams 1979*c*), and from the marginal drawdown region dominated by *Cyperus immensus* at Lake Naivasha (Gaudet & Muthuri 1981*a, b*).

Animal excretion is another source of mineralized products, especially NH_4-N and PO_4-P. Three studies of their production in pelagic environments by zooplankton have been considered in Section 3.2. This source was estimated to be quantitatively minor in lakes Titicaca and Calado, but highly significant in Lake George. Some animals are a means of transfer to water of products derived from terrestrial plant production. Examples that have been studied quantitatively in Africa are the hippopotamus at Lake George (Viner 1975*a*; Chapter 5.2i) and herbivorous ungulates at Lake Kariba (McLachlan 1971). The occurrence of the excretory product urea, and its relation to phytoplankton activity, have been followed by Mitamura *et al.* (1995, 1997) in some Brazilian lakes. There uptake and decomposition of urea appeared to be mainly by phytoplankton rather than bacteria and led to estimated turnover times as short as two days in the dry season. Excretion of organic material by migrating zooplankton may increase microbial activity during the night, measured by thymidine uptake, in the Ebrié lagoon (Torréton *et al.* 1994). Another and more drastic form of migration, the emergence of adult insects through the water surface (e.g., in clouds of lakeflies), constitutes an *export* of biomass and nutrients.

Recycled products may be taken up rapidly, or accumulate locally before use. Accumulant regions include most hypolimnia (examples in Fig. 2.36) and the interstitial fluid of sediments. In deeper lakes, re-use from deep accumulations is favoured by transport in vertical mixing. Lewis (1987, 1995) has suggested that a greater frequency of partial mixing (atelomixis) in deep tropical lakes will increase the effectiveness of recycling and hence primary production.

A tropical condition of 'endless summer', i.e., indefinitely long growing season, was considered by Kilham & Kilham (1989) to imply that recycling was more continuous and preponderant in tropical than in temperate lakes. In other terminology, 'new production' from externally introduced nutrients was relatively less important than 'regenerated production' based on recycled nutrients. This generalization would be further supported if the available nutrient stock (e.g., the elements N and P) were more bound up as biomass in tropical waters that have fewer physical limitations to seasonal growth. However, the factors of hydrological input and water throughput with reduced retention time, and growth limitation in turbid waters, would locally negate the supposed relationship.

Specific rates of respiration and of microbially mediated decomposition are positively temperature-dependent, and given an adequate capa-

making the generalization of ~10% no longer tenable for the most efficient food chains, although otherwise it may still be representative (e.g., Pauly & Christensen 1995).

Food webs in the pelagial of lakes Tanganyika and Malawi also suggest high efficiencies for P_2/P_1 and/or P_3/P_2. For Lake Tanganyika the evidence centres on the high yields and implied production (P_3) of zooplanktivorous clupeid fishes, relative to that likely to be achieved by the phytoplankton (P_1) and zooplankton (P_2). Unfortunately, the last two quantities, and especially P_2 (Burgis 1984, 1986; Hecky 1984, 1991), are not well known. However Hecky (1991) believed that the ratio of production rates between planktivorous fishes and the precursor zooplankton (i.e., P_3/P_2) must be ~15% or more. For Lake Malawi there has been more extended study of P_1 (Degnbol & Mapila 1985; Bootsma 1993*a*; Patterson & Kachinjika 1995) and P_2 (Irvine 1995*a*), together with that of fish production at trophic levels P_3 to P_5. The interrelations have been modelled quantitatively as a food web (see Fig. 3.45 and Section 3.5d). In essence the web embodies four or five trophic levels, with upward transfer from herbivorous zooplankton (especially *Thermocyclops cunningtoni*) being partly by planktivorous fishes (especially *Engraulicypris sardella*) and subsequent tertiary+ levels of predatory fishes, and partly by invertebrate predators (mainly *Mesocyclops aequatorialis aequatorialis* and *Chaoborus edulis*) and subsequent planktivorous fishes and their later fish predators. The number of levels is sufficiently great to prevent a markedly high ratio between aggregated fish production and aggregated zooplankton production. However, the high value of 23.9% was estimated for the ratio of tertiary to secondary (herbivore) production. The absolute fluxes involved are much restrained by the relatively low densities of phytoplankton and zooplankton.

Less documented food webs, with some important transfers unquantified or very indirectly estimated, have been described from other tropical lakes. Hecky (1984) gives a comparative, and in parts speculative, survey that includes lakes George, Chad and Lanao, as well as Tanganyika and Malawi. The first three illustrate the relevant and varying base of biomass distributions, but do not obviously raise issues of high efficiencies as production ratios.

(d) Quantitative models of food webs

A food web can be modelled as a network of biomass stocks linked by fluxes of production and consumption, with contributions to detritus (some re-utilized) and respiratory costs. The fewer the species-stock components, the more realistic this exercise becomes. As virtually all fluxes are determined by the biomass-stocks, many input parameters are expressed per unit biomass. Examples are specific rates of photosynthesis (ϕ) and respiration (R), and ratios of production to mean biomass ($P/\overline{B}$), consumption to mean biomass ($C/\overline{B}$) and production to consumption (P/C). Unit area of habitat is usually adopted for estimates of biomass and fluxes; these estimates can be expressed in units of mass (e.g., dry weight, fresh weight, carbon) or energy.

A single computerized model, ECOPATH II, has recently been applied to several tropical lakes – including Malawi, Tanganyika, Chad, George, Kariba and Turkana (Christensen & Pauly 1993; Allison *et al.* 1995). Aspects of transmission to higher trophic levels are emphasized, with top-down controls incorporating measures of 'ecotrophic efficiency' – the proportion of production by a component that is utilized by further consumers. Trophic level is assigned according to the proportions of food from different sources; heterogeneous aggregates like 'zooplankton' – that can be very misleading – are sometimes adopted by default. As steady-state conditions are assumed, the network of fluxes must be inter-compatible.

Two applications have been made to the pelagial of Lake Malawi. The later one, by Allison *et al.* (1995), based on unusually full data from a multidisciplinary survey, is illustrated in Fig. 3.45. Of note are the central role of herbivorous zooplankton, the division of the next trophic level between one fish and two invertebrate zooplanktivores, and the multiple food sources used by most subsequent predators. A special study was made of food consumption per unit mean biomass ($C/\overline{B}$) and per unit area for the main component fishes and their aggregate (Allison *et al.* 1996). It was calculated that only about 3% of production by the crustacean zooplankton was directly consumed by the fish community, but that consumption rose to >80% of production of late instars of *Chaoborus* larvae and of young of the planktivorous fish *Engraulicypris sardella*.

An application of the model to Lake Kariba (Machena *et al.* 1993; Moreau *et al.* 1997) involves a wider range of organisms, including submerged macrophytes and mussels, and both pelagic and benthic communities. The web diagram depicts a large consumption by mussels, involving both phytoplankton and detritus, three primary to secondary fish consumers that include cichlids and the introduced clupeid

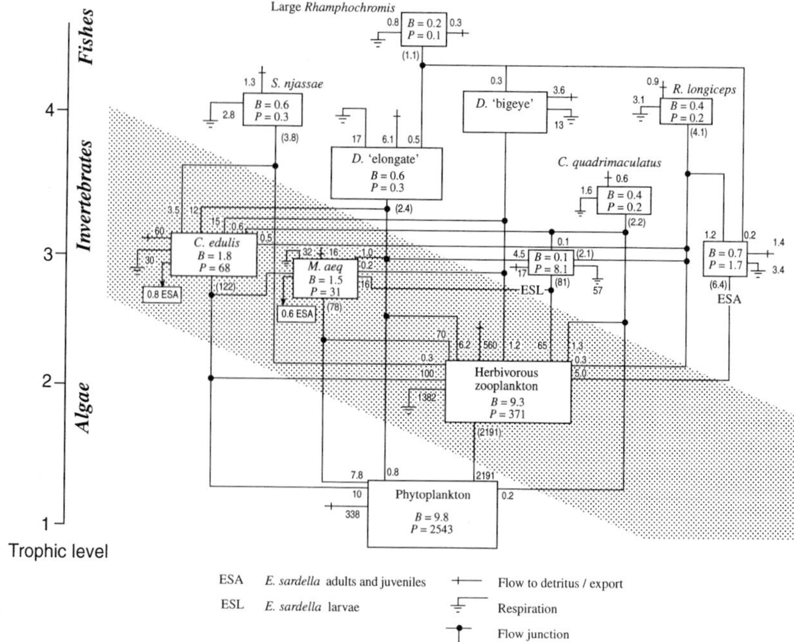

Fig. 3.45. ECOPATH quantitative model of the food web in the pelagial of Lake Malawi. For each biological component numbers indicate biomass (*B*) and fluxes of production (*P*), respiration, loss rate and consumption (in parentheses). Quantities are in g fresh weight per m^2 and, for fluxes, per year. C, *Chaoborus (edulis)* and *Copadichromis (quadrimaculatus)*; D, *Diplotaxodon*; E, *Engraulicypris*; S, *Synodontis*; R. *Rhamphochromis*; *M. aeq*, *Mesocyclops aequatorialis*. From Allison *et al.* (1995).

Limnothrissa miodon, and the piscivore *Hydrocynus vittatus* as a top-predator.

Both these flow representations are idealized from inevitably incomplete data, but do bring interrelationships of biomass stocks and fluxes into focus.

One component flux is the yield to fisheries, denoted by m_c as a component of total specific mortality m (units time^{-1}) of the species-stock in question. In an intensely exploited fishery m_c is an appreciable fraction of m with – at steady state – an upper fractional limit of 0.5. Actual values are lower, and though arguable are probably often above 0.1. The estimation of this fraction is obviously of practical importance. A notable early study was that of Garrod (1963) on a population of the cichlid *Oreochromis esculentus* (now very rare) in northern Lake Victoria. In this area an extensive use of gill nets introduced a large catch mortality m_c for intermediate age-groups, which could be distinguished by a reading of scale-rings. If absolute mortality in a cohort or age-group is exponential with time, the specific mortality rate can be estimated as the slope of a plot of log numerical abundance per unit age class against age. Examples for *Limnothrissa miodon* from Lake Kariba appear in Fig. 3.46. Here, for 1983, the estimate of total specific mortality was 1.15 month^{-1} and its natural component 0.7 month^{-1} (Marshall 1987).

(e) Fish yield related to primary production

Various attempts have been made to correlate commercial fish yields with biological and environmental factors on a world-wide scale. Among these factors, daily rates of photosynthetic production per unit area by phytoplankton have been widely considered. The foundation study for the

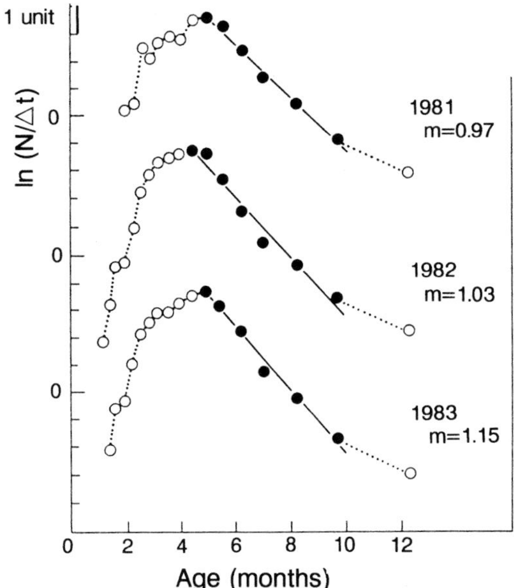

Fig. 3.46. Catch-age relationships over successive years for the clupeid fish *Limnothrissa miodon* in Lake Kariba. Age was calculated from successive 3 mm-incremental length classes; catch is calculated as numbers per unit age interval, $N/\Delta t$, and plotted logarithmically. For the solid points, with age-classes most reliably sampled, linear fitted gradients were used as estimates of total mortality m (month^{-1}). Modified from Marshall (1987).

tropics is by Melack (1976), who combined information from Africa and India. The result (Fig. 3.47) was a strong positive relationship. Here the logarithm of fish yield increases linearly with gross photosynthetic production, implying that the yield (now based upon exceptionally high values from waters near Madras) rose sharply at higher levels of primary production. Viewed in relation to photosynthetic biomass, these levels would be restrained by self-shading behaviour and the derived net production by increased respiratory loss. Prowse (1964) demonstrated these two restraints in fertilized fishponds at Malacca, Malaysia and evaluated fish production (largely tilapiine) as 1.0–1.8% of net photosynthetic production when based upon energy content.

Later correlation studies have partly considered further tropical situations (e.g., Lake Bangweulu: Toews & Griffin 1979; Lake Tanganyika: Hecky *et al.* 1981) and partly a wider range of lakes from high to low

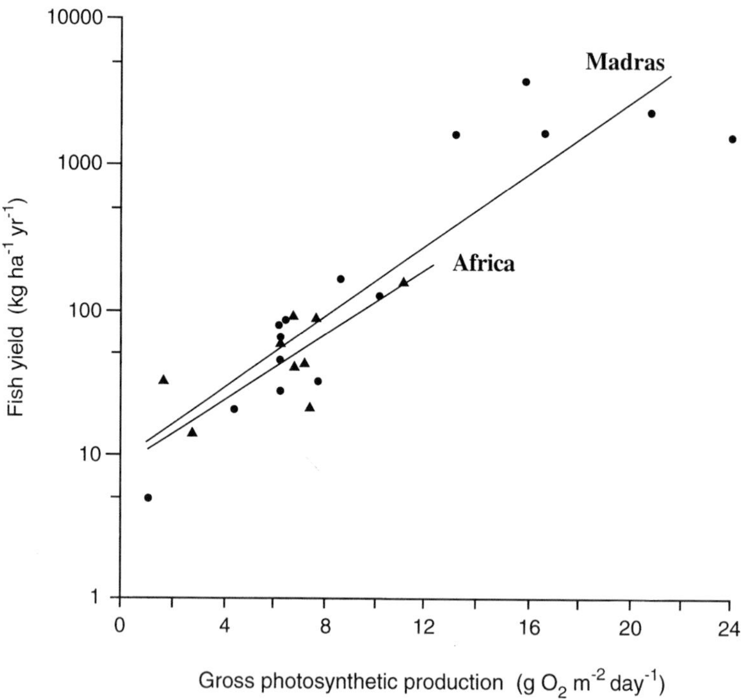

Fig. 3.47. Relationship between annual fish yield (fresh weight, logarithmic scale) and gross photosynthetic production by phytoplankton in tropical lakes and reservoirs of Africa (▲) and of India near Madras (•). Regression lines are inserted. From Melack (1976).

latitudes (Downing *et al.* 1990; Downing & Plante 1993). The Tanganyika data pointed to an exceptionally high ratio of fish yield to photosynthetic production as already discussed (Section 3.5c). World-wide the correlation of yield (FY, units kg wet weight ha^{-1} yr^{-1}) and photosynthetic production (PP, units g C m^{-2} yr^{-1} – interpreted as net rates) had a greater scatter because of other climatic variables, but could be summarized by a log/log regression:

$$\log_{10} \mathrm{FY} = 0.600 + 0.575 \log_{10} \mathrm{PP} \qquad (3.15)$$

that differs from the log–linear relationship of Melack.

As a predictive guide, the relationship to planktonic photosynthetic production will be weakened if other sources of primary production are heavily involved. Examples could be aquatic macrophytes that are consumed directly or contribute detritus that sustains invertebrate or fish detritivores.

The previous positive relationships generally suggest a predominant control by 'bottom-up' relationships. The opposite possibility of control by 'top-down' grazing relationships cannot be excluded in specific cases. Experimental enclosures in Brazilian reservoirs with and without planktivorous fish have provided some evidence, albeit scanty (Northcote *et al.* 1990; Starling & Rocha 1990). Enforced exclusion of fishes can thus reduce phytoplankton through enhanced grazing by zooplankton. The latter effect can also be obtained by manipulating concentrations of zooplankton, as in the field experiments by Weers & Zaret (1975) in Gatún Lake, Panama.

(f) Chemical tracers

The sources of food to consumers cannot always be recognized visually and distinctive chemical features can provide useful evidence. Among these rank the relative depletion or enrichment in the heavy isotopes ^{13}C and ^{15}N, expressed as $\delta^{13}C$ and $\delta^{15}N$ in parts per thousand. Applications have been made to potential foodstuffs and animals in the Amazon and Orinoco rivers and in two African lakes.

Both rivers bear abundant stands of macrophytes, especially as floatings mats, that have distinctive and varied levels of $\delta^{13}C$ according to whether C_4 or C_3 photosynthetic metabolism is involved. This vegetation will contribute substantially to organic detritus. In the Amazon, detritivorous fishes are strongly represented, but the $\delta^{13}C$ values of one major group – the Characiformes – points to a predominance of micro-algae rather than macrophytes as the ultimate C-source (Araujo-Lima *et al.*

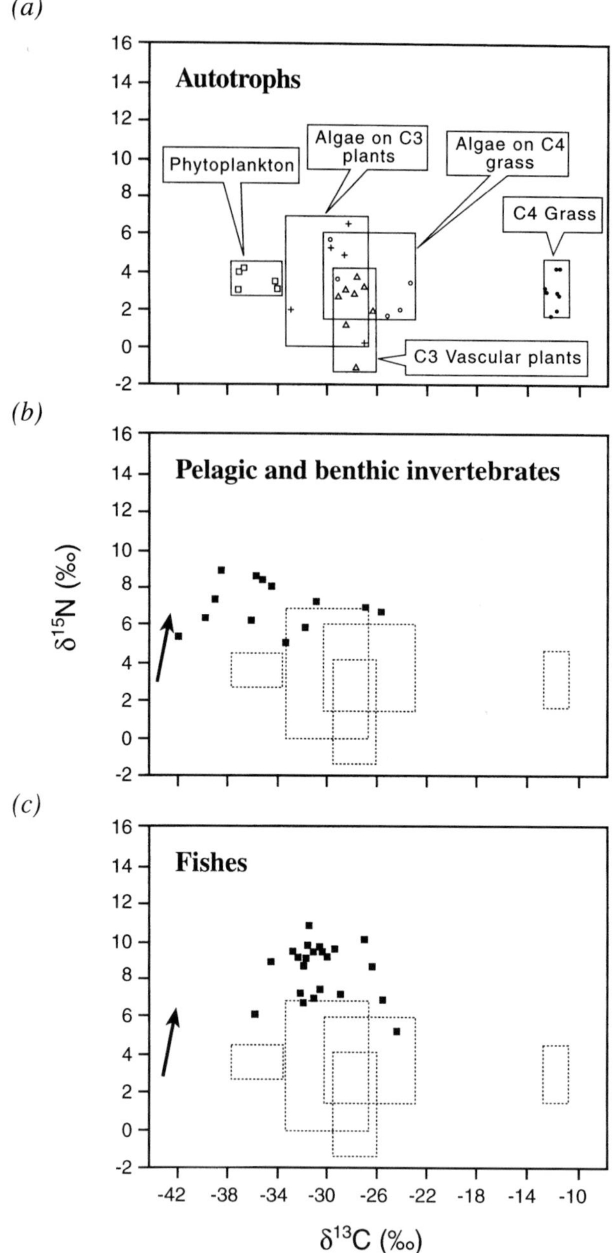

Fig. 3.48. Combinations of the isotope enrichment indices $\delta^{15}N$ and $\delta^{13}C$ in (*a*) groups of algae and vascular plants of the Orinoco river system, (*b*) various associated groups of invertebrates, and (*c*) fishes. Regions occupied in (*a*) are indicated in (*b*) and (*c*); also, by arrows, shifts expected during metabolic assimilation by the animals. From Hamilton *et al.* (1992).

1986). The same conclusion was reached on still stronger isotopic evidence from the Orinoco (Hamilton *et al.* 1992), involving both $\delta^{13}C$ and $\delta^{15}N$ values and a wider range of animals, invertebrate and vertebrate. Correspondence and non-correspondence in isotopic composition are shown in Fig. 3.48; some allowance is required for a small metabolic elevation of $\delta^{15}N$ that occurs during N-assimilation by consumers. Later work on the Amazon floodplain (Forsberg *et al.* 1995) has shown, however, that both C_3 and C_4 macrophytes can contribute considerably to the carbon of invertebrate and fish consumers – although the contribution from C_4 grasses was small relative to their abundance. However, this source of carbon appeared to provide the major part (mean 89%) of that metabolized by planktonic heterotrophic bacteria in an Amazonian floodplain lake (Waichman 1996). The application of isotope analysis to Lake Malawi has demonstrated, or confirmed, the importance of benthic plant and animal production for the feeding of nearshore fishes (Hecky & Hesslein 1995; Bootsma *et al.* 1996). Diffusive limitations of CO_2 fluxes by unstirred boundary layers, and bulk depletion by photosynthesis, might increase the relative incorporation and hence later transfer of the heavier ^{13}C isotope, here and in Lake Kyoga.

A quite different isotope, radioactive ^{32}P, was used by Walker *et al.* (1991) to label vegetable litter with fungal decomposers in a backwater river of the Amazon system. A transfer was followed to invertebrate consumers, including decapod shrimps and fishes; their movements from the source of input were limited. Lastly, there has been transmission and unwelcome persistence in food webs of chlorinated hydrocarbons (e.g., the insecticide DDT) introduced by man. Accumulation occurred especially in the tissues of top consumers, such as the African Fish Eagle (*Haliaeetus vocifer*) in the ecosystem of Lake Kariba, where concentrations and fluxes have received detailed study (Berg 1995).

In such ways, chemical tracers can provide a direct indication of material flow along food webs. The webs themselves are a summary of qualitative relationships (e.g. Fig. 5.56) and potentially an integration of compatible quantitative (flux) relationships (e.g., Fig. 3.45). The latter, rarely available, would ideally resolve the relative importance of 'bottom-up' (or nutrient-regulated) versus 'top-down' (or predation + grazing regulated) control. These two controls can be separately illustrated by tropical situations of considerable economic interest – as in the linkage of primary production and fish-yield characteristics, and the impact of the introduced predator *Lates niloticus* on its cichlid prey in Lake Victoria.

4

Patterns of environmental change with time

Here we examine time-variability at various levels in tropical inland waters, and explore the proposition that a tropical location can confer distinctive possibilities for such variability. We seek to provide a general survey of ecosystem variability over a broad spectrum of time scales, with reference to environmental driving variables and to innate characteristics of biological response. Most ecological studies have included some account of time-relations at individual sites. More generalized treatments of tropical time-variability are scanty. In them are represented general freshwater seasonality (Payne 1986); population-time relationships over wide geographical regions for phytoplankton (Melack 1979*a*; Ashton 1985*a*; Talling 1986), macrophytes (Mitchell & Rogers 1985), invertebrates (Hart 1985) and fishes (Lowe-McConnell 1975, 1987); and the latitudinal control of environmental seasonality in lakes (Talling 1969, 1992; Lewis 1987, 1995).

Time-variability involves two components, relating to period/frequency and to amplitude or range. Analyses of the latter are normally framed in absolute units, as of energy flux or stock density, but for comparative purposes relative ratios or derived logarithmic units (e.g., Talling 1986) can be more useful. Frequency is important in two ways: its absolute magnitude determines whether or not processes with circumscribed response times will be involved (e.g., the limnological spectrum illustrated by Harris 1986), and its high or low variability distinguishes between near-random or regular cyclic behaviour. Either of these short-term behaviour patterns can be compatible with progressive or longer-term change; the latter can also arise from unique, singular events such as the creation of a reservoir or a biological introduction.

The treatment of time-variability begins here with primary environmental causes and their expression on various time scales. Responses are then considered, that are expressed in chemical, biological and system variability. Besides response, biological innovation must also be taken into account.

It is often useful to distinguish between immediate or *proximate*, and *ultimate*, causes and factors. This is straightforward in relatively simple environmental systems of variability, such as for lake level and preceding rainfall as determinants of an outflow discharge. More complex cases arise in biological systems involving higher organisms, wherein co-ordinations of sensory responses allow a normally reliable (i.e., well correlated) proximate factor to act as cue or 'trigger' for subsequent behaviour, often in reproduction. Examples in tropical freshwaters have been discussed by Chutter (1985) and Payne (1986).

4.1 Quantitative characterizations of time-variability

Although the analysis of time-series is now a well-established discipline (e.g., Chatfield 1984) with ecological applications (e.g., Platt & Denman 1975), there have been few applications to tropical freshwater ecology other than in the purely hydrological background. For the latter, reference can be made to the books of Balek (1977, 1983) and Bonell *et al.* (1993). Tropical records exist of up to 100 years for rainfall, and river and lake levels. Issues of wide concern include long-term trends, periodicities by spectral analysis, coherence between records at widely separated sites and the probability for specified values.

Examples of the time-variability of lake- and river-levels appear in Chapters 2.2 and 4.3b. Balek (1983) considered that evidence of periodism in river discharge with periods > 1 year increases towards the equator.

In tropical freshwater biology, data sets for long-term series (> 5 years) are mainly represented by fish-catch statistics and palaeolimnological studies. Although the former are generally of low accuracy, some have been used to derive parameters of population change, including long-term changes of mortality components (e.g., Garrod 1963).

Some intensive studies have been made of quantitative interrelationships within pelagic lake environments sampled for one or a few years. Thus various annual environmental and planktonic sequences in Lake Titicaca were examined by Richerson *et al.* (1986, 1992) and Richerson & Carney (1988) for *auto-correlations* (i.e., tendency to repeat similar values) given time lags of 0–12 months. Some results are illustrated in Fig. 4.1. Year-to-year differences in the time-variation of phytoplankton

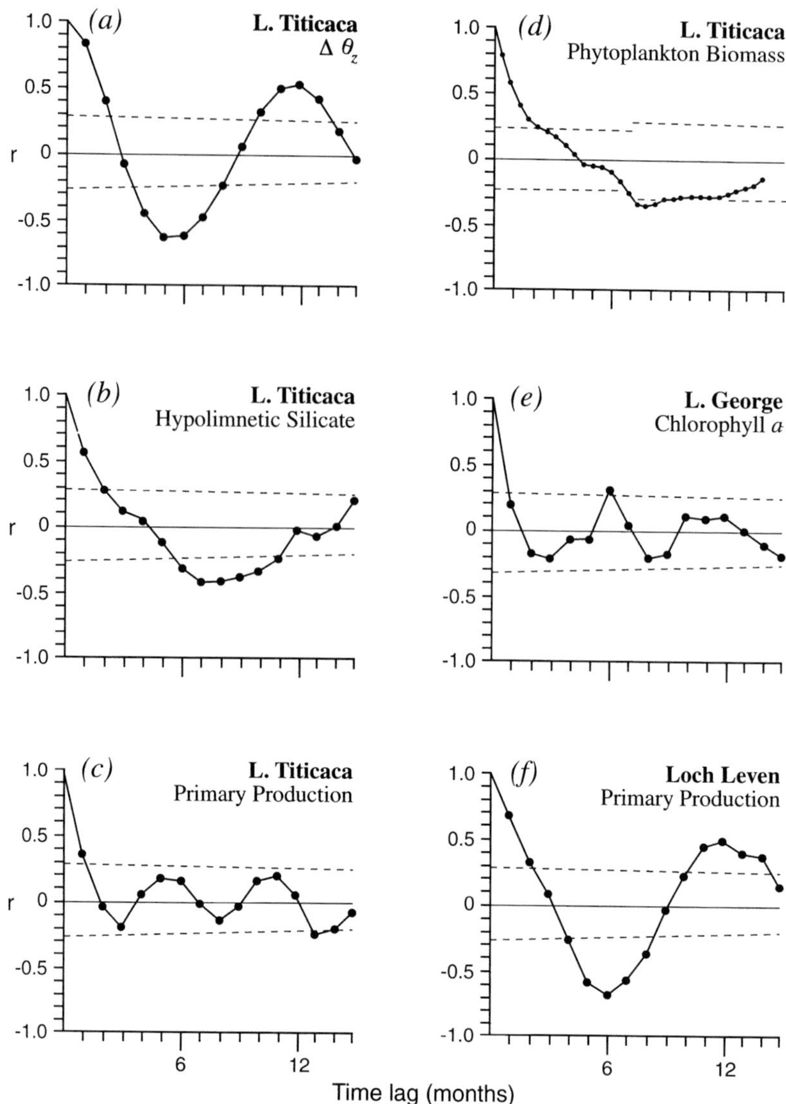

Fig. 4.1. Plots to test for positive or negative auto-correlation (r) with increasing time separation for four variables in Lake Titicaca (a–d), with comparable examples from the equatorial Lake George (e) and the temperate Loch Leven (f). Envelopes enclosed by broken lines indicate non-exclusion from random origin at 95% confidence limits. $\Delta\theta_z$ signifies vertical difference in temperature. From Richerson *et al.* (1986) and Richerson & Carney (1988).

biomass were high and the corresponding auto-correlation low. The outstanding auto-correlations were positive for solar radiation income and vertical temperature difference at a lag of 12 months, implying a predominant role of the former in the annual stratification cycle. In contrast, the auto-correlation between radiation and rates of primary (photosynthetic) production was low, unlike the situation in some temperate lakes similarly analysed. Later data from the more frequently mixed Puno Bay of Lake Titicaca (Vincent *et al.* 1986) also indicated a minor relationship between mean monthly variances of primary production rates and solar radiation, unlike the stronger relationship from the temperate lakes that were compared.

Records for phytoplankton of one or a few years duration have been treated to derive various quantitative measures of species-succession rate. Data for the tropical lakes Victoria, Lanao and Titicaca have been so used. Limited comparisons with north temperate lakes indicated lower values of overall succession rate in Victoria and Titicaca (Williams & Goldman 1975; Richerson & Carney 1988). In Lanao, and probably elsewhere, the index of succession rate is clearly correlated with rates of absolute change in biomass (Lewis 1978*b*). Lewis (1978*a*, 1986*a*) has also analysed records of up to five years to obtain trends of successional sequence, following vertical mixing, of diatoms → green algae → blue-green algae → dinoflagellates that he considered to be of probable wide application among tropical lakes. For African waters, the sequence diatoms → blue-green algae is widely recognizable (Talling 1986); under some circumstances relatively invariant green algae might – passively – form an intermediate stage.

4.2 The diel (24 hour) cycle: radiation control and environmental consequences

The diel 24-h cycle is established by the predictable variations of solar elevation and photoperiod (daylength), that of the latter being minimal at the equator (± 1 min). The diel regularity of the solar radiant flux density is naturally greatest under clear-sky conditions. The highest value, and hence diel amplitude, of the solar radiant flux density at perpendicular incidence is approximately 1.09 kJ m^{-2} s^{-1} (= 1.09 kW m^{-2} or 1.56 cal cm^{-2} min^{-1}), assuming a transmission factor per unit air mass of 0.8 and mean solar constant of 1.36 kJ m^{-2} s^{-1} (1.95 cal cm^{-2} min^{-1}).

For tropical freshwaters, the most widespread and generally influential correlate of the diel radiation cycle is the diel temperature cycle. It was first given limnological study in Africa by Worthington (1930) and has been a component of virtually all later diel work. The relationship between the diel radiation and water-temperature cycles is complex, although gross correlations that may include a time-lag of temperature are often recorded. Temperature is related to a 'stock' quantity of heat content per unit area, changes in which are the resultant or residual from energy flux densities that also include long-wave radiative exchange upwards and downwards (resultant = net 'back-radiation'), the conductive + convective transfer of sensible heat down an air–water temperature gradient and the latent heat of evaporation. A comprehensive diel energy budget that includes all these terms has rarely been attempted for tropical freshwaters. The information available is summarized in Chapter 2.1.

Except in completely mixed shallow water-bodies, the diel temperature cycle is liable to evoke a diel cycle of density stratification near the water surface, with a diel reduction of vertical eddy diffusivity – quantified for Lake Titicaca by Powell *et al.* (1984). For a given heat storage, such stratification is enhanced by the more rapid changes of water density with temperature at high tropical temperatures. If the modifying weather factors listed above change, corresponding differences between successive stratification cycles will result (examples in Talling 1957*a*; Melack & Kilham 1974; Montenegro-Guillén 1991). Of these modifying factors, wind especially is liable to its own diel cycles that may differ from, and interact with, the radiation cycle. Thus the strongest wind action usually occurred in mornings at Lake Chad (Carmouze, Chantraine & Lemoalle 1983), late afternoon at Lake George (Viner & Smith 1973) and Lake Nakuru (Vareschi 1982), and by night at the Jebel Aulia reservoir (Talling 1957*a*). Otherwise, in wind-sheltered situations that include crater lakes (example in Fig. 4.2) and forest lakes, stratification is enhanced and nocturnal mixing reduced. Results for one Amazonian forest lake, Lake Calado 'suggested that penetrative convection [induced by surface cooling and of widespread significance: see Chapter 2.3] makes a major contribution to the diurnal mixing cycle' (MacIntyre & Melack 1988).

In most tropical lakes of moderate depth, the stratification of diel period is superimposed upon another of annual period. Interrelations between the two are exemplified by the systematic observations of Kannan & Job (1980*b*) on a reservoir in southern India. Thus night-time mixing can destroy a diel thermocline but leave a deeper seasonal thermocline. In some well-studied crater lakes the diel stratification is superimposed upon a persistent, possibly inter-annual, density stratification that is

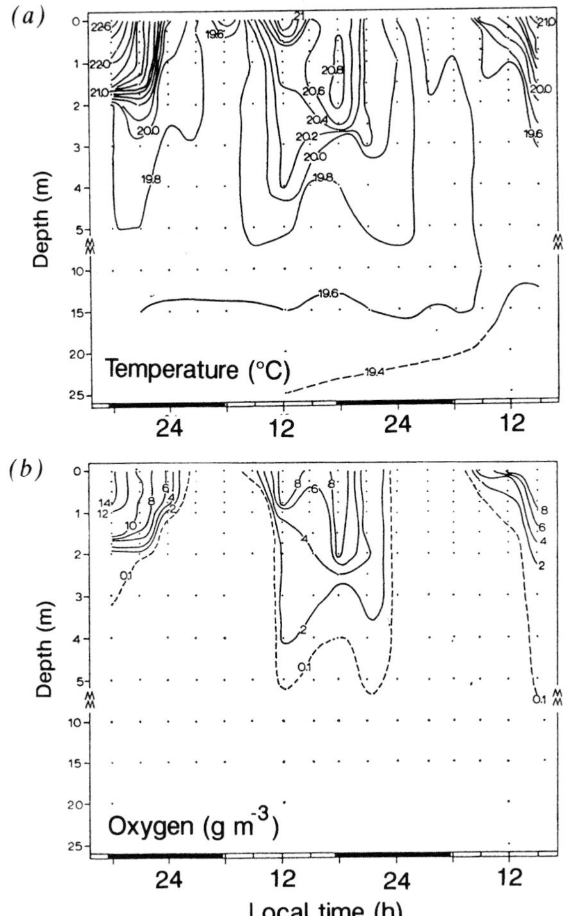

Fig. 4.2. Depth-time variation of (*a*) temperature and (*b*) dissolved oxygen (contours) in the crater lake Aranguadi, Ethiopia, 30 October–1 November, 1964, under conditions of diurnal thermal stratification. From Talling *et al.* (1973).

stabilized by increased solute concentration at depth. Examples are lakes Sonachi (Melack 1981; MacIntyre & Melack 1982) and Simbi (Melack 1979*b*; Ochumba & Kibaara 1988) in Kenya, and the subtropical Pretoria Salt Pan (Ashton & Schoeman 1983, 1988) in South Africa.

4.3 The annual cycle: control by radiation, water and wind regimes

The annual cycle, now of period 365.24 days (but varying over geological time), is *ultimately* derived from earth–sun relationships. These operate through incident radiation as the ultimate agent of intra-annual change or seasonality. However, *proximate* control of tropical freshwater environments can be conveniently divided between more direct effects of solar radiation (e.g., temperature) and more indirect effects associated with air-mass circulation and water-balance.

(a) Radiation regime dominance

The world-wide seasonality of solar radiation input as a function of latitude can be precisely calculated for a horizontal plane above the atmosphere (e.g., List 1951). It is only slightly modified below clear-sky conditions, which can be approximately represented (neglecting spectral variation) by a transmission coefficient per unit air-mass of 0.8; resulting contours for daily radiation income related to latitude and season are shown in Fig. 4.3. Here the seasonal patterns for the Northern and Southern Hemispheres are not exactly symmetrical, and the seasonal range of monthly mean radiation is least not at the equator but about latitude 3.4° N (Linacre 1969). Even at 20° N, near the limit of the tropics, the seasonal range of daily radiation income is low, indicated by a ratio between maxima and minima of 1.73:1 (cf. 10.7:1 at 50° N); the corresponding difference in daylength is 2.4 h. The seasonal maxima and minima of daily solar radiation differ in their relations to latitude. The maxima are relatively insensitive to latitude, whereas the minima decrease markedly with increasing latitude.

Under actual tropical conditions, seasonal and irregular variations in the interception of solar radiation by cloud and also dust (Monteith 1972 gives African examples) can greatly modify the ideal latitude-dependence. In Fig. 4.4, a latitudinal series of recorded seasonal changes in monthly means of daily radiation income is shown, from sites in Africa. Reduced values at the times of cloudy-rainy seasons (marked by * in Fig. 4.4) are widespread. These may accentuate latitudinal 'winter' minima or – more often – introduce new minima at other seasons, as occurs around June (summer solstice) near Addis Ababa, Ethiopia and in the southern Sudan (Griffiths 1972*a*).

As in the diel cycle, higher radiant flux density generally leads to a positive energy balance of a water-body and hence, with some time-lag, higher water temperature. The correlation between seasonal cycles of radiation income and surface water temperature is considerable over wide latitudinal ranges that include the subtropics; an example for Africa is shown in Fig. 4.4. This, with other latitudinal sequences of

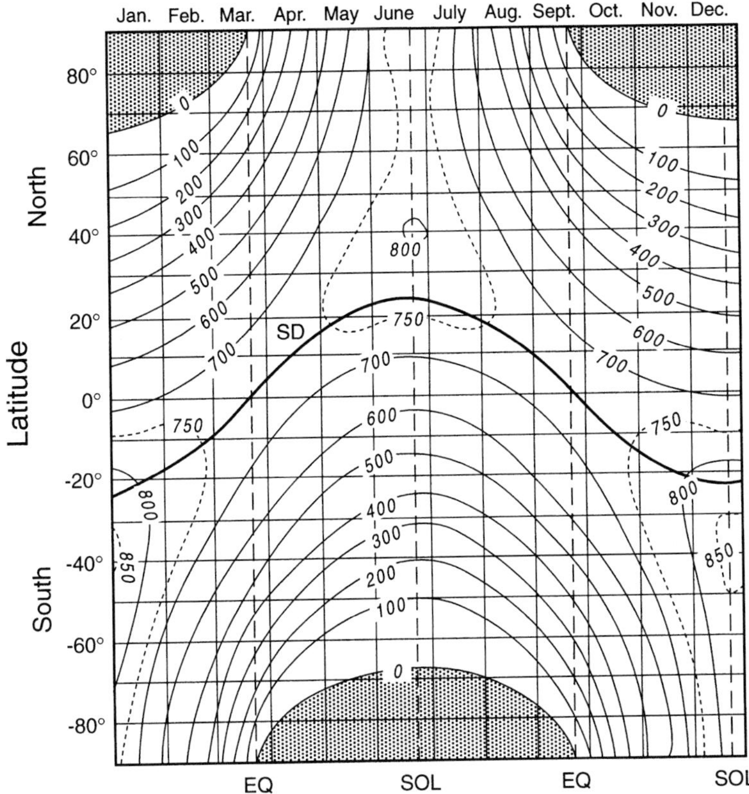

Fig. 4.3. Distribution with latitude and season of flux densities (in cal cm^{-2} day) of solar radiation under clear-sky conditions, calculated for an atmospheric transmission factor of 0.8 per unit air-mass. SD, solar declination; EQ, equinox; SOL, solstice. From Deacon (1969).

temperature from the neotropics and Australasia (Fig. 4.5), show the inevitable inversion of seasonal pattern in the Southern Hemisphere. Within the tropics the seasonal amplitude of temperature change is reduced and the relationship with solar radiation income more uncertain. For example, Wood *et al.* (1976) recorded the annual *minimum* temperature in some Ethiopian crater lakes during a season of *maximum* solar radiation but low humidity and suggested that increased evaporation was an important factor for the cooling. Regular seasonal cooling can be associated with dry cool winds. One is the North East harmattan blowing near the winter solstice from semi-desert regions of West Africa, which led to a sudden fall of daily minimum air temperature and mixing in a small Nigerian savanna lake (Hare & Carter 1984: see Fig. 2.14). By

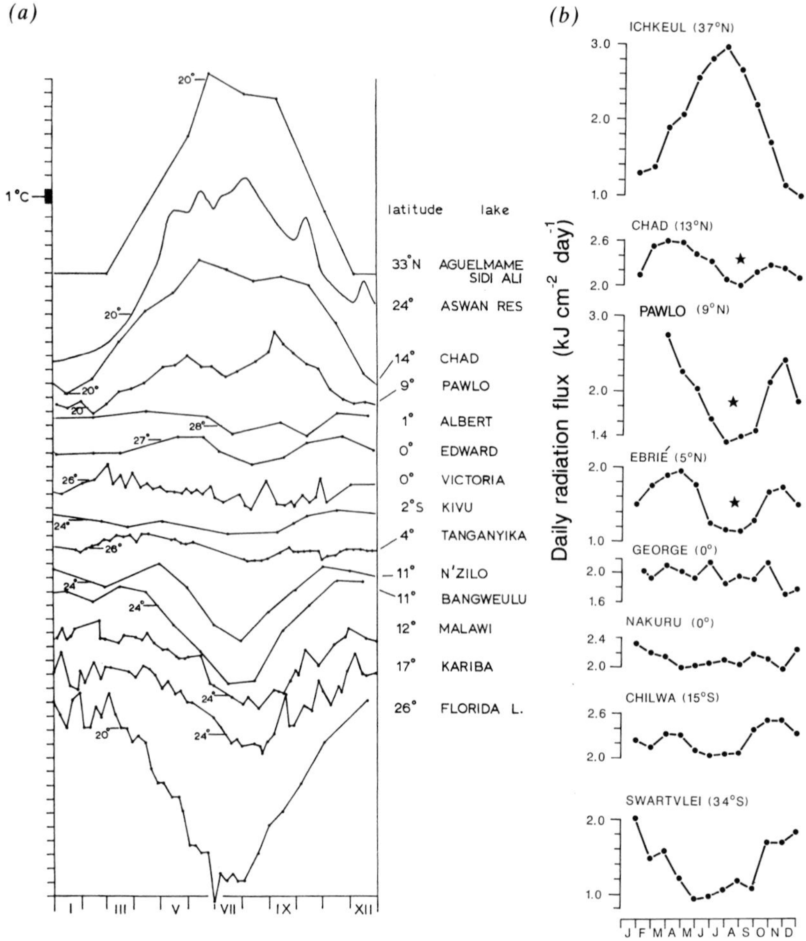

Fig. 4.4. Annual variation of surface water temperature (left) and solar radiation (right) for lakes and reservoirs of differing latitude in Africa. *denotes marked reduction during the rainy season. From Talling (1990).

contrast, in the humid tropics – where net back-radiation is reduced – the amplitude of seasonal temperature variation in surface water is typically less. In the drier tropics, a cloudy-rainy season will reduce both incoming solar radiation and outgoing net back-radiation, so that the seasonal pattern of surface water temperature is likely to vary more regularly with latitude than does solar radiation (Figs. 4.4, 4.5). However, monsoonal wet seasons can lead to a regular depression of surface water temperature, as in the Indian reservoir illustrated in Fig. 4.5. There are

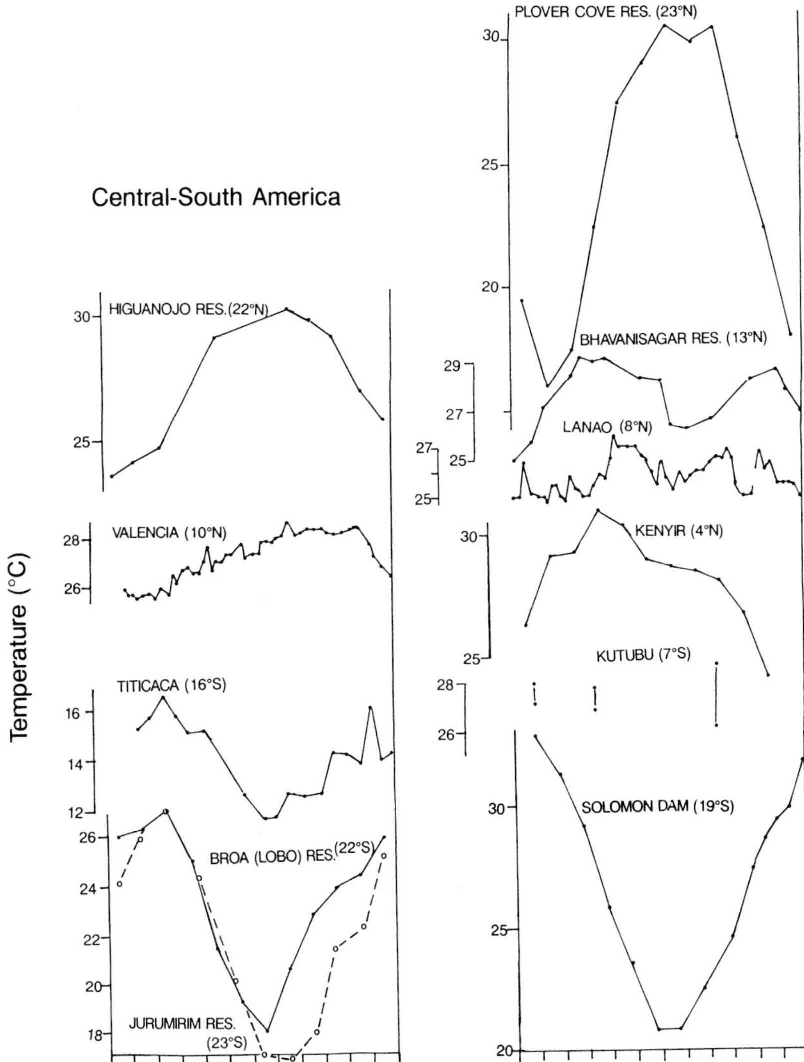

Fig. 4.5. Annual variation of surface or near-surface water temperature in lakes of differing latitude in (*a*) Central and South America, (*b*) Australasia. Data from Laiz *et al.* (1993*b*), Lewis (1983*a*), Richerson (1992), Tundisi *et al.* (1978), Henry (1993); Hodgkiss (1974), Sreenivasan (1974), Lewis (1973), Fatimah M. Yusoff (personal communication), Osborne & Totome (1992), Hawkins (1985). The years selected were, respectively, 1989, 1977, 1973, 1975, 1988–9; 1969, 1962, 1970–1, 1992, 1988–9, 1982. For Lake Kutubu vertical bars indicate diel ranges.

then two minima of temperature over the annual cycle, corresponding to the hemispheric winter and to the wet (e.g. monsoon) season. These are analysed in terms of surface energy fluxes by Townsend *et al.* (1997) for two reservoirs in northern Australia, where the second minimum may be lacking in years when the monsoon period is brief. In equatorial regions, this depression may be the lowest of the entire year, as in Cameroon lakes of West Africa where vertical mixing is favoured in a season remote from the hemispheric winter solstice (Kling 1987).

The solar radiation–temperature relationship is open to further modification when the spectrum of oceanic to continental climates is considered, as well as the rather regular and now much discussed relationship (Fig. 4.6) between deep-water temperature and altitude (Löffler 1968*a, b*; Lewis 1973, 1987; Talling 1990, 1992; Kling *et al.* 1991). Figure 4.6 shows the decline of deep-water temperature with altitude, both in deep lakes world-wide and in a series of shallow African lakes. In deeper tropical

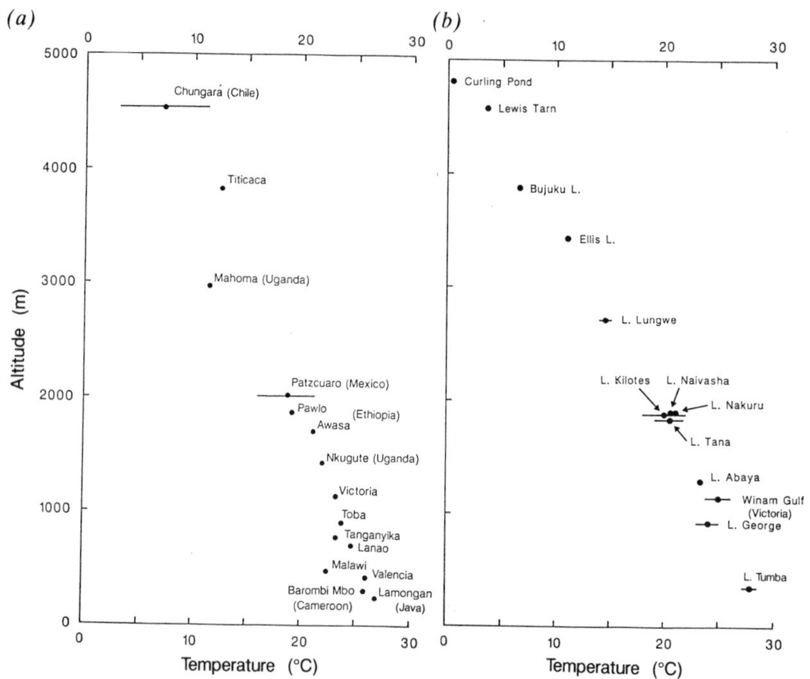

Fig. 4.6. Decrease with altitude of near-bottom temperature measured in (*a*) tropical lakes exceeding 10 m depth, (*b*) various shallow tropical African water-bodies. Bars indicate annual ranges. Data in *a* from various authors, *b* from Talling (1992).

lakes, the corresponding temperature in bottom-water tends to be lower, because of longer isolation. Seasonal changes of temperature and stratification have rarely been followed in tropical lakes at very high altitude (>4000 m); Banderas Tarabay *et al.* (1991) describe an example from Mexico subject to surface cooling in winter and complicated by a volcanic heat source at depth. Mühlhauser *et al.* (1995) describe another from northern Chile. In general, the amplitude of seasonal change in temperature increases towards the centre of a continent ('continentality'). In the tropics, this is less marked than at higher latitudes, especially (Ratisbona 1976) in the humid Amazon basin. Increases in altitude can lead to considerable amplitudes of temperature change, especially on a diel basis, about lowered mean levels that are otherwise atypical of tropical freshwaters. Lake Titicaca is the most studied example. Here the temperature of surface water, ranging between *c.* 11.5 and 16 °C, has a regular seasonality (see Fig. 4.7) influenced by wet and dry seasons and a wind regime, with net back-radiation and evaporative loss as major and seasonally variable components of the energy balance (Carmouze 1992).

Just as the net access of thermal energy to a water-column by day can lead to a diel temperature/density cycle of stratification, so can net access

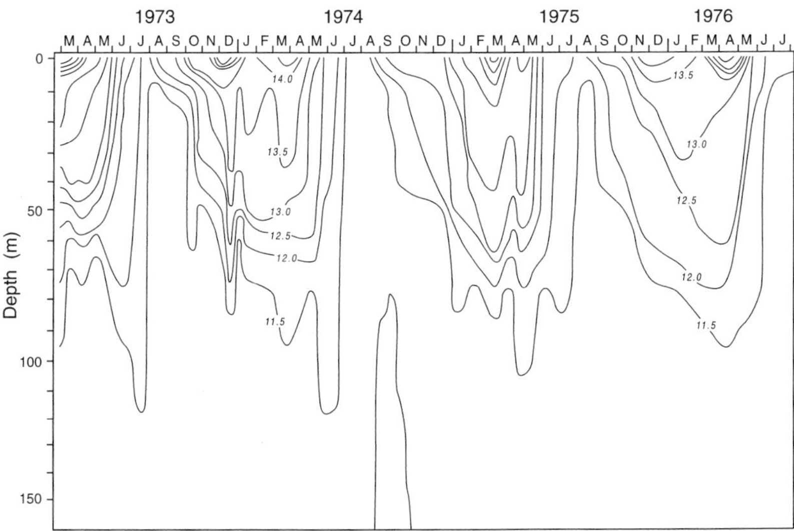

Fig. 4.7. Lake Titicaca, Andes. Depth-time diagram, with isotherms in °C, showing successive annual cycles of thermal stratification in the upper part of the lake during 1973–76. From Taylor & Aquize (1984).

over longer periods (as the summed residuals of diel balances) lead to seasonally persistent stratification. The last is not likely to develop in most very shallow lakes, which mix frequently under the influence of wind stress and convection. However, it is inevitable in all sufficiently deep water-columns and favoured in climatic regimes of low wind stress or marked local shelter (e.g., forest lakes, sunken crater lakes). Thus exposure to occasional strong winds, with long over-water action or 'fetch', is probably mainly responsible for the limited development of thermal stratification in the African rift lakes Albert (Talling 1963; Evans 1997) and Turkana (Hopson 1982; Liti *et al.* 1991) of mean depth >20 m. Conversely, seasonally persistent thermal stratification is possible in small shallow lakes, of mean depth sometimes <6 m (e.g., Opi Lake, Nigeria: Hare & Carter 1984), that include sheltered crater lakes in Africa (e.g., Melack 1978) and forest lakes in Brazil (e.g., Barbosa & Tundisi 1980, 1989; MacIntyre & Melack 1988). Thermocline depth then varies from day to day under the influence of preceding meteorological conditions; MacIntyre & Melack (1988) provide a detailed two-year study for one Amazonian floodplain lake (Fig. 4.8).

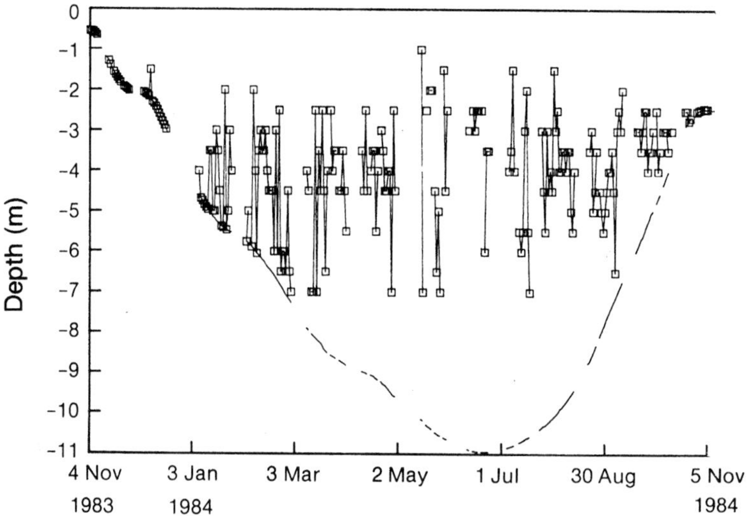

Fig. 4.8. Lago Calado, Amazonian floodplain, November 1983–November 1984. Time-variability in the depth measured at *c.* 06.30 h of the thermocline, shown in relation to the maximum depth of the lake – discontinuous line. Points superimposed on the latter indicate a vertically mixed state. From MacIntyre & Melack (1988).

The origin of the relatively cool deep water is usually traceable back to a preceding overall cool state with more complete vertical mixing (e.g., Lake Victoria: Talling 1966; Lake Titicaca: Richerson 1992). However, in some examples (represented in Fig. 2.16) it has been suggested that local cooling in shallow areas might lead to profile-bound density currents that descend and originate or enhance a deep colder layer (Talling 1963, 1969; Eccles 1974; Coulter & Spigel 1991; Patterson *et al.* 1998). In lakes or reservoirs with inflows that are large and often very seasonal, laterally derived (advected) water may, if cool, contribute to deep stratification at a suitable density level (e.g., Begg 1970; Gliwicz 1976*a*), or otherwise – as in the uppermost basin of Lake Kariba (Coche 1974) – destabilize a pre-existing stratification.

The seasonal incidence of thermal stratification in tropical lakes is illustrated in Fig. 4.9, using the divergence in temperature between surface and deep water. This incorporates the latitudinally related variation of surface temperature, already discussed and illustrated (Figs. 4.4, 4.5), with seasonal range ($\Delta\theta_t$) that is a potential determinant of vertical temperature difference ($\Delta\theta_z$) and hence the stability of stratification. Towards the limits of the tropics values of $\Delta\theta_t$ reach 5–10 °C (e.g., lakes Kariba, Malawi, Chad in Africa, and reservoirs in Cuba (Laiz *et al.* 1993*a, b, c*, 1994: example in Fig. 5.7) and those of $\Delta\theta_z$ a similar magnitude in deep lakes (e.g., Kariba, Malawi) but lesser in shallower lakes more susceptible to wind-mixing (e.g., Lake Kilole, Ethiopia). Near the equator both $\Delta\theta_t$ and $\Delta\theta_z$ are generally <5 °C if the near-surface amplitude of diel cycles is discounted, with <3 °C in some examples (e.g., Lake Albert: Talling 1963; Evans 1997) that approach 'constant-temperature baths'. Least variation is found in the lowland humid tropics with equatorial climates, where the seasonal variation of air temperature is also low. At the high altitude Lake Titicaca (Fig. 4.7) $\Delta\theta_t$ is ~5 and the maxima of $\Delta\theta_z$ ~5.5 °C, both values not atypical for the latitude of 16–17° S. Lewis (1987, 1995) has summarized graphically and by regression equations the trends of variation with latitude (referred to as a 'meteorological equator' of minimal annual variation in irradiance at 3.4° N) for deeper lakes in maximum θ_z, bottom temperature and surface or mixed-layer temperature.

In duration, the seasonal stratified phase occupies more than 70% of the year in most deep *monomictic* tropical lakes, that have one period of complete vertical mixing in the annual cycle. Intervals of de-stratification after surface cooling, with complete or near-complete vertical mixing, are then correspondingly short. A typical example is the Brazilian lake of

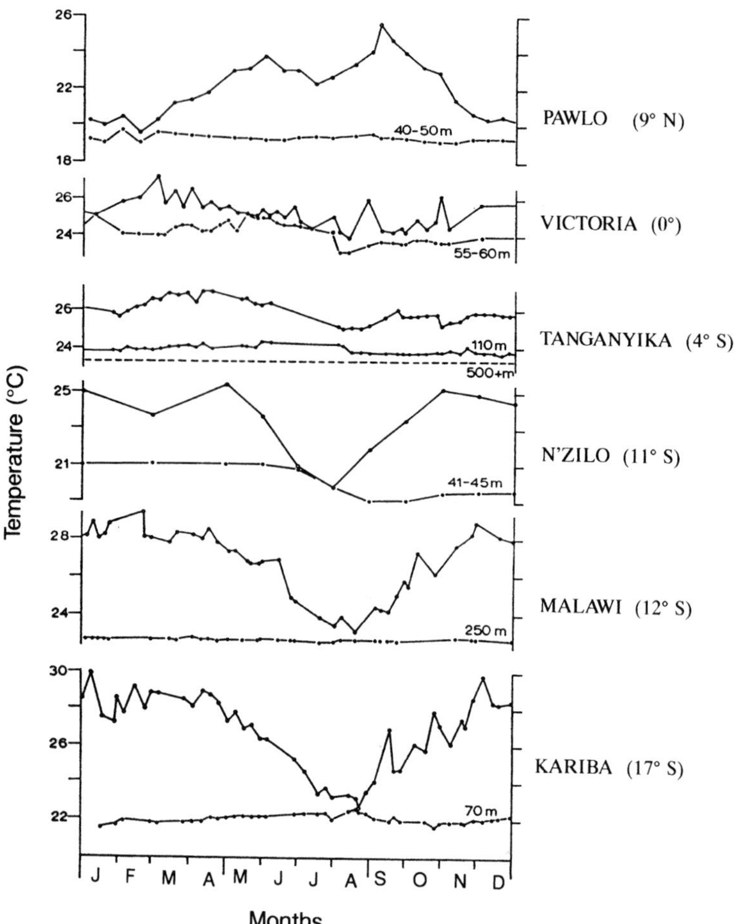

Fig. 4.9. Annual variation of the temperature difference between surface and deep water (depths indicated) in a north to south series of lakes in Africa. From Talling (1969).

Dom Helvécio (Tundisi *et al.* 1981). De-stratification tends to occur in the hemispheric 'winter', around the timing of minimum surface water temperature that tends to change abruptly near the equator (Fig. 4.10). Interception of radiation by cloud-cover can modify this timing (e.g., Cameroon lakes: Kling 1988) as can change of air-mass and wind regime. Thus the South East Trade Winds – part of a monsoon system – are probably a synchronizing influence over a series of East and Central African lakes (Talling 1969). However, periods or episodes of entrainment or partial mixing (*atelomixis* of Lewis 1973), with enlargement of

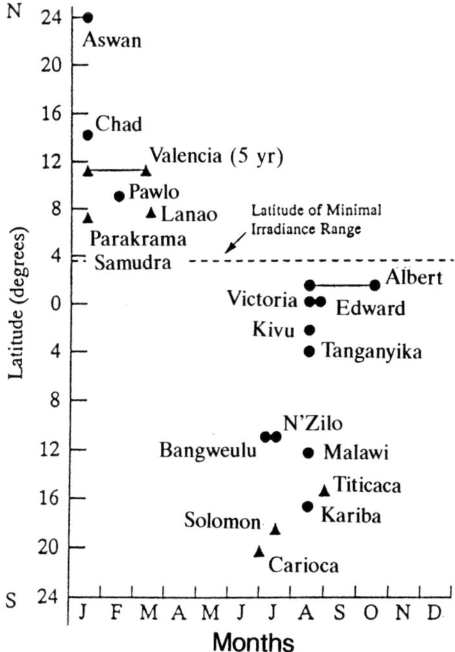

Fig. 4.10. Timing of the annual minimum of surface temperature in a latitudinal series of lakes over the tropics. Symbols indicate different sources of data. From Lewis (1995).

the upper mixed layer, are more frequent and more unpredictable. They also occur in permanently stratified lakes (e.g., Tanganyika, Malawi) and can be ecologically influential, affecting the depth of the upper mixed layer and nutrient return thereto. Irregular incidence of partial or complete mixing is more likely in tropical areas with travelling cyclones as in the Philippines, Indonesia and the Caribbean or with travelling cold polar fronts as in Brazil (Brinkman & Santos 1973; Froehlich & Arcifa 1984; Arcifa *et al.* 1990; Domingos & Carmouze 1995), Bolivia (Ronchail 1989) and North West Africa (Leroux 1983); also when the temperature ranges $\Delta\theta_t$ and $\Delta\theta_z$ are low.

In a hydrographically complex lake like Lake Victoria, deeper offshore and shallower gulf regions are characterized by stratification cycles of markedly different timing (Fish 1957: see Fig. 5.17) although they share almost the same overall temperature cycle (Talling 1966). In Lake Kariba, a more linear but multi-basin lake, successive basins differ in the timing of stratification and de-stratification by virtue of another factor, the input and downstream travel of floodwater (Coche 1974). Inputs of cooler water can also descend and create stratification, as in the Gatún lake system of Panama (Gliwicz 1976*a*), the Lagartijo reservoir of Venezuela (Lewis & Weibezahn 1976) and the Guma Dam of Sierra Leone (Mtada 1986). Another possibility is local cooling at one extremity of an elongate tropical lake that spans a wide range of latitude. The outstanding example, now well-documented, in Lake Malawi (9.5–14.5° S), where about the winter solstice cooler water develops near the southern tip (see cover) and possibly contributes to descending density currents (Talling 1969; Eccles 1974; Patterson & Kachinjika 1995; Patterson *et al.* 1998).

In such ways an annual stratification cycle can be modified both temporally and spatially. Its wider environmental significance lies in the direct physical impact of vertical mixing for the suspension of particulates, including plankton, and the vertical redistribution of chemical quantities that include dissolved O_2 and plant nutrients. Links to biological cycles are considered in Chapter 5.2.

(b) Water regime dominance

Reference has already been made (Chapter 2) to distinctive features of the latitudinal and temporal distribution of rainfall in tropical regions. Amplification can be found in text-books of tropical meteorology (Riehl 1979; Dhonneur 1985) and in volumes of the *World Survey of Climatology* edited by Landsberg (1972–81). The equatorial belt is largely a region of high annual rainfall (Fig. 2.7), with principal subcentres in Amazonia, West Africa and Indonesia that are marked by a stronger prevalence of rising air-masses (Dhonneur 1985) – although this is subject to seasonal change, especially in Africa. The wet equatorial climate in the strict sense has no extended dry season, so that there are < 3 months of less than 50 mm rainfall $month^{-1}$. However, most of the tropics has markedly seasonal rainfall influenced by the seasonal latitudinal movement of the equatorial low-pressure trough and inter-tropical convergence zone (ITCZ), that over land-masses typically lags by about two months behind the sun's zenithal position. There are, however, many local modifications – some associated with mountainous regions, as in the Horn of Africa. The movement is illustrated world-wide for the months of January and August in Fig. 4.11a. An example from eastern Africa, showing other correlated variables of solar zenithal timing (declination) and seasonal rainfall, appears in Fig. 4.11b. Wide areas are subject to a monsoon-type climate (details in Ramage 1971; Nieuwohlt 1981)

(a)

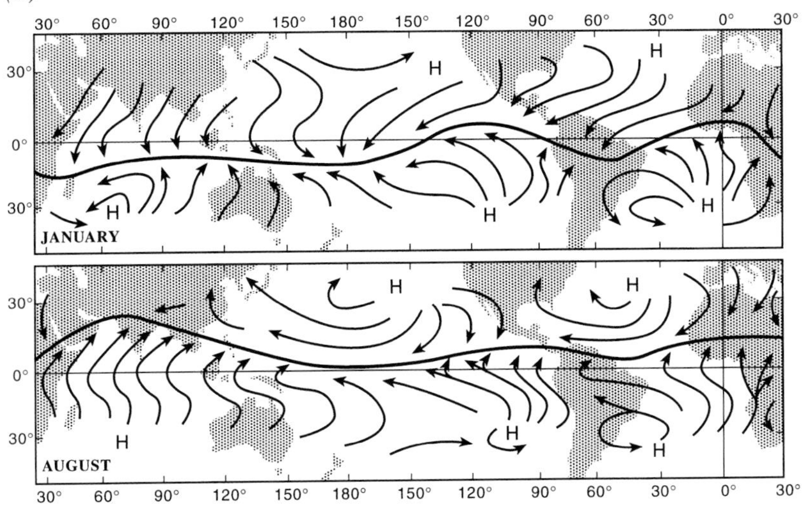

(b)

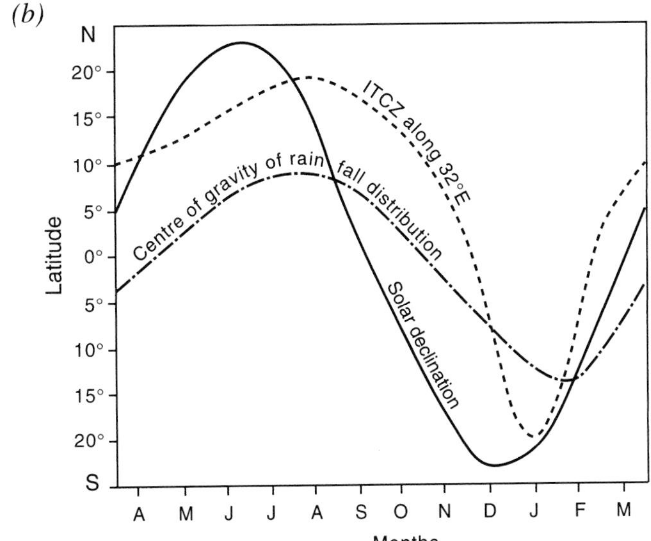

Fig. 4.11. Latitudinal distribution and seasonal movement of the inter-tropical convergence zone (ITCZ), (*a*) at all longitudes (H = high pressure), (*b*) at longitude 32° E, with associated shifts over latitude of solar zenithal position (declination) and distribution of rainfall. Modified from Dhonneur (1985) and Griffiths (1972*a*).

with seasonal alternation of wind regimes allied with a cycle of precipitation – with one or sometimes two maxima each year – of oceanic origin. Further, there is a linked seasonal cycle of atmospheric humidity that affects both radiative and evaporative components of the limnological energy budget, and the fractional run-off factor (run-off/precipitation) from land to water-bodies. Variation in the run-off factor, as well as in the rainfall itself, contributes to the marked seasonal variation of discharge in most tropical rivers. If the net balance between precipitation and evaporation is computed on a global basis and plotted against latitude (Dhonneur 1985), the positive peak of equatorial regions is flanked by negative peaks of the often arid subtropics.

Latitudinal sections down the three main tropical regions give some sampling of the seasonal distribution of monthly rates of precipitation and evaporation. One-peaked (unimodal) rainfall patterns are generally found, with 'dry season' minima tending to become more prolonged at the higher tropical latitudes. They are also extensive in a few lower-latitude regions in the shadow of atmospheric upwelling areas, as in North East Brazil and Somalia. Bimodal rainfall can be expected near the equator from well-separated crossings of the equatorial low-pressure trough and ITCZ (Fig. 4.12), but local modifications by highlands are frequent. A bimodal pattern occurs in equatorial East Africa where there are

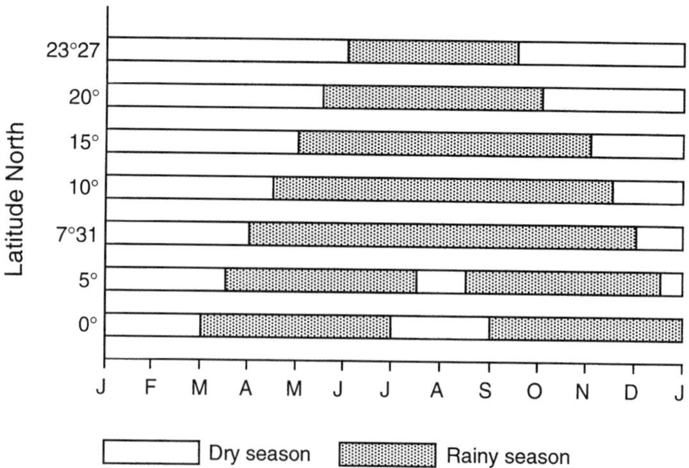

Fig. 4.12. Seasonal-latitudinal changes showing the expected distribution of rainy seasons deduced from the solar zenithal position (declination) and its theoretical lag by the equatorial low-pressure trough or ITCZ. From Dhonneur (1985).

numerous important and relatively well-studied lakes. Evaporation rates (from the meteorological literature, usually based on pan-evaporation) are generally inversely related to rainfall with the humidity correlate. They may be raised by the cooler seasonal winds (e.g., South East Trades in Central and East Africa, North East Trade or harmattan in West Africa) that are hydrodynamically important for lake stratification-mixing cycles. These cycles can then be influenced by heat loss linked with additional evaporation, as postulated by Talling (1966) for Lake Victoria and more directly indicated by the analysis of Lewis (1983*a*) for Lake Valencia (Figs. 2.5, 4.17). Pouyaud (1986, 1987*a*) has analysed, in terms of the energy budget of a small West African lake, the difference of evaporation under a marked alternation of wet and dry seasons.

However, water input related to rainfall, rather than water loss related to evapotranspiration, is typically the most variable factor determining the hydrological seasonality of tropical water-bodies. Water budgets are the key to understanding, and have been presented on an annual basis in Chapter 2.2. Some aspects of their seasonal variations are analysed here.

The seasonal discharge of **rivers and streams** reflects the rain regime on the watershed, especially for the smaller ones which typically show more day-to-day variability. Two non-dimensional indices may be used to compare different regimes (Frécaut 1982). The monthly discharge coefficient (C) is the ratio of the discharge for a given month to the mean annual monthly discharge. The ratio (R) of the extreme (largest/smallest) monthly discharge coefficients, hence also of the extreme monthly discharges themselves, is a relative measure of the annual range. It will be used here to present examples of the main types of seasonal variations in discharge. These can be described as:

true equatorial regime, with two maxima (bimodal)
altered equatorial regime, with one maximum (unimodal)
humid tropical, with one strong maximum
complex regime of large rivers fed with tributaries of different types.

The following discussion relates mainly to African examples. Others from South America and South East Asia are provided by Lewis *et al.* (1995) and Dudgeon (1995), respectively.

Bimodal regimes, as a direct image of rainfall seasonality, are characteristic of the equatorial region, but are more frequent in Africa than in Indonesia or equatorial America. The R ratio is usually between 2 and 6. An example is the River Ogoué at Lambaréné in Gabon (West Central Africa) (Fig. 4.13). The bimodal regime of the lower Niger, with 'white'

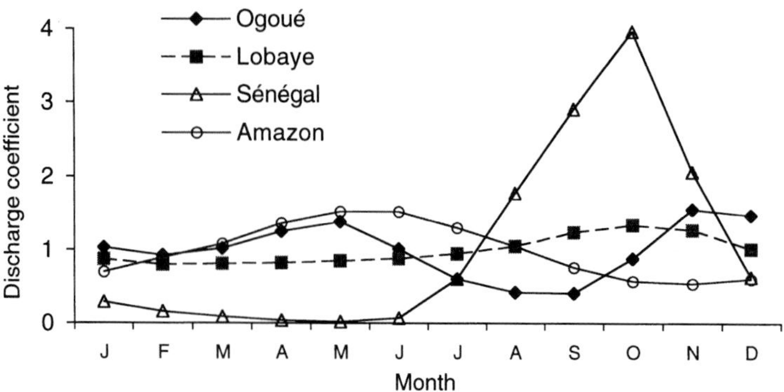

Fig. 4.13. Within-year variation of the monthly discharge coefficient C (monthly discharge/mean annual monthly discharge) in four tropical rivers. Based on data in Frécaut (1982).

and 'black' floods, results from the different timing of rainfall and flood levels in the two main upstream drainage systems (Grove 1985). An altered unimodal equatorial regime may result either from specific soil and drainage conditions, or from local modifications of the rain distribution: thus a greater permeability of both soil and basement rock (sandstone) smooths the first flood peak of the River Lobaye, also in West Central Africa. Although near-equatorial by its situation, the River Maroni (French Guiana) has a unimodal regime as a result of local rain distribution (Hiez & Dubreuil 1984), as has the Rupununi District of nearby Guyana (Lowe-McConnell 1964).

Unimodal tropical regimes are widely distributed in the humid tropics. They are usually characterized by a higher R ratio: 9 for the River Oubangui (Central Africa), 22 for the Upper Niger at Koulikoro (Mali), 18.5 for the River Chari (Chad) and up to 132 for the River Senegal at Dagana (Senegal). The last has a very low minimum discharge and may be considered as dry tropical, with less than 1000 mm yr^{-1} rainfall on most of its watershed.

These closely associated regimes of rainfall and discharge are no longer apparent for large watersheds and rivers in which different affluents experience different climates, or when the river course is hydrologically damped. This happens for the Amazon (R = 2.8) and the Congo (R = 2.4) which have major tributaries from both hemispheres, for the Zambezi (upstream of Maramba) with its internal delta, or for the River Ganges where glacier and snow-melt upstream contribute to R = 18.7, lower than it would be for a purely monsoon-fed tropical river. The White Nile regime (R = 1.6 at Mongalla) is largely smoothed by lakes Victoria and Albert, and then by the Sudd swamps. The influence of the other Ethiopian tributaries and of the Blue Nile, which is little damped by major lakes and is more directly dependent upon strongly seasonal rainfall, resulted in R = 11.7 at Aswan before the High Dam construction in 1962. Below this point, the resulting large reservoir now much reduces seasonal variations in the lower Nile (Table 4.1).

The most extreme influence of seasonality in rainfall and river discharge occurs with temporary water-bodies as in **floodplains, oxbow lakes** or **rainpools**. In these, the input components of the water budget expressed relatively (normalized) to water storage have large numerical values.

Fringing floodplains often contain a number of depressions, including oxbows and Australian billabongs, which are usually permanent water-bodies with a seasonality closely linked with that of the river, and a chemically and biologically active transition zone (ecotone) between aquatic and terrestrial vegetation (Loubens *et al.* 1992: see Fig. 2.9). Various classifications of these water-bodies, according to the importance and the timing of their relation with the river, have been proposed (Junk 1982, 1997; Junk *et al.* 1989; Drago 1989). The proportion of permanent to seasonal flooded areas is also an ecologically significant variable. On one extreme, the Sudd (Nile Basin) and the Okavango (Botswana) have a permanent to seasonal ratio of 0.6:1. These are extensive permanent marshes of seasonally variable area (Bullock 1993). A ratio of 6:1 was given for the floodplains of the rivers Niger and Senegal before dam constructions (Fig. 4.14), and much greater values apply for those of the Yaéré (Chari-Logone Basin, Chad and Cameroon) or the Orinoco (Vásquez 1992).

Discrepancies of estimates on the extent of floodplains arise from their hydraulic functioning. In a number of situations there is a close coincidence between the rainy season and the river-overspill season so that direct rainfall on large flat areas with poor drainage may result in floodplain-like situations. This is the case in the Orinoco–Apure internal floodplain where the standing water in contact with the river is given as 4920 km^2 (Lewis 1988) whereas the whole flooded area amounts to 70 000 km^2 (Welcomme 1979; Vásquez 1992). The Yaéré (Chari-Logone) and the Llanos de Mojos (Rio Mamoré, Upper Amazon: Fig. 2.9) also receive direct rainfall just before the river overspill through channels in the natural levées. Their respective maximal extent is 12 500 and 150 000 km^2, with an inter-annual variability linked to both rainfall and river flood.

Table 4.1. *Mean annual discharge, and index (R ratio) of its within-year variation, at locations on various tropical rivers*

River	Gauge Location	Catchment area 10^3 km^2	Mean discharge m^{-3} s^{-1}	Period	R ratio
Ogoué	Lambaréné	203 000	4730	1929–65	3.7
Lobaye	M'Bata	30 000	336	1950–66	1.7
Oubangui	Bangui	131 500	4400	1910–66	9
Chari	N'Djamena	600 000	1270	1936–66	18.5
Senegal	Dagana	268 000	1687	1903–66	143
Amazon	Obidos	5 000 000	157 000	1928–46	2.8
Zambezi	Maramba	1 236 000	730	1908–65	6.8
Ganges	Harding Bridge	770 000	11 650	1934–62	18.7
White Nile	Khartoum	1 759 000	2500	1912–27	2.5
Zaïre	Brazzaville	3 380 000	38 900	1987–88	2.4

Source: Data from Frécaut (1982) and Moukolo *et al.* (1990)

A particular example of a side-lake connected by a channel to the main river is that of the Grand Lac (Cambodia) which is linked to the River Mekong by the Tonle Sap channel and acts as a regulation storage. During the flood there is a net inflow towards the Grand Lac and a reversal as the Mekong recedes. The water level variation is about 8 m and the seasonal area change from 3000 to 16 000 km^2. Direct tributaries, flooding from July to October, bring about 24 km^3 yr^{-1} to the lake, while 48 km^3 yr^{-1} are fed through the Tonle Sap. Direct precipitation on the lake is roughly equal to the evaporation (Carbonnel & Guiscafré 1965).

In the Amazon basin, 'white-water' rivers have developed large fringing floodplains known as *várzea* (Sioli 1984), whereas black-water rivers, with lower pH, ionic and nutrient contents, inundate a corresponding *igapó*. Floodplains of the white-water rivers are the most densely colonized areas in the whole Amazon basin (Junk 1982, 1984).

The huge area of the Amazon floodplain (180 000 km^2) results from the combination of a very strong seasonality in the river water level and a very flat landscape. In Manaus, the water level of the Amazon rises for 6–8 weeks after the end of the local rainy season. The flood peak always

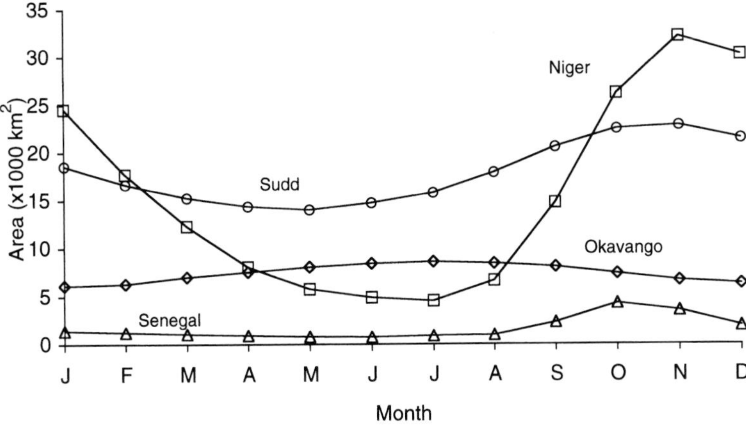

Fig. 4.14. Seasonal changes in the extent of four African floodplains. Modified from Sutcliffe & Parks (1989).

happens at the end of June or the beginning of July, whereas the lower levels occur at some more irregular dates, between September and December. The yearly fluctuation of water level may reach 11 m. Such a flood height results in a very large flooded area, as the Amazon, after descending from the Andes, traverses some 5000 km over a vast low lying depression, with an average slope of only 15 mm per km for the last 1500 km. The vegetation cover of the floodplain depends on the periodicity of flooding, and ranges from herbaceous annuals or semi-perennial grasses like *Paspalum* and *Echinochloa* to tall permanent floodplain forests.

Contrasting with temporary systems, **large deep lakes** are characterized by small numerical values of the components of their volume-normalized water budget (Table 2.1). Although a strong seasonality may occur in the water inputs, this does not significantly alter the aquatic environment.

A well known example of a large lake with atmospheric control, where the main contributors to water input and output are, respectively, rain and evaporation, is Lake Victoria (see Fig. 2.8). On a monthly basis, with rather low and relatively invariable surface inflows and outflows, the water level depends very much on the rainfall seasonality. A maximum develops at the end of April and a minimum in September, the total amplitude being about 0.4 m (Kite 1981, 1982). Inter-annual variations in rainfall have also resulted in long-term changes of level (see Section 4.5).

Also similar is the regime of Lake Tanganyika, with inflow R_i as 37% and precipitation P as 63% of input, and where output is dominated by evaporation E at 94%, in a relatively small catchment area. Lake Malawi presents similar characteristics in its annual budget, although with a much longer residence time of the water.

Another large lake with important atmospheric exchange is the high-altitude Andean Lake Titicaca. Its maximum level is generally centred on April, at the end of the rainy season (47% of total inputs are by direct rainfall) and the period of high river inputs (53% of total). Evaporation, about 1630 mm yr^{-1}, constitutes 91% of the annual losses, while riverine output via the Desaguadero accounts for only 9% (period 1956–89: Roche *et al.* 1992). The minimum level usually occurs in December, just before the start of the rains, the mean annual range of level being 0.6 m.

In all these deep lakes, the seasonality resulting from the hydrological regime applies mainly to a small fraction of the water-body, the immediate shoreline, whereas radiation and/or wind regimes are the main control variables for the structure of the water-mass.

As these large lakes have long residence times, an equilibrium has been reached between solute input and output, and the total dissolved concentration is little affected by the input seasonality. The situation is quite different in **shallower water-bodies**, which also have shorter residence times and thus higher numerical values in their volume-normalized water budget. In these waters, the annual input volume may be of the same order of magnitude as the lake volume; the seasonality in inputs creates a significant seasonality in relative water level, in ionic concentrations, in suspended matter as a result of susceptibility to bottom turbulence and more generally in the aquatic environment. Also, in shallow lakes, a moderate seasonal variation in water level implies a large relative variation in area with its associated changes in flora and fauna.

This applies also to some **endorheic lakes**, with no surface outflow, such as lakes Chad, Chilwa, Naivasha, Nakuru and Turkana in Africa. The difference between direct precipitation (P) or surface inflow (R_i) as dominant fluxes of input is here often reflected in horizontal gradients in concentrations of ionic or suspended matter, as a result of variable inflow from the rivers.

(c) Interaction between the radiation, water and wind regimes

Some interactions between the radiation and water regimes, already noted in part, influence the seasonal variation of environmental features controlled by both. In the atmosphere, water vapour and hydrosols intercept short-wave and long-wave radiation; at the water surface, the energy budget and hence water temperature is sensitive to a large evaporation term; in the water-mass, light penetration and chemical concentrations are influenced by water-borne inputs and indigenous biotic responses to these two resources. There is therefore some wider significance in the relative time-phasing of seasonal cycles of the two factor regimes.

Interesting interactions between the two regimes have been studied on man-made lakes near the southern margin of the tropics in North to North East Australia (Farrell *et al.* 1979; Finlayson *et al.* 1980; Walker & Tyler 1984; Hawkins 1985; Boland & Imberger 1994; Townsend 1998). Here the short and monsoon-related rainy season occurs in the warmer season near the summer solstice (Fig. 4.15). At this time water input can greatly modify, or in a shallow lake destroy, the seasonal stratification that would otherwise be then most developed (Fig. 4.15d). A further interaction occurs between the chemical consequences of inflow and of the vertical stratification/mixing sequence.

At many sites in the tropics, especially at lower latitudes, the two dominant seasonal variables of solar radiation input and precipitation input are inversely related. At higher tropical and subtropical latitudes the stronger annual radiation minimum is less sensitive to cloud-rain conditions, and the main rains may be near the winter solstice (e.g., North Africa) or – more usually – the summer solstice (e.g., North Australia). Some of the most important effects of the two inputs – solar radiation on water temperature, rainfall on river discharge and water levels – follow cumulatively with time-lags. In both the main regimes, loss or output fluxes exist that are generally less seasonal than the main input fluxes. These output fluxes may be *positively* related to the main input factor of the other regime (e.g., evaporative water loss to insolation), so further modify the interaction between regimes. In the example of the Ethiopian crater lake Pawlo, as interpreted by Wood *et al.* (1976), the cloudy-humid conditions of the rainy season reduced both solar-radiation input and evaporative energy loss (probably also net back-radiation); there was a combined effect upon water temperature and heat content whose seasonal cycle reached a maximum that was displaced in time from that of radiation input (Fig. 4.16). Evaporative energy loss was also shown by Lewis (1983*a*) to be a major variable factor affecting the heat content and seasonal loss of stratification in Lake Valencia, Venezuela (Figs. 2.5, 4.17).

Wind seasonality is a further interacting factor that has no simple relation to either radiation or water regime at a given site. However, there is a strong latitudinal differentiation; easterly winds are predominant at low latitudes beyond the equatorial belt, including the North East and South East Trade Winds that alternate seasonally (as in many mon-

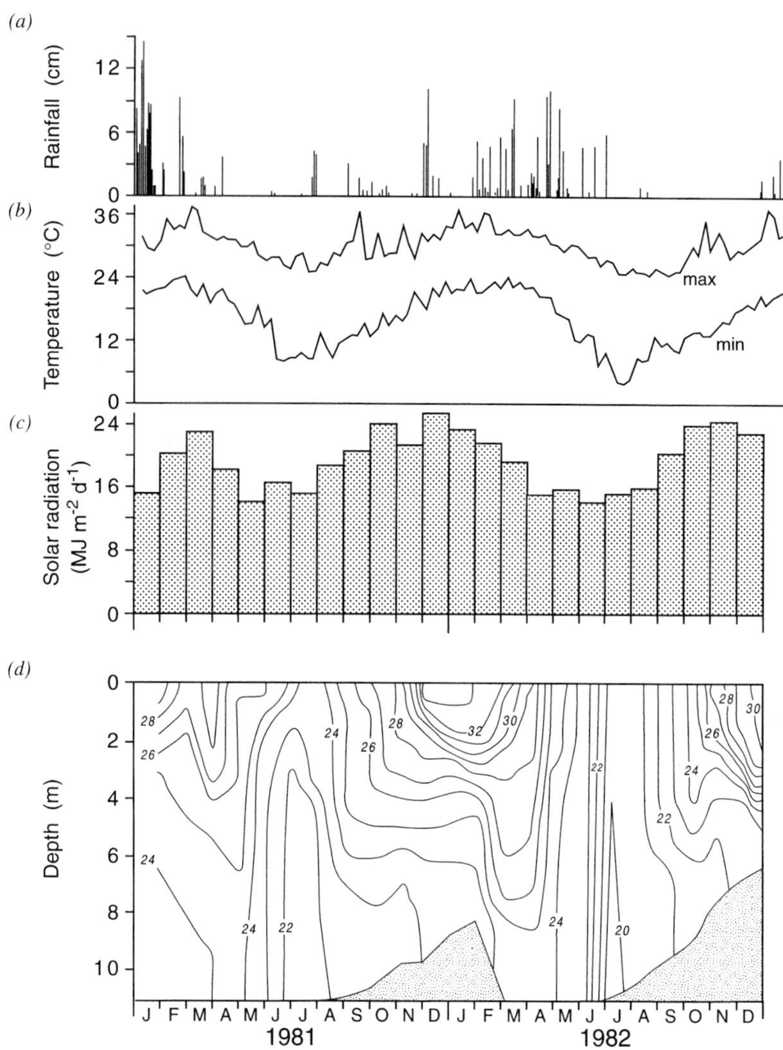

Fig. 4.15. Townsville region, latitude 19° S, tropical North East Australia, 1981–82. Seasonal variability in (*a*) rainfall, (*b*) maximum and minimum air temperature, (*c*) monthly-average daily solar radiation, (*d*) temperature stratification (contours in °C) in the Solomon Dam reservoir. From Hawkins (1985).

soon climates – although the South *West* monsoon is important in India). Their strength, greatest in the hemispheric winter, and regularity are important for cycles of stratification in non-equatorial tropical lakes. Here direct energy transfer is by kinetic movement, with water currents

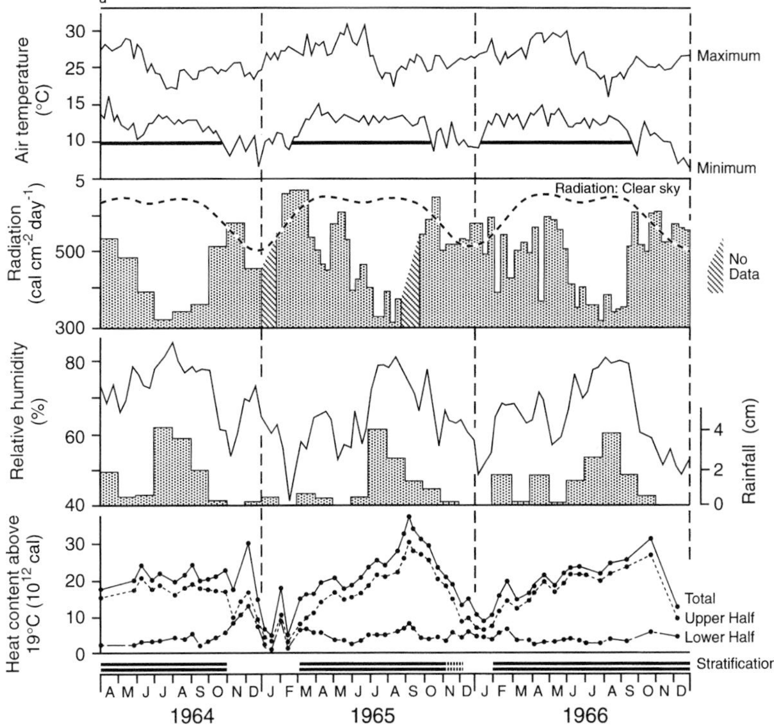

Fig. 4.16. Lake Pawlo, Ethiopia, 1964–66. Annual variation of four meteorological variables in relation to that of heat content in the lake. Besides the measured income of solar radiation (histograms, centre), its variability estimated for clear-sky conditions (transmission coefficient 0.8 per unit air-mass) at the latitude concerned is also shown (broken line). From Wood *et al.* (1976).

and turbulence. In addition, surface fluxes that involve transfers of sensible heat and latent heat (evaporation) are also affected. The limnological effects are non-linearly related to wind velocity, with higher velocities more strongly influential; they interact most strongly with those of radiation regime, influencing temperature distribution in depth (lake stratification) and in time. Examples, discussed in other sections, include the seasonal effects of the harmattan winds on Opi Lake and Lake Asejire, Nigeria, and of the South East Trade Winds on lakes Malawi (Fig. 4.18), Tanganyika and Victoria. There are also effects upon suspended particulate material, both living and non-living, and so upon turbidity. Such effects are especially marked in shallow turbid lakes, exemplified by Lake Xolotlán in Nicaragua (Hooker *et al.* 1991; Montenegro-Guillén 1993) and Lake Chapala in Mexico (Lind *et al.* 1992).

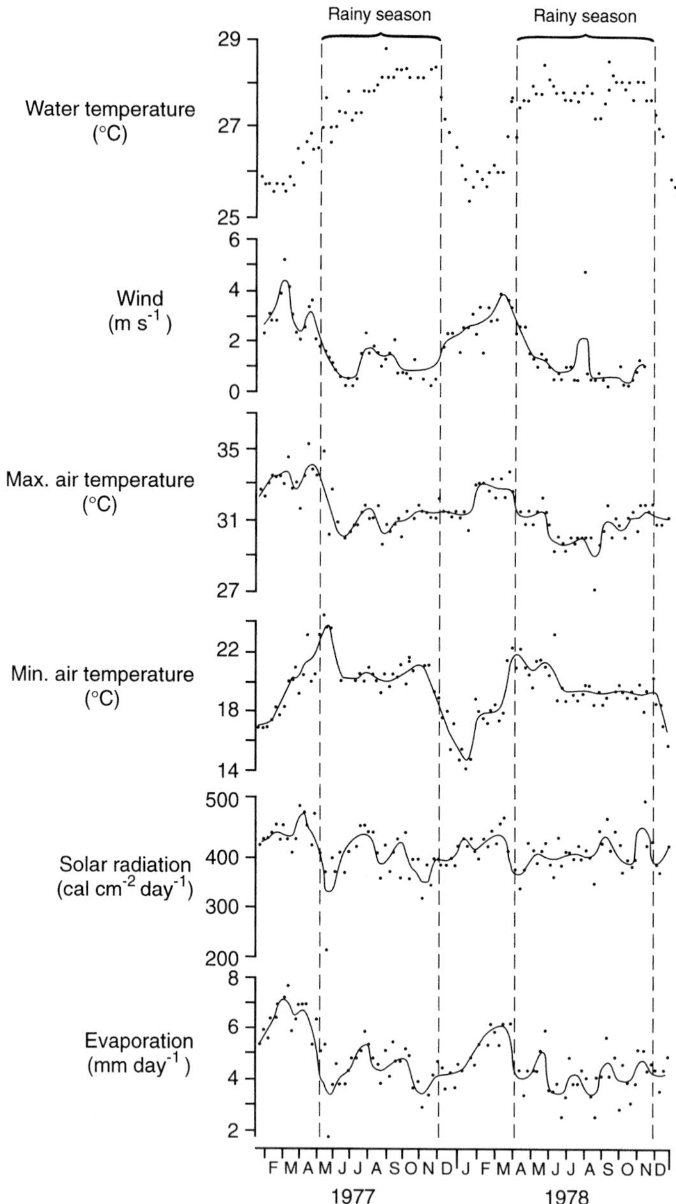

Fig. 4.17. Lake Valencia, Venezuela. Annual variation of five meteorological variables in relation to, top, the annual variation of surface water temperature. Inserted lines indicate running means of three weeks. From Lewis (1983*a*).

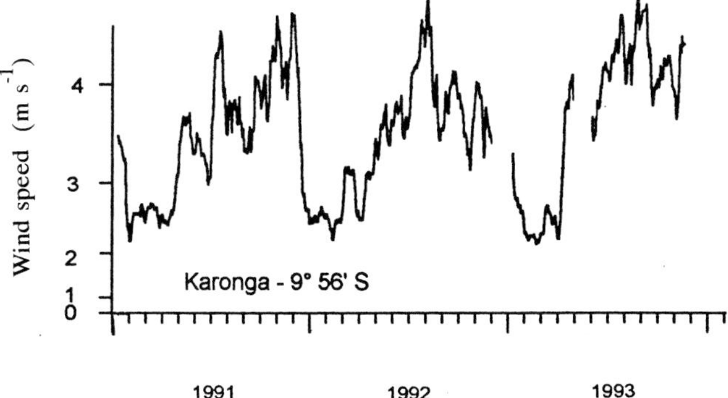

Fig. 4.18. Example of a strongly seasonal wind regime, from a location on Lake Malawi. A time-series for mean weekly wind speed is plotted on a square-law scale, roughly indicative of stress on a water surface. From Patterson & Kachinjika (1995).

(d) Approach to the aseasonal aquatic environment

The aseasonal aquatic environment can be approached in two ways, by the reduction in amplitude of intra-annual variation (conceivably to below diel amplitude) and by the irregularity of factor-variations. Although gradations exist, the absolute state is possibly never completely realized even in the equatorial tropics – as was concluded during a symposium (Chutter 1985) on seasonality/aseasonality in freshwaters of the Southern Hemisphere.

A near-equatorial location for aseasonality is favoured by the potential occurrence there of considerable year-round rainfall and small variation of the seasonal solar elevation (geometric) factor for radiation income. The most investigated freshwater site is the equatorial Lake George in Western Uganda, summarized in Greenwood & Lund (1973), Burgis (1978) and Talling (1992). Because of its shallowness a prolonged (> diel) stratification cycle is absent; the records of solar radiation income indicate a relative annual range of < 2:1 as monthly means; rainfall is moderately bimodal but inflow is locally augmented by rain on nearby mountains; and lake level is buffered by an outflow channel of considerable size. Local hydrological factors are therefore of importance for low amplitudes of variation, which also extend to bottom-water temperature, and most major and minor solutes.

There are apparently few described examples of tropical rivers that approach such a state of minimal environmental seasonality, which might be sought in the outflow from a large equatorial lake. However, the best described examples are from the zoobenthos of running waters on Pacific islands. A largely aseasonal stream on the Palau Islands had an invertebrate community with all growth stages present at all times; however, two dominant species showed some seasonality, possibly related to rainfall, in their population densities (Bright 1982). An apparent completely aseasonal pattern is described by Yule (1995) and Yule & Pearson (1996) from the short Bovo River and Konaiano Creek on Bougainville Island, Papua New Guinea. Some seasonality seems to appear in deep lakes with persistent thermal stratification, which invariably changes in some systematic way throughout the year, partly in relation to the almost ubiquitous seasonal wind regimes. Some independence from the consequences of a seasonal rainfall is possible in very large lakes of long retention time such as Lake Victoria, although these generally do not escape the variability from thermal stratification. Hydrological near-constancy of water level can be preserved within large lakes by shallow bays and gulfs that also lack persistent stratification, such as the Nyanza (Winam, Kavirondo) Gulf of Lake Victoria, or have frequent mixing such as the two major bays of Lake Titicaca (Lazzaro 1981; Vincent *et al.* 1986).

Tropical aseasonality that incorporates oscillations of wide amplitude but irregular timing does not seem to be established for lakes, but irregular spates of flow probably so qualify in the two running water sites of Bougainville Island (Yule 1995; Yule & Pearson 1996). In some regions, but not all, the thermal stratification of lakes can suffer large changes of irregular timing (e.g., Lake Lanao, Philippines: Lewis 1973; Lake Nyos, West Africa: Kling *et al.* 1987; Kling 1988) but an annual cycle is still recognizable. Exceptions with substantial aseasonality may yet be found in lakes of Indonesia and Papua New Guinea, where episodic mixing with fish-kill is known (Green *et al.* 1976; Osborne & Totome 1992; Lehmusluoto *et al.* 1995); also with shallow saline pans or rainpools in regions of low and irregular rainfall.

4.4 Cycles with other periodicities

Other environmental cycles of well-defined period exist but their influence in freshwaters is more limited.

Lunar cycles (period 28 days) can operate via gravitational or light-related influences. Of the former, tidal levels – that also include periods of 12.4 and 24.8 h – govern the entry or retreat of seawater into numerous coastal lagoons and estuaries of intermediate salinity. Tropical examples are the much-studied Ebrié lagoon (Ivory Coast, Africa: e.g., Durand & Chantraine 1982) and the Saquarema lagoon (Brazil: Costa-Moreira & Carmouze 1991). Here the lunar-tidal cycles interact with longer-period cycles of freshwater discharge to determine the extent of influence by seawater (Fig. 5.2). Far from the sea, in Lake Kariba, semi-diurnal tides of low amplitude (see Fig. 4.19b) have been traced by Ward (1979). Their period is 12.4 h and amplitude 3–6 cm, the last being greater at the west end of this elongate basin and also near phases of the new and full moon, especially at the spring and autumn equinoxes.

Light-related lunar cycles of biological importance are more distinctive of tropical freshwaters. Their role as a cue or proximate factor for the emergence of many aquatic insects has been established in various tropical regions since the original description from Uganda by Hartland-Rowe (1955, 1958). An example for chaoborids in the West African Lake Opi was described by Hare & Carter (1986). Here there was also some evidence that a lunar rhythm of abundance could be impressed upon the crustacean prey of the chaoborids (Hare & Carter 1987). One main period of emergence per lunar month is usual (e.g., MacDonald 1956; Corbet 1958; Fryer 1959; Fukahara *et al.* 1997), but two – at first and last quarters – were recorded by Tjönneland (1962) for one species of chironomid at Lake Victoria. A different influence is that upon visual predation by zooplanktivorous fishes and associated vertical migration in the pelagial of lakes, demonstrated for a clupeid fish and zooplankton by Gliwicz (1986*a, b*) in Lake Cahora Bassa, Mozambique. In this lake the lunar influence via fish predation could be recognized in the abundance of Copepoda, Cladocera and Rotifera (Fig. 4.20), and even the varying density of other particulate matter subject to ingestion (Gliwicz 1986*b*).

Seiches and internal waves (Chapter 2.3), of various types, are set up as oscillations in lakes after wind stress, with periods determined by travel-mode and basin dimensions. The longest periods, of several weeks or more, can be expected in the largest lakes. Latitude can have an influence on internal waves via earth-rotational (geostrophic) effects, but these are minimal at the equator. Internal seiches or waves are propagated in the water-mass on density-temperature boundaries. The best known, from Lake Tanganyika, is the predominantly unimodal longitudinal seiche. In this lake (Coulter 1988; Coulter & Spigel 1991) it is set up by the seasonal south wind and is prominent during the season of maximal stability of stratification, approximately November to May. The amplitude of 30–40

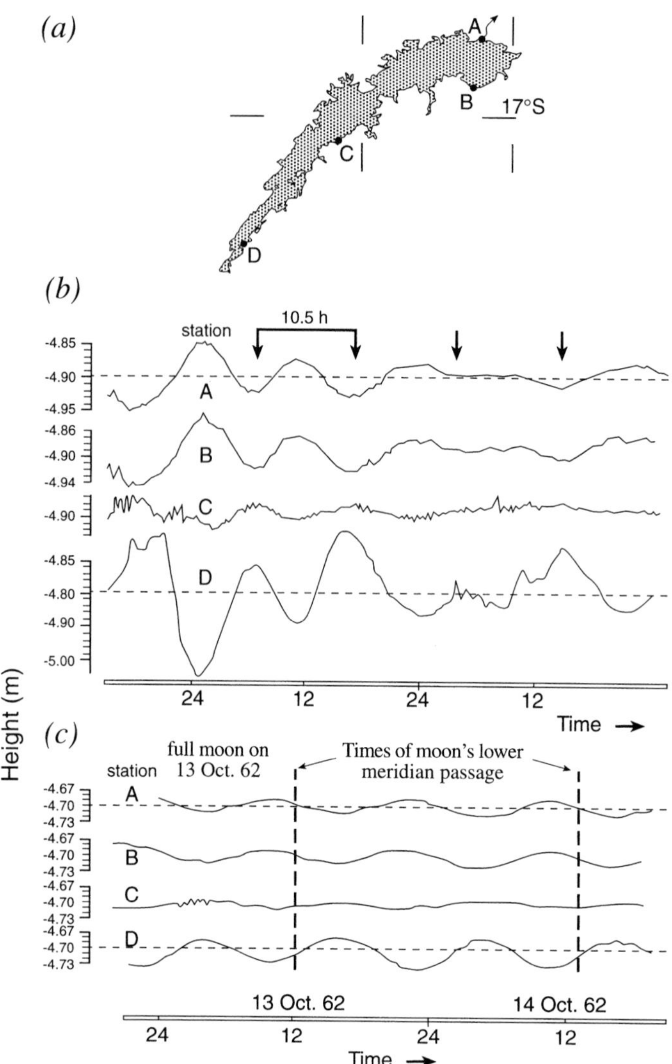

Fig. 4.19. Lake Kariba, Zimbabwe-Zambia, Africa. Time-variations of water level at four stations A–D along the axis of the lake (*a*) associated with a surface seiche (*b*) semi-diurnal tides (*c*) From Ward (1979).

m, period of 25–30 days and operative duration of *c.* six months are apparently larger than established in any other lake. The oscillations have significance for both horizontal and vertical exchanges in the water-mass. In the neighbouring rift lake of Malawi there is less complete evidence for internal waves of large amplitude (Beauchamp 1953*a*; Eccles

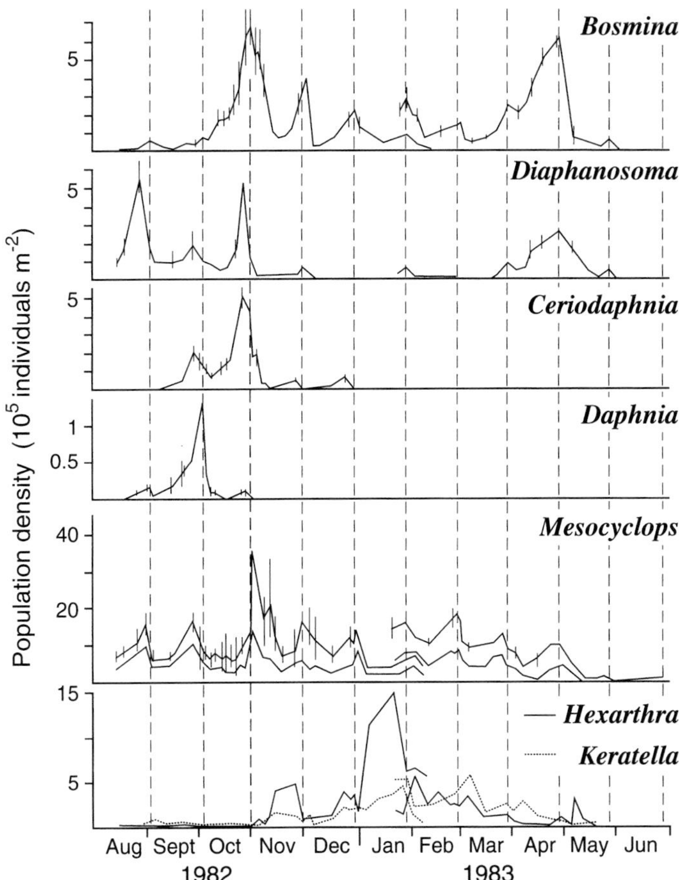

Fig. 4.20. Lunar periodicity in the population density per unit area of seven zooplankters in Lake Cahora Bassa, Mozambique. Dates of full moon are marked by vertical broken lines. From Gliwicz (1986*a*).

1962, 1974; Patterson & Kachinjika 1995), that are likely to contribute to upwelling of nutrients and hence to lake fertility.

Less regular oscillations of the deeper isotherms occur in offshore Lake Victoria (Fish 1957; Talling 1966), and for 1951–2 were interpreted by Fish as indicative of a unimodal internal seiche with a period of about 40 days. This interpretation is insecure in view of the difficulty of distinguishing direct wind-displacements from later oscillations, but its rejection by Newell (1960) is also open to criticism (Talling 1966). Whatever their status, the offshore movements transmit their effects to some inshore waters that are seasonally stratified, as in the Buvuma and Rosebery Channels studied intensively during 1952–53 by Fish (1957).

Direct wind-displacements of surface water level are a common form of time-variability and may set up **surface seiches** of much shorter period than the internal seiches. The amplitudes involved are small in deep lakes, usually of the order of a few centimetres. Surface seiches in Lake Tanganyika have been analysed mathematically by Servais (1957). In Lake Kariba (Fig. 4.19) the first mode of seiche oscillation, of period 9.7 ± 0.4 h and amplitude 10–30 cm, can approach near-resonance with the smaller semi-diurnal lunar tide (Ward 1979). In the large shallow Lake Chad examples of wind set-ups of ~40 cm are known. Here they could be damped by stands of macrophytes (Carmouze, Chantraine & Lemoalle 1983).

Solar (sun-spot) cycles (period *c.* 11 years) were formerly correlated with a similar periodicity in recorded lake levels of major African lakes (Hurst 1952; Pike 1965), but the correlation broke down after the 1920s. Nevertheless, the long-term variation of these lake levels (Fig. 4.22) shows at least some repeated oscillations, as well as coherence between lakes resulting from shared climatic change (see Section 4.5).

4.5 Long-term and aperiodic environmental change

Non-periodic environmental change has many origins, in which chance plays a large part. There may be large irregularities between successive oscillations, long-term trends superimposed on some short-term annual variations, or unpredictable perturbations of short duration with either short- or long-term effects.

The very long-term includes a thousand-year+ time scale and implies climatic and environmental changes which resulted in the present day tropical conditions. They are also accompanied by evolutionary change (Chapter 5.4). Very long-term changes have been recognized from geological surveys and study of the record preserved in lake sediments (palaeolimnology). Examples are available for a number of tropical lakes among which are Lake Valencia (Bradbury *et al.* 1981; Binford 1982) and Lake Titicaca (Wirrmann *et al.* 1992: Ybert 1992) in South America, Lake Victoria (Kendall 1969), Lake Naivasha (Richardson & Richardson 1971), Lake Chad (Maley 1981; Servant & Servant 1983) and the Ethiopian Rift lakes (Gasse *et al.* 1980) in Africa, and the highland lakes of Papua New Guinea (Oldfield *et al.* 1980) in

Australasia. At the time of writing there are active and promising, ongoing investigations of the larger African Rift lakes (Johnson & Odada 1996).

In these tropical regions, the major long-term changes are more readily connected with changes in rainfall and related hydrology than to those in temperature that predominate at higher latitudes. For African lakes, alternations between wetter and drier conditions have been widespread over the last 15 000 years (reviewed by Livingstone 1975). These can be recognized in corresponding sediment horizons in cores from a number of lakes and in reconstructions of changing lake level (Street & Grove 1979). There is no consensus as to the geophysical origin of these alternations, but hypotheses include earth orbit-related changes in solar radiation income (Milankovitch fluctuations: Kutzbach & Street-Perrott 1985) and in Atlantic Ocean circulation (Street-Perrott & Perrott 1990; Lamb *et al.* 1995).

We shall focus here on shorter time-spans, with different time scales depending on the duration of the oscillations which are considered.

A predisposition towards more irregular oscillations in the tropics might be expected to follow from the much reduced amplitude of the annual radiation and temperature cycles. Small irregular changes – e.g., of heat income – could then exert considerable overall effects. This expectation was modelled by Lewis (1987) in relation to the stability of lake stratification and was borne out by observed thermal cycles in tropical lakes such as Lake Lanao. However, some qualification may be needed. Irregular change in tropical freshwaters can also be favoured by some atmospheric patterns. These include unpredictable cyclones and the irregularly periodic (*c.* 2–7 yr) El Niño–Southern Oscillation (ENSO) phenomenon. The latter involves changes in atmospheric circulation that (as the Walker circulation) forms a tropical series of west–east cells, the strongest between Indonesia (upwelling) and the Eastern Pacific (downwelling), with widespread consequences for surface sea temperature and climate – including rainfall – in the southern tropics. Some ENSO-related effect seems to exist in long-term records (1903–85) of the Amazon discharge, on the between-year variability that mainly appears at a 2–3 year time scale (Richey *et al.* 1989). Also relevant is the reduced magnitude of the Coriolis force in tropical atmospheric circulation, which lessens the likelihood of travelling fronts and favours more persistent weather patterns of wind and rainfall.

(a) Climate and hydrological balance

Environment changes related to some modification in the elements of the water budget are generally results of climatic change or of human activity. The usual hydrological balance is based on an annual period. Hence, irregularities apply to between-year differences and trends extend over several to many years.

Large departures from the mean annual rainfall are a common feature for wide tropical areas and mean values over a moving time-span of several years may indicate the occurrence of a dry or wet period. However, auto-correlations in time-series (Section 4.1) seem a particular feature of the Sahelo-Sudanian region of Africa: dry or wet years often occur not interspersed but as a series of successive years, as shown in Fig. 4.21 (from Olivry *et al.* 1993). No comparable persistent trends or series have been observed since 1900 in other tropical regions such as India (Fontaine 1991), eastern or southwestern Africa (Hulme 1992) or North East Brazil (Hastenrath *et al.* 1984).

As a result of an annual rainfall deficit since 1968, total river input to the Atlantic Ocean by West African rivers from the Congo (Zaïre) to Senegal has decreased, when compared to a 40-year mean (1951–90) which was estimated as 2585 km^3 yr^{-1}. This total input was 2395 km^3 yr^{-1} during 1971–80 and 2155 km^3 yr^{-1} during 1981–90, that are, respectively, 93 and 83% of the long-term average (Olivry *et al.* 1993). When

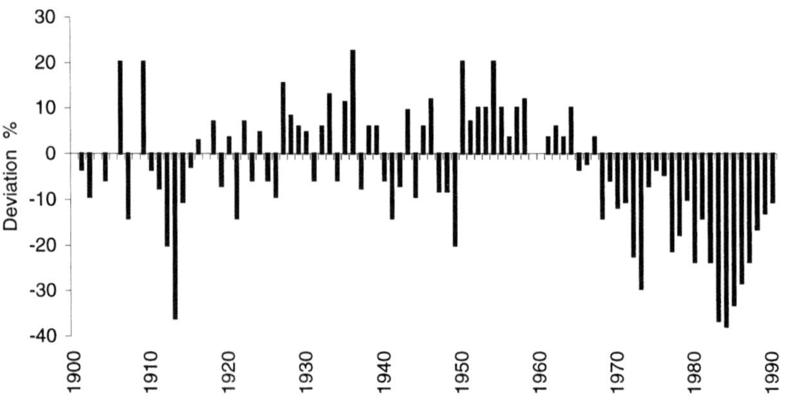

Fig. 4.21. Variation of relative annual rainfall, as % deviation from the long-term mean, in dry tropical West Africa. Based on Nicholson *et al.* (1988) updated by Olivry *et al.* (1993).

the component Sahelo-Sudanian region only is considered, these proportions are 87 and 73%, respectively, for the two periods.

The effects on river discharge of departures from the mean for rainfall are accentuated in regions of high evaporation or evapotranspiration losses. In the total catchment budget, these losses are approximately constant and large compared to rain inputs, so that the remaining water for river discharge is highly variable: in the River Niger, at Koulikoro, a ± 15% change in annual rainfall results in ± 33% variations in annual discharge (Grove 1985).

The fact that dry and wet groups of years alternate has led to the question of **long-term periodicity**. But, although there may be pluri-annual periodicities in some tropical aquatic features, it seems that the direct time-series of data available are generally too short to allow for significant determination of such periods. This is particularly true for hydrological variables such as river flow or lake level. Indirect time-series can sometimes be deduced from a long sedimentary record. Thus fine-resolution of change in $\delta^{13}C$ and magnetic susceptibility in cores from Lake Turkana has suggested periodicites at 11, 16, 18.6, 22 and 32 years in the river inflow. The first four periods have reported equivalents elsewhere (Halfman *et al.* 1994).

The Senegal River discharge from 1903 to 1979 was used by Faure & Gac (1981) to illustrate a return period of about 30 years. This has not been corroborated since, due to the persistence of below-average rainfall over most of the sahelian region. On a shorter time-span (1896–1922) there appeared to be some correlation between the sunspot number and the level of Lake Victoria, with a period of about 11.5 years. The correlation, however, did not hold for the period 1923–50, as discussed by Hurst (1952). Other hydrological correlations with the solar cycle have been proposed and debated (Section 4.4).

Two main tropical rivers are included in an analysis of periodicity by Andel & Balek (1971): the Nile and the Niger. The 1871–1954 data for the Nile reveal periods of 84, 22.6 and 7.3 years, and for the Niger (1906–1957) a 'highly significant' period ranging from 20.4 to 34.2 years was found. When comparing the length of the data records and of the periods computed, the inference is that the calculated periods are accidental. The term pseudo-periodicity was purposely used by Dyer (1979) for the level of Lake Chilwa.

Variations in rainfall have direct effects on lakes where surface or seepage outflow is a small component of the water budget compared to a relatively constant evaporation. Seasonal variations of direct rain input have been shown to be responsible for within-year variations of level in a number of lakes, such as Lake Victoria or Lake Titicaca in which river outflow accounts, respectively, for 24 and 9% of the losses. Between-year variations of level occur in these lakes as a result of one or more years of abnormal rainfall.

A decrease of 5 m has been observed in Lake Titicaca (Roche *et al.* 1992) between 1933 and 1943, and an increase of 2 m occurred between 1961 and 1964 in Lake Victoria (Piper *et al.* 1986) (Fig. 4.22). This last increase of level had considerable downstream effects along the Upper Nile (Sutcliffe 1988). It was related to two above-normal rainy years (1961 and 1963) in East Africa, where other large lakes also increased in level: Lake Albert (Mobutu) by 1.7 m, Lake Tanganyika by 1.5 m and Lake Malawi by 0.8 m (Kite 1981, 1982) (Fig. 4.23). The causes of increase in Lake Malawi have been shown to be, in decreasing order: rainfall in the catchment, catchment runoff and rainfall on the lake. For the same lake, Drayton (1984) showed that man-made changes in runoff and outflow have been comparatively unimportant for the level increase over the period 1976–80. A similar analysis has been made for Lake Toba, which covers an area of 1120 km^2 within a volcanic caldera in a rather small basin of 3650 km^2 among the northern mountains of Sumatra (Bullock 1993), and which has undergone a fall in level during the 1980s (Fig. 4.22).

Lake Malawi also underwent large changes of level at the beginning of the twentieth century. As a result of 15 years of below-normal rainfall, the water level steadily fell by about 1.7 m until 1915. By that time the outflow, the River Shire, had dried up and sand bars and vegetation growth had notably increased the sill height. With more normal climatic conditions, it took about 20 years before the lake reached a new equilibrium level about 3 m higher than the former one (Beadle 1981: see Fig. 4.22).

An extrinsic shift towards a drier climate, with its effects magnified by human activity and water use, has resulted in a change in the regime of Lake Valencia, Venezuela. This lake was last reported to have reached its outlet in the early eighteenth century. It has since been a closed (endorheic) basin, and steadily contracting for more than 200 years, with an associated increase in salinity and change in phytoplankton and other communities (Bradbury *et al.* 1981; Binford 1982).

Lake Titicaca, and large African lakes such as Victoria, Malawi, Tanganyika and Kivu, have in common a long residence time and a relatively large proportion of their inputs through direct rainfall. As a

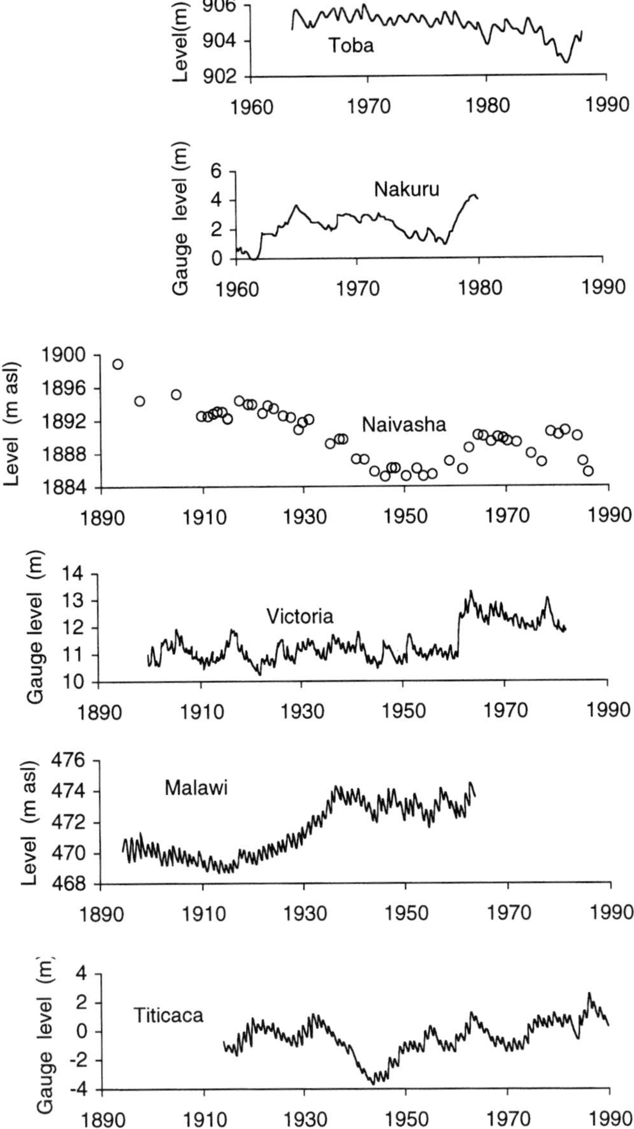

Fig. 4.22. Some long term variations in lake levels for lakes in Indonesia (Toba), Africa (Nakuru, Naivasha, Victoria, Malawi) and South America (Titicaca). Respectively based on Meigh *et al.* in Bullock (1993), Vareschi (1982), Harper (1987), Piper *et al.* (1986), Pike & Rimmington in Beadle (1981), and Roche *et al.* in Dejoux & Iltis (1992).

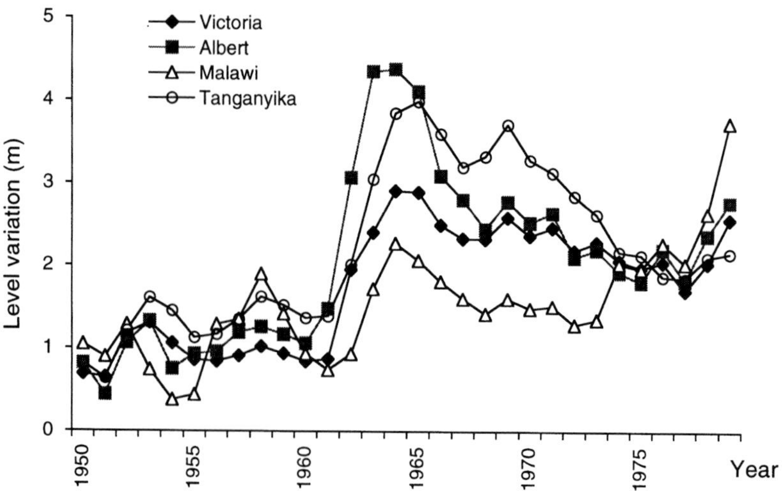

Fig. 4.23. The synchronous increase in level of some East African lakes as a result of increased rainfall in the region. Beginning-of-the-year levels are given on an arbitrary common scale. Modified from Kite (1981).

result, changes in water level of a few metres are not accompanied by other significant changes in water chemistry or other environmental variables. The situation is quite different for lakes of short residence time with small or no outflow, and shallow depth. Such lakes respond to changes in inputs by environmental, chemical and level fluctuations which are large relative to their volume or mean depth, although the correlated change in inundated area often serves to dampen the level variation.

Most tropical lakes with little or no outflow, and usually shallow, are found in arid areas and are particularly sensitive to changing water balance. They have been described as amplifier lakes by Street (1981). Documented examples for Africa are those of lakes Turkana, Chilwa, Chad, Naivasha and Nakuru, and a number of other saline lakes widespread in the arid tropics.

Lake Chad, lying in a largely closed (endorheic) basin, depends mainly on the River Chari for water input (87%) which is compensated for by evaporative loss (92%) and seepage-out through the sandy northeastern shores (8%). The lake level (or area) is a good descriptor of regional climatic fluctuations. It has varied greatly during the past century, with periods of very low level in 1904–17, around 1940 and from 1973 onwards. The 'Normal Chad', as it was in the late 1960s, had an area of about 19 000 km^2 and a surface altitude of 281.5 m. As a result of rainfall deficits since 1968 (see Fig. 4.21) the level decreased rapidly to less than 279.5 m, at which the lake split into several basins in April 1973. Since then, the northern basin of the lake has only been fed intermittently by overspills from the southern basin which is fed by the River Chari. This state of the lake has been named 'Lesser Chad' (Carmouze, Chantraine & Lemoalle 1983) (Fig. 4.24). Significant overspills to the northern basin occur only when the inflow of the Chari exceeds 15 km^3 yr^{-1}, and the whole basin is temporarily inundated if the inflow reaches 28 km^3 yr^{-1}. This has occurred only once (1988–9) between 1976 and 1996. The northern basin thus behaved as a seasonal lake with a size changing both seasonally and annually (Lemoalle 1991).

(b) Climate and stratification regime

Aperiodic climatic events may modify the heat content of the upper layer and thus the vertical structure of a lake.

Strong winds, gales or typhoons are obvious destabilizing events, but there seem to be few documented direct observations on tropical deep lakes. An exception is that of Lake Valencia, at the end of November 1977, when an increase in wind velocity was associated with a rapid and complete vertical mixing of this 36-m-deep lake, and with a fish-kill (see Fig. 5.41).

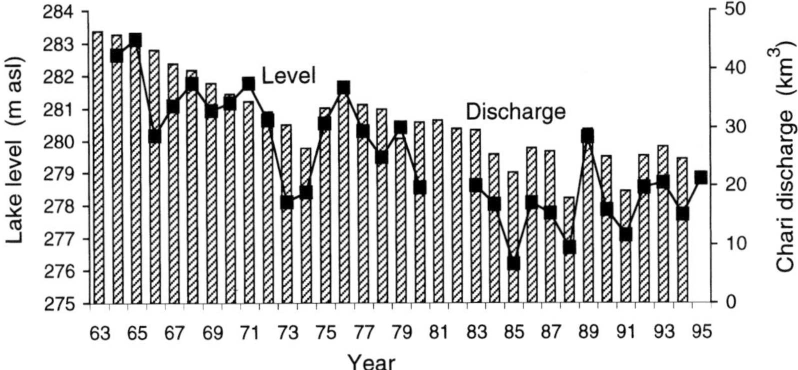

Fig. 4.24. Co-variation in the maximum annual level at Bol, Lake Chad Southern Basin (squares, in m above sea level) and annual input by the River Chari (bars, km^3). From 1973 onwards, the lake behaved as a 'Lesser Chad'. Based on Olivry *et al.* (1996).

Heavy storms, occurring in a period of lower stability, may also induce partial mixing and sudden stirring up of sufficiently anoxic water to cause mass fish-kills in the equatorial African lakes Nkugute and Bunyoni (Denny 1972). For shallow lakes, an account of selective fish-kills linked with bottom stirring during storms is given by Bénech *et al.* (1976) for Lake Chad. The South American cold spells of Antarctic air, which may suddenly lower the air temperature by about 10 °C, are also aperiodic stimuli that suddenly induce vertical mixing and fish-kills in Brazilian lakes (see Chapter 5.2h).

Further, a short series of a few climatically peculiar days may strongly modify the diurnal stratification regime of shallow lakes. Anoxia may thus develop in shallow eutrophic lakes when nocturnal cooling is lessened by a strong cloud cover (which reduces back-radiation) associated with low wind. A lack of nocturnal mixing resulting in mass fish-kills has occurred in Lake George (Ganf & Viner 1973) and in the Ebrié Lagoon (Dufour *et al.* 1994). On all these occasions, a sequence of 2–3 days was responsible for the fish mortality. Such short-term events, although with long-lasting effects, would not be detected through consideration of the mean monthly rainfall or wind-run; continuous recording or intensive monitoring is needed in accordance with the time scale of these rapid phenomena.

On an inter-annual time scale, there may be a tendency to infrequent complete mixing (oligomictic behaviour) in deep tropical lakes (Lewis 1987). Ecuadorian lakes, which lie close to zero latitude and experience little seasonal variation in irradiance, provide an example (Steinitz-Kannan *et al.* 1983).

Climatic trends, or atypical years, may alter the stability of lakes and increase their mixing depth. Although long-maintained records of temperature profiles are scarce for tropical lakes, a number of other indicators may be used.

Massive CO_2 release from crater lakes in Cameroon have been described for Lake Monoun (15 August 1984) and especially Lake Nyos (21 August 1986, causing the death of 1700 people) (Kling *et al.* 1987). In these lakes, the hypolimnion was strongly enriched with CO_2 of magmatic origin. Stable stratification prevented mixing and had allowed gas accumulation to a concentration close to the local in-depth saturation such that any upward displacement of hypolimnetic water would generate oversaturation and CO_2 gas bubbles. According to one theory (Kling *et al.* 1987), a predictable seasonal interval of reduced stability may have been enhanced by recent trends of decreasing air temperatures and insolation relative to long-term means. Another theory favours a tectonic triggering of the degassing (Freeth & Kay 1987). Oral tradition indicates that the possibility of degassing is known by populations living close to the crater lakes, although not reaching the amplitude of the Nyos catastrophe. There is thus a clear indication of aperiodic occurrence of destratification in these crater lakes.

Between-year differences in the vertical circulation of lakes usually result from either more or less pronounced cooling seasons, although differences in wind stress or deep water salinity can also be involved. The factor of variable salinity is well illustrated by a long-term study of Lake Sonachi, a Kenyan crater lake (MacIntyre & Melack 1982). The difference between three successive stratification cycles in Lake Titicaca from 1980 to 1982 has been described through the vertical distribution of O_2, NH_4^+ and NO_3^-. The deep anoxic hypolimnion decreased in thickness from 1980 to 1981 and was completely eliminated during deep-mixing only in 1982 (Vincent *et al.* 1985).

(c) Other trends and singular changes

More examples of these types, together with related changes in other lake characteristics, are described in later sections. Progressive long-term trends can be manifest in salinization (Chapter 5.1a) and in nutrient enrichment or eutrophication (Chapter 5.5b). Unique events that set off long-term changes are reservoir creation and biotic introductions or invasions (Chapter 5.5b).

Looking back at the patterns of differing period, it is a truism to say that the shorter coexist with – and are 'nested' within – the longer. The diel is always regular in length, as it is linked to the dependable factor of solar elevation. Its further possibilities depend on superimposed environmental cycles (e.g., of wind) with different diel phasing and the presence of biological systems of sufficiently rapid response. The long-term are the least regular, with aspects of water balance and of singular events most strongly represented in the tropics. The intermediate lunar period has impact limited by the weak forces or energy fluxes involved, although it is a cue of considerable influence in tropical aquatic biology. The annual period is, overall, less muted in tropical freshwaters than might be anticipated from the reduced amplitude of variation in the main driving variable, daily solar radiation income. It is particularly liable to be influenced by hydrological events of water input.

5

Reactive components of time-variability

So far our survey of tropical time-variability has centred upon primary controls by physical factors and the resulting spectrum of periodic behaviour. We now turn to responsive or *entrained* components that are chemical and biological in nature. Especially in biological communities, many new response mechanisms are *initiated.* Over very long periods there is evolutionary change. Components are first considered individually, later as making up *systems* of interactive behaviour.

5.1 Chemical components

The time-variations of chemical components are variously imposed by the radiation and hydrological factor-complexes, plus the effects of changing biological activity. All these can generate cycles of chemical flux and concentration, whose timing and amplitude may be set by distinctive tropical circumstances. Thus a physical stratification cycle, regular or irregular in occurrence, typically leads to a vertical chemical layering and its periodic destruction. Pronounced cycles of net water input and water loss, that often relate to rainy and dry seasons, are unevenly accompanied by cycles of solutes and particulates. These cycles are especially pronounced in rivers.

Chemical variability is usually known from time-series of *concentrations* in unit water volume; such provide most of the examples illustrated below. However, interpretation may be better served in lakes by considering *contents* per unit area, as of a water-column with sediment or surface-related inputs (e.g., photosynthesis: Talling 1957*a*; Talling *et al.* 1973), or of a complete water-body (e.g., Lake Pawlo, Ethiopia: Wood, Baxter & Prosser 1984) for which budget-terms may be required (e.g., Lake Chad: Carmouze 1983; Lake Turkana: Yuretich & Cerling 1983). A budget-scale understanding requires knowledge of chemical *fluxes*, at least as boundary inputs and outputs.

Time-variability in concentrations will reflect that of boundary chemical transfers (inflow–outflow, air–water, sediment–water), internal transfers (e.g., nutrient uptake or release) and the changing extent of compartment zones (e.g., hypolimnia). All these, and especially the first, can be altered by change in the regional water balance. There may be passive effects of water balance independently of chemical fluxes: dilution by flood water input and evaporative concentration. Time-change of temperature also has implications for rates of chemical reactions and of constituent concentrations linked by equilibria, as in the CO_2-system.

For purposes of general survey, it is convenient to distinguish between major ionic constituents (normally Na^+, K^+, Ca^{2+}, Mg^{2-}; HCO_3^-, Cl^-, SO_4^{2-}), major plant nutrients that are available forms of the elements N, P and Si; and the metabolically involved gases O_2 and CO_2. These three categories typically form a series with increasing susceptibility to short-term biological modification of environmental concentrations. In periodic reducing environments, also generally induced by biological activity, a further group of reduced chemical species can accumulate (e.g., Fe^{2+}, Mn^{2+}, CH_4, H_2S). Finally, as a product of biological activity, there is added organic matter in dissolved and particulate forms.

(a) Major ionic constituents

Listed above, these may show no significant within-year variation of concentration in the surface water of a lake with long retention time. This can also be favoured by a similarity between amounts of evaporative loss and on-lake rainfall, as for offshore water of Lake Victoria, where variation with depth is also minimal (Talling & Talling 1965; Talling 1966). Even with more prolonged stratification, and significant depth-differences, the within-year variation in surface water can remain small (e.g., Lake Tanganyika: Coulter 1991*a*). However, it can become considerable in lakes with high and seasonal net loss of water by evaporation (e.g., Lake Ihotry, Madagascar: Moreau 1982), and in those of relatively short retention and with a seasonally large inflow of dissimilar (usually lower) concentration. Examples are provided by some floodplain lakes of the Amazon (Schmidt 1972, 1973*a*; Furch 1982, 1984; Furch *et al.* 1983) and Orinoco (Hamilton & Lewis 1987, 1990); that of Lago Camaleão is shown in Fig. 5.1. Such seasonal impact is greater if the lake water has

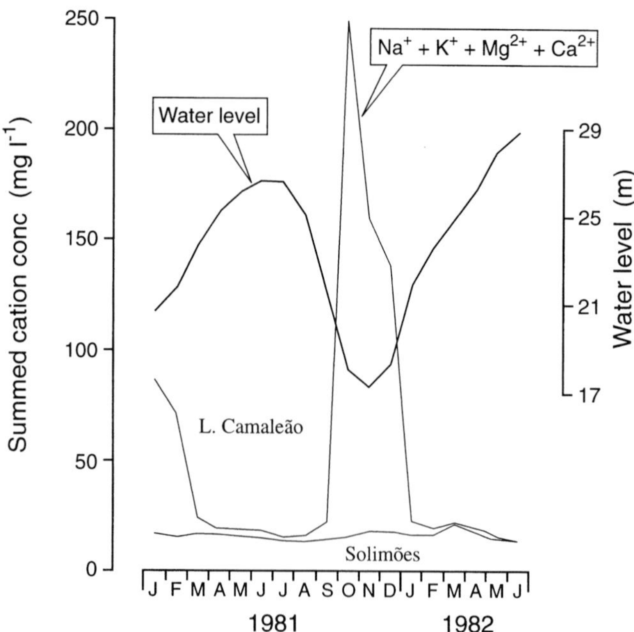

Fig. 5.1. Lago Camaleão, Amazonia, 1981–2. Seasonal variability in the summed content of major cations in relation to corresponding variability in the adjacent Rio Solimões (Amazon) and to water level. From Furch (1984).

had a history of evaporative concentration, as in the large African lakes of Chad (Gac 1980; Carmouze, Chantraine & Lemoalle 1983) and Turkana (Ferguson & Harbott 1982). In both these lakes the seasonal floodwater input leads to changing horizontal patterns of ionic concentration.

More generally, shallow lakes of the semi-arid tropics show bulk-concentration changes in time, within and between years, that are induced by periods of greater evaporative concentration alternating with those of greater water income. The ranges of concentration, and the biological consequences, are particularly large in closed saline lakes such as Nakuru (Fig. 5.21), Elmenteita and Bogoria in the Rift Valley of East Africa (Tuite 1981; Vareschi 1982; Melack 1988). A similar alternation with concentration changes can occur in floodplain lakes and pools that are isolated at low level. Examples are the savanna ‘pans’ of changing alkalinity in southern Africa (Weir 1968), and water-bodies in floodplains such as Lake Murray in Papua New Guinea (Osborne *et al.* 1987) and wetlands of the Pantanal, Brazil (Heckman 1994). In some lakes, such as the large but shallow Lake Chapala in Mexico, removal of water for irrigation is a major factor (Limón *et al.* 1989). In coastal lakes with marine connection (e.g., Lake Maracaibo, Venezuela: Gessner 1956; Redfield & Doe 1964; Lake Songkla, Thailand: Limpadanai & Brahamanonda 1978; Ebrié, Ivory Coast: Fig. 5.2) a varying salinity can develop in relation to tidal and mean sea levels on one side and to varying land runoff on the other. Exceptionally, evaporative concentration can also lead to periodic conditions of high salinity, even above that of seawater, in a river where marine ingress is involved. Seasonal examples have been described from the Casamance and Saloum rivers in West Africa (Pagès *et al.* 1987; Pagès & Debenay 1987; Pagès & Citeau 1990), where a maximum of salinity developed further upstream and changed position with season.

Flowing waters with rapid renewal are more likely than lakes to show short-period fluctuations of ionic (and silt) content. Strong seasonal changes follow marked seasonal rainfall in many tropical regions. Other well-studied examples include the Amazon (Stallard & Edmond 1981; Furch 1982; Devol *et al.* 1995), Orinoco (Lewis & Saunders 1989), Caura (Lewis *et al.* 1987) and Gambia (Lesack *et al.* 1984) rivers and – on a smaller scale of discharge – tributaries of the Rio Tempisque in Costa Rica (Newbold *et al.* 1995). High flows after rain are usually marked by reduction of total ionic content (and hence conductivity) plus increase in silt content (and turbidity). These features are illustrated by a four-year record with fine daily resolution for a small river in Kenya (van Someren 1962), subject to two rainy seasons each year; also by seasonal studies of the mouth of the Mwenda River into Lake Kariba (King & Lee 1974; Bowmaker 1976; King & Thomas 1985). Time-variability is increased in sectors below the confluence of chemically dissimilar tributaries with variable discharge and reduced in sectors fed predominantly from a large lake or reservoir. Exceptionally, a variable lake contribution may result from changes of flow between inflow and outflow when these are close together. Thus on the upper White Nile below the salt-rich Lake Albert there is evidence for travelling short-term pulses of water of higher conductivity (Beauchamp 1956; Talling 1957*c*; Prosser 1987).

(b) Major plant nutrients

These supply the elements C, N, P and (for diatoms) Si, and are influenced by phases of net biological depletion followed by net replenishment in tropical lakes and rivers – although few locations have been studied in

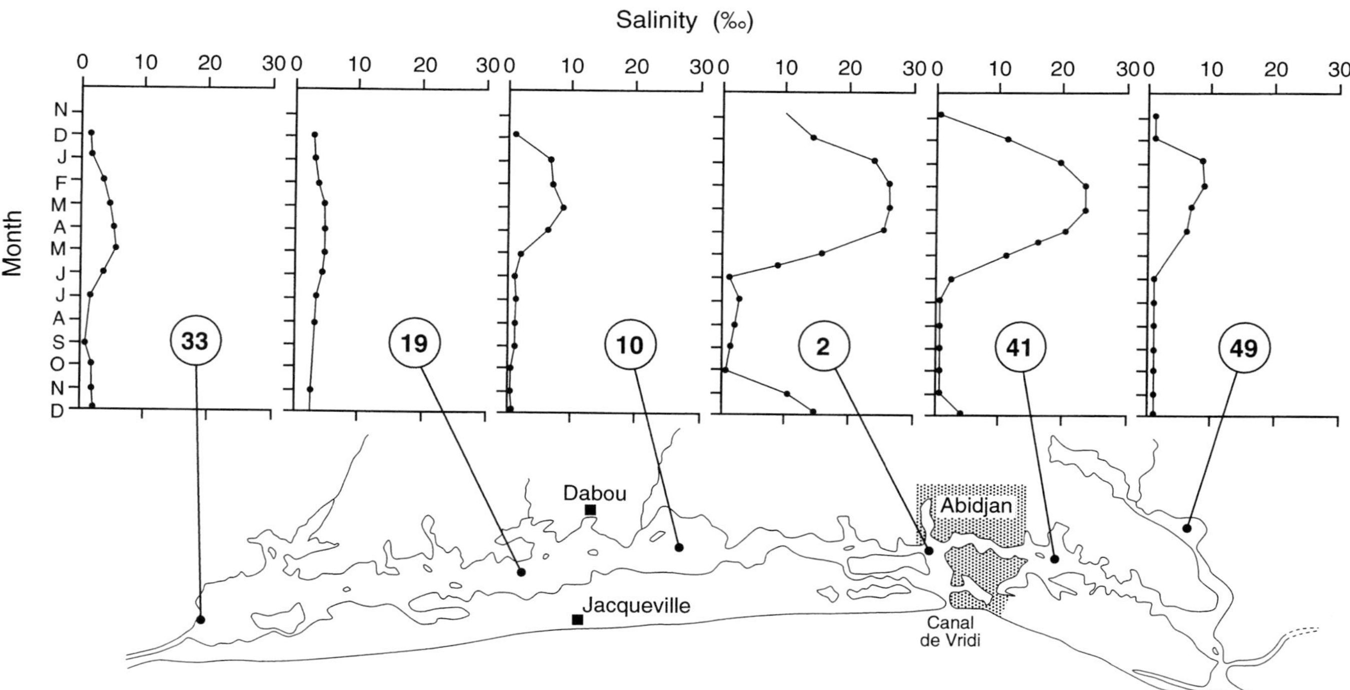

Fig. 5.2. Ebrié Lagoon, Ivory Coast. Seasonal variation of salinity during 1975, measured at six stations whose locations are shown below. From Pagès *et al.*, after Durand & Chantraine (1982).

detail. The time scale is mainly annual, although in productive lakes the high fluxes of carbon, nitrogen and phosphorus combined with diel cycles of stratification plus physiological activity can lead to diel fluctuations of concentrations. Increases of inorganic N and P after nocturnal mixing are known from Lake George in Uganda (Viner 1973) and, for P, increases with phosphatase activity during the afternoon in the lake Parakrama Samudra in Sri Lanka (Gunatilaka & Senaratna 1981; Gunatilaka 1983, 1984). Also relevant is a seasonality in the contributions of nutrients from atmospheric precipitation, wet and dry, that is likely to be pronounced in many tropical localities (as for Lake Valencia: Lewis 1981); and the possible role of forested catchments as 'buffers' in reducing the variability of nutrient (N, P) concentrations that are susceptible to short-term change of discharge (Lewis 1986*b*).

Examples of Si depletion with abundance of the main abstractors, diatoms, are described from the White Nile (Prowse & Talling 1958), Lake Victoria (Talling 1966), Lake Kainji (Adeniji 1977) and Lake Chad (Lemoalle 1978: see Fig. 5.3). For Lake Victoria there is also recent evidence for a long-term depletion over the last two decades, accompanying increased phytoplankton production (Hecky 1993: see Fig. 5.55).

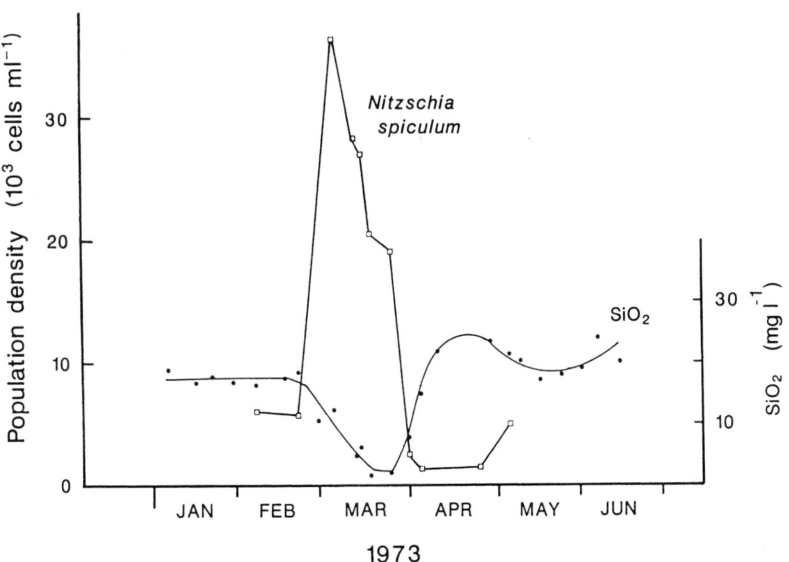

Fig. 5.3. Lake Chad. Large depletion of soluble reactive silicon (expressed as SiO_2) from lake water during an episode of diatom growth (*Nitzschia spiculum*) at the station Bol. From Lemoalle (1978).

The circumstances controlling the depletion of nitrate are generally difficult to delimit in tropical freshwaters. Compared to most temperate freshwaters, prevalent concentrations are usually low. Plant growth may mainly utilize the alternative source of ammonium-nitrogen (e.g., Fisher *et al.* 1988; Morrissey & Fisher 1988) or even N_2-fixation whose seasonality is linked to that of heterocystous blue-green algae (Wurtsbaugh *et al.* 1985; Moyo 1991, 1997). Bacterial denitrification is probably influential (Viner 1982*b*; Vincent *et al.* 1985; Kern *et al.* 1996) as well as nitrification. Perhaps the most conspicuous pulses of nitrate occur in river floodwater, as illustrated by seasonal or spatial studies of the Blue Nile (Talling & Rzóska 1967; Sinada & Abdel Karim 1984*a*), the Ganges (Lakshminarayana 1965*a*), a floodplain tributary of the Amazon (Lesack 1993*a*) and a tributary of the Paraná River (Pedrozo & Bonetto 1987). However, mainstem regions of the Orinoco and Amazon (Solimões) showed lower concentrations at higher discharge (Lewis & Saunders 1989; Devol *et al.* 1995). Phased transfer from surrounding soils will be promoted by the increase of nitrate content widely found in tropical savanna soils early in a rainy season. There can also be a nitrate-flush from re-wetted marginal soils of swamps or lakes after a dry phase (e.g., at Lake Chilwa: Howard-Williams 1972, 1979*a*). In at least some stratifying tropical lakes (e.g., Lake Valencia: Lewis 1986*a*; Lake Titicaca: Vincent *et al.* 1984, 1985) nitrate increased in surface water during and following phases of vertical mixing (Fig. 5.4). Its derivation may then be from accumulations previously built up in deep- or mid-water, or by transformations of ammonium-nitrogen similarly accumulated. In Lake Victoria and Lake Malawi the supply of inorganic nitrogen from below was probably critical for the production of phytoplankton, but rapid uptake apparently led to a state of almost year-long depletion in the upper and more populated zone (Talling 1966; Patterson & Kachinjika 1993, 1995: see Fig. 2.37). In Lake Titicaca nitrate depletion after re-stratification was related to a well-established limitation of phytoplankton growth by inorganic nitrogen (Vincent *et al.* 1984; Wurtsbaugh *et al.* 1985), a feature often suspected from less rigorous evidence in other tropical freshwaters (see Chapter 3.2).

In the water-column of Lake Tanganyika, as interpreted by Edmond *et al.* (1993), the *ultimate* supply of N from above is currently in balance with that of P from below, although this balance would be sensitive to climatic change. However, a significant atmospheric source of phosphorus cannot be generally discounted (Lewis 1981; Lewis *et al.* 1987, 1990); even such a low concentration as $10\,\mu g\,l^{-1}$ could locally dominate

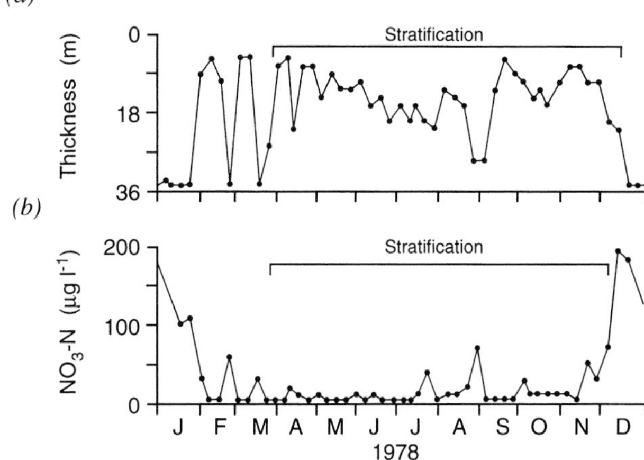

Fig. 5.4. Lake Valencia, Venezuela, 1978. Within-year variability in (*a*) the thickness of the upper mixed layer, (*b*) the average concentration of NO_3-N in the 0–5 m layer, with durations of thermal stratification indicated. From Lewis (1986*a*).

input (Lewis *et al.* 1995). The existence of internal waves of large amplitude in lakes Tanganyika and Malawi probably ensures a discontinuous or 'pulsed' transfer of nutrients from below (Bootsma 1993*a*, *b*; Plisnier *et al.* 1995).

The time-courses of phytoplankton abundance and concentrations of soluble reactive phosphorus are often (as elsewhere) inversely related, and may reflect a redistribution of the element. Within such redistribution, a unidirectional time-course from solution to particulates has been followed by radio-tracer additions to water samples from some Kenyan lakes (Peters & MacIntyre 1976; Kalff 1983) and Lake Titicaca (Wurtsbaugh *et al.* 1992). Analyses of total phosphorus content are highly desirable; few existed before the work of Ida Talling in 1960–61 (Talling & Talling 1965). Like nitrate-nitrogen, concentrations of phosphate are often relatively high in river floodwater; appreciable quantities may be carried adsorbed on suspended silt (Viner 1982*a*; Grobbelaar 1983). Floodplains, such as those of the Amazon and Orinoco rivers, can show an annual cycle within which much phosphorus and nitrogen are transferred from the river at a high level by overspill to lateral standing waters within which plant growth induces progressive nutrient stripping (Fisher 1979; Fisher & Parsley 1979; Hamilton & Lewis 1987; Furch & Junk 1993). Within a river channel itself, the concentrations of different chemical components typically bear varied relationships to discharge (e.g., Saunders & Lewis 1988*d*, 1989*a*, for a Venezuelan river) and to rising or falling levels. Thus negative, positive and near-neutral relationships to discharge can be distinguished. Time-courses may have rising levels with bias towards high concentrations ('clockwise hysteresis loop': e.g., particulates), or falling levels with such positive bias ('anti-clockwise hysteresis loop': e.g., dissolved organic carbon). These were well-marked in an annual study of the Gambia River, West Africa, by Lesack *et al.* (1984) (see Fig. 5.5). Differing time-courses can be expected in rivers with differing degrees of inundation in the drainage basis (Saunders & Lewis 1988*d*).

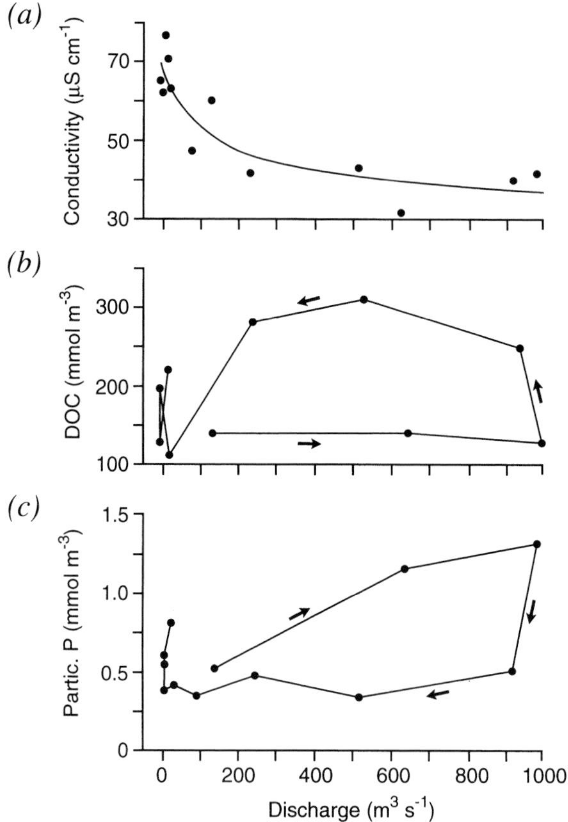

Fig. 5.5. Gambia River, West Africa. Dependence upon water discharge of (*a*) conductivity, (*b*) dissolved organic carbon, with time relation as anti-clockwise hysteresis loop, (*c*) particulate phosphorus, with clockwise hysteresis loop. Modified from Lesack *et al.* (1984).

Increased nutrient inputs over long periods lead to trends of enrichment or *eutrophication*; there are few well-studied examples from the tropics (Section 5.5b).

(c) The gases O_2 and CO_2

These undergo variations of concentration with time that have some unique characteristics. Involvement in photosynthetic metabolism leads to light-induced diel cycles that entrain other components of the CO_2-system – concentrations of HCO_3^- and CO_3^{2-}, and pH. There is also interaction with diel and annual cycles of temperature/density stratification, leading to phases of vertical compartmentation.

The amplitude of such diel O_2 cycles in warm tropical waters rich in phytoplankton of high photosynthetic capacity, and subject to diel temperature/density stratification, such as the African crater lakes of Aranguadi (Talling *et al.* 1973; see Fig. 4.2) and Simbi (Melack 1979*b*), is rarely if ever surpassed in other natural waters, weed-beds excepted. Strong diel changes (and the sampling time) should be borne in mind when interpreting less frequent sampling of supposed seasonal changes. Diel cycles can include complete or near-complete nocturnal anoxia (Baxter *et al.* 1965; Talling *et al.* 1973; see Fig. 4.2), an uncommon condition that also developed from decaying organic matter in the newly forming Brokopondo Reservoir of Suriname (van der Heide 1978; see Fig. 5.6).

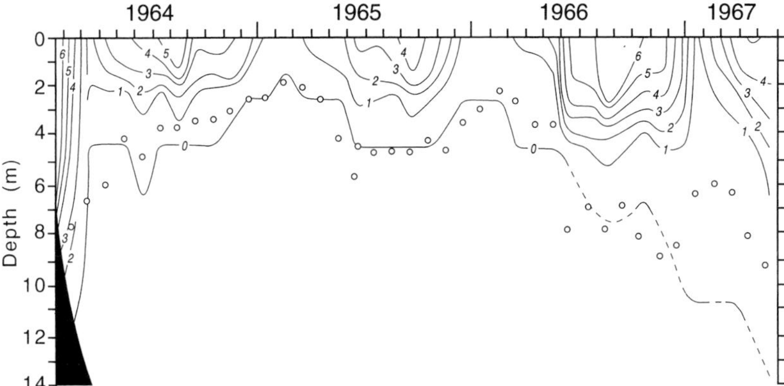

Fig. 5.6. Lake Brokopondo, Surinam. Depth-time distribution of dissolved oxygen during successive early years after filling, indicated by contours in mg l^{-1}. Open circles show the depth of the illuminated, euphotic zone. From van der Heide (1982).

Less extreme but strong diel cycles are described by Marzolf & Saunders (1984) from ponds in southern India. Further analyses of rates of gaseous change per unit area can give estimates of photosynthetic production per unit area (Chapter 3.1e).

The sensitivity of rates of O_2 consumption (respiration, decomposition) to temperature, with the tropical characteristic of warm hypolimnia, promotes the development of deep O_2 depletion in productive waters well supplied with organic substrates. Examples with seasonal anoxia have been described from water-bodies over a range of latitudes. We can instance the Higuanojo Reservoir in Cuba, 22° N (Fig. 5.7); Lake Pawlo, Ethiopia, 9° N (Wood, Baxter & Prosser 1984); Lake Awasa, Ethiopia, 7° N (Gebre-Mariam & Taylor 1989*a*); Lake Kariba, Zimbabwe-Zambia, 17° S (Fig. 5.32); Solomon Dam, Australia, 19° S (Hawkins 1985). In older records from Lake Victoria (Fish 1957; Talling 1966), anoxia was of limited deep occurrence in offshore water. At a northern station the vertical distribution of O_2 concentration showed a sequence of three seasonal phases, which were conditioned by corresponding phases of the temperature/density stratification and correlated with patterns of the CO_2-controlled pH stratification.

The organic substrate may be partly of external terrestrial origin, as in Amazonian floodplain lakes that develop a seasonal deep anoxia during the phase of higher water level (Schmidt 1972, 1973*a*; Rai & Hill 1982*b*; Melack & Fisher 1983, 1990; Tundisi *et al.* 1984; MacIntyre & Melack 1988). The rates of consumption may show trends with time on the diel, annual and inter-annual time scales. In the shallow Lake George, for example, Ganf (1974*a*) found that specific rates per unit quantity of phytoplankton varied systematically with diel time and depth (Fig. 3.10). In the long-term records from Lake Kariba (Fig. 5.32), seasonal anoxia of deep water developed most rapidly and completely during the earlier years when plankton was most abundant and flooded terrestrial vegetation (see in Fig. 5.13) was newly available for decomposition (Harding 1964; Coche 1974). Similar long-term trends have been observed in other tropical man-made lakes (e.g., Lake Volta in Ghana, Lake Brokopondo in Suriname, Samuel Reservoir in Brazil) after first filling (McLachlan 1974; van der Heide 1978, 1982; see Fig. 5.6; Matsumura-Tundisi *et al.* 1991).

Although the depletion of O_2 in deep water occurs in a regular seasonal manner in some tropical lakes, depletions of apparently irregular incidence are not uncommon and contribute to the phenomenon of 'fish-kills' (Section 5.2h). Their causes are varied, but probably include the excep-

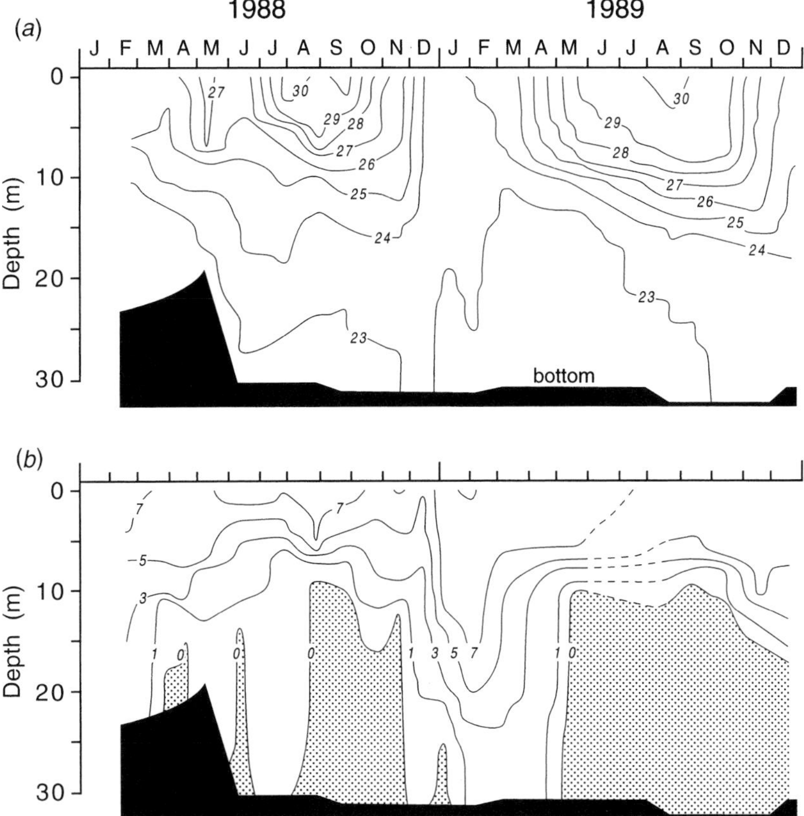

Fig. 5.7. The annual cycle of stratification in Higuanojo Reservoir, Cuba, as depth-time diagrams over two years that show (*a*) temperature, contours in °C, (*b*) dissolved oxygen, contours in $mg\,l^{-1}$. Cooler conditions around the winter solstice do not preclude some residual stratification. From Laiz *et al.* (1993*b*).

tional extension of daytime stratification under prolonged calm weather (e.g., Lake George: Ganf & Viner 1973), the lateral transfer (advection) of pre-existing deoxygenated water at depth (e.g, Lake Albert: Eccles 1976), vertical mixing with upward transport of O_2-depleted but H_2S-rich water (e.g., Lake Valencia: Infante *et al.* 1979) and decomposition associated with periodic phytoplankton blooms (e.g., Lake Victoria: Ochumba & Kibaara 1989). Still poorly documented are the possible biological implications of CO_2 accumulated in deep layers or of elevated pH in surface layers induced by photosynthetic consumption of CO_2. Dunn (1967) noted that a varied zooplankton survived diurnal episodes of a surface pH rise of two units in fishponds at Malacca bearing dense phytoplankton, but not a later mass death of the bloom. As mentioned in Chapter 4.5, an exceptionally large accumulation of CO_2 of deep-seated (magmatic) and non-biological origin in the crater lake Nyos (Cameroon) was followed by catastrophic de-gassing on 21 August 1986 and mortality in human settlements nearby. Although pre-existing information is inadequate, the interruption of a long-persistent density stratification can be deduced (Kling 1982, 1988; Kling *et al.* 1987, 1989), completed by energy from buoyancy resulting from the de-gassing.

(d) Reduced chemical species

Oxygen depletion, periodic or aperiodic, is universally influential for the formation and accumulation of reduced chemical species such as CH_4 (see Chapter 2.4a vii), H_2S, NH_4^+, Fe^{2+} and Mn^{2+} ions. In lakes, deep accumulations typically mark prolonged temperature/density stratification. They are likely to be dispersed or oxidatively eliminated by extensive vertical mixing, so that an annual periodicity of stratification – common in tropical lakes – is likely to be imposed on the reduced species. There are relatively few tropical examples described with depth-time resolution over several years, although shorter series are not uncommon (e.g., for Lake Victoria: Talling 1966). One two-year and another 2.5-year series from reservoirs in tropical Australia illustrates the seasonal accumulation of Fe^{2+} in the O_2-depleted hypolimnion (Hawkins 1985; Townsend 1995). Two long series from Africa that include H_2S (or sulphide) are especially notable. That from Lake Kariba (see Harding 1964; Coche 1974; McLachlan 1974) showed a strong accumulation of sulphide in the deep hypolimnion of the early productive phase (with incidental corrosion of turbine metal at the dam), accompanied by CH_4, but lessening in later years (Fig. 5.32). The other and uniquely intensive study is from Lake Pawlo, a crater lake in Ethiopia (Wood, Baxter & Prosser 1984). Here the annual occurrence of sulphide was somewhat variable, corresponding to the varying extent of annual mixing and thermocline formation. In Amazon floodplain lakes a seasonal stratification, linked to changes in water level, is often accompanied by deep oxygenation with accumulation of H_2S, as in the example from Lago do Castanho (Santos 1973) shown in Fig. 5.8.

Numerous tropical lakes are indefinitely stratified or meromictic. Their anaerobic lower layers are typically rich in the reduced species listed above, with a vertical extent that is governed by changes in the depths

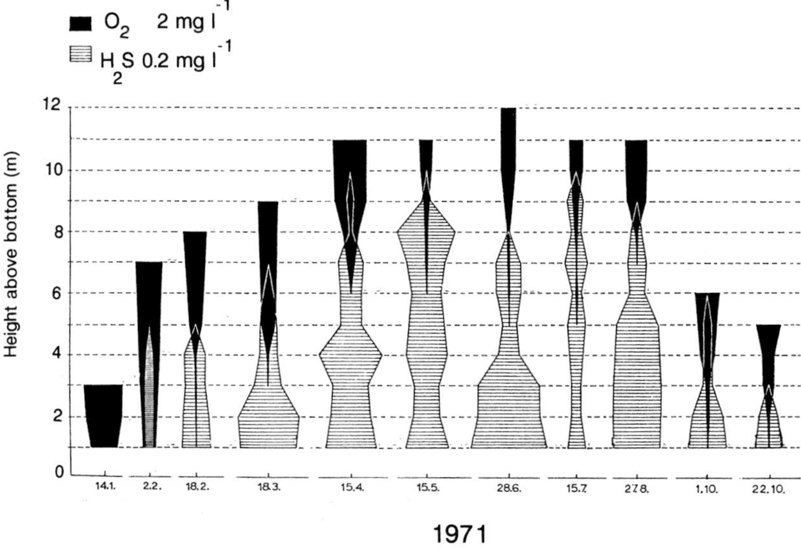

Fig. 5.8. Lago do Castanho, Amazonia. Vertical profiles of concentrations of oxygen and hydrogen sulphide in relation to the seasonal rise and fall of water level and of temperature stratification. From Santos (1973).

of the upper mixed layer and lower persistent layer or monimolimnion. Examples include the African lakes of Tanganyika (Hecky *et al.* 1991) and Sonachi (MacIntyre & Melack 1982).

Reducing conditions are also widespread in waters below swamp vegetation. Here within-year changes are likely to depend mainly on local hydrological conditions, as described by Gaudet (1979*b*) for a swamp fringing Lake Naivasha in Kenya.

(e) Organic matter

Dissolved organic matter, usually reported as dissolved organic carbon (DOC), is universally present in freshwaters, generally at carbon concentrations in the range 2–10 mg l^{-1} (Chapter 2.4a ii). Its time-variability is not well known for tropical lakes, but has attracted attention for some large tropical rivers – including the Amazon (Richey *et al.* 1980, 1990) and in other rivers of the 'black-water' type of low ionic content draining forested catchments, such as the Rio Negro, Orinoco and Zaïre. The variation of concentration with time relates to that of discharge, often with a time-lag and suggestive of purging, as described for the rivers

Orinoco (Saunders & Lewis 1988*d*; Lewis & Saunders 1989; Paolini 1991, 1994) and Gambia (Lesack *et al.* 1984). However, the amplitude of variability is typically not large considering the strong environmental changes involved, indicating a degree of stabilization with respect to runoff (Schlesinger & Melack 1981).

5.2 Biological components

(a) General

The compatibility of particular modes of biological response to environmental variability depends upon the innate response time compared to the frequency of the external change. Vegetative or somatic growth is a fundamental characteristic, that in organisms which are microbial or with repeated structural units (modular) can produce relatively rapid response which is directly expressed in demographic change of population numbers. In most higher organisms the generation length or duration of a life history sets lower time-limits for demographic response. Here the adverse growth or survival prospects of the higher-latitude winter make the distinction between a generation length of more or less than one year important, as are various strategies for reproduction and feeding involving diapause, resistant reproductive bodies, and migrations. Tropical equivalents of this stoppage or low-survival season, if present, generally have other environmental origins; further, the prevalence of temperature as a master-factor for potential growth rate and stage-duration opens possibilities for higher frequencies of cyclic response. At all latitudes possibilities of rapid response are generally high for mobile behaviour patterns, as exemplified in diel feeding behaviour, and vertical migration of zooplankton and fishes. There are further possibilities for endogenous rhythms of biological activity, mainly around daily (circadian) or annual (circannual) periodicities.

Another type of biological variability concerns the changing frequencies of different structural forms in a population. This feature is represented in the zooplankton Cladocera, as with the seasonal occurrence of parthenogenetic and sexual females, and of form-variations (e.g., helmet form) that constitute **cyclomorphosis**. The latter phenomenon, possibly induced by predation, is apparently infrequent and little recorded from tropical freshwaters even though helmet-dimorphism is known (e.g., in *Ceriodaphnia* and *Daphnia*: Rzóska 1956; Green 1967; Robinson & Robinson 1971; Zaret 1969; 1972*a*, *b*; Infante 1982). There is a report by Arcifa-Zago (1976) of cyclomorphosis from a Brazilian reservoir; Egborge & Ogbekene (1986) provide another for a rotifer in a Nigerian reservoir (Fig. 5.9). An expression that involved seasonal variation of animal size is described by Masundire (1991) for the cladoceran *Bosmina longirostris* in Lake Kariba.

If biological variation is analysed in terms of a stock or population quantity subject to inputs and outputs (losses), treatment can be mainly demographic (as population dynamics) or biogeochemical (as chemical dynamics). Linkage between the two aspects has been explored in only a few tropical freshwater systems; examples appear below in relation to the macrophyte papyrus (Zaïre), phytoplankton (lakes George, Chad, Victoria and Titicaca), zooplankton (Lake George, Lake Chad) and zoo-

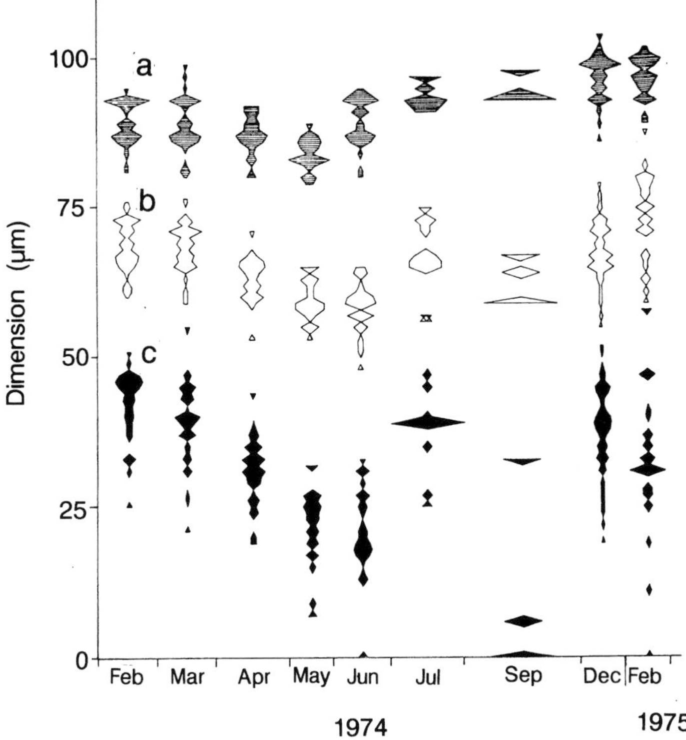

Fig. 5.9. Seasonal variation in three measures of form (cyclomorphosis) for the rotifer *Keratella tropica* in Lake Asejire, Nigeria. The dimensions, in μm, were lengths of (*a*) lorica, (*b*) right posterior spine, (*c*) left posterior spine. Modified from Egborge & Ogbekene (1986).

benthos (Lake Chad). Stock inputs are dominated by nutrient/food incorporation and demographic recruitment; outputs by biological interchanges (grazing, predation, competition) and mortality from environmental stress, of regular or irregular incidence. The environmental *distribution* of a population, rather than its size, can vary cyclically in time as a result of passive transport or volumetric dilution (e.g., plankton), active migration (e.g., zooplankton, fishes), water/air phase alternation (insect emergence) and social aggregation patterns.

In natural communities, many time-relationships are determined by interactions between the component organisms as well as by direct responses to the abiotic environment. The interaction can minimize competition, as between related species that occupy niches which differ with respect to diel or seasonal time; it can also result from prey–predator relationships that involve time-dependent exposures to predation.

Illustrative examples can be drawn from the relatively well-studied behaviour of freshwater fishes. Niche differences between numerous closely related species of endemic haplochromine cichlids in Lake Victoria can involve spawning times and depth/time aspects of foraging for zooplankton prey (Goldschmidt & Witte 1990; Goldschmidt *et al.* 1990). Although some haplochromine species in both lakes Victoria and Malawi breed continuously, there is a gradation to breeding seasonality at discrete and different times for others in these lakes (D.S.C. Lewis 1981). There are uses of space and hiding places to escape larger predators that hunt at different times of day or night. During the wave of migration in seasonal rivers and floodplains, a sorting out of species in time-succession often occurs (Welcomme 1985), typically linked to time-differences in spawning and juvenile stages.

In temperate freshwaters there is normally periodic interruption of recruitment to animal and plant populations, so that discrete cohorts or year-classes can be recognized in population structure. This is also true of markedly seasonal tropical freshwaters. Most described examples are from fish populations repeatedly sampled for length-frequency distribution, as from the El Beid River near Lake Chad (Fig. 5.10), Lake Malawi (Fig. 5.33) and Lake Valencia (Fig. 5.36). Benthic invertebrates provide other instances, especially larval stages of insects with recruitment determined by phased oviposition, as in the lunar cycles of Lake Victoria (MacDonald 1956; see Fig. 5.11) and the seasonal cycles in a West African stream (Hynes 1975*a*).

For the less seasonal tropics a potential for continuous, unphased reproduction and survival might be anticipated, yielding a population

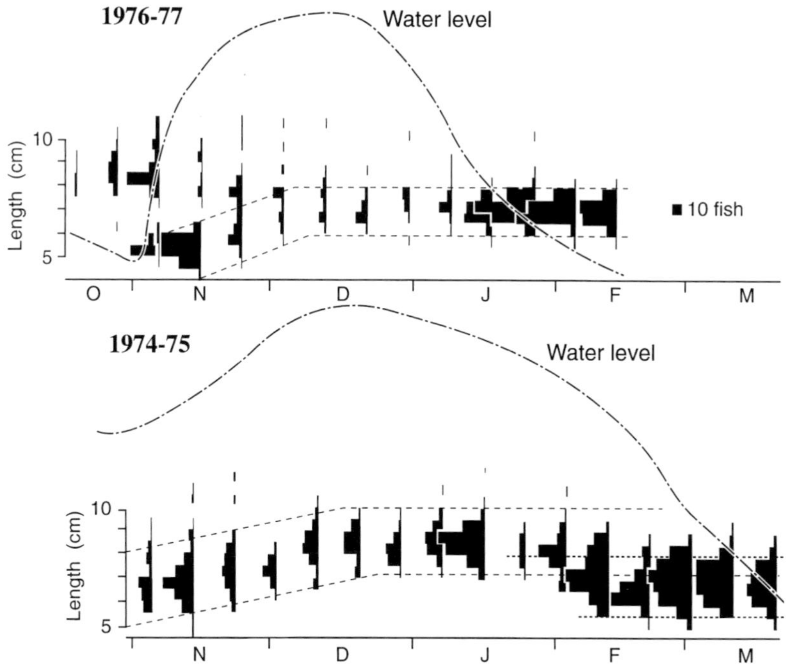

Fig. 5.10. Development over time of one (1977) or two (1975) cohorts – indicated by length-frequency distributions – of the fish *Siluranodon auritus* according to flood magnitude of the El Beid River inflow near Lake Chad. Cohorts are framed by dashed or dotted lines. From Bénech & Quensière (1985).

structure without pronounced discontinuities of age-frequency. This was apparently true of the zooplankton (Burgis 1971), zoobenthos (McGowan 1974; Darlington 1977) and some cichlid fishes (Gwahaba 1978) of the equatorial Lake George. For planktonic copepods worldwide, an almost unchanging or 'stationary' age-frequency distribution of developmental stages has been very rarely recorded in freshwaters; Wyngaard *et al.* (1982) cite only examples from Lake George and the subtropical lakes of Kinneret (Israel) and Thonotosassa (Florida). Continuous reproduction is also shown by a minority of fishes in some large tropical lakes that stratify seasonally but where the within-year ranges of water level and temperature are low. Examples include some cichlids of Lake Victoria and clupeids of Lake Tanganyika (Lowe-McConnell 1987), and the cyprinodont *Orestias agassii* of Lake Titicaca (Loubens & Sarmiento 1985). In a rare analysis of age structure for a species of tropical swamp vegetation, Thompson *et al.* (1979)

showed that this varied considerably for stands of papyrus (*Cyperus papyrus*) in African sites of differing seasonality, even though the resultant biomass density was virtually identical.

If discontinuous and phased reproduction and recruitment is combined with high fecundity or propagule formation, time-oscillations of population size will follow for non-environmental reasons. As high fecundity is one of the supposed characteristics of species with *r*-selected reproductive strategy (Ssentongo 1988 gives examples for African fishes), and survival at high density a characteristic of *K*-selected species, the *r–K* spectrum of 'strategy' has some relationship to the amplitude of temporal population change.

The long-term stability of fluctuating populations implies some degree of tolerable density-dependent, negative feedback at high densities and some assurance of recovery – rather than extinction – from low densities. Unfortunately most population-time studies on tropical freshwaters do not resolve the periodic minima, nor enable the relative or logarithmic range (Talling 1986; Kebede & Belay 1994) to be assessed. The recovery phase may owe much to resting stages or immigration – including insect oviposition and water-borne transfer (advection) of plankton – as well as the re-growth and reproduction of surviving individuals.

As is widely recognized, adaptation during evolution has allowed communities to survive and cyclically re-establish after environmental changes of great magnitude but regular incidence. Those of irregular incidence can be more damaging. Examples are provided, respectively, by the floodpulse of the central Amazon, level range approximately 10 m, and the de-stratification with fish-kill in *várzea* lakes of the same region induced by travelling cold fronts.

Some features of the time-variability of populations relate to forms of life history. Successive stages in a lengthy and differentiated life history often have different environmental requirements and tolerances. Compatibility of the stage-succession with environmental (abiotic + biotic) time-sequences is particularly important in variable environments; tropical floodplains provide examples. Aids to such compatibility include a response to well-correlated proximate factors as 'cues', insertion of a resting stage (diapause), differentiation of storage organs or stages and restriction of most active growth to specific stages. In insects with aquatic larvae and flying adults, the emergence event is often synchronized between individuals on the diel and – sometimes – lunar time scales, and may induce a synchrony of origin for the next generation or cohort. An example for benthic chaoborids of Lake Victoria appears in Fig. 5.11.

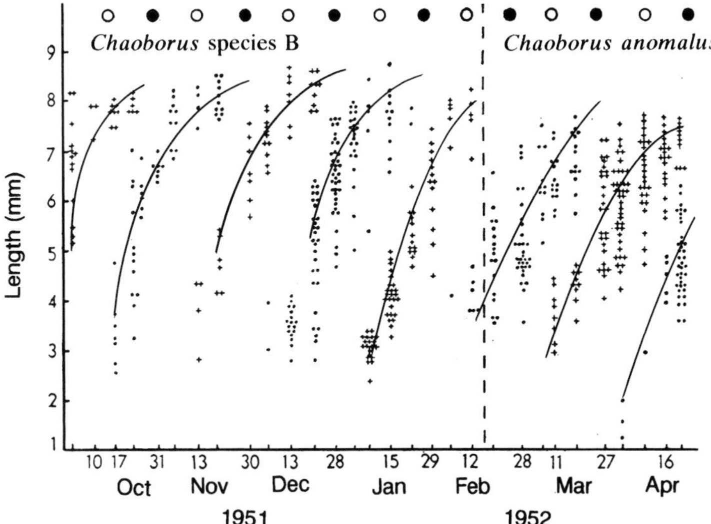

Fig. 5.11. Growth in length of larvae of two *Chaoborus* spp. collected periodically from the benthos of Ekunu Bay, Lake Victoria. Lines indicate trends of cohort development. Different symbols (•, +) differentiate coexisting cohorts of each species. Lunar phases are also shown. From MacDonald (1956).

In terrestrial environments, the predominant function of diapause changes from a cold- or winter-resistant stage in temperate regions to a drought-resistant stage in the subtropics and tropics. However, in persistent tropical freshwaters drought-resistance may be irrelevant and the diapause stage can bridge other unfavourable periods. These include adverse conditions for growth during seasonal stratification of lakes. Here one might recall the summer encystment of the dominant dinoflagellate in the subtropical Lake Kinneret, Israel (Pollingher 1986).

In some groups of freshwater organisms, notably algae, protozoa, Rotifera, Cladocera and macrophytes, there is a potential for both sexual and asexual modes of reproduction. The latter include vegetative and parthenogenetic increase, and can enable rapid rates of increase of population biomass under favourable conditions. In temperate waters the two modes of reproduction can be at least partly linked to season. In the tropics such linkage to season is apparently not well known, and in the Upper Nile was not evident for Rotifera and Cladocera of the zooplankton (Rzóska 1976). However, at Lake Chilwa in Malawi there was a conspicuous appearance of male cladocerans and then resting eggs on the first major fall in level during a drying sequence (Kalk 1979*a*), as shown in Fig. 5.26. Among planktonic algae, episodes of sexual reproduction and auxospore-formation are known for the diatoms of many temperate waters and have recently been described by Jewson *et al.* (1993) for one tropical and subtropical species, *Aulacoseira herzogii*, in Bangladesh. Here the time-sequence of auxospore-formation was related to a check to population growth and probably to reduced solar radiation in the cloudy monsoon climate.

Especially for smaller (including micro-) organisms, present knowledge in tropical freshwater ecology is limited by the scarcity of supporting studies with cultured populations. The contrast with temperate situations is obvious in work with bacteria, protozoa, rotifers and especially algae. Among common tropical representatives of the last, perhaps only the blue-green *Spirulina 'platensis'* (in part *S. fusiformis*) has received close attention regarding light utilization (e.g., Kebede & Ahlgren 1996), sodium tolerance (Kebede 1997) and nutrient requirements relevant to ecology. A rare example for a plankton diatom appears in Fig. 3.27.

(b) Heterotrophic bacteria

No general account is possible of the time-variability of bacteria (excluding cyanoprokaryotes or Cyanobacteria) in tropical freshwaters. There are few studies of total numbers by reliable methods of direct counting; components are difficult to distinguish, the classical quantitative plate counts being of very limited value; indices of cellular activity only sporadically investigated over time; and micro-environments are of great significance. Below examples are given from some sustained investigations of bacterial variability, mostly for several types of standing and flowing waters in South America with considerable inputs of organic matter.

Seasonality in the Amazonian floodplain is ensured by the annual floodpulse, but considerable differences in bacterial concentrations exist between organic-rich 'black' and 'white' river waters, and between them and floodplain lakes (Schmidt 1970; Rai 1979). Much in the longitudinal downstream flux of organic material is refractory and persistent (Richey *et al.* 1980, 1990; Hedges *et al.* 1986; Ertel *et al.* 1986), and the highest concentration of bacterial numbers (Rai & Hill 1981, 1984) and activity with added glucose (Rai 1979; Rai & Hill 1982*a*, 1984) showed some correlation with time-changes of phytoplankton abundance and photosynthetic production. Highest values of these characteristics were reached towards the low-level phase in lakes of the floodplain, such as Lago Tupé, from which the main river channel can be enriched. Thus at high level and discharge Benner *et al.* (1995) found in the river relatively low bacterial numbers but maximum growth rates, as deduced from the uptake of tritiated thymidine and leucine.

In Lake Valencia, not in a floodplain but highly productive, the variation in concentrations of planktonic bacteria has been studied with depth and time over five years (Lewis *et al.* 1986). Time-variability in both near-surface and deep water was more related to changes in vertical stratification than to those of phytoplankton abundance. During prolonged periods of stratification bacterial numbers declined in both upper and lower layers, with increases during entrainments to upper layers and transfers of O_2-bearing water to lower layers. By indirect calculation, it was estimated that in this lake bacterial activity was not large enough to process a major part of the products of primary production by phytoplankton. Seasonally, in a lake of the Amazonian floodplain where heterotrophy appears to dominate over autotrophy (Melack & Fisher 1990), size-fractionation experiments by Fisher *et al.* (1988) suggested the opposite conclusion for nutrient (N, P) regeneration.

Lewis *et al.* (1986) point out that in very shallow tropical lakes, such as Lake Nakuru in Africa that is recorded as supporting exceptionally high bacterial concentrations (Kilham 1981), planktonic bacteria may be partly derived by resuspension of sediments. There may also be a bias towards higher bacterial concentrations in saline lakes, and especially soda lakes, such as those in Ethiopia examined by Zinabu & Taylor (1997). Numbers may also respond to irregular phytoplankton mortalities as indirect evidence from rotifer abundance in Lake Nakuru suggests (Vareschi & Jacobs 1985). Measurements of bacterial numbers and cellular growth on the diel and seasonal scales are available for only a few tropical lakes including Lake Awasa, Ethiopia (Gebre-Mariam & Taylor 1989*a*, *b*). Here specific growth rates were estimated by three methods: incorporation of tritiated thymidine, frequency of dividing cells and increase in cell numbers during bottle incubations. These generally agreed, with a range of $0.006–0.026\,h^{-1}$ and mean $\sim 0.013\,h^{-1}$. Estimates of opposing grazing rates, probably chiefly from ciliates, were of similar magnitude. Diel variations were also found to be significant in the Ebrié coastal lagoon of West Africa (Torréton *et al.* 1994), and in Lake Xolotlán, Nicaragua (Vammen *et al.* 1991). For the former, the variation may be due to the resuspension of sediment by stronger winds with a diel incidence, bearing bacteria of larger size attached to particles (Bouvy *et al.* 1994; Arfi & Bouvy 1995). A similar influence of particulate resuspension on bacterial numbers and size has been found in the shallow Lake Chapala, Mexico (Lind & Dávalos-Lind 1991). Time-variability in total bacterial numbers over one year was followed by Freitas & Godhino-Orlandi (1991) for the surface sediment of an oxbow lake in southern Brazil. Concentrations, expressed per g dry weight of sediment, were strongly seasonal and hydrologically influenced. They ranged from 1.26×10^{10} cells g^{-1} dry weight in the dry season to 8.6×10^{10} cells g^{-1} in the rainy season during which the main decomposition of allochthonous material occurred.

In some tropical waters, as elsewhere, photosynthetic sulphur bacteria are known to occupy a water stratum immediately above a sulphide-containing hypolimnion, where sufficient light is available. Examples are described from Borneo (Brunei Darussalam) by Booth & Choy (1995) and Papua New Guinea by Vyverman & Tyler (1995). The latter relate short-term changes in these communities to those of density stratification linked to incursions of seawater.

(c) Aquatic macrophytes

These plants are very varied in their relations to the phase boundaries of air–water and water–sediment, to which their communities have an overall depth-zonation. Consequently, besides responses as primary producers to changes in the radiation/temperature regimes affecting atmosphere and water, they are generally susceptible to tropical hydrological regimes that produce marked changes in water level. Further, as most are higher plants, they tend to show internal controls of time-related events of life cycle such as flowering (phenology). Their large biomass in many shallow freshwaters makes their time-variations important as environmental characteristics for smaller organisms, independently of trophic links. These time-variations can be very different for various parts of the same plant, such as leaves, erect shoots and rhizomes.

Relevant studies-in-depth from the tropics, which are not numerous, mainly concern major distributional expansions of a few species (e.g., *Eichhornia crassipes*, *Salvinia molesta*) and effects of seasonal or long-term change in water level. Wider comparative surveys of note are the book edited by Denny (1985) on the ecology of African freshwater macrophytes and the review of low-latitude seasonality/aseasonality by Mitchell & Rogers (1985).

Free floating plants are richly developed in tropical freshwaters. The combination of extensive asexual vegetative reproduction, potentially exponential population increase and horizontal mobility has enabled some spectacular range-extensions in nutrient-rich waters. The originally South American species of *Eichhornia crassipes* and *Salvinia molesta* have become pan-tropical, like the earlier-established and comparatively innocuous *Pistia stratiotes*, and expanded to pest-proportions in various man-made lakes (Little 1966; Gaudet 1979*a*; Mitchell & Gopal 1991), some natural lakes and in river-systems. Examples are the expansion of *Salvinia molesta* in the early and more nutrient-rich phase of Lake Kariba (Mitchell 1969) followed by later decline (Marshall & Junor 1981: Figs. 5.12, 5.13); the corresponding but lesser expansion followed by decline of *Pistia stratiotes* in Lake Volta (Okali & Hall 1974; Hall & Okali 1974); and the invasion of *Eichhornia crassipes* in the river-systems of the Zaïre (Congo) (Berg 1959, 1961) and, later, the Nile (Gay 1960; Obeid 1975; Batanouny & El-Fiky 1975).

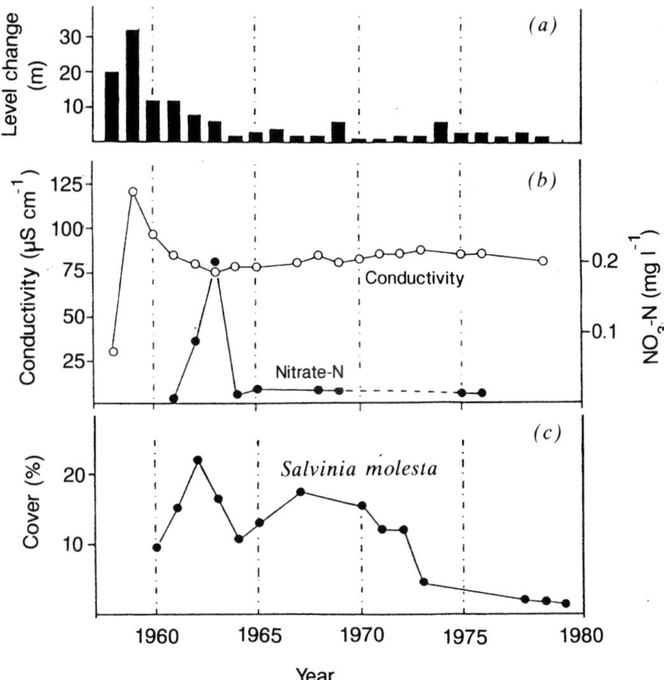

Fig. 5.12. Lake Kariba. Physical, chemical and biological changes after dam-completion in 1958, including (*a*) annual change of water level, (*b*) concentration of nitrate-nitrogen and conductivity of surface water, (*c*) percentage of lake area covered by the floating plant *Salvinia molesta*. From Marshall & Junor (1981).

Fig. 5.13. Lake Kariba. A shoreline in October 1983, with dead flooded trees and an accumulation of the floating water-fern *Salvinia molesta*.

Here the intrinsic rates of increase (= relative or specific or instantaneous growth rates, *g*: Chapter 3.3, Section 5.3) are of obvious importance. They have been studied for the species of *Salvinia* (Mitchell & Tur 1975; Bond & Roberts 1978; Room & Thomas 1986*a*), *Eichhornia* (Obeid 1975; Bond & Roberts 1978; Junk & Howard-Williams 1984; Gopal 1987) and *Pistia* (Hall & Okali 1974; Junk & Howard-Williams 1984). Doubling times under near-optimal conditions can be as low as 3.4–5.3 days (*Salvinia molesta*: Gaudet 1973; Mitchell & Tur 1975; Sale *et al.* 1985), but *in situ* generally lie in the range 8–15 days under favourable conditions. Crowding, via local nutrient release after decay, can stimulate as well as retard growth rates (Mitchell & Tur 1975). Rates were also recorded as lower in the cooler season for *S. molesta* at Lake Kariba and for *Eichhornia crassipes* in a reservoir at 23° N in northern India (Sen *et al.* 1990). Besides temperature, the nitrogen status of *S. molesta* was found by Room & Thomas (1986*a*) to be an important factor in the seasonal variation of relative growth rate (Fig. 5.48), which they modelled for latitudes between 0° and 50° S.

Changes of water level may lead to marginal stranding but are not generally of major significance. Decrease in the shade from pre-existing forest occurred during the newly flooded phase of Lake Brokopondo in

Suriname. Afterwards, with leafless trees, expansion by *Eichhornia crassipes* was favoured, but not that of shade-plants which included the submerged water-fern *Ceratopteris pteroides* – or of the duckweeds *Lemna valdiviana* and *Spirodela biperforata* (van Donselaar 1968).

Completely submerged plants have rarely been studied for time-variation in the tropics. In temporary waters a seasonal dry phase can produce near-complete vegetative destruction by desiccation or preceding elevation of salinity, with survival by seeds or turions, as described for *Potamogeton crispus* in a subtropical (27° S) floodplain pan of South Africa by Rogers & Breen (1980). However, in an equatorial Malaysian swamp, without desiccation, the dry seasons were the main periods of vegetative growth of the bladderwort *Utricularia flexuosa*, contrasting with two periods of wash-away during the South West and North East monsoonal rains (Lim & Furtado 1975). For the principal species in Lake Titicaca, *Schoenoplectus tatoram*, there was an increase of biomass cover during the warmer season (October–March), around the summer solstice, but not for the still more abundant species of *Chara* (Collot *et al.* 1983).

Emergent reedswamp plants are exposed to atmospheric factors such as air temperature and also to variation in water level. Stock density and composition may vary seasonally in relation to phasing in the generations (e.g., annual) of shoots, as is conspicuous in temperate regions. A tropical example is provided by the sedge *Eleocharis interstincta* in a Venezuelan lagoon (Gordon Colón & Velásquez 1989), where the seasonal cycle was pronounced and involved a maximum of biomass and culm height in November at the transition from the rainy to dry season. Near the edges of the tropics growth may largely cease in the coolest 'winter' season, as for *Typha angustata* at latitude 26°49′ N in northern India (Sharma & Pradhan 1983). In Africa the well-studied *Typha domingensis* of the swamps of Lake Chilwa is a near-constant perennial, as at another tropical site in Cuba (Plasencia & Květ 1993), but its growth rate – judged from experimental cutting – varied with season with a correlation to water level (Howard-Williams 1979*b*). Its resistance to high ionic concentrations (e.g., up to ~44 mmol $Na^+ l^{-1}$) was a factor in its survival, if not growth (Howard-Williams 1975), through annual or long-term increases of salinity during low water levels. Long-term changes of water level in Lake Chad, that also involved salinity change, altered the relative abundance and spatial distribution of major reedswamp species (Iltis & Lemoalle 1983). In a newly created African lake of variable level, Lake McIlwaine, the small emergent *Polygonum senegalense* was outstandingly successful in littoral colonization; contributing factors were its abundant production of seeds suited to germination on wet mud, rapid vegetative growth, structural adaptability to changing water level and wide tolerance of chemical conditions (Jarvis *et al.* 1982).

Long-term littoral colonization with transition from lake to swamp constitutes the classic *hydrosere*. Osborne & Polunin (1986) describe an apparently 'reversed hydrosere' from Waigani Lake in Papua New Guinea, where the transition of emergent swamp → floating-leaved species → open water was apparently caused by a combination of water level change and nutrient enrichment by urban effluents.

Attached or semi-attached floating plants are typified by grasses that are attached to the sediments but bear subaerial shoots from a ramified floating mass. Such 'floating meadows' are a major feature of the seasonally inundated *várzea* region of the Amazonian floodplain. Notable species here are *Paspalum fasciculatum* and *Echinochloa polystachya*; *Echinochloa stagnina* and *Vossia cuspidata* are common species in African floodplains. Sites on the Amazonian floodplain may have a seasonal amplitude of 11 m in water level, such as Lago do Castanho where the seasonal development of a stand of *Paspalum fasciculatum* was studied in 1974–5 (Junk & Howard-Williams 1984: see Fig. 5.14). In this time three annual generations of shoots could be distinguished, each with a growth–senescence cycle. After seed germination on the sediment at low level, very rapid extension growth (e.g., 20 cm per day) normally allowed vegetative development to keep up with rising water level, with maximum biomass near maximum flood level.

Another form of floating vegetation is that of papyrus (*Cyperus papyrus*) which consists of mats of interlaced rhizomes bearing erect culms. Culm growth can be continuous, with the replacement of growth stages in time, at equatorial locations (e.g., Thornton 1957; Thompson *et al.* 1979), but with a marked minimum of growth rate during the cool season at subtropical locations (e.g., Breen & Stormanns 1991). The mats can accommodate smaller amplitudes of level variation, as at the equatorial African Lake Naivasha where small falls can extend the community (Gaudet 1977*a*) although larger long-term changes to high level are unfavourable (Harper 1984, 1992). Detached mats of this and other species form 'sudds' that encroach on open water at rates much faster than reedswamp in the traditional hydrosere. Floating examples in the Okavango Delta, Botswana, varied in number seasonally by a unique mechanism probably linked to seasonal gas production – predominant sinking in the cooler season, predominant rise in the warmer season (Ellery *et al.* 1990).

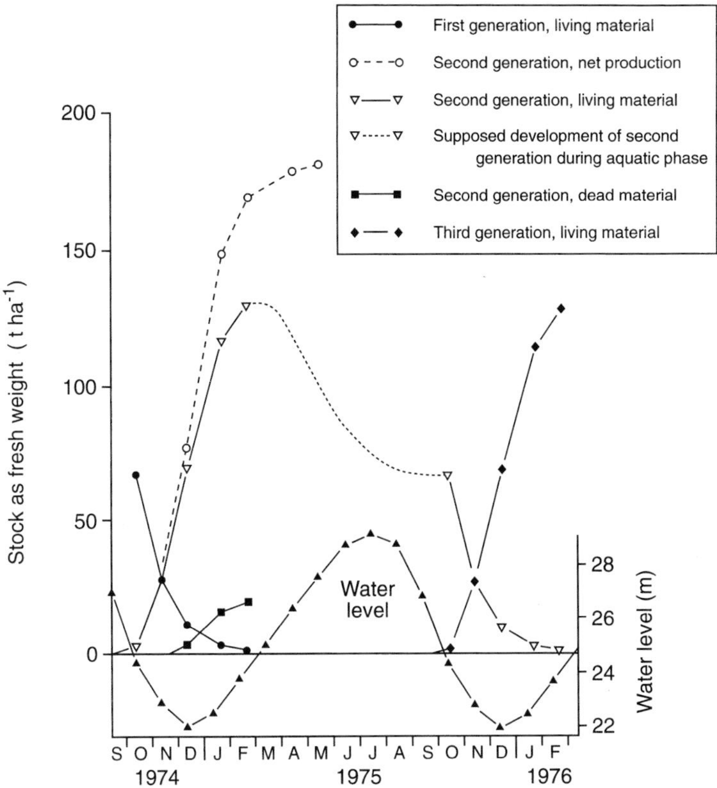

Fig. 5.14. Seasonal biomass, as fresh weight, of successive generations of the grass *Paspalum fasciculatum* in the Amazonian lake, Lago do Castanho, in relation to changes of water level. From Junk & Howard-Williams (1984).

An adverse reaction to media of markedly low or high ionic content has consequences in some time-series involving evaporative concentration, and was prevalent during phases of falling level at Lake Chilwa (Howard-Williams 1979*b*) and Lake Chad (Iltis & Lemoalle 1983).

Rock-encrusting plants include the unique tropical Podostemaceae that inhabit swiftly running waters, with a life cycle in which flowering and seed formation are triggered by falling water level (Payne 1986).

(d) Phytomicrobenthos

Phytomicrobenthos (or periphyton, *Aufwuchs* in part) develops as algal growth on submerged surfaces, of stone (epilithic), sediment (epipelic) or plants (epiphytic). It may be a competitor with phytoplankton and submerged macrophytes for nutrients and light. Its study in tropical freshwaters has been minimal, although it serves as food for many invertebrates and – as in lakes Malawi and Tanganyika – rock-frequenting cichlid fishes. In general its biomass density per unit area is determined by availability of substratum, light and nutrients, by dispersal losses from currents and turbulence, and by animal consumption.

Influences from all these factors are illustrated in the study by Engle & Melack (1990) of epiphytic material on floating meadows of an Amazonian floodplain lake, Lake Calado. Here recovery after disturbance was rapid (< 7 days). The seasonal pattern of abundance per unit area was related to the hydrological cycle, with high values during the phase of high-level, nutrient-rich, turbid water conditions that was unfavourable for plankton development. Seasonal increase was conspicuous in several rock-attached (epilithic) components – *Cladophora*, *Calothrix*, diatoms – in southern Lake Malawi during the cool season, when vertical exchange probably increased the availability of nutrients (Haberyan & Mhone 1991). Seasonal change of biomass was established for a community attached to submerged macrophytes in a temporary water-body of flooded savanna in Venezuela (Cruz & Salazar 1989). Here the biomass maximum (August) was phased before that of the host plant, from which an unfavourable shading might result. The reverse, an unfavourable shading of the host plant, is also possible. Seasonal change was also marked for the periphyton attached to the trunks of submerged trees in the Volta Lake of Ghana (Obeng-Asamoa *et al.* 1980), where change in water level – by as much as 5 m during the year – was a dominant factor. This could be eliminated by following sequences on wooden blocks suspended at fixed depths (John *et al.* 1981). In both cases diatoms were the main primary colonizers, with green filamentous algae following later.

The time required for recolonization can also vary with season. In experiments of Fernandes & Esteves (1996) with immersed leaves of *Typha domingensis* in a Brazilian coastal lagoon, maximum biomass was reached in 14 days during the warmer season (~28 °C) and in 21 or 28 days during the cooler season (~21 °C). Other experimental exposures were used to follow seasonal changes in the colonization of artificial substrata by Ho (1976) in a Malaysian stream and by Iltis (1982) in two West African rivers. Colonization was generally heaviest near the end of low-water periods, due to adverse scouring effects in high-level periods after rains.

Longer-term changes have rarely been followed quantitatively. In very shallow waters a fall of water level, and evaporative concentration, may promote a rich flora of benthic diatoms – as described for Lake Chilwa in Malawi by Moss & Moss (1969), and for several Kenyan soda lakes by Tuite (1981) and especially Melack (1988). The last study extended over two years (1973–75) on Lake Elmenteita, where reduced rainfall caused mean depth to decrease from 1.1 to 0.65 m, and conductivity to rise above 21 mS cm^{-1}. Possibly related to the salinity change, populations of major phytoplankters fell abruptly whereas those of several bottom-living diatoms rose.

(e) Phytoplankton

As a community of primary producers in the form of dispersed cell suspensions, phytoplankton is inherently susceptible to change in both the radiation/temperature and water balance complexes of environmental factors. Its own reactions upon the physical and especially the chemical environment can also be profound, and often cyclic in time. Biotic interactions that include grazing can introduce further temporal change.

Diel variability originates at several levels. The day–night cycle of photosynthetically available radiation will induce cycles of relative carbohydrate content in algal cells. These have apparently not been studied in the tropics. Diel cycles of photosynthetic activity have been followed very widely with varying degrees of time-resolution (Chapter 3.1, Section 5.3). A corresponding cycle of light-dependent nitrogen fixation by blue-greens has been measured in Lake George (Ganf & Horne 1975) and in Lake Valencia (Levine & Lewis 1984: see Fig. 3.26). The diel period is too short for cycles of population growth to be recognized; effects on a cycle of frequency of dividing cells can be anticipated but do not appear to have been investigated in tropical freshwaters for phytoplankton (unlike bacteria: Gebre-Mariam & Taylor 1989*b*). There is, however, evidence for bursts of cell division at longer intervals for the diatom *Aulacoseira* (formerly *Melosira*) *italica* in a Brazilian reservoir (Nakamoto *et al.* 1976).

The diel cycle of temperature/density stratification can have large effects on the vertical distribution of blue-green algae (cyanobacteria) with varying positive or negative buoyancy. In the most intensive tropical studies, a predominant rise during the day was found in the Jebel Aulia reservoir on the Nile (Talling 1957*a*) and a predominant sinking in Lake George (Ganf 1974*b*, *d*). This varying behaviour can be encompassed by known mechanisms of buoyancy regulation in these organisms, but applications must be speculative pending more measurements for the situations concerned (Ganf 1974*d*). Vertical near-uniformity was restored during nocturnal mixing, sometimes aided by diel wind cycles as well as nocturnal cooling. Day-to-day variation of wind speed was positively related, through turbulence and resuspension, to the abundance of the diatom *Aulacoseira italica* in a shallow Brazilian reservoir (de Lima *et al.* 1983). Sectors of a population isolated for a time above a diel thermocline can show increased effects of a near-surface light-inhibition of photosynthesis (Vincent, Neale & Richerson 1984: see Fig. 3.8). Conversely, at depth in darkness, reduced rates of respiration per unit biomass have been demonstrated in Lake George (Ganf 1974*a*: see Fig. 3.10).

Vertical redistribution of populations with diel cycles can also result from active migrations of flagellates. Although largely unexplored in the tropics, an outstanding example of large amplitude (to 18 m) by *Volvox* is described by Sommer & Gliwicz (1986) from an African man-made lake, Cahora Bassa.

Within-year (annual) variability, seasonal or aseasonal, provides most examples of temporal variability for phytoplankton. Population dynamics and productivity have been central themes, and wide-ranging studies are accumulating from an increasing – although still very limited – number of tropical sites. A sampling is contained in Munawar & Talling (1986); others appear in the survey of Serruya & Pollingher (1983).

General surveys of the extent of annual variability with latitude, including the tropics, have been made by Melack (1979*a*) and Ashton (1985*b*), using as index the coefficient of variation (= standard deviation/mean) applied to rates of photosynthetic production per unit area or concentrations of chl-*a*. Results relating to photosynthetic production, which partly reflect biomass concentration, are shown in Fig. 5.15. Although there is a trend towards minimum variability at the equator, the scatter there is wide and will be influenced by hydrological variability. The coefficient of variation is loosely related to relative annual range, a parameter which was used by Talling (1986), Kalff & Watson (1986) and Kebede & Belay (1994) to compare variations of surface concentrations of total biomass in three series of African lakes. This range was much less than that of most component species, indicating a degree of species-replacement. It was also lower in the shallower lakes – a feature correlated by Kalff & Watson (1986) with more extensive contact between the upper mixed layer and sediments. Later Lewis (1990) made a further comparison of seasonal variability of biomass as chl-*a* concentration in

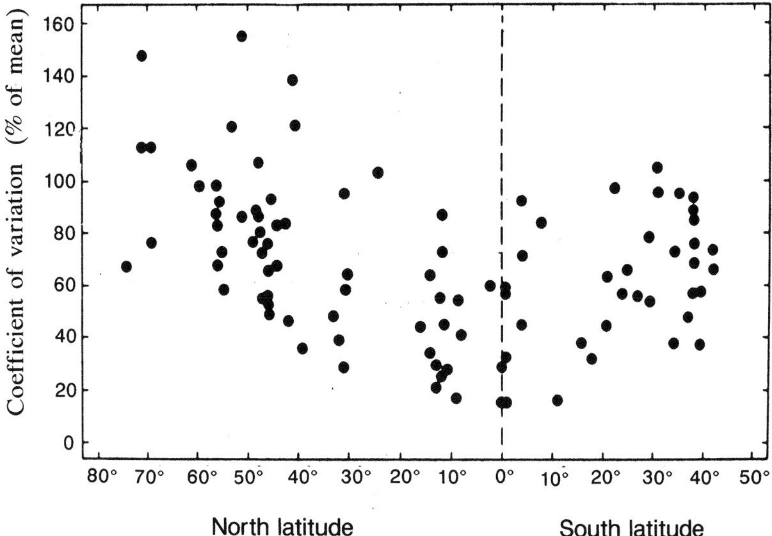

Fig. 5.15. Scatter diagram showing coefficients of variation for within-year variability of phytoplankton photosynthesis rates per unit area in relation to latitude. Based on Melack (1979*a*) and Ashton (1985*b*).

stratifying lakes, five tropical and 15 temperate, in which seasonal range was taken between the 5 and 95 percentiles, and maximum concentrations were taken in ratios with both annual and seasonal mean concentrations. These ratios differ in their sensitivity to the period-duration of biomass limitation. On the annual basis, the mean concentration was a much higher fraction of the maximum in the tropical than the temperate lakes, whereas on the seasonal basis it was less different. These features can be interpreted as suggesting that minima in the tropical lakes are relatively less depressed and that support might derive from more extensive recycling of nutrients there.

Figure 5.16 shows examples of time-variability for phytoplankton abundance in a series of tropical lakes, arranged by latitude. Periods of reduced stratification and greater vertical mixing are indicated. Many deeper tropical lakes, such as Victoria, Lanao, Valencia, Tanganyika, Malawi and Titicaca, have an annual cycle of thermal stratification that includes a short phase of extensive, sometimes near-complete, vertical mixing. This phase is often accompanied or immediately followed by a peak of algal abundance, to which diatoms make a major contribution. However, a decline of phytoplankton density on mixing is not

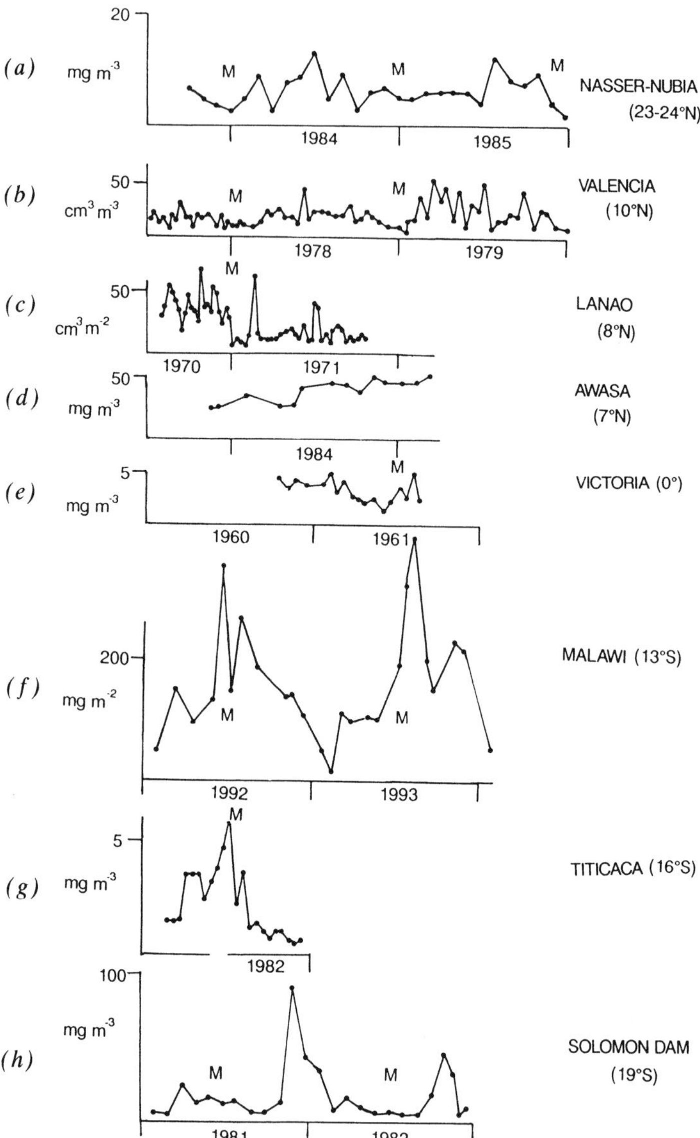

Fig. 5.16. Time-variation of total phytoplankton density in the upper layers, or entire water column (examples *c*, *g*), of a series of deep tropical lakes arranged by latitude. Time-scales are aligned according to the winter and summer solstices. Phytoplankton density is assessed by chlorophyll *a* (examples *a*, *d–h*), or by cell volume (*b*, *c*). Periods of stronger vertical mixing (M) are indicated. Sources are respectively, Habib *et al.* (1987), Lewis (1986*a*), Lewis (1978*a*), Kifle & Belay (1990), Talling (1966), Patterson & Kachinjika (1995), Vincent *et al.* (1984) and Hawkins & Griffiths (1993).

uncommon, especially where the euphotic zone becomes only a small fraction of the total mixed depth, as in lakes Titicaca (Vincent *et al.* 1984), Lanao (Lewis 1978*a*) and Valencia (Lewis 1986*a*). In Lake Lanao a negative response for the total phytoplankton biomass was compatible with a positive response by some diatom (*Aulacoseira*) components. In several lakes, such as Victoria and Lanao, and probably Awasa (Kebede & Belay 1994), multiple partial mixings occur with some phytoplankton response. A second major peak may develop after re-stratification and is often dominated by blue-greens; examples have been described from lakes Tanganyika (Symoens 1956; Hecky & Kling 1981) and Victoria (Talling 1966). Such cycles, one- or two-peaked, seem to be primarily determined by changes of hydrographic structure in the water-mass related to the radiation/temperature and wind complexes of environmental change. In large lakes some effective changes of hydrographic structure can be local. Examples include responses to upwelling at the southern ends of lakes Tanganyika (Coulter 1963, 1968, 1991*a*) and Malawi (Eccles 1974; Bootsma 1993*b*), and the differentiation seen between bays and 'open lake' environments of lakes Victoria (Fish 1957; Talling 1966, 1987; Akiyama *et al.* 1977: see Fig. 5.17) and Titicaca (Lazzaro 1981; Vincent *et al.* 1986).

Especially in shallow lakes that lack a cycle with prolonged stratification, and in rivers such as the Nile (Talling & Rzóska 1967), Orinoco (Lewis 1988; Carvajal-Chitty 1993) and Ganges (Lakshminarayana 1965*b*; Singh *et al.* 1983), there is more opportunity for response to major, usually seasonal, water inputs (i.e., the hydrological complex). Examples for lakes and reservoirs in Africa are surveyed by Talling (1986) and in the Southern Hemisphere generally by Ashton (1985*b*). The water input may influence phytoplankton by deepening the water-column, by reducing light penetration through introduced silt, by wash-out effects in basins of short retention time, and – more favourably – by injecting nutrients. These influences are seen seasonally and spatially, especially near inflows, in the large African lakes or reservoirs of Turkana (with a pervasive North–South polarization: Ferguson & Harbott 1982; Harbott 1982; Ferguson 1982), Albert (Evans 1997), Volta, Kariba, Nubia and Chad; also in smaller reservoirs and ponds of India and Bangladesh where a plankton minimum can be characteristic during wash-out and raised turbidity in the wet monsoon season (e.g., Sugunan 1980; Kannan & Job 1980*a*; Zafar 1986; Khondker & Parveen 1993). This or other influence of seasonal hydrology is recorded for lakes and reservoirs of Malaysia (Fatimah *et al.* 1984) and of Sri

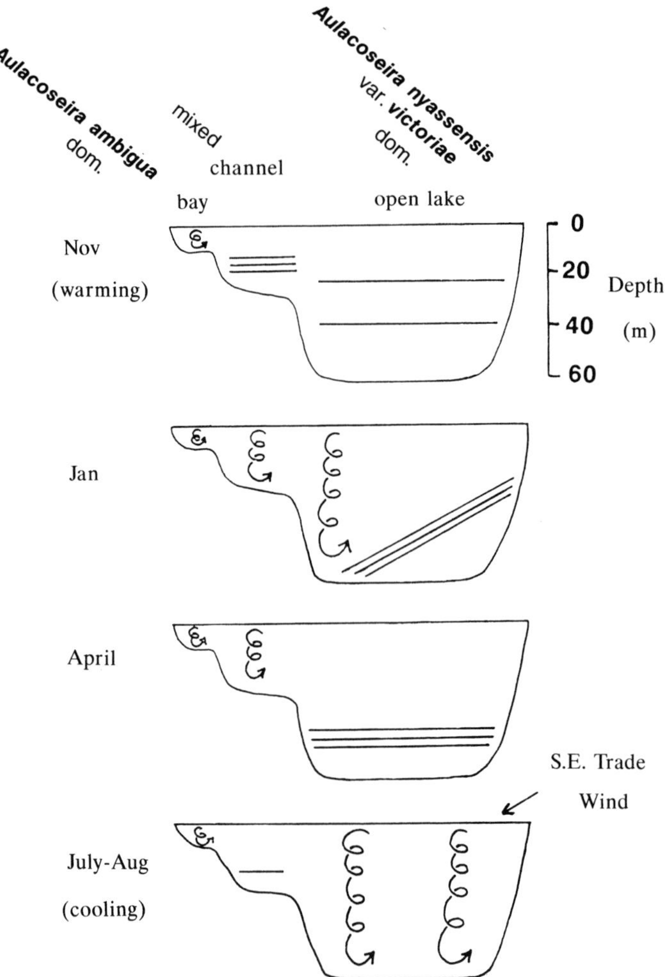

Fig. 5.17. Lake Victoria. Diagrammatic representation of a seasonal sequence of stratification and vertical mixing, showing differences of timing between bays, inshore channel and open lake areas that influence their respective populations of two species of the planktonic diatom *Aulacoseira*. Based on Fish (1957) and Talling (1966).

Lanka, where studies of plankton seasonality have a long history (Apstein 1907, 1910; Holsinger 1955; Schiemer 1983). However, a minimum in Sri Lanka during the wet North East monsoon, affecting photosynthetic production, appears to be not simply explicable from wash-out effects (Silva & Davies 1987).

In Central and South America a seasonal water-input influence on phytoplankton is described for the large lakes of Xolotlán, Nicaragua (Hooker *et al.* 1991) and Chapala, Mexico (Limón *et al.* 1989; Lind *et al.* 1992), and numerous small reservoirs of which those in South East Brazil are surveyed by Tundisi (1983, 1994). That at Brasilia (Paranoá Reservoir) is heavily enriched by urban effluents and the dry season corresponds with a seasonal minimum of the dense phytoplankton dominated by the cyanophyte *Cylindrospermopsis raciborskii* (Branco & Senna 1994, 1996). Especially notable are small lakes of the Amazon floodplain (reviewed by Melack & Fisher 1990) that are seasonally refilled. Lago Jacaretinga and Lake Calado are well-studied examples of a class where the flooding river channel contributes to the lake with nutrient-richer water that in subsequent months deposits silt, develops a dense phytoplankton and is progressively stripped of soluble inorganic nitrogen and phosphorus (Fisher & Parsley 1979; Forsberg *et al.* 1988). Low penetration of light in floodwater is an important seasonal limitation (less so for the accompanying emergent macrophytes: Forsberg 1984). It also retarded the seasonal growth of phytoplankton in the Blue Nile during a phase when measured major nutrients were most abundant (Talling & Rzóska 1967). For this reservoir–river system, the seasonal increase of retention time by dam closure and also ponding-back near Khartoum at low level (Rzóska *et al.* 1955; Talling & Rzóska 1967; Hammerton 1972) was decisive for determining a two-peaked annual cycle of algal abundance. Further downstream the phytoplankton in the upper reaches of the elongate High Aswan Dam reservoir, Lake Nasser-Nubia, has shown large increases during July–August that were lacking from the lower reaches near the dam (Habib & Aruga 1988) and were probably responses to the input of nutrients in seasonal floodwater.

There remains a few studied equatorial lakes in which strong seasonal inputs of floodwater are not marked and which are too shallow to develop a long-maintained thermal stratification. Lake George, with local hydrological buffering, has already been mentioned (Chapter 4.3d); here low amplitudes of environmental factors are matched by low amplitudes of annual change in phytoplankton abundance and composition (Burgis *et al.* 1973; Ganf 1974*b*; see Fig. 5.21). A small positive response to seasonal rainfall is discernable. The Kenyan Lake Naivasha shows greater seasonal change in the phytoplankton (Kalff & Watson 1986) that may be influenced by variable resuspension of sediment, with nutrient exchange, induced by the wind regime (Kalff & Brumelis 1993).

There is here a possibility of buffering against seasonal nutrient inputs by interception in marginal swamps, investigated by Gaudet (1979*b*).

Environmental seasonality in West African freshwaters related to another wind regime, the harmattan (Chapter 4.3c), is apparently connected with one of two seasonal phases of major changes in the phytoplankton of Lake Volta (Biswas 1972*a*; Talling 1986). Also of wide influence in East and Central Africa are the South East Trade Winds (part of a monsoon system), which promote seasonal mixing in lakes that include Tanganyika, Malawi and Victoria. In these lakes, and in the man-made Lake Kariba (Ramberg 1987; Cronberg 1997), this mixing elicits positive growth response from the diatom components of the phytoplankton. It is an essential factor in the annual cycle of sedimentation and resuspension that is known for *Aulacoseira* (*Melosira*) *nyassensis* var. *victoriae* in Lake Victoria during the years 1950–52 (Fish 1957), 1956 (Talling 1957*b*) and 1960–61 (Talling 1966: see Fig. 5.18).

The annual variation of component species is rarely known for more than two years in tropical water-bodies, so the regularity or otherwise of annual cycles is not well established. A considerable degree of seasonal regularity was found for stretches of the White and Blue Niles near Khartoum (Prowse & Talling 1958: see Fig. 5.19; Talling & Rzóska 1967; Hammerton 1972; Sinada & Abdel Karim 1984*b*), Lake Lanao, Philippines (Lewis 1978*a*), Lake Valencia (Lewis 1986*a*), and the 'normal' phase of Lake Chad (Compère & Iltis 1983; Lemoalle 1983). In Lake Victoria the changes in population density of many species, resolved over one year, fell into three main categories characterized by positive response to vertical mixing (diatoms), negative response (most blue-greens), and fluctuations of low amplitude (green algae) (Fig. 5.20). A corresponding resolution of species changes in Lake George (Ganf 1974*b*) showed mainly fluctuations of low amplitude, excepting the blue-green *Anabaena flos-aquae* with the large relative range of $>10^5$:1.

Temporal changes in species-populations that differ by time-shifts, growth and loss rates, and inoculum levels give rise to patterns of species succession. Generalized sequences of algal classes or morphotypes, and of predominant factors, have been proposed with application to stratifying tropical lakes as discussed by Lewis (1978*a*, 1986*a*) and Ashton (1985*b*). Much depends upon the different responses of diatoms and blue-greens to mixing and stratification, illustrated from Lake Victoria in Fig. 5.20, but the patterns shown by other algal groups (e.g., greens, dinoflagellates) are more variable between lakes, as is the succession of smaller- and larger-celled species.

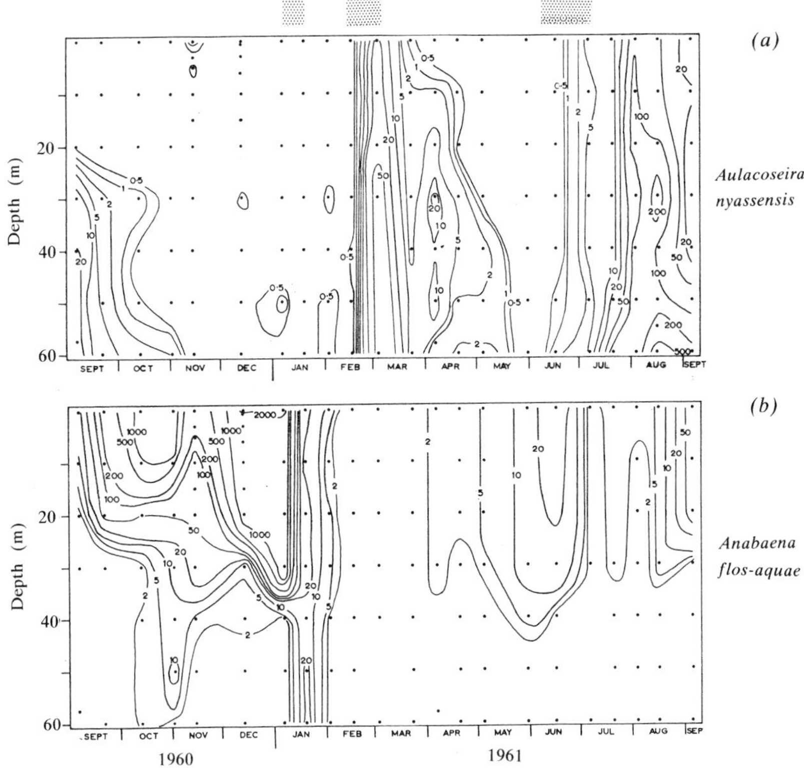

Fig. 5.18. Lake Victoria. Depth-time diagrams with contours of algal concentration in cells ml^{-1}, showing the complementary patterns of occurrence of two principal phytoplankters, (*a*) diatom and (*b*) blue-green, over an annual cycle of stratification. Stippled blocks indicate onset of extended vertical mixing. From Talling (1966).

Growth cycles or fluctuations of phytoplankton often induce corresponding cycles or fluctuations of nutrient concentrations subject to depletion. Perhaps the best established examples are of diatom–Si relationships as described by Adeniji (1977) for Lake Kainji and Lemoalle (1978) for Lake Chad (Fig. 5.3). Depletions are often countered or obscured by recycling and horizontal transfers (advection), especially of the elements nitrogen and phosphorus.

Depletions in time of the populations themselves occur by various processes, including sedimentation as well as grazing by zooplankton and some planktivorous fishes. The quantitative role of grazing is virtually unexplored for tropical freshwater phytoplankton. However,

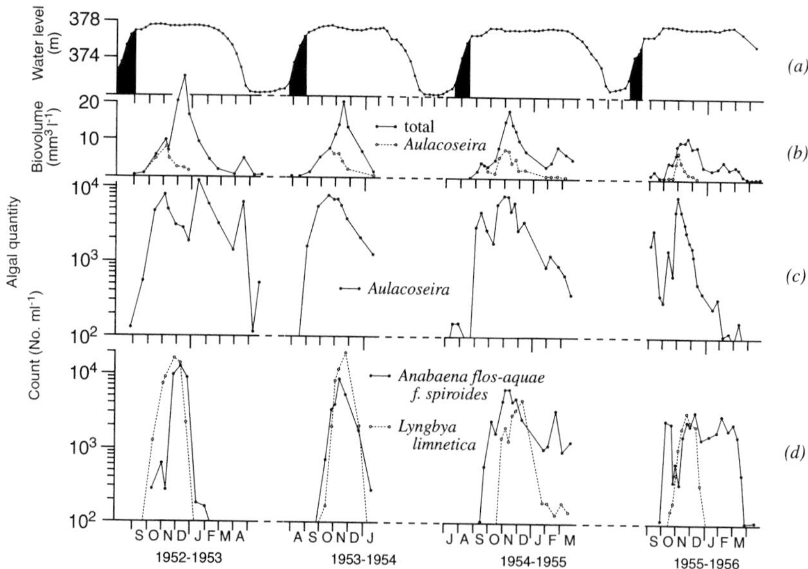

Fig. 5.19. White Nile, 1952–56. Seasonal changes downstream of the Jebel Aulia dam for the abundance in surface water of three major phytoplankters (*c*, *d*) and of the total phytoplankton and *Aulacoseira granulata* assessed by cell volume (*b*), in relation to change of water level above the dam indicative of annual cycles of water storage and release (*a*). Modified from Prowse & Talling (1958).

Gliwicz (1976*b*), working on lakes in Panama, obtained an index of grazing rate (in % day^{-1}) from the removal of plastic micro-beads. Some idea of bounds to grazing possibilities can be obtained from general experience with specific filtration or consumption rates. Thus Lewis (1978*a*, 1985) applied an upper bound of 2000 ml per mg zooplankton dry weight and day to Lake Lanao, and a consumption rate of 30% of body weight per hour to estimates of protozoan biomass in lakes Lanao and Valencia. In all these cases the impact upon phytoplankton was judged to be small. Fungal parasitism and ingestion by protozoa are, however, little known from direct observations; examples are illustrated by Talling (1987) and Finlay *et al.* (1987), respectively. Possible seasonal increase by excystment or germination of resting stages is largely unknown; this can be of great importance in the subtropical Lake Kinneret, Israel (Serruya & Pollingher 1983).

Few examples of time-variability in tropical phytoplankton have been analysed in terms of absolute or specific (relative) rates of population change that are the resultant of corresponding rates of gain and loss as

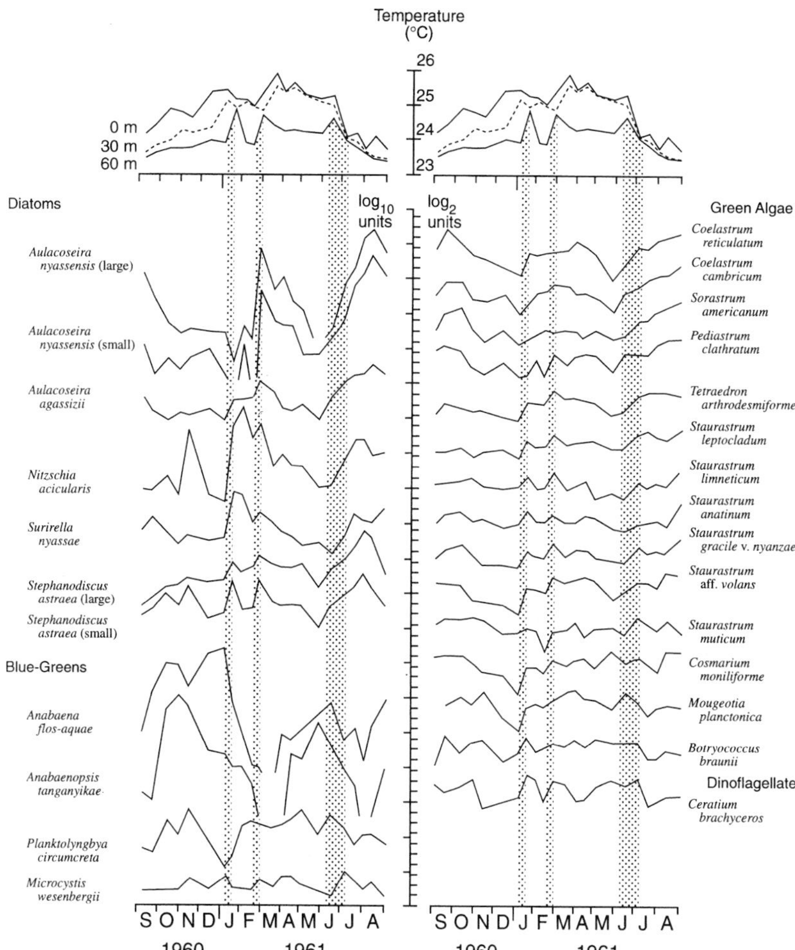

Fig. 5.20. Annual variation in the abundance of four groups of phytoplankters in offshore water of Lake Victoria, shown successively displaced on a relative logarithmic scale in relation to temperature stratification and (stippled) onset of periods of vertical mixing. From Talling (1966).

by grazing and sedimentation. One has been attempted for Lake Malawi (Bootsma 1993*a*). In another, involving a short-term sequence in a West African fishpond (Arfi & Guiral 1994), one large loss component was from grazing by protozoa and rotifers, and a second by sedimentation after the depletion of ammonium-nitrogen.

Long-term (or inter-annual) variability has been documented only sporadically in the tropics. Year-to-year variability is likely to be pronounced in shallow water-bodies susceptible to occasional drought, such as the two ponds in India where depression of photosynthetic productivity during a drought year was studied by Kundu & Jana (1994). In deeper waters one of the best examples of large year-to-year (and within-year) differences in abundance is that of the major diatom *Aulacoseira* (formerly *Melosira*) *nyassensis* in Lake Malawi, where they are shown by historical records (Hecky & Kling 1987) and by sediment stratigraphy (Owen *et al.* 1990; Pilskaln & Johnson 1991; Owen & Crossley 1992). However, the abundance of accompanying *Nitzschia* spp. is much under represented in the sediments (Haberyan 1990), due to dissolution in the water-column. Sedimentary remains of diatoms have indicated changing water level and climate in the remote past from various tropical lakes, including Lake Victoria (Kendall 1969; Stager *et al.* 1997), Lake Abhé, Ethiopia (Gasse 1977; Gasse & Street 1978), Lake Chad (Servant & Servant 1983; Gasse 1987), Lake Naivasha (Richardson & Richardson 1971; Richardson & Dussinger 1986), Lake George (Haworth 1977), Lake Rukwa (Haberyan 1987), Lake Texcoco (Bradbury 1971) and Lake Valencia (Bradbury *et al.* 1981). Thus in tropical Africa an alternation of wetter and drier periods has been extensive over the last 15 000 years. Stratigraphic correlations between African lakes are illustrated by Beadle (1981).

Long-term changes in phytoplankton can be expected in closed basin, saline lakes with fluctuating water level. In the African lakes Nakuru (Tuite 1981; Vareschi 1982), Elmenteita (Melack 1988), Chilwa (Moss & Moss 1969; Kalk *et al.* 1979) and Chad (Iltis & Lemoalle 1983) they can partly be correlated with trends, or rates of change (Melack 1988), of salinity (see Fig. 5.21). Changes in the quantity and species composition also occur in tropical lakes with long-term enrichment (eutrophication), such as Lake McIlwaine, although few are well documented. It is now established that phytoplankton concentrations have increased considerably in Lake Victoria during recent decades, together with chemical evidence of eutrophication, although the cause(s) remains speculative (Hecky 1993; Mugidde 1993; Lehman 1996). Species-sequences are to be expected during the early development of man-made lakes; examples are recorded for Lake Volta in Ghana (Biswas 1969, 1972*b*, 1975), Lake Brokopondo in Suriname (van der Heide 1973), Lake Asejire in Nigeria (Egborge 1974, 1979) and Lake Kariba (Cronberg 1997).

There are a few records of rapid spread of a prominent species in a river system, attributable to accidental introduction or environmental change. Hammerton (1972) has described the downstream spread of a new dominant, the blue-green *Microcystis flos-aquae*, below a new

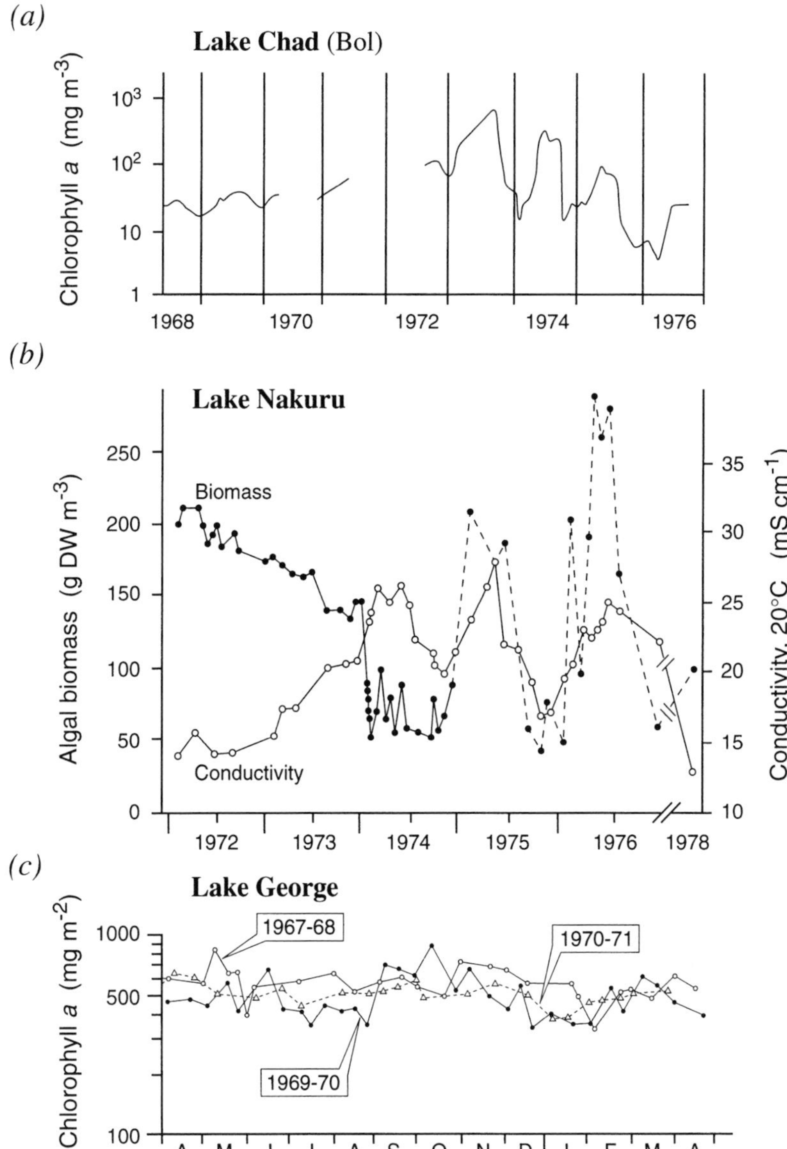

Fig. 5.21. Long-term and annual variation in the abundance of phytoplankton in three shallow African lakes. In Nakuru, a soda lake, the concomitant variation of electrical conductivity is also shown. From Lemoalle (1979*a*), Vareschi (1982) and Ganf (1974*b*).

reservoir on the Blue Nile. In the same river, and in Lake Nubia below, there was a rise to abundance in the 1970s of a previously unrecorded dinoflagellate, *Ceratium hirundinella* (A.I. Moghraby, personal communication).

(f) Zooplankton

This community comprises secondary producers, often predominantly herbivores, in a range of complexity that includes flagellate and ciliate Protozoa, Rotifera, Crustacea – Copepoda, Cladocera plus a few Decapoda – and some insect larvae, notably chaoborids that are facultatively benthic and important predators in many tropical zooplankton assemblages (e.g., Lake Lanao: Lewis 1975, 1979; Lake Malawi: Irvine 1995*b*, Allison *et al.* 1995). Most ecological studies have concentrated on the crustacean components, which account for most of the behaviour described below. Least well known are the protozoa, chiefly flagellates and ciliates: variations in their gross abundance with time has been followed in lakes Tanganyika (Hecky & Kling 1981), Turkana (Ferguson 1982), Lanao and Valencia (Lewis 1985), and in a small Brazilian reservoir (Barbieri & Godhino-Orlandi 1989). Among these were distinctive features of community composition. In Tanganyika the ciliate *Strombidium* sp(p). contained symbiotic algae or possibly sequestered plastids and as biomass could exceed the accompanying phytoplankton; in Lanao and Valencia the protozoans were chiefly represented by small flagellates and ciliates, respectively.

Reproductive rates of zooplankters are influenced by developmental sequences of varying length; the rates of rotifers and cladocerans can be high during prevalent phases of asexual reproduction that often generate population maxima. The usual strong positive dependence of specific growth rates (and, inversely, of stage-duration) upon temperature has been widely studied and is relevant to tropical dynamics (Chapter 3.3). However, a general trend to reduced individual size in tropical forms tends to be correlated with reduced fecundity. Community patchiness and vertical migrations are often pronounced and reduce the accuracy of stock census. Relationships to phytoplankton abundance and composition are complicated by size-selectivity of food intake, occasional antagonistic effects, alternative food sources (e.g., bacteria) and predation.

Diel variability is most conspicuous in vertical diel migrations that are widespread at all latitudes, especially with the Crustacea and chaoborid larvae. Typically there is descent to deeper levels by day and ascent to nearer the surface by night. Quantitative tropical studies began in 1927 with that of Worthington (1931) on Lake Victoria (Fig. 5.22). Later work from Africa included varied lakes in Uganda and Kenya (Worthington & Ricardo 1936; Ferguson 1982; Mavuti 1992), a Nile reservoir (Rzóska 1968), Lake Kainji in West Africa (Adeniji 1978, 1981), Lake Kariba in Central Africa (Begg 1976) and the downstream Cahora Bassa reservoir in Mozambique (Gliwicz 1986*a*: Fig. 5.23). From Central America there are studies on Gatún Lake (Zaret & Suffern 1976); from South America on Brazilian lakes (Fisher *et al.* 1983; Matsumura-Tundisi *et al.* 1984, 1997) and a reservoir (Arcifa-Zago 1978); from the Indo-Pacific work on a crater lake (Ranu Lamongan) in Java (Ruttner 1943) and Lake Lanao (Lewis 1975). Vertical migration is not marked in shallow turbid waters readily mixed by wind action, such as the Nile reservoir. In Lake Calado, on the Amazonian floodplain, it was probably restrained by anoxia below 3 m depth. A diel wind regime on Lake Turkana, Kenya, clearly affected vertical migration of the dominant zooplankter, *Tropodiaptomus banforanus* (Ferguson 1982). This movement usually appeared – as in earlier work of Worthington & Ricardo (1936) – as a *reversed* migration, with ascent during relatively calm afternoons and descent during windy late-night periods. In some other deep stratified waters the centre of gravity of species-populations could shift vertically by about 30 m (Cahora Bassa) or 25 m (Lake Lanao).

Diel cycles of food intake and of excretion are probably widespread in tropical zooplankton; they have been established quantitatively for the copepod *Thermocyclops hyalinus* (= *crassus*) in Lake George (Ganf & Blažka 1974). Saunders (1980) has shown that in Lake Valencia there is a diel reproductive pattern in rotifers that affects egg release.

Annual variability probably often involves a regular, seasonal, component. Annual patterns for tropical (and Southern Hemisphere) water-bodies are reviewed by Hart (1985), from whom the summary in Fig. 5.24 is taken. The series of deep water-bodies (Fig. 5.24a) involves stratified waters subject to periods of de-stratification followed by re-stratification, as indicated. There is no consistent *general* relationship between these and the maxima or minima of gross zooplankton abundance, although a relationship is not excluded for individual lakes and species components.

For example, in Lake Malawi Twombly (1983) and Irvine (1995*a*) found an initial positive reaction to de-stratification, whereas in Lake Valencia the initial response could be negative perhaps as a result of an adverse chemical admixture from below (Infante 1982; Saunders & Lewis

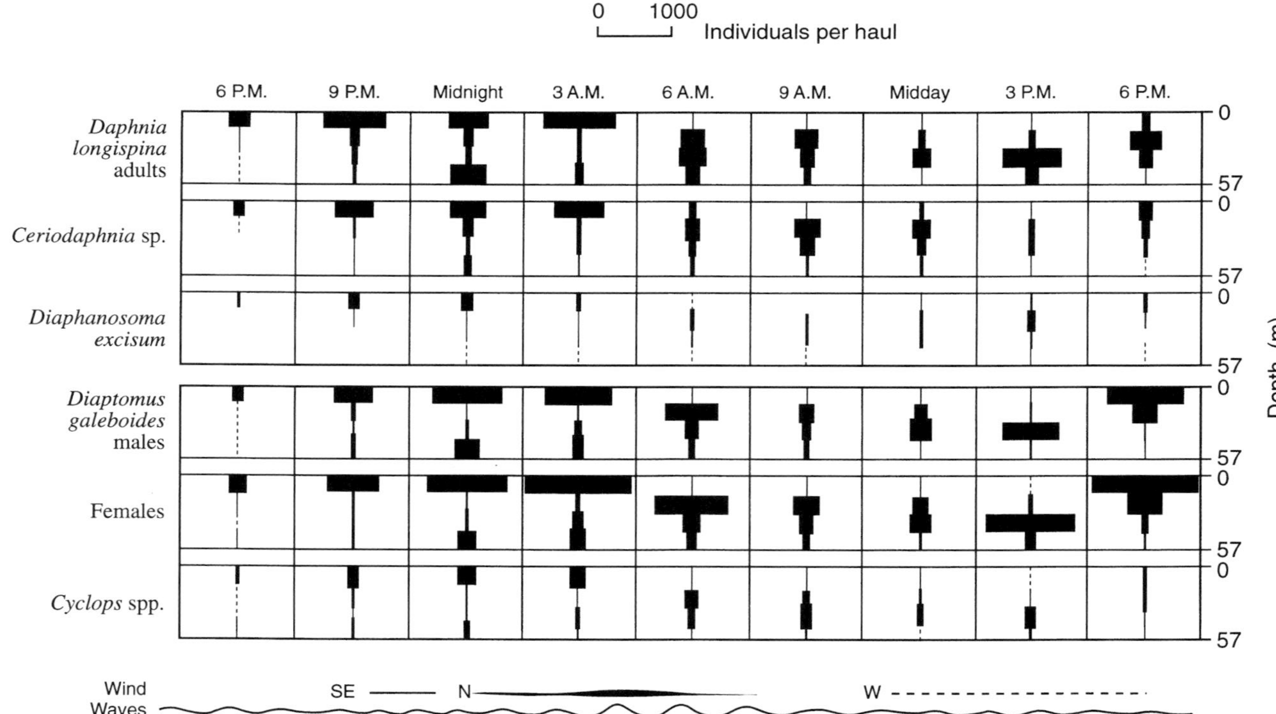

Fig. 5.22. Lake Victoria, 22–23 September 1927. Diel vertical migration sequences of five components of crustacean zooplankton, from sampling within the depth-range 0–57 m by closing vertical net hauls at 3-h intervals. Subjective estimates of wind and surface waves are indicated. Modified from Worthington (1931).

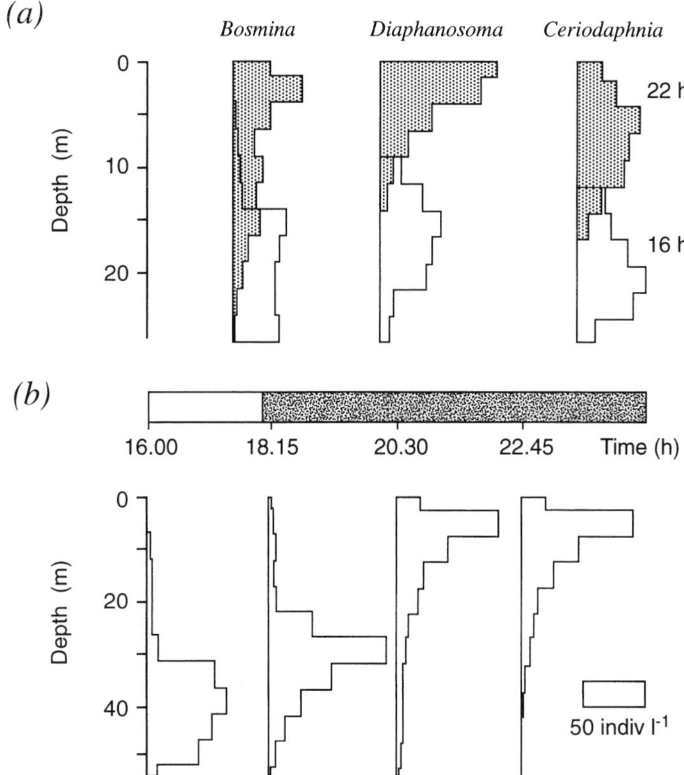

Fig. 5.23. Diel vertical migration of zooplankters in Lake Cahora Bassa, Mozambique: (*a*) changes in the relative depth-distribution of three species between 16.00 and 22.00 hours (shaded), post-full moon, (*b*) details of the evening rise of a *Mesocyclops* sp. on 25 February 1983. Modified from Gliwicz (1986*a*).

1988*a*). There was evidence of an annually recurrent bimodal pattern, with one of the minima possibly induced by a seasonal abundance of planktivorous fish fry (Infante 1982) and/or *Chaoborus* larvae (Saunders & Lewis 1988*a*). A brief annual phase of vertical mixing was believed by Matsumura-Tundisi & Okano (1983) to be the predominant influence behind the seasonality of zooplankton in the small lake Dom Helvécio of South East Brazil, one of a group of forest lakes in which hydrological (water balance) seasonality is minimal (Tundisi 1983). Although a seasonal stratification-mixing cycle is well developed in the high altitude Lake Titicaca, it does not seem to be closely related to the changes in zooplankton populations, which show much irregularity (Pawley & Richerson 1992).

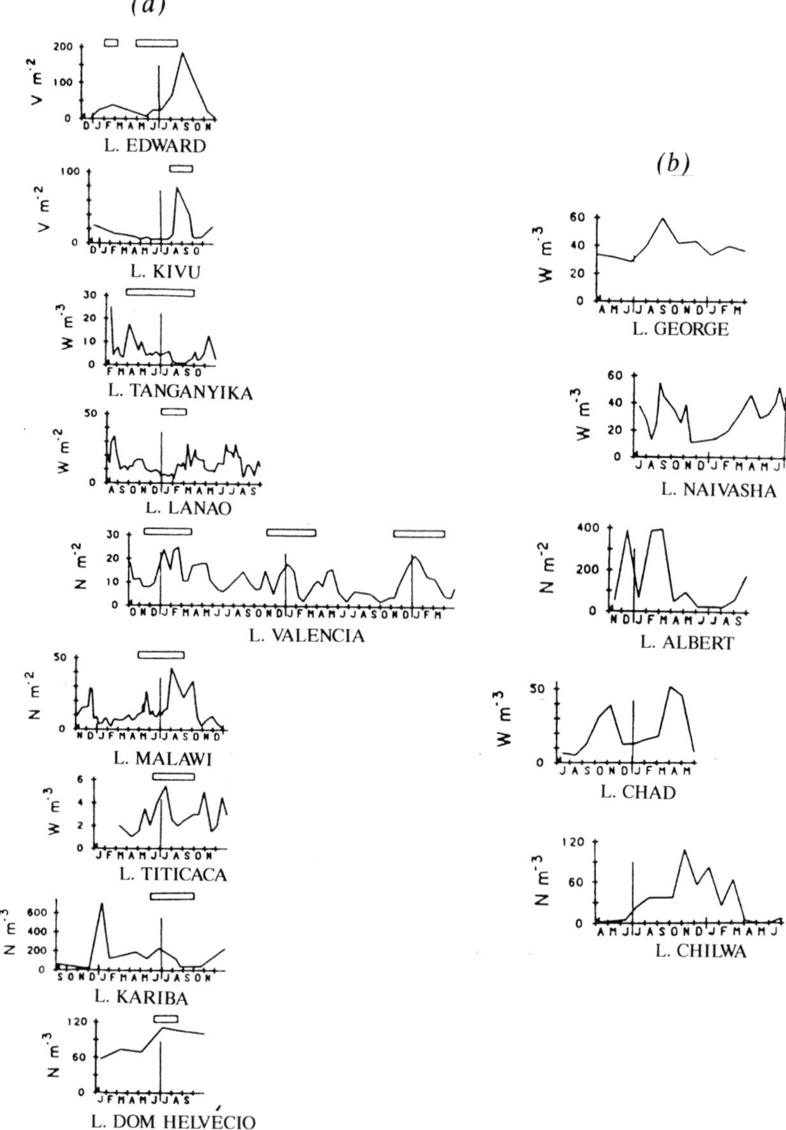

Fig. 5.24. Annual variation of various measures of zooplankton abundance (V, volume; W, weight; N, numbers) in two series of tropical lakes arranged by latitude, that are (*a*) stratified with periodic vertical mixing (bars), (*b*) without persistent stratification. Vertical lines mark the winter solstice. From Hart (1985).

Predation losses to larvae of *Chaoborus* spp. are probably influential in many of the lakes illustrated (e.g., Lake Lanao: Lewis 1975, 1979; Lake Valencia: Saunders & Lewis 1988*a*, *b*) and contribute to the seasonal variability in numbers of herbivores. Such variability in Lake Valencia is unusually great (range > 10/1) for a large tropical lake. In a floodplain lake of the Orinoco River, Venezuela, predation probably strongly influences the seasonal succession (Twombly & Lewis 1987, 1989), as of the cladocerans shown in Fig. 5.25. General issues of planktonic predation by fishes in tropical water-bodies are discussed by Lazzaro (1987). One of particular interest concerns the differential impact of a predator upon different forms of a polymorphic prey species. This situation has been deduced for the horned and non-horned forms of the cladoceran *Ceriodaphnia rigaudi* (= *C. cornuta*: Rzóska 1956) and helmeted and non-helmeted forms of *Daphnia lumholtzi*. The less vulnerable helmeted and horned forms appeared stable or relatively favoured under fish predation in lakes Albert (Green 1967) and Gatún (Zaret 1969). Finally, there is the special case of lunar cycles in the Cahora Bassa reservoir (Gliwicz 1986*a*), induced by fish predation (Chapter 4.4).

Sequential and causal relationships between the abundance of zooplankton and that of phytoplankton have often been postulated for tropical waters, but regulating aspects such as size-compatibility for ingestion have received less attention. For Lake Valencia, Infante & Riehl (1984) obtained evidence of a strongly antagonistic influence of filamentous blue-greens upon zooplankton – as did Hawkins (1988) for the blue-green *Cylindrospermopsis raciborskii* in Solomon Dam, a tropical Australian reservoir. In a fishpond at Malacca the zooplankton coexisted with very dense blooms of *Cylindrospermopsis* (*Anabaenopsis*) *philippinensis*, but not with the mass death and decay of the bloom; later recovery is described by Dunn (1970). Other phytoplankton–zooplankton interactions may involve the 'microbial loop'. For example, in Lake Nakuru the irregular mortality of a dense phytoplankton appeared to be correlated with increases in planktonic rotifers, possibly via an induced abundance of bacteria that served as food for the rotifers (Vareschi & Jacobs 1985).

Most lakes in the second series of Fig. 5.24b are shallow and, though lacking a long-persistent thermal stratification, are more likely to show time-changes related to the water budget. The hydrologically buffered Lake George is an exception, in which the amplitude of annual change in zooplankton density is low and reproduction is continuous (Burgis 1971, 1973). Changes of large amplitude and hydrological determination, related to water level, discharge and retention time, are well known from the middle reaches of the Nile and Amazon rivers, and from the lower Orinoco. However, evidence of population decline by wash-out is not reliably shown by population density alone, owing to dilution effects and the influence of water discharge on population transport and recruitment from backwaters and floodplain in a flowing system (e.g., Saunders & Lewis 1988*e*). Such recruitment on rising levels is the main factor for varying density in the Caura River within the Orinoco system (Saunders & Lewis 1988*c*). On the two Niles near Khartoum dense populations of rotifers, copepods and cladocerans generally built up, together with phytoplankton, in annual cycles that corresponded to the filling and retention of reservoirs (Rzóska *et al.* 1953; Brook & Rzóska 1954; Talling & Rzóska 1967) – although some instances of increase under purely running-water conditions were also recorded (Rzóska 1976). Stages in diapause may possibly aid or enable recolonization after a seasonal phase of turbid floodwaters as occurs in the Blue Nile (Moghraby 1977).

The Amazon floodplain includes numerous lakes filled by seasonal overspill, such as Lago Jacaretinga and Lago Camaleão in which the seasonal peak of total zooplankton is developed at low water before levels begin to rise (Brandorff & Andrade 1978; Hardy *et al.* 1984). During the rising phase of February–April in Lago Camaleão, a well marked succession of four cladocerans occurs (Hardy 1993). In another lake, Lago Grande, highest populations of a principal zooplankter (*Daphnia gessneri*) were developed earlier under high-level and relatively clear water conditions, but were apparently then successively reduced from predation by a planktivorous fish and the unfavourable effect of very high turbidity occasioned by wind disturbance of sediments at low water level (Carvalho 1984). Further examples of population cycles and species succession are known from three floodplain lakes of the Orinoco River (Twombly & Lewis 1987, 1989: see Fig. 5.25; Hamilton *et al.* 1990). Again the overall hydrological control was clearly shown, with interaction between lake retention time and the durations of species development (egg to egg) that broadly increase in the series rotifers–cladocerans–copepods. There was also evidence for an early contribution from diapause-stages and later losses from predation by *Chaoborus*; also for other losses by interception from flow across mats of macrophytes. It was shown that dilution effects could produce a decline in population *density* even at times of increase in population *size*.

At tropical latitudes of 10° or more, an annual cycle of surface temperature that exceeds 8 °C in range is often combined with a strong

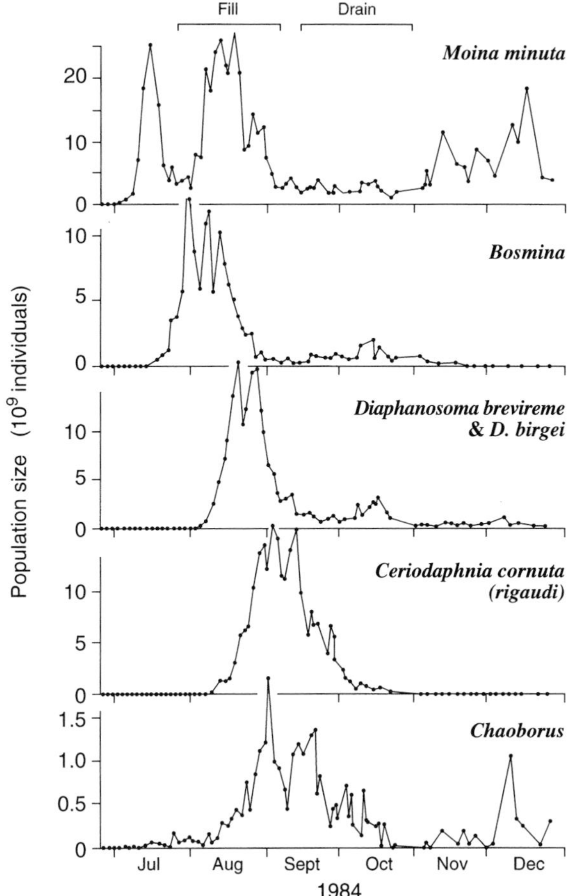

Fig. 5.25. Laguna la Orsinera, Orinoco system floodplain, Venezuela. Changes in the estimated total population size of four cladoceran components and of the insect predator *Chaoborus* associated with phases of lake fill and drain. From Twombly & Lewis (1987).

hydrological influence in shallow water-bodies of varying levels. Lake Chad is an example in which at 'normal' high level the seasonal abundance of the total larger zooplankton was bimodal near the southern inflow (Gras & Saint-Jean 1983; Saint-Jean 1983) and probably also in the more distant northern basin (Robinson & Robinson 1971) – where, however, there were some relatively invariable species like *Thermocyclops neglectus* with continuous reproduction. Saint-Jean (1983) considered that the main determining seasonal factors in Lake Chad were turbidity related to wind disturbance, floodwater injection, water level, temperature (for Cladocera) and possibly zooplanktivorous fishes. Lake Chilwa, at 15° S, is another example in which these factors influence seasonal change in the zooplankton. As at Lake Chad, abundance of *Daphnia barbata* is regularly associated with the cool season (Kalk 1979*a*, *b*). In reservoirs of southern Brazil there is also evidence that water level and retention influence the temporal variation of zooplankton (Rocha *et al.* 1982).

At lower latitudes three shallow lakes in the Eastern Rift of Africa – Naivasha in Kenya, Abijata and Langano in Ethiopia – show considerable annual variation in the abundance and composition of the zooplankton (Mavuti & Litterick 1981; Wodajo & Belay 1984; Mengestou & Fernando 1991*a*, *b*; Mengestou *et al.* 1991). In Naivasha, maxima occurred in the two rainy seasons but with different dominant species; in Awasa, the seasonal incidence of stratification was a principal influence. A strong link between zooplankton changes and rainfall, acting via available food, was also suspected in the Central American lake of Xolotlán (Cisneros & Mangas 1991) and for rotifer populations in two Ugandan crater lakes (Kizito & Nauwerck 1995, 1996).

As with phytoplankton, the annual patterns of abundance in large lakes often differ considerably between offshore and shallow inshore regions. An example, for the cladoceran *Ceriodaphnia rigaudi* (= *C. cornuta*) in Lake Kariba, was described by Masundire (1994). However, there was, at least in the lowermost basin of this lake, a generally positive response of crustacean zooplankton density to the seasonal de-stratification, and probably to inputs of river water (Masundire 1997).

A still more extreme situation of hydrological controls is provided by temporary rainpools bearing often dense populations of phyllopod Crustacea. Tropical examples at Khartoum, studied by Rzóska (1958, 1984), formed in the rainy season with a duration typically of about 1–3 weeks. The crustacean populations developed rapidly from resting stages that survived in hot dry soil for most of the year.

Long-term changes are chiefly documented for the zooplankton of hydrologically unstable shallow lakes in closed basins subject to interannual change of water level and area. The African lakes of Chad (Saint-Jean 1983) and Chilwa (Kalk 1979*a*, *b*) are the major examples, in which there were influences at low level from higher salinity as well as desiccation. At Lake Chilwa the open water dried up completely during 1967–68, but the original main zooplankton components were re-established after the refilling during 1969 (see Fig. 5.26). Contributing to this were the survival in diapause of resting eggs and probably of copepodids of an early recolonist *Mesocyclops* sp. (Kalk & Schulten-Senden 1977; Kalk 1979*a*). In the 1970s evaporative concentration to much higher levels of salinity led to a disappearance of the copepod *Paradiaptomus africanus* (= *Lovenula africana*) from Lake Elmenteita (Melack 1988) and its decline in Lake Nakuru (Vareschi & Vareschi 1984). Populations of various species of rotifers may also change considerably with

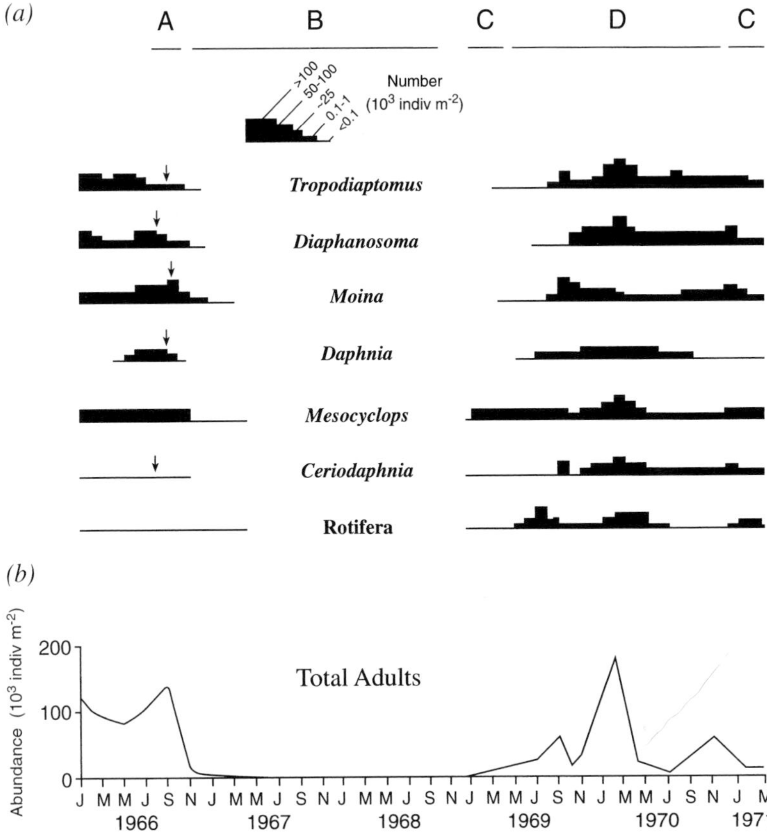

Fig. 5.26. (*a*) Relative abundance of zooplankters in a western bay during the decline and recovery of Lake Chilwa, 1966–71. A, critical months of the 'pre-drying period'; B, 'drying period'; C, dilute months of the 'filling period'; D, 'post-filling period'. Arrows indicate the first appearance of resting eggs. Organisms are listed in ranked order with the dominant species above. (*b*) Total adult zooplankton estimated from horizontal net hauls (1966–67) or vertical hauls (1969–71). Modified from Kalk & Schulten-Senden (1977).

natural increases of salinity, as in Lake Nakuru (Nogrady 1983; Vareschi & Vareschi 1984) and the Kanem soda lakes, Chad (Iltis & Riou-Duwat 1971).

The closed-basin Lake Valencia in Venezuela has shown large year-to-year differences in the occurrence of Cladocera, that have been ascribed to interference from filamentous blue-green algae of intermittent abundance (Infante 1982; Infante & Riehl 1984).

Few hydrologically stable tropical lakes have been studied over periods of decades. Similarity of early and more recent records of faunistic composition is exemplified by Ranu Lamongan, Indonesia (1928 and 1974: Green *et al.* 1976), but Green (1976) encountered considerable differences of species between collections of 1962 and 1975 from Lake Mutanda, and of 1931 and 1962 from Lake Bunyoni, in western Uganda. Mavuti (1990) commented on the apparent similarity of the zooplankton assemblages found in Lake Naivasha, Kenya, in 1929–31 and 1978–80. Another similarity is recognizable in the records from Lake Victoria since 1953, despite a major eutrophication there (Branstrator *et al.* 1996). However, a correlated entry, or rise to detectable numbers, may have been that of *Daphnia lumholtzi* var. *monacha*.

(g) Zoobenthos

As the community of animals closely related to a bottom substratum, zoobenthos is recruited from diverse groups of invertebrates with oligochaetes, molluscs and insect larvae generally the most prominent. In lakes and reservoirs, littoral, sub-littoral and profundal communities can be distinguished. These often differ with respect to type of substratum and the likelihood of deoxygenation that is a widespread restrictive factor. Quantitative time-studies in the tropics are not numerous, the most notable being from African lakes that include Kariba, McIlwaine, Chilwa, George, Chad and Volta. An early landmark was set by the work of MacDonald (1956) on Lake Victoria. Similar quantitative studies of tropical stream and river benthos are even fewer.

The differing mobility of the benthic components – some with an aerial stage in their life history – influences patterns of variation in time. The often rich but complex communities in fringing submersed vegetation are not considered here in any detail. Their time-variability is greatly influenced by that of the vegetation they inhabit, often in consequence of varying water level – as in the hydrologically unstable Lake Chilwa (McLachlan 1975) and Lake Chad (Dejoux 1983*a*), the Amazonian floodplain (Junk 1980), and the billabongs of northern Australia (Marchant 1982). In Lake Chad the seasonal patterns of species abundance were varied, reflecting the diversity of the animal groups involved.

Diel variability is most prominent in three types of movements. Active vertical migration in the water-column can be sporadic and lead to adventitious occurrence, but is systematic and of 24-h periodicity with the predatory larvae of chaoborid flies that are especially common in tropical lakes. These typically rise at night as a planktonic component after a deeper, sometimes benthic, existence during the day (Fig. 5.27). Descriptions of these diel cycles include Worthington & Ricardo (1936) and McGowan (1974). Second, the last aquatic instars of these and other insects, notably chironomids, often give rise by synchronous emergence to swarms of aerial adults with a diel rhythm (e.g., Corbet 1964; Elouard & Forge 1978). Third, in running waters a small fraction of benthic individuals often detach and are found as a *drift* component with a diel rhythm, numbers being generally larger at night (e.g., Bishop 1973; Hynes 1975*b*; Elouard & Lévêque 1977, Statzner *et al.* 1984, 1985*a*, *b*;

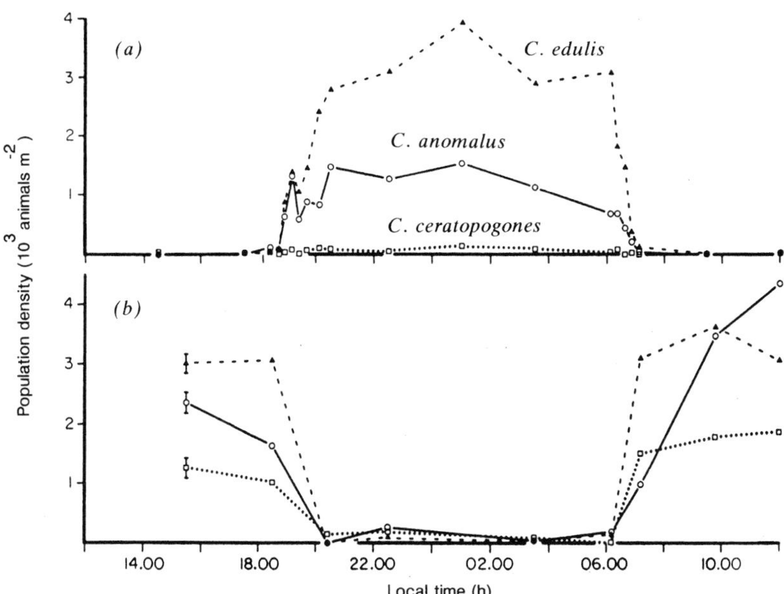

Fig. 5.27. Diel abundances of (*a*) planktonic and (*b*) benthic individuals of three species of *Chaoborus* in Opi Lake A, Nigeria, on 27–28 January 1979. From Hare & Carter (1986).

Barnes & Shiozawa 1985; Benson & Pearson 1987) but not always so (Turcotte & Harper 1982). There are implications of drift for fish feeding (Chapter 3.3a), downstream dispersal and the possible existence of a compensating upstream movement. The last two features appeared as quantitatively minor in a study by Benson & Pearson (1987) of a stream in North East Australia – although incidence varied strongly between wet and dry seasons.

Annual variability can be induced by various factors of the radiation–temperature and hydrological complexes. There may, in lakes, also be periodic disturbance by wind of shallow-water sediments and phases of renewed supply of food by sedimentation of detritus. Temperature can control not only growth rates but, in deeper lakes, the seasonal incidence of stratification that often involves O_2 depletion in deep water. For the shallow Opi Lake, Nigeria, depletion is relatively mild and is survived by a zoobenthos dominated by chaoborid larvae that, as we have seen, migrate diurnally (Fig. 5.27). More severe anoxia with corresponding depth-restriction of zoobenthos is illustrated in the early phase of three African man-made lakes, Kariba (McLachlan 1970*a*, 1974: see Fig. 5.32), McIlwaine (Marshall 1978, 1982*a*) and Volta (Petr 1972). Simultaneously, all these lakes were subject to considerable changes of level, both short- and long-term, by varying water input–output. A recent rise in level created a habitat extension that was rapidly colonized by insects, notably chironomids, with flying and egg-laying adults.

In the shallower Lake Chad at 'normal' higher water level, a lake-wide seasonal variation of zoobenthic biomass could be recognized (Lévêque *et al.* 1983). Of the three main components, oligochaetes and chironomids showed highest biomass during the coolest season with higher water level (Fig. 5.28), whereas the biomass of most molluscs was not seasonally variable. In the similarly shallow but hydrologically stable Lake George, the very limited zoobenthos (other than *Chaoborus*) of fluid sediments was of near-constant and non-seasonal biomass, and showed continuous reproduction (Darlington 1977). These and other examples of annual variation in African lakes of varying latitude are reviewed comparatively by Hart (1985).

Reproduction by caridean (decapod) shrimps can be continuous and year-round in some tropical regions, but elsewhere can be synchronized by the monsoon or seasonally delimited by periods of low temperature near the edge of the tropics (e.g., Hong Kong: Dudgeon 1985: see Fig. 5.30). For Lake Zwai, Ethiopia, marked seasonal change is shown by the abundance of benthic ostracods; there is a correlation with hydrological

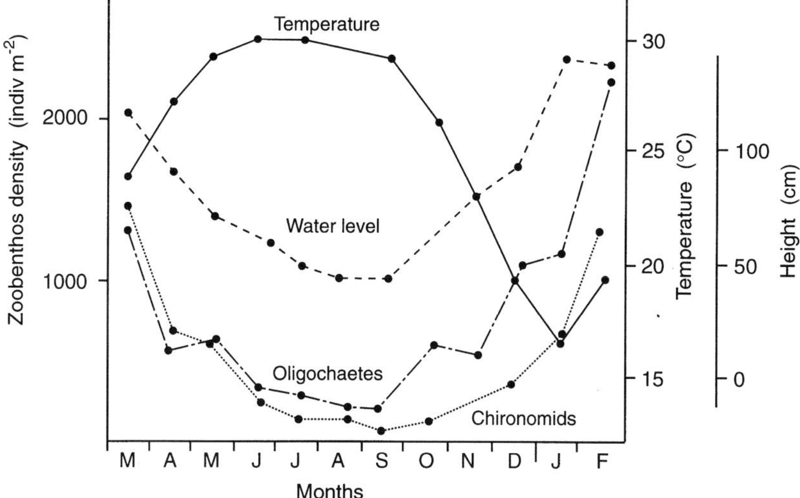

Fig. 5.28. Seasonal variation in mean density of benthic chironomids and oligochaetes for eastern Lake Chad during 1967, in relation to water temperature and water level on the Bol gauge. From Dejoux *et al.*, after Lévêque *et al.* (1983).

seasonality, numbers being reduced after the onset of the long and short rains (Martens & Tudorancea 1991). More widely in Africa, breeding incidence and numbers of freshwater snails often vary considerably in strongly seasonal higher-latitude environments but not in more equable ones. The difference is seen, for example, from studies of *Bulinus globosus* in Zimbabwe and Zaïre (Brown 1994).

For seasonally filled lakes in the Amazon floodplain, deoxygenation is a principal restriction on the development of zoobenthos. In rivers and streams here the floodpulse dominates and induces lateral extensions of range, as of abundant species of shrimps (Walker & Ferreira 1985). In the vertical dimension the level change is coped with by the vertical migration of some species and by the association of others with surface-bound vegetation (Reiss 1976, 1977). However, the populations of many less mobile species are reduced to low numbers, from which recovery is enabled by high reproductive rates (Junk 1984).

In contrast to the large river with a regular annual floodpulse, smaller streams often experience short irregular spates of high flow after rainstorms, that are *disturbances* to the populations of benthic animals. Consequent reduction in numbers, and later recovery, have been analysed by Flecker & Feifarek (1994) for two Andean streams. In a Brazilian stream of very variable flow, the abundance of mayfly nymphs on floating litter was responsive to spates in the rainy season; it also showed temporary decline on the transition from rainy to dry and dry to rainy seasons when considerable change occurred in the species-composition (Nolte *et al.* 1997). The general feature of seasonal abundance of larval blackflies (Simuliidae) in a Venezuelan stream was a maximum near the end of the rainy, high flow, season (Grillet & Barrera 1997). However, the consequences of variable flow are also influenced by the local stream gradients, as illustrated by aquatic Hemiptera in two Costa Rican streams (Stout 1982). There can be further interaction by modification of fish predation on the zoobenthos, as found in a stream by Hong Kong (Dudgeon 1993). High flows may also wash away detritus of significance as food, as in a river of South India during the monsoon (Arunachalam *et al.* 1991). The 'drift' component of living organisms, a flux recruited from the zoobenthos, is generally likely to increase with discharge. Examples have been described from the Naro Moru River on Mount Kenya (Mathooko & Mavuti 1992, 1994; Mathooko 1996).

For the zoobenthos of many lakes and streams dominated by insects, the number of generations per year (i.e., voltinism) is an important factor of intra-annual change. Its value is likely to increase, above the usual temperate univoltine state, in tropical waters of higher temperature and shorter developmental times. However, at the edge of the tropics, a stream near Hong Kong contained two *Ephemera* species that were univoltine (Dudgeon 1996). Of the few sites studied at lower latitude, a mountain stream in Central Africa contained species of Trichoptera with generation times between two and four months (Statzner 1976). A stream in Ghana yielded examples of complete growth of aquatic stages of insects in about 2.5 months (Hynes 1975*a*). Here seasonal faunal change was related to water flow and development of a vegetational substratum of mosses and algae. After a dry phase faunal renewal was probably mainly by eggs laid by flying adult insects rather than by a drought-survival strategy of a resting stage. Both strategies of renewal appear in other temporary streams (e.g., Harrison 1966) and are shown by the insects that colonize temporary pools, often of brief duration (cf. zooplankton Crustacea, Section 5.2f). Thus larvae of *Polypedilum vanderplanki* can survive in a dehydrated state, whereas those of its frequent regional associate *Chironomus imicola* cannot but are associated with reinvasion from eggs and a larval development time that can be as short as 12 days (McLachlan & Cantrell 1980; McLachlan 1983). Cohort succession in these species is illustrated in Fig. 5.29 for pools in Malawi.

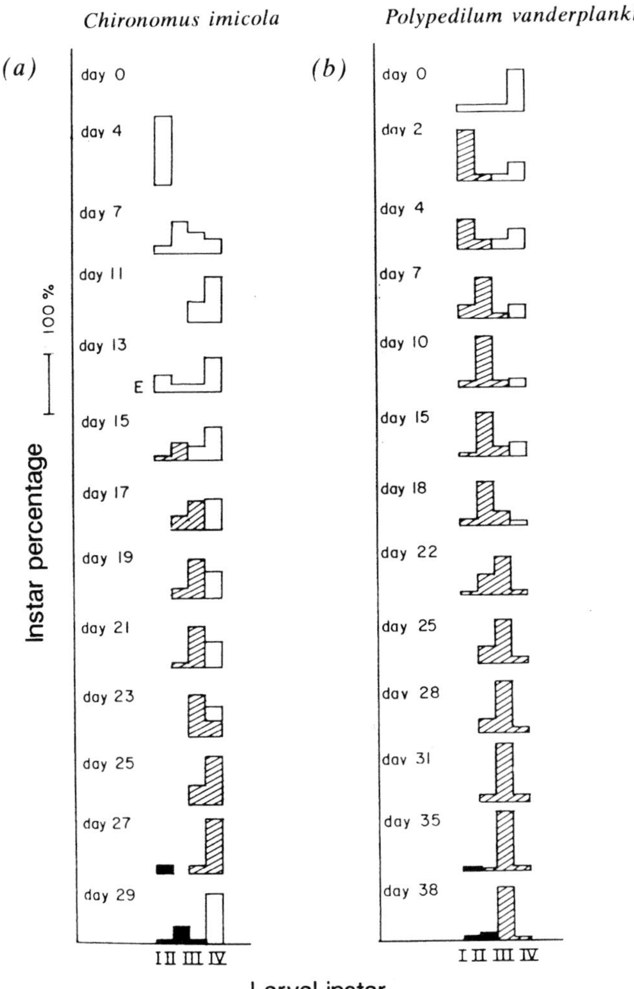

Fig. 5.29. Temporary pools, Malawi. Time-progression of the percentage representation of successive instars (I–IV) in counts of the larvae of two chironomids, (*a*) *Chironomus imicola* and (*b*) *Polypedilum vanderplanki*, occurring as separate pool populations. Successive cohorts are distinguished by shading. Modified from McLachlan (1983).

Streams by Hong Kong, especially the Lam Tsuen River studied by Dudgeon (1992), illustrate a range of seasonality in components of the zoobenthos at the northern edge of the tropics. Here there is a relatively cool 'winter', with a water temperature of 15–20 °C and a monsoon period of heavier rainfall. Breeding of pulmonate snails and the prosobranch snail *Melanoides tuberculata* is interrupted by the cooler season, unlike the year-long continuity found for the latter species in Malaysia by Berry & Kadri (1974). A similar temperature control in freshwater decapod shrimps (Fig. 5.30) has already been mentioned. Aquatic insects such as Odonata, Trichoptera and Ephemeroptera vary in their voltinism, and emergence can be seasonally synchronized or substantially aseasonal – although often timed to precede the summer monsoon during which sharp spates of flow may reduce population densities. The fauna is recruited from both tropical and north-temperate forms, and this is relevant for temperature responses in seasonality. Comparison can be made with a lower-latitude stream of central Malaysia, 3° N, where temperature seasonality and monsoonal impact are small (Bishop 1973). There the reproduction of invertebrates was largely continuous and aseasonal; changes in population numbers were primarily caused by instability of the substratum during spates and not to the consequences of synchrony in life cycles. In a Malaysian swamp, instability of the substratum – here the macrophyte *Utricularia flexuosa* – occurred during two monsoonal rainy seasons with accentuation by an annual senescence of endogenous origin. It caused a seasonal depression in the abundance of the associated macro- and micro-fauna (Lim & Furtado 1975).

In bivalve molluscs, intermittent checks to growth can often be recognized from rings induced in the shells (cf. scale-rings of fishes: Section 5.3h). A tropical example from the Nile near Khartoum is described by Moghraby & Adam (1984). Rings on *Corbicula consobrina* corresponded to a resting state during the flood season, when growth was arrested by an unfavourably high silt content.

Collections of the flying adults of aquatic insects by light-traps have also been used to assess seasonal activity and abundance, although certain biases are possible. In some cases a lunar periodicity has been found (e.g., Corbet 1958, 1964), as is illustrated in Fig. 5.31. Results of McElravy *et al.* (1982) for Trichoptera, from a relatively non-seasonal environment in Panama, showed statistically significant seasonal changes but with active seasons longer and seasonal peaks less sharp than in temperate regions. An emergence trap above a Central African hill stream in Zaïre showed year-round emergence of various insect groups with little influence of rain or water level (Böttger 1975), although emergence rates of component species could vary cyclically due to the varying reproductive success of antecedent generations (Statzner 1976).

Long-term variability has been most studied in relation to hydrological change, affecting level and depth, in deep man-made lakes (e.g., Kariba,

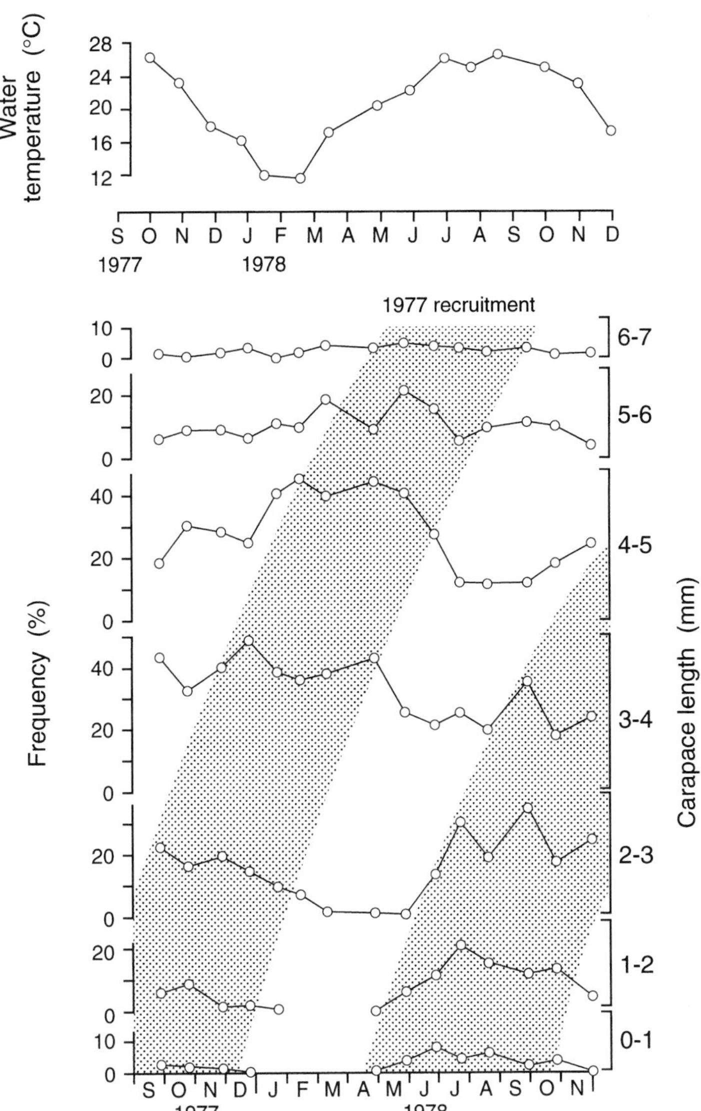

Fig. 5.30. Evidence of seasonal recruitment to two successive cohorts (shaded) of the shrimp *Neocaridina serrata* in a Hong Kong stream, from changes in the % frequency of individuals in seven size classes shown in relation to water temperature. From Dudgeon (1985).

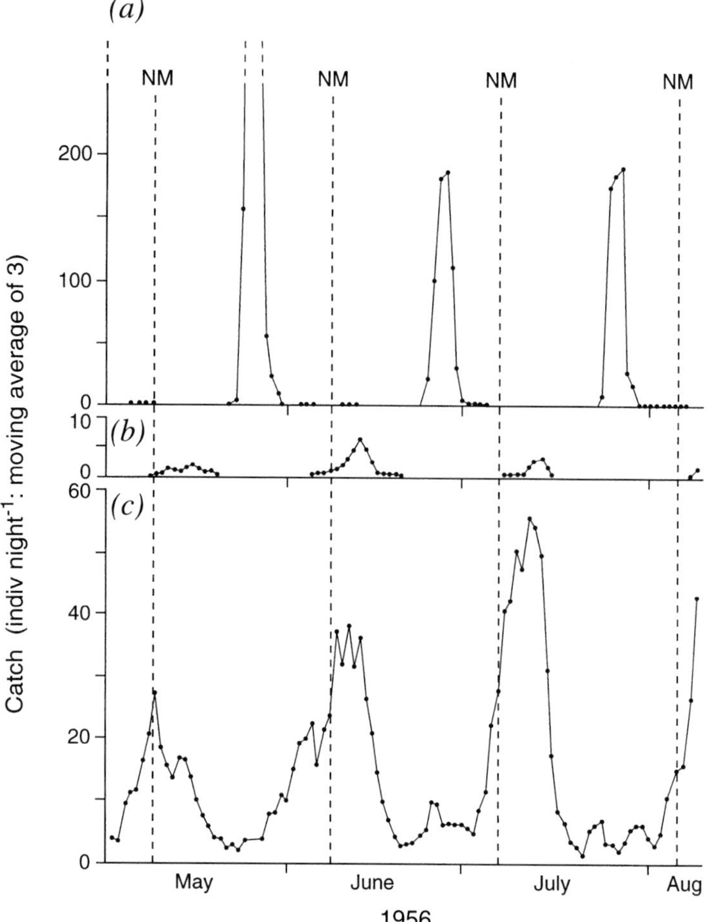

Fig. 5.31. Lake Victoria at Jinja, Uganda. Three examples of lunar phase-related emergence of aquatic insects, based on night catches in light traps: (*a*) *Povilla adusta* (Ephemeroptera), (*b*) *Clinotanypus claripennis* (Chironomidae), (*c*) *Tanytarsus balteatus* (Chironomidae). NM, new moon. From Corbet (1964).

Volta, McIlwaine) and shallow unstable lakes (e.g., Chad, Chilwa). As during annual change, various animal components differed in their response to long-term changes of level/depth. A new substratum appeared in the form of submerged trees and bushes (Fig. 5.13), that were colonized by surface-living and – especially after death – by wood-boring animals such as the mayfly *Povilla adusta* (McLachlan 1970*b*, 1975; Petr 1970). Chironomids often rose to abundance at the mud–water interface; the flying adults appeared in pest-like numbers at some localities on the Nile after the creation or extension of reservoirs either nearby or even far upstream (D.J. Lewis 1956; Rzóska 1964). Some species are rapid colonizers of newly flooded shallow areas and in depth-distribution also readjust to falling levels, as at lakes Kariba (McLachlan 1970*a*, 1974), Volta (Petr 1972, 1974), McIlwaine (Marshall 1982*a*) and Chilwa (McLachlan 1974; Cantrell 1988). In the Central African lakes the principal pioneer species, *Chironomus transvaalensis*, was only temporarily abundant during the phase of refilling.

The deeper populations of molluscs were less tolerant of level change in Lake McIlwaine, where severe drops in level during 1968–69 and 1972–73 decimated the mussel population (Marshall 1982*a*). Nevertheless in Lake Kariba abundant and ecologically influential bivalve populations have developed in the littoral despite considerable changes of level (Machena & Kautsky 1988). Molluscan components in the hydrologically unstable lakes Chad (Lévêque *et al.* 1983) and Chilwa (McLachlan 1979) were much reduced by low-level phases of the 1970s. Elsewhere bivalves suffered from the deep deoxygenation that occurred after the Aswan reservoir on the Nile was extended to over-year storage in the 1960s (Entz 1976). The early post-filling phase of the Volta and Kariba lakes was marked by deep deoxygenation that later lessened (Fig. 5.32), but was originally inimical to a development of deep-water zoobenthos (Petr 1972, 1974; McLachlan 1974).

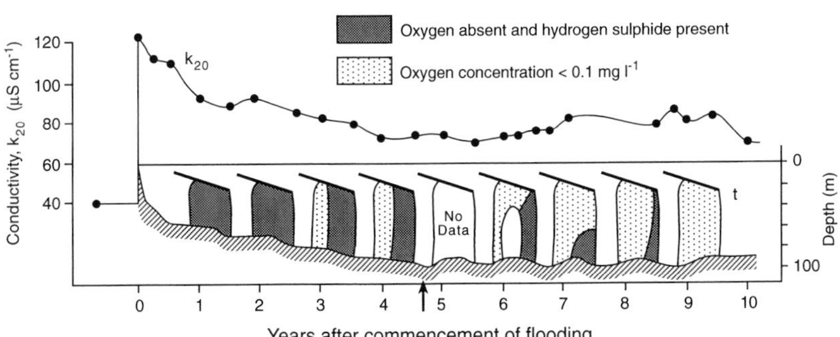

Fig. 5.32. Lake Kariba, Zimbabwe-Zambia. Diagrammatic representation of changes in the early years after dam-closure, including the conductivity (k_{20}) of surface water. The development of oxygen depletion with or without H_2S in deep water below the seasonal thermocline (t) is represented by successive depth-time diagrams. Arrow, end of primary filling. From McLachlan (1974).

In temperate water-bodies, long-term enrichment or eutrophication often has considerable effects on the abundance and composition of the zoobenthos, especially in the profundal of lakes, but long-term studies in the tropics are rare. One of the best-studied examples is Lake McIlwaine (Lake Chivero), Zimbabwe (Thornton 1982), but here the effects of enrichment upon the zoobenthos – that probably included an increase in the oligochaete *Limnodrilus hoffmeisteri* – were overshadowed by the effects of changing water level (Marshall 1978, 1982*a*). Comparison between surveys in 1973 and 1992–3 showed a pronounced increase in prosobranch snails, which may have been aided by a three-fold increase in Ca^{2+} concentration to 2 mmol l^{-1} (Marshall 1995). The apparent enrichment of the largest tropical lake, Lake Victoria, in the 1970s and 1980s was probably accompanied by more extensive deep deoxygenation (Ochumba & Kibaara 1989; Hecky 1993; Hecky *et al.* 1994) that would be unfavourable to the profundal zoobenthos. In rivers and streams also, the zoobenthos is susceptible to organic and nutrient enrichment. This was seen in an organically polluted river at Hong Kong, studied during 1976–7 by Dudgeon (1984), where the most polluted zone shifted upstream or downstream according to rainfall and combined loss of diversity with increased numbers of some species.

(h) Fishes

Fishes are distinctive as the subjects of much work on populations and behaviour, being mostly long-lived organisms for which cohorts (age-classes) and growth checks can often be distinguished (Figs. 5.33, 5.36). Their time-variability is frequently influenced by seasonal migrations and reproductive phases. These temporal aspects of their ecology in tropical freshwaters are outlined briefly by Payne (1986), in more detail by Lowe-McConnell (1975, 1979, 1987), and – for Africa – by Lévêque *et al.* (1988) and Lévêque (1997).

Of the two main factor-complexes, the hydrological is most prominent for this group. Most tropical fishes inhabit rivers of seasonally variable discharge, and even for lacustrine species the majority – other than cichlid species flocks of African lakes – retain some features of riverine ancestors by seasonal migration and spawning in inflow streams. Nevertheless, the radiation/temperature complex has its effects on diel variability and on annual variability of growth and reproduction especially at latitudes $> 15°$. At low latitudes there is less potential for control by photoperiod.

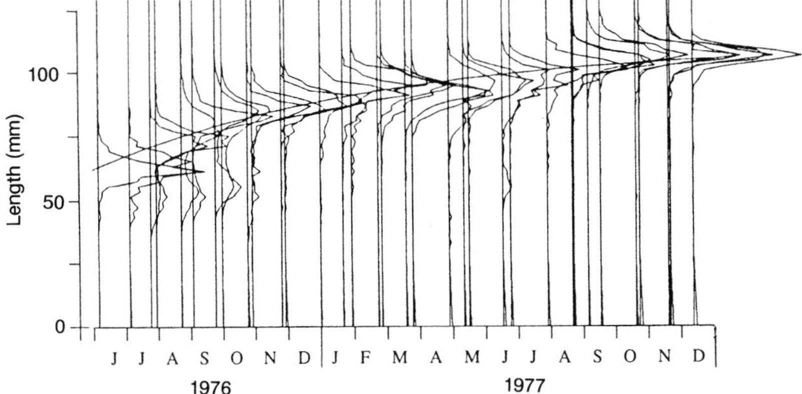

Fig. 5.33. Shift with time, 1976–77, of relative length-frequency distribution in a cohort of the fish *Engraulicypris sardella* collected at Monkey Bay, Lake Malawi. From Thompson *et al.* (1995).

In the tropics there is a prominent development of fishes adapted to feed on lower trophic-level food, as detritivores (Bowen 1984) and filter-feeding planktivores (Lazzaro 1987). Feeding over long periods is then usually possible on this low-quality food, providing sustained nutrition that is reflected in within-year patterns of growth.

Diel variability is often prominent in behavioural traits associated with feeding. For this a refuge may be vacated. Many species specialize as daytime or night-time, or dawn or dusk, feeders and perform diel movements accordingly. These movements may be correlated with corresponding movement of prey, as in the vertical migration patterns described by Begg (1976) that involve nocturnal ascent of the cladoceran *Bosmina longirostris* and its predator clupeid fish *Limnothrissa miodon* in Lake Kariba. Comparable, linked, vertical migrations of prey and predator are known, involving planktivorous fishes, in lakes Cahora Bassa (Gliwicz 1986*a*) and Tanganyika (Hecky 1991; Coulter 1991*b*). Even when migrations are absent, ingestion and digestion by planktivores can follow a marked diel rhythm. This has been quantified from stomach contents for the cichlids *Oreochromis niloticus* and *Haplochromis nigripinnis* in Lake George (Moriarty & Moriarty 1973*a*: see Fig. 5.54), with daytime feeding. A general account of diel feeding patterns in African fishes is given by Lévêque (1997).

It is not uncommon for some tropical lake fishes, and notably cichlids, to alternate between deep and shallow water in a day–night rhythm. This may possibly reduce loss by predation, but additionally the temperature changes involved may have implications for consumption and metabolism that lead to enhanced growth rates. Such implications were modelled by Caulton (1978) from experimental data on *Tilapia rendalli*.

Lunar cycles of behaviour are well known as endogenous rhythms in fishes generally (Leatherland *et al.* 1992) but not for tropical freshwater species. The lunar spawning synchronicity of some cichlids in Lake Tanganyika (Nakai *et al.* 1990; Rossiter 1991) is a rare example, shown in Fig. 5.34. Lunar phase is also known to influence the timing of spawning migration by some fishes; tropical examples are known from the Mekong and Niger rivers (Welcomme 1985). In the Mekong River fishes move downstream in a definite order and are caught during only part of each lunar month, mainly between the first quarter and full moon from October to February. The largest fishes migrate down in the first lunar period (October), the siluroids travelling by night, the large cyprinids by day. The migration builds up to a maximum of species in the third lunar period (December) (Blache & Goosens 1954). Directly light-controlled (and so exogenous) periodicity of grazing on zooplankton by pelagic clupeids in Lake Cahora Bassa has already been described (Chapter 4.4). Prey availability, and hence intake, may also depend upon the phase of the moon (see Fig. 5.35).

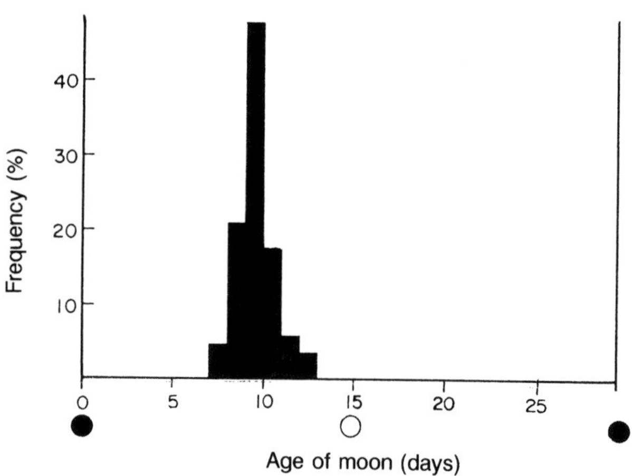

Fig. 5.34. Spawning in Lake Tanganyika of the cichlid fish *Lepidiolamprologus elongatus* in relation to lunar phase, as percentage frequency of sampling occasions. From Nakai *et al.* (1990).

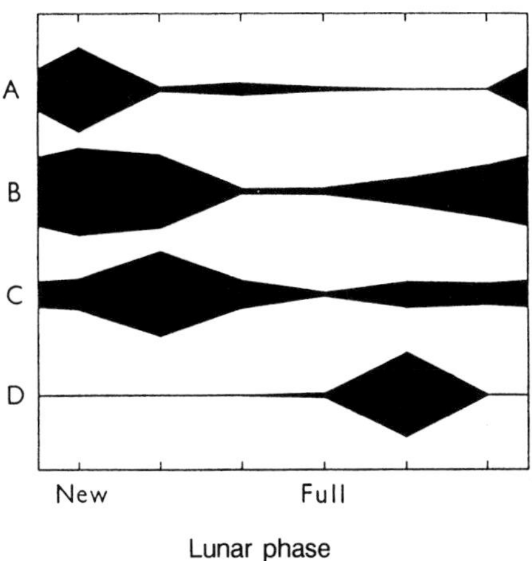

Fig. 5.35. Lunar periodicity of feeding, over one lunar cycle, in a pelagic fish, *Alestes jacksoni*, in Lake Victoria. Feeding is indicated by the percentage occurrence (various scales) of five foods in the gut contents. A, chironomid pupae (main contents); B, pupal exuviae; C, chironomid adults; D, *Povilla* adults. From Corbet (1961).

Annual, usually seasonal, variability is abundantly represented in the feeding, growth, migration, reproduction and age-structure of tropical fish populations. The expression here of the hydrological factor complex is partly related to its general influence on environmental variability (Chapter 4.2), for which there is a graded sequence from lakes of long retention and stable level, lakes of short retention or unstable level, rivers of small to large amplitude for discharge, seasonal floodplains, and temporary waters.

In equatorial examples of the first group, such as lakes Victoria and George, are found some cichlid species with year-round breeding and populations of relatively continuous age-structure. Nevertheless sources of annual variability in the fish populations are not lacking. Year-round breeding can vary in intensity according to the incidence of rainy and drier seasons, as reported by Gwahaba (1978) for the cichlid *Oreochromis niloticus* in Lake George. Variable food supply is exemplified by planktonic diatoms under a hydrographic (mixing) control, whose cichlid consumer *Oreochromis esculentus* in Lake Victoria appeared to show one or two main breeding seasons in relation to differences in the frequency of mixing between northern and southern areas of the lake (Lowe-McConnell 1956, 1987). A single annual phase of mixing also seems to determine the seasonal period of greater abundance of rotifers in Lake

Where high river levels are combined with extensive lateral overspill, as in tropical floodplains, there develops a cycle, normally seasonal from seasonal rainfall, in which the greater opportunities for dispersal and feeding of fishes at higher water levels are central features. These, and other correlates with the cycle of rising and falling level, have been summarized by Welcomme (1979, 1985), Lowe-McConnell (1987: see Fig. 5.37) and Junk *et al.* (1989). There is, for many fishes, a sudden access to food supplies of terrestrial origin that is reflected in higher growth rates. Goulding (1980) provides dramatic illustration for Amazonian forests. For detritivores, changes in growth rate are not marked (Bayley 1988). However, a phase of higher mortality may follow as levels fall and as areas of water become small and often isolated. Such mortality, and its contributing factors, have been studied quantitatively by Chapman & Kramer (1991) for *Poecilia gillii* in a Costa Rican stream.

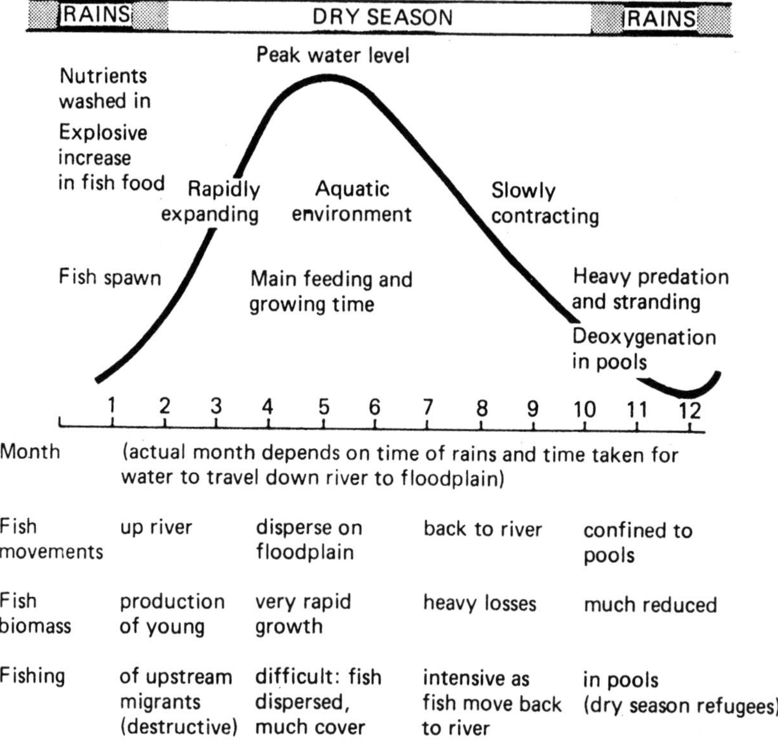

Fig. 5.37. Seasonal cycle of events in a floodplain river related to the biology of fishes. From Lowe-McConnell (1987).

Predation by fishes is typically more intense at low water levels, with reciprocal disadvantage and advantage for prey and predator. It is also seen, independently of river overspill, in the concentration phase of pools in flooded Venezuelan savanna studied by Prejs & Prejs (1987, 1992). For most floodplain fishes the hydrological, feeding and reproductive cycles are interconnected, with spawning early under rising levels and consequent opportunities for the feeding of young during high levels. Gonad activity must therefore begin before the main hydrological events. Overall, total biomass production is increased in years with greater extent and duration of flooding (Welcomme 1979).

Between the flood periods there is an unfavourable dry season in many tropical river systems and small water-bodies, over which survival of some cyprinodontids in South America and Africa is by resting eggs buried in the bottom mud ('annual fishes': Lowe-McConnell 1987). Here there is an embryonic or pre-embryonic stage of diapause, followed by rapid growth (e.g., Bailey 1972). In rivers where flow ceases, the bed may bear isolated pools in which fishes persist. An extensive series in the River Sokoto, Nigeria, was studied by Holden (1963). Here it appears (Chapman & Chapman 1993) that there is year-to-year regularity as regards the species present, but much variability in their relative proportions – perhaps partly due to chance factors.

Fish response to the radiation/temperature complex of within-year variability is likely to be most marked at the higher tropical latitudes, where the environmental amplitude is greatest. The Nile Perch, *Lates niloticus*, of wide distribution in the northern half of Africa, provides an example from the variable indication of growth checks as rings on its scales. In Lake Chad (13–14° N) these are relatively well defined and correspond to the period of seasonal low temperature (Hopson 1968, 1972; Loubens 1974), whereas in more equatorial African lakes such as Lake Albert they are ill defined or irregular. The colder season at Lake Chad also appears to reduce growth rates and induce scale-rings in *Alestes baremoze* (Hopson 1972), and to inhibit reproductive activity in a variety of fishes. Such inhibition may also occur in the Okavango internal delta, latitude 19–20° S, where the flood arrives at the coldest season; here breeding of fishes is related not to it but to the warmer season (Lowe-McConnell 1987). Elsewhere a dry season can be the equivalent of a 'physiological winter', with checks in breeding activity and in growth, the latter with visible scale-rings, as in a marshy savanna region of Guyana in Central America (Lowe-McConnell 1964). Another situation is provided by the annual scale-rings of an introduced salmonid,

Rainbow Trout, in elevated equatorial regions of East Africa. These are not related to environmental temperature or food supply, but to the maturation of gonads (van Someren 1950).

In deeper lakes the annual radiation–temperature cycle often induces a stratification cycle (Chapter 4.3a), and sometimes local upwelling, that in turn may govern the abundance of phytoplankton and zooplankton (Section 5.2e–f). Thus the food intake and growth of planktivorous fishes may respond to this complex of environmental factors. A contributing influence lies in the more ready digestibility of diatoms, the phytoplankton component that typically reacts most positively to cooling with vertical mixing. Probable examples for African fishes are described for cichlids in lakes Victoria (Lowe-McConnell 1956) and Malawi (Eccles 1974) and for a clupeid (sardine) in Lake Tanganyika (Chapman & van Well 1978; Coulter 1970, 1991*b*). Spawning of the latter occurs in the northern end of the lake some months later than in the southern end, 'probably adapted to timing of wind induced nutrient enrichment and plankton production', so that young fry recruits are synchronized or 'matched' to the plankton maximum. This is a special case relevant for the match-mismatch hypothesis of Cushing, the possible application of which to freshwaters is discussed by Harris (1986).

Long-term variability has been most studied in four contexts.

(i) Quantitative changes with time in the fish populations of water-bodies subject to maintained fishing pressure. For the tropics an unusually long record is available for Lake Victoria, where gill nets have had an exceptionally long use. Their later use, with progressively smaller mesh sizes that took younger fishes, has been related to the decline of a major indigenous species, *Oreochromis esculentus* (Fryer 1973; Craig 1992: see Fig. 5.38). After 1965 bottom trawling extended fishing methods and caused a reduction in the abundance of haplochromines. In Lake Tanganyika both sides of a predator (perch)–prey (sardine) system have come under fishing pressures, but in the 1960s that on the predators reduced their numbers and led to an increased abundance of sardines (Fig. 5.39a). One species of sardine, *Limnothrissa miodon*, has been introduced to Lake Kariba, for which Fig. 5.39b shows rising long-term records of catch that also include a marked seasonality.

(ii) Sequences following introduction of alien species. These have altered the fish communities of many tropical lakes throughout the world, especially since about 1930. Regional surveys are available for Africa (Moreau *et al.* 1988; Craig 1992; Pitcher & Hart 1995) and South West Asia (Fernando 1991); for Central to South America exam-

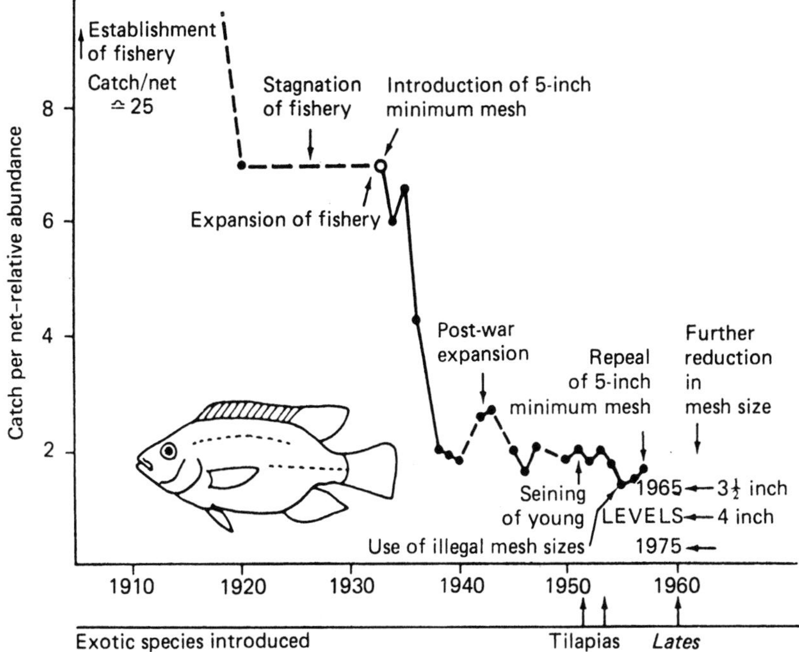

Fig. 5.38. The decline of the fishery for the native tilapia *Oreochromis esculentus* in Kenyan waters of Lake Victoria, in relation to fishing methods and the timing of introductions. Based on Fryer (1973), from Lowe-McConnell (1987).

ples from Gatún Lake, Panama (Zaret & Paine 1973) and Lake Titicaca (Loubens and Osario, in Dejoux & Iltis 1992) may be cited. In Africa, transfers from the planktivorous clupeid ('sardine') populations of Lake Tanganyika have led to new pelagic fisheries in lakes Kariba (Fig. 5.39b) and Kivu. Another outstanding but controversial example is the introduction and spread of *Lates niloticus* in Lake Victoria. This led to a considerable increase, two decades later, in the total fish catch, but with near-complete elimination of many prey species that included endemic cichlids and haplochromines (Barel *et al.* 1985; Hughes 1986; Ogutu-Ohwayo 1988, 1990*a*, *b*, 1992; Witte *et al.* 1992; Kaufman 1992; Goldschmidt *et al.* 1993; Gophen *et al.* 1993). These changes are reflected in the long-term fishery statistics (Fig. 5.40) and in food webs (Fig. 5.56). Examples of geographically widespread and generally additive changes in fish fauna are provided by the introductions of temperate salmonids (e.g., Rainbow Trout) to high altitude, cool tropical lakes such as Titicaca (South America), and – on a larger scale – of the African lacustrine

(a)

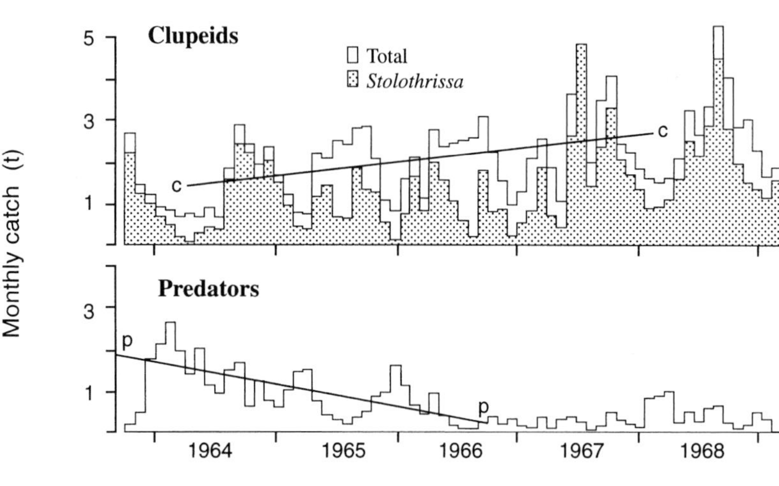

(b)

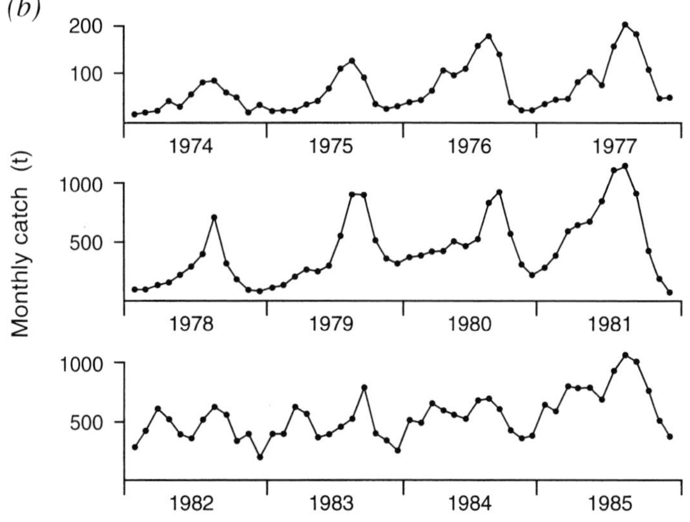

Fig. 5.39. Long-term records of monthly commercial catches of pelagic clupeid fishes (sardines): (*a*) in southern Lake Tanganyika, 1964–68, showing evidence of rising abundance (line c–c) in relation to a declining abundance of centropomid predators (line p–p); (*b*) in Lake Kariba, 1974–85, following the clupeid introduction in 1967–68. Modified from Coulter (1970) and Marshall (1988).

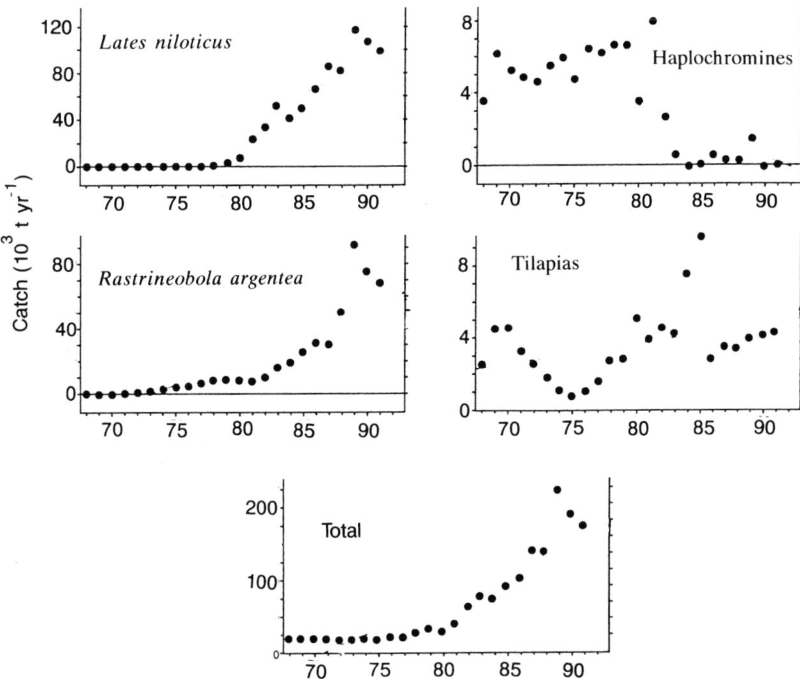

Fig. 5.40. Long-term changes in annual commercial catches of fishes from the Kenyan waters of Lake Victoria, 1968–1991, accompanying the increase in numbers of Nile Perch, *Lates niloticus*. From Gophen *et al.* (1995).

cichlids *Oreochromis niloticus* and *O. mossambicus* into small artificial water-bodies of South East Asia. In both cases under-exploited feeding niches were occupied (Piet *et al.* 1994).

(iii) Changes conditioned by altered chemical composition of the water. Nutrient enrichment (eutrophication) is a widespread cause, with effects upon food availability and oxygenation, and consequences in species representation and biomass. In dry tropical climates sequences of salinization have also affected fish faunas, as the long-term record for Lake Chilwa (Kalk *et al.* 1979) has demonstrated. Drought can lead to a greater penetration of seawater into coastal lagoons with effects on their fish fauna. In a West African lagoon studied between 1962 and 1982, a relatively stable group of about 20 species could be recognized (Albaret & Ecoutin 1990).

(iv) Developing populations in man-made lakes. These are essentially consequences of local hydrological revolutions whose environmental consequences have already been outlined. Impacts upon the ecology of fishes are reviewed by Lowe-McConnell (1987) and, for the African Great Lakes, by Craig (1992). The widespread early-phase upsurge of productivity, influenced by nutrient inputs of terrestrial origin, involves species of the original riverine fauna that are capable of adapting to lake conditions, though not necessarily for reproduction therein or for feeding on a newly developed plankton. This upsurge and its trophic support is a parallel with that occurring seasonally in floodplains, although the time scale is longer and the consequences for population increase more species-selective. Of great importance is the changing availability of varied types of food, among which Petr (1975) emphasizes the development of periphyton or *Aufwuchs* on submerged terrestrial vegetation. During the development of Lake Kainji (Nigeria) the mormyrids declined with the loss of the original chironomid-dominated zoobenthos, the detritivore *Citharinus citharus* increased strongly but briefly, the piscivore *Hydrocynus forskalii* became more abundant as clupeid prey increased, as did the food-opportunist *Alestes baremoze* (D.S.C. Lewis 1974). Fishes that are especially successful in lake conditions include cichlids and clupeids, with reproductive phases that are extended in duration and not dependent on inflowing streams. Thus, in West Africa, cichlids like *Oreochromis niloticus* and *Sarotherodon galilaeus* rose to prominence within a few years of the impoundments of Lake Volta (Petr 1967, 1968*b*) and Lake Kainji (Blake 1977). In Lake Volta the most successful clupeid *Pellonula afzeliusi* has adopted a relaxed and seasonally extended breeding pattern (Reynolds 1974); it can be traced to the antecedent river fauna (Lowe-McConnell 1987), but the species in lakes Kariba and Cahora Bassa – *Limnothrissa miodon* – was introduced by man from Lake Tanganyika. Year-to-year differences in its catches from Lake Kariba appear to be influenced by the antecedent river flow with presumed nutrient replacement for phytoplankton (Marshall 1982*b*, 1988). The assessment of the time-period involved for attaining a measure of faunistic and ecological maturity is somewhat subjective (McLachlan 1974), but seems to be considerably shorter (possibly ten years or less) than for Northern Hemisphere man-made lakes of latitude 50° or more. Delay may be introduced by such special features as submerged woody vegetation.

A sequence of colonization can also arise from irregular natural causes, as when an exceptionally high lake level induces or extends a series of marginal lagoons. Such extension occurred on Lake Victoria during the high levels of the 1960s (Fig. 4.22), when the new habitats were colonized by several species of introduced cichlids (Welcomme 1970).

There is also a class of irregular and episodic events known as 'fish-kills'. Although preceding environmental conditions are rarely well documented, the onset of O_2 deficiency in space or time is usually suspected. This ill-characterized phenomenon is widespread in tropical freshwaters. For the equatorial Lake George it is suggested (Ganf & Viner 1973) that over prolonged periods of calm weather the O_2 depletion of deeper water is less offset by nocturnal mixing, and the normally stable conditions of the diel cycle give way to phytoplankton increase and more extensive deoxygenation below. These last features were also associated with greater incidence of fish-kills in North East Lake Victoria during the 1980s (Ochumba 1987; Ochumba & Kibaara 1989). In Lake Albert (Uganda–Zaïre) a deep-water deoxygenation that is variable in horizontal and vertical extent (Talling 1963) appears to be responsible (Eccles 1976). A rare instance of intensive study of limnological conditions over some period (21 days) after a fish-kill exists for the Indonesian crater lake of Ranu Lamongan (Green *et al.* 1976). The onset of wind-induced mixing with deoxygenation has been linked to fish-kill at a location in Lake Chad (Bénech *et al.* 1976), in the Nyanza Gulf of Lake Victoria (Ochumba 1990), and in Lake Valencia (Infante *et al.* 1979: Fig. 5.41) where the mixing formed part of a normal seasonal sequence during November–December. In 1977, however, it was preceded by an unusually prolonged calm period during which accentuated deep anoxia was accompanied by accumulation of toxic hydrogen sulphide. Upward movements of this gas, with extended anoxia, also lie behind heavy fish-kills in lakes of the Central Amazonian floodplain. These occur sporadically when cold fronts or 'friagems' travel to the region from southern Brazil and destroy pre-existing stratifications (Brinkmann & Santos 1973, 1974).

Low-O_2 conditions need not be involved in fish-kills within acid water-systems, where toxic aluminium concentrations can be high – as in some tropical waters of northern Australia (Fig. 5.42) during the transition from dry to wet seasons (Morley *et al.* 1985; Brown *et al.* 1985; Townsend 1994).

(i) Air-breathing vertebrates

Although often neglected by limnologists, many air-breathing vertebrates – reptiles, amphibians, birds and mammals – are associated with fresh-

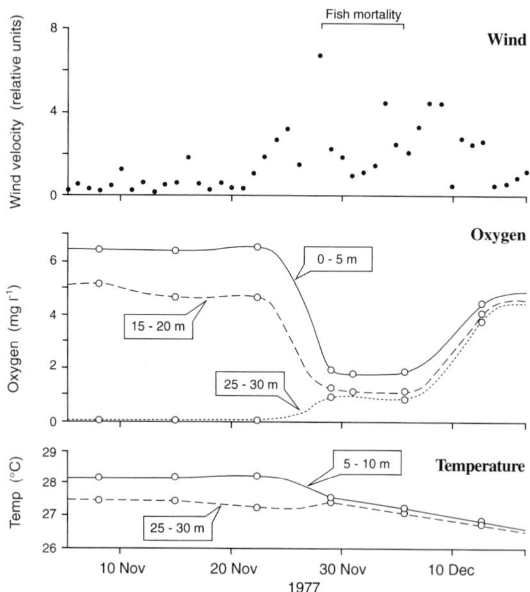

Fig. 5.41. Lake Valencia, Venezuela, 1977. The occurrence of a fish-kill in relation to increased wind velocity and evidence of de-stratification with vertical mixing from the distribution of dissolved oxygen and temperature. From Infante *et al.* (1979).

waters for support, food or shelter, can be important in food webs and have distinctive tropical representation. Their out-of-water mobility influences their distributions on several time scales. Breeding behaviour has opportunity for extended occurrence over time in the less seasonal tropics, a feature illustrated by Baker (1938) for birds in general (Fig. 5.43). However, the potential for year-round breeding in equatorial regions may be opposed by other factors, as Marshall & Roberts (1959) describe for African cormorants whose nests were damaged by seasonal winds. For physiological reasons, very large reptiles are predominantly tropical. Three distinctive tropical species – a reptile, a bird and a mammal – are taken as case-examples below.

The Nile Crocodile (*Crocodylus niloticus*) exemplifies the large reptiles, that also include the alligators and caymans of regions other than Africa (e.g., Amazonia: Best 1984). As a poikilotherm ('cold-blooded'), it responds to temperature on both diel and seasonal scales. The latter is illustrated by Hutton (1987) from the seasonal climate of Zimbabwe, where young animals showed growth in the hot season only, so that a three-year record of increase in length was stepped. The maintenance of

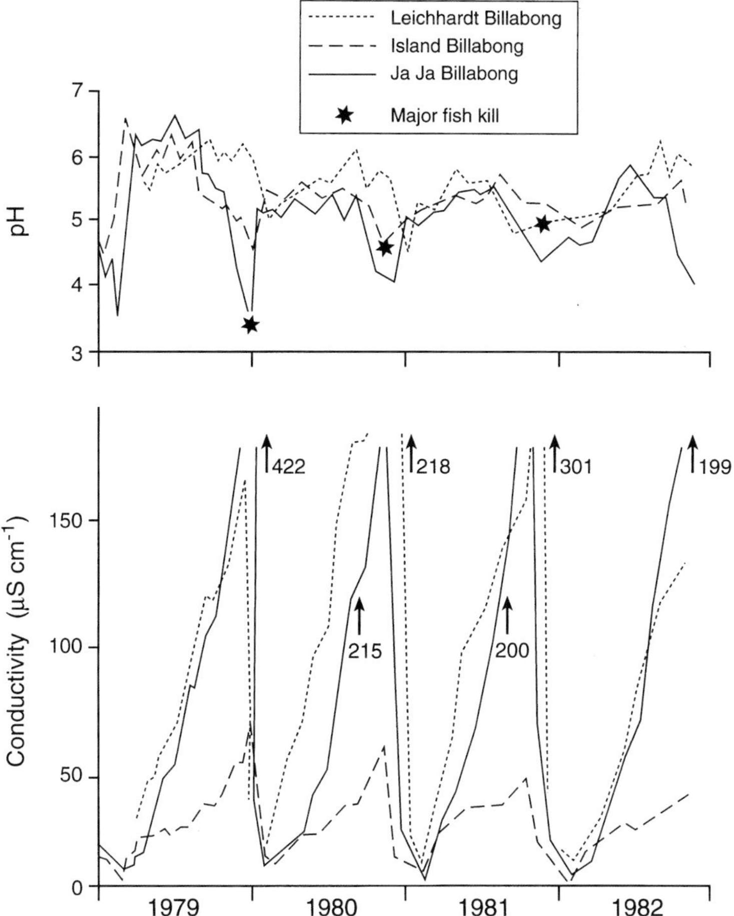

Fig. 5.42. Billabongs, Northern Australia. Within- and between-year changes in chemical characteristics of water, showing extreme values of pH and salinity (as conductivity) reached near the end of the dry season. (October–December), and the association with fish-kills. From Brown *et al.* (1985).

body temperature within a favourable range is often achieved by a diel rhythm of water-to-air movements with basking ashore. This frequently involves two maxima of animals ashore, in morning and afternoon with avoidance of too high midday temperature (Cott 1963; Hutton 1987: see Fig. 5.44); at Khartoum, the water-to-air movements were observed mostly when the air temperature was near 24 °C (Cloudsley-Thompson 1964). Foraging for food is linked with these diel cycles. Breeding is seasonal, with local timing determined by water level (Cott 1963). Eggs

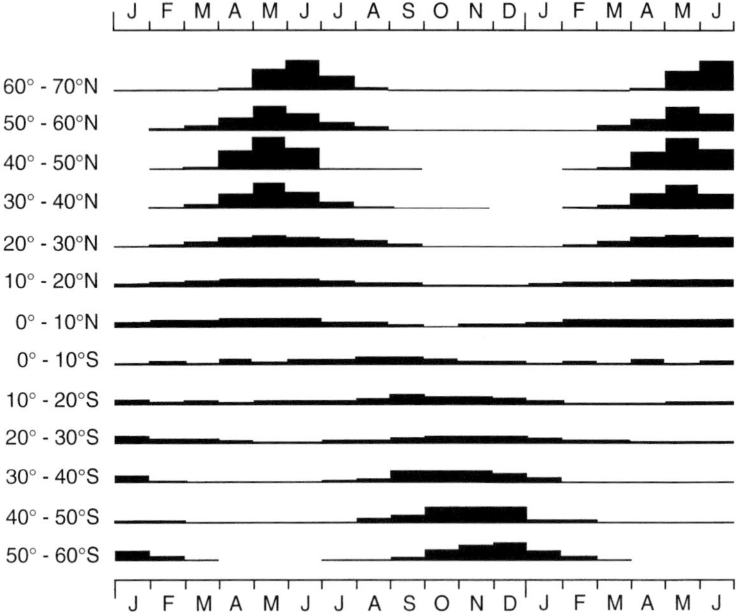

Fig. 5.43. The distribution with latitude of breeding seasons in birds: the number of times each month occurs in the records of egg-seasons in each 10° interval of latitude. From Baker (1938).

are laid at low level, with hatching just after increased seasonal rainfall begins with subsequent rise in level. In much of tropical Africa this results in one breeding season per year, but in the equatorial climate of northern Lake Victoria the two seasonal maxima of rainfall are linked with two breeding seasons (Fig. 5.45). On the long time scale, many African populations have been greatly reduced by man during the twentieth century (Cott 1954, 1963).

The Lesser Flamingo (*Phoeniconaias minor*) reaches high densities on some shallow productive tropical lakes, such as those (e.g. Lake Nakuru: Fig. 2.31) along the Rift Valley of East Africa. It, and a few relatives in other continents, are rare instances of filter-feeding birds, utilizing planktonic algae large enough to be compatible with a unique pharyngeal filtration mechanism (Fig. 3.31). Of this, Jenkin (1957) and Vareschi (1978) give details. Vareschi also calculated (see Chapter 3.3, 3.5) that the food removed daily from a highly productive lake (Nakuru) could amount to the larger part of the known primary production. The requirement for year-round abundant phytoplankton would be incompatible

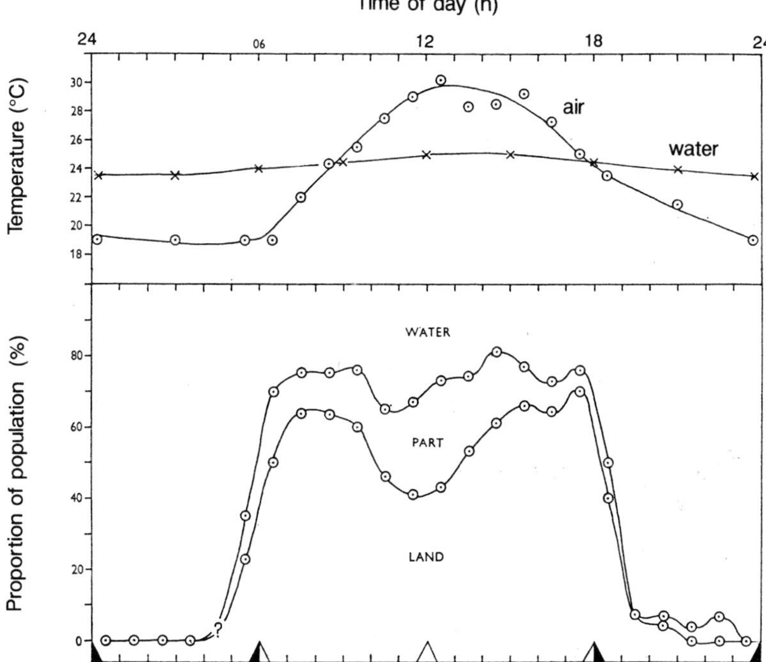

Fig. 5.44. Diel activity in the Nile crocodile, *Crocodylus niloticus*, in Uganda: changes in the mean percentage distribution of animals on land, partly on land, and in water, shown in relation to mean air-shade (○) and water (×) temperature. From Cott (1963).

with the large seasonal changes inevitable at high latitudes; thus, as with fishes, the tropical location encourages planktivory. Even in Lake Nakuru, the quantitative and qualitative availability of suitable food can be interrupted by long-term changes of water level and salinity (Section 5.1b), during which the flamingoes migrate to other lakes of the region (Vareschi 1978: Fig. 5.46). This mobility, combined with that associated with breeding behaviour centred upon one lake (Brown & Root 1971), produces a varying representation, seasonally and year-to-year, of populations in the East African series of shallow soda lakes (Tuite 1979).

The hippopotamus (*Hippopotamus amphibius*), by contrast, is a large herbivorous mammal that obtains its food by grazing on land beside its habitat of rivers and lakes. This grazing is mainly by night, so that a diel rhythm of movement occurs. As the animal is locally abundant, and will defaecate in water by day, it is an agent for significant transfers of nutri-

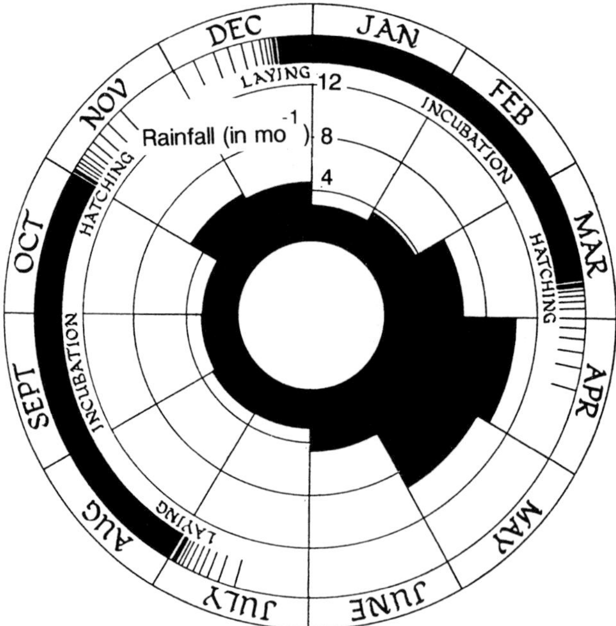

Fig. 5.45. Seasonal distribution of the two breeding seasons of *Crocodylus niloticus* in northern Lake Victoria, in relation to (centre) the bimodal equatorial pattern of rainfall as monthly averages (in inches) over 55 years. From Cott (1963).

ents from land to water (examples in Viner 1975*a*; Kilham 1982). The timing of reproductive phases can be influenced by seasonality of climate. On the equator in Uganda, breeding was year-round but conceptions and births showed two peak periods, those of births corresponding to the two seasonal maxima of rainfall. As a result a better quality, protein-rich herbage was available at a critical time (Laws & Clough 1966).

5.3 Rates of biological production

Rates of change in biomass are of significance for time-variability in two respects: as relating biomass quantities separated by a time interval, and as variables subject to time-variability themselves. In nature they are fundamentally divisible (Appendix C) into *absolute rates* (e.g., $\mathrm{g\,m^{-2}}$ $\mathrm{day^{-1}}$) and *relative or specific or instantaneous rates* that can have the dimensions of time only (e.g., $\mathrm{day^{-1}}$). The latter are most familiar in

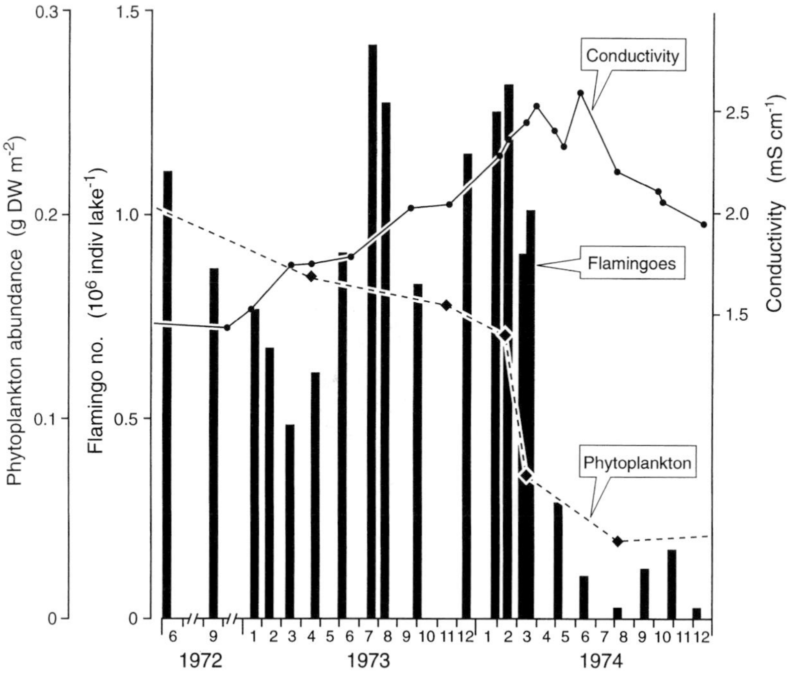

Fig. 5.46. Lake Nakuru, Kenya, 1972–74. Long-term changes in the estimated numbers of flamingoes (histograms), in relation to the abundance of phytoplankton in g dry weight per m^{-2} and to the conductivity of the lake water as an index of salinity. Modified from Vareschi (1978).

biology as specific rates of growth (g), mortality (m) or other loss; examples are discussed in several earlier sections.

For practical reasons, there are few sustained studies of seasonal variation in the specific growth rates of tropical freshwater organisms. Free-floating macrophytes such as *Salvinia*, *Pistia* and *Eichhornia* spp. (Section 5.2c) are perhaps the most easily studied. Common representatives of these three genera were studied comparatively, using confinement in floating cages, in an Amazonian floodplain lake by Junk & Howard-Williams (1984). Seasonal change in specific growth rate was marked, and correlated with the changing water level (Fig. 5.47). For *Salvinia molesta* the seasonal variability increased with latitude in the region of Papua New Guinea and northern Australia, with reduced rates associated with seasonal lower values of temperature and nitrogen status deduced from the plant's percentage N content (Room & Thomas 1986*a*: see Fig. 5.48). The same factors also appeared to be rate-limiting in an earlier

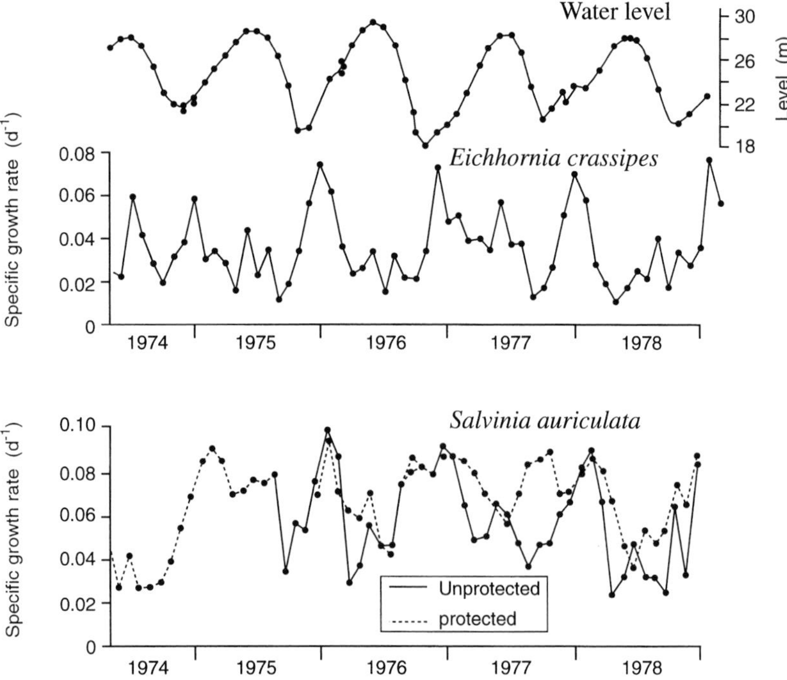

Fig. 5.47. Comparison of yearly and seasonal changes in the specific (or relative) growth rate of two species of free-floating aquatic plants, maintained in floating cages with and without protection above, on a floodplain lake of the Amazon near Manaus. Annual cycles of water level are also shown. From Junk & Howard-Williams (1984).

study of the same species in Lake Kariba (Mitchell & Tur 1975). Records of seasonal fluctuations in absolute growth rates are more numerous. One example has been noted in connection with an attached grass in the Amazonian floodplain (Section 5.2, Fig. 5.14), and there are many others from the length–weight–time relationships observed in tropical fish populations (Section 5.2h).

The time-variability of population or community production rates (e.g., as $\mathrm{g\,C\,m^{-2}\,day^{-1}}$) has been studied in a number of tropical freshwaters. Variability within the diel scale is mainly known for the photosynthetic production of phytoplankton. Example-studies are marked in Table 3.1. A well-resolved example appears in Fig. 5.49. Diel changes are primarily conditioned by solar radiation flux density, but – as already noted (Chapter 3.1) – can be skewed relative to solar noon by changing

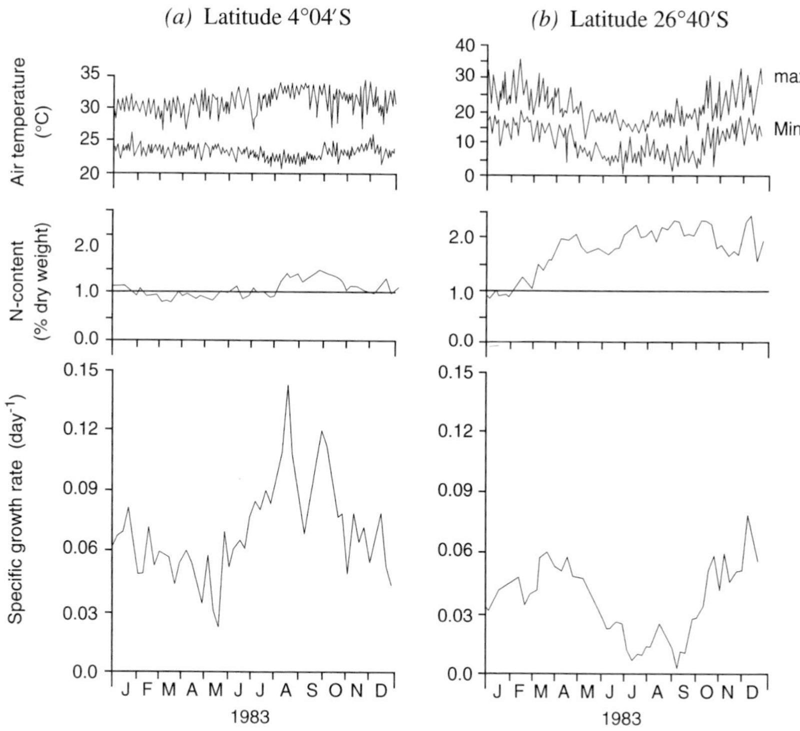

Fig. 5.48. Seasonal changes in the specific (or relative) growth rate *g* of the free-floating aquatic plant *Salvinia molesta*, based on leaf numbers of uncrowded plants, at sites of varying latitude in (*a*) Papua New Guinea, 4°04′ S, (*b*) northern Australia, 26°40′ S. At each site the concomittent changes of maximum and minimum air temperature, and nitrogen content of the plant (indicative of N-status), are also shown. Modified from Room & Thomas (1986*a*).

vertical distribution of biomass or the accentuated effects of photoinhibition later in the day (Fig. 3.13). In only one example, from Lake George, has the diel course of net C-assimilation been followed by direct analyses of phytoplankton-C, and expressed as sequential changes of carbon-based specific growth rate (Ganf & Viner 1973). The diel cycle of CO_2-fixation may be accompanied by the similarly light-dependent cycle of N_2-fixation given the presence of heterocyst (= heterocyte) bearing blue-green algae (cyanobacteria). The diel variation of rates of N-fixation has been measured in lakes George (Ganf & Horne 1975) and Valencia (Levine & Lewis 1984: see Fig. 3.26). In the N_2-fixation a diel variability of heterocyst abundance can be a subordinate factor, as demonstrated by the study of Levine & Lewis (1984) on Lake Valencia.

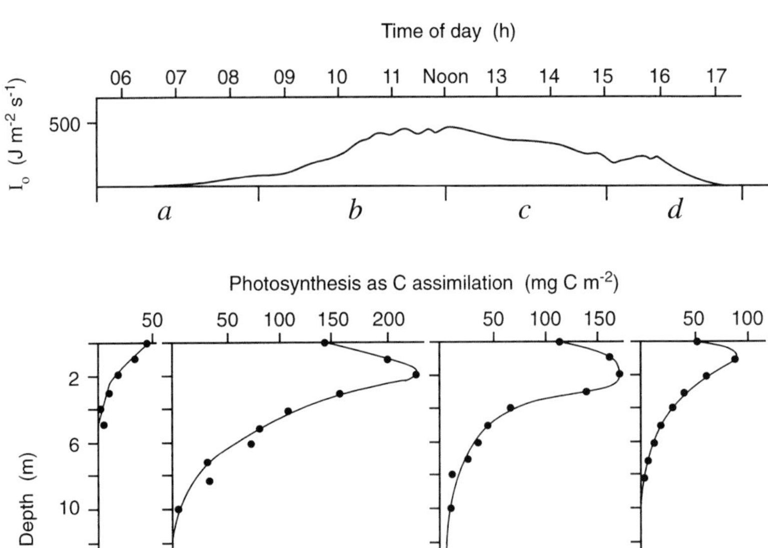

Fig. 5.49. Lake Lanao, Philippines. A diurnal sequence of depth-profiles of photosynthesis by phytoplankton, in $\mathrm{mg\,C\,m^{-3}}$, measured over periods related to the variation of surface-incident solar radiation (I_0). Modified from Lewis (1974).

Annual variability of primary, photosynthetic production is documented from studies on various tropical lakes (also marked in Table 3.1). The amplitude of within-year variation generally relates to that of euphotic biomass; responses of both to major episodes of vertical mixing can be positive (e.g., Lake Victoria, 1960–1) or negative (e.g., Lake Lanao). Variation of photosynthetic capacity, expressed as a specific activity per unit biomass, is then limited, unlike the seasonal situation in most lakes of higher latitude. Thus, in Lake Titicaca, absolute rates (Fig. 5.50) reflect biomass. An instructive exception is Lake McIlwaine, Zimbabwe, where the photosynthetic production rates per unit area measured by Robarts (1979) during 1975–6 were greatest within the warmest season around December–January when phytoplankton abundance (chlorophyll *a*) was low, light penetration increased and photosynthetic capacity high. Another instance of high photosynthetic capacity during a seasonal (here monsoonal) low abundance of phytoplankton is described by Khondker

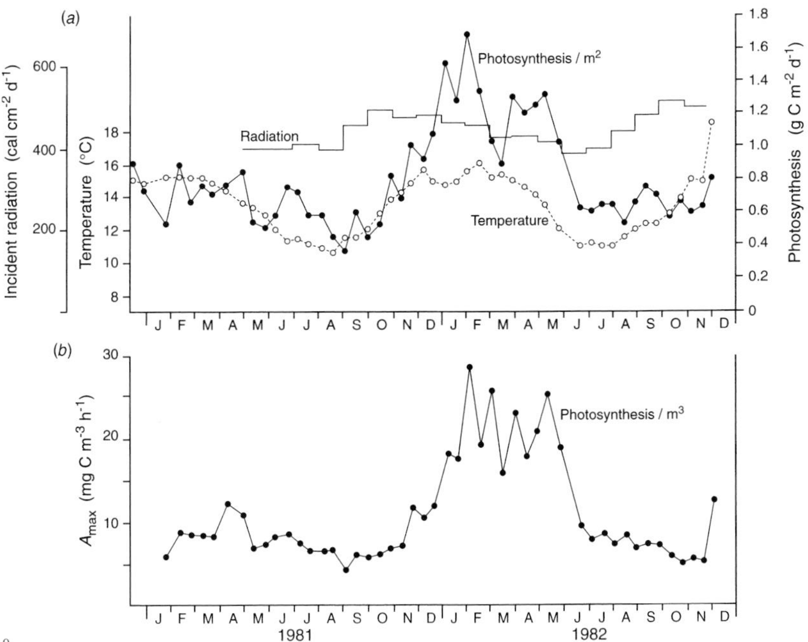

Fig. 5.50. Lake Titicaca, Andes. Within-year and between-year variation of (*a*) estimated daily rates of photosynthesis per unit area by the phytoplankton, in relation to corresponding variation in the temperature of near-surface (4 m) water, monthly average incoming radiation and (*b*) maximum rates of photosynthesis per unit volume down the water column (A_{max}). From Vincent *et al.* (1986).

& Parveen (1993) from a lake in Bangladesh. Such a combination tends to reduce the seasonal range of absolute photosynthetic activity.

In tropical, as opposed to temperate, lakes the linkage of within-year variability of areal production rates with that of solar radiation income is slight (Vincent *et al.* 1986; France 1992). However, the influence of variable light penetration underwater can be strong. The overall variability tends to decrease with the shift from temperate to tropical latitudes as assessed by the coefficient of variation (standard deviation/mean) (Melack 1979*a*: see Fig. 5.15) and by time-courses showing averaged deviations from annual mean rates in temperate and tropical lakes (Alvarez Cobelas & Rojo 1994: see Fig. 5.51). Within-year variability is notably low in the shallow equatorial lakes George and Naivasha. Conversely, it becomes high in reservoirs of short retention time, as Jebel Aulia, Sudan (Prowse & Talling 1958), Lubumbashi, Zaïre

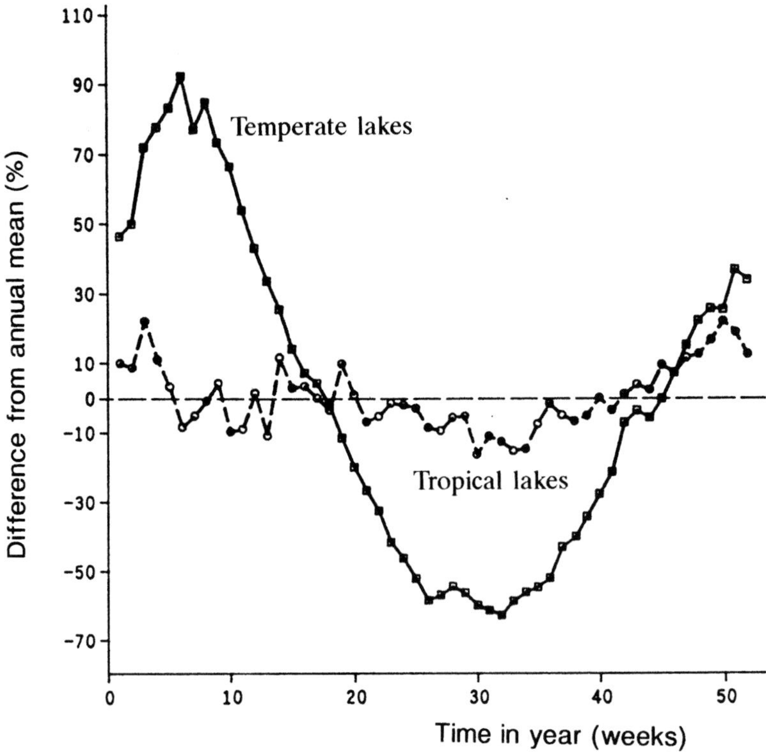

Fig. 5.51. Within-year variation of areal photosynthetic production by phytoplankton in two sets of temperate and tropical lakes, for each expressed as the mean percentage difference from annual mean values. Time is measured from the summer solstice. From Alvarez Cobelas & Rojo (1994).

(Freson 1972), and reservoirs in India and Sri Lanka during the wet monsoon (e.g., Kannan & Job 1980*c*; Saha & Pandit 1987; Silva & Davies 1987) where increased turbidity and an annual wash-out of the phytoplankton can occur.

Annual variation in rates of secondary production is more difficult to estimate (Downing & Rigler 1984). The few year-long examples available, some illustrated in Figs. 5.52 and 5.53, include the zooplankton of lakes Lanao (Lewis 1979), Valencia (Saunders & Lewis 1988*a*), Chad (Gras & Saint-Jean 1983; Lévêque & Saint-Jean 1983) and George (Burgis 1971, 1974), and the molluscan zoobenthos of Lake Chad (Lévêque 1973*a*; Lévêque & Saint-Jean 1983). Variability of biomass density is generally a major source of time-variability in community or group production rate, as shown for the zooplankton of Lake Chad (Fig. 5.53). In these,

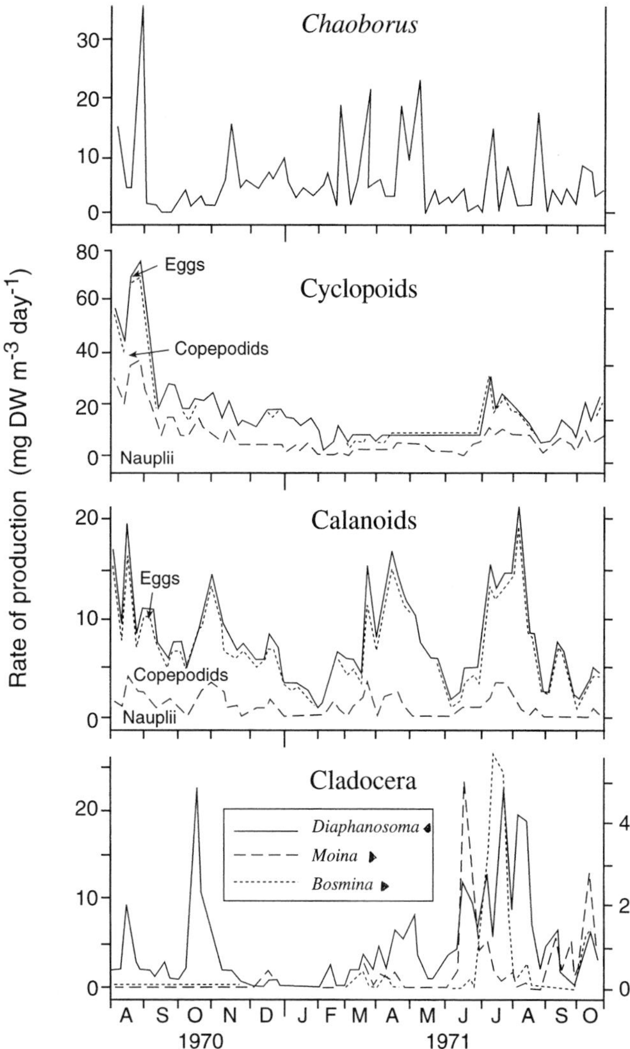

Fig. 5.52. Lake Lanao, Philippines, 1970–71. Annual variation in estimates of daily rates of production, as mg dry weight m^{-3} day^{-1}, of four component groups of zooplankton. From Lewis (1979).

as in other patterns, the equatorial Lake George shows the least variability. There the recruitment rate of the main copepod was found to be higher at the beginnings of the two rainy seasons, although total copepod numbers decreased during those seasons. Some relation of production parameters for zooplankton to rainy seasons also seemed likely from a one-year study of another shallow equatorial lake, Naivasha (Mavuti 1994).

At higher tropical latitudes the temperature factor is typically responsible for a marked seasonal reduction of specific growth rate and increase of stage-duration in the cooler season. This is true, for example, at Lake Chad for both zooplankters and molluscs. In all cases the biomass density per unit area is a principal factor for areal production rates, and was most variable for zooplankton in Lake Chad (seasonal inflow, wider temperature range) and lakes Valencia and Lanao (marked stratification cycle). In Lake Chad, central archipelago region, estimates of the production to biomass quotient (P/B) for zooplankton followed the seasonal temperature–time curve (Fig. 5.53) but not production rate per unit area as this was affected by a maximum of biomass in the cool season (Gras & Saint-Jean 1983). At the high temperature of 30 °C typical of the hot season, specific growth rates of copepod nauplii reached values of 0.79 day^{-1} (*Thermocyclops*) and 0.87 day^{-1} (*Mesocyclops*), with similar high

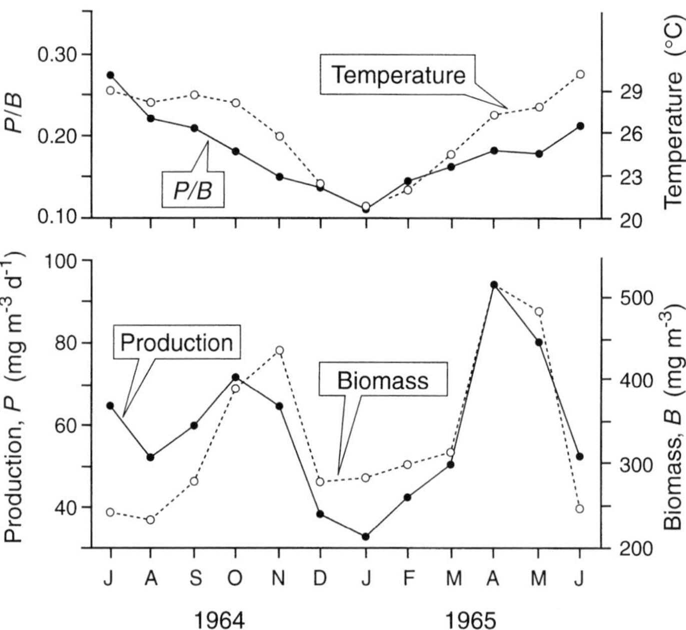

Fig. 5.53. Lake Chad. Annual variation in water temperature and three production-related characteristics of the micro-crustacean zooplankton: biomass per unit water volume (B); production in dry weight per unit water volume and day (P); and production per unit biomass and day (P/B). From Gras & Saint-Jean (1983).

maxima in the cladoceran *Moina micrura* (Gras & Saint-Jean 1981, 1983). These values exceed a doubling per day (i.e., ln 2 or 0.69 day^{-1}) and are of a magnitude common among the much smaller planktonic algae. In the deep Lake Malawi temperature seems to influence seasonal production rates of zooplankton and a principal planktivorous fish mainly through its connection with vertical mixing, nutrient entrainment and phytoplankton production (Allison *et al.* 1995).

A reduction of growth rate during the cool season is often found in tropical fish populations; the Nile Perch, *Lates niloticus*, and *Alestes baremoze* provide examples from Lake Chad (Section 5.2h). Conversely, increased growth rates of fishes in floodplains are often found in the hydrological phase of rising water level – a feature discussed quantitatively by Bayley (1988). Detritivores show much less response than omnivores. However, mean growth rates of fishes in natural populations in Africa are apparently not higher than those common in temperate regions (de Merona *et al.* 1988), although the very high crop yields per unit area from densely stocked tropical fish ponds are outstanding (e.g., Delincé 1992).

5.4 Biological diversity

Further perspectives of biological change emerge when species diversity – rather than abundance or activity – is the focus and very long time periods are considered. Higher levels of biological diversity in the tropics have attracted attention in terrestrial faunas and floras, but for inland waters are less evident and the generalization even denied (Lewis 1987, 1995). Also questionable (as by Serruya & Pollingher 1983) is the concept of equable tropical conditions allowing relatively uninterrupted long-term accumulation of biological novelty. The combination of localized evolutionary origins with barriers to dispersal have led to some distinctive patterns of distribution of freshwater organisms in the tropics. These range in scale from entire continents to single lakes.

In the geological past, tropical conditions prevailed in land-masses whose derivatives are now remote from the equator and which once bore very different biota. Some large-scale patterns of distribution have resulted from continential separation and drift during the Mesozoic and Tertiary eras. Over these periods some major groups of freshwater fishes evolved, with subsequent distributions that reflected the barriers between separated land-masses (see Lowe-McConnell 1975). South America and Asia differ radically, for example, in the natural absence of characoids from the latter and of cyprinids from the former. In Africa both coexist. Adaptive radiation is seen in various animal groups at this continental scale down to that of individual drainage basins and lakes. Fryer (1969) has surveyed examples from Africa.

The outstanding small-scale feature is the presence of numerous endemic species – mostly animals – in some ancient tropical lakes of tectonic (crustal) origin. Of these the African lakes Tanganyika, Malawi and Victoria are pre-eminent. Their respective origins – although not in the present forms – were approximately 20, 2 and 0.5–1.5 million years ago; past work has indicated the presence of > 500 endemic species in each of these lakes (Coulter 1991*a*; Martens *et al.* 1994). These endemics are clearly associated with the lakes and not with communicating waters in their drainage basins, as pointed out by Coulter (1991*c*) for Lake Tanganyika. Their number alone, as species, is an incomplete measure of the biological diversity involved: all species are different, but some are more different than others. Cichlid fishes are major components in all three lakes, and number between ~200 and > 600 species – but are probably now reduced in Lake Victoria by recent extinctions from predation by the introduced Nile Perch. In each of the lakes Malawi and Victoria most if not all form a *species flock* of numerous related species with a common (monophyletic) origin; in Lake Tanganyika at least seven older lineages have been distinguished (Meyer *et al.* 1994). Within the last and particularly ancient lake endemics are numerous in several other groups of fishes and also invertebrates, most notably the gastropod molluscs (Coulter 1991*a*). No other tropical lake approaches the three above in number or diversity of endemics. An example in the Phillippines, Lake Lanao, has been recorded to contain about 25 endemic fishes, mainly cyprinids of the genus *Barbus* (Lowe-McConnell 1975).

The association of endemic species or species-groups with ancient lakes serves to emphasize the time-dimension in cumulative evolutionary change. However, complexities arise from the variable *rate* of speciation, for which there are strong indications from tropical lakes. Rapid recent speciation seems to have taken place in the cichlid species flock of Lake Victoria. Five locally endemic species have been found in a shallow marginal water, Lake Nabugabo, that was probably isolated from the main lake only ~4000 years ago. Recently evidence has extended to the entire species flock, as work on the deepest sediments has been interpreted as indicating that the entire lake dried up in the late Pleistocene (Johnson *et al.* 1996). Reflooding was dated to ~12 400 years ago, so allowing a remarkably short time for the species flock to develop – a scientific dilemma (Fryer 1997), unless one invokes survival of earlier stocks in marginal water-bodies. Rapid speciation is consistent with evidence of but slight divergence from biochemistry (isozymes, electrophoresis) and molecular genetics (mitochondrial DNA sequences) between these cichlids within lakes Victoria and Malawi (Meyer *et al.* 1994, 1996).

It is conceivable that short periods with rapid rates of evolutionary change might alternate with long periods of little change or 'stasis'. A general application has been much discussed, as the theory of 'punctuated equilibria', for which supporting evidence has been described from a tropical African lake, Lake Turkana. Here fossil shells of gastropods (many still extant) are well preserved in unconsolidated lake sediments, with characteristics in dated sequences that Williamson (1981) studied by biometrical and statistical methods. However, his deduction, that there was 'long term stasis in lineages punctuated by rapid episodes of major phenotypic change', is probably not firmly established in view of likely discontinuity in environmental conditions that could affect fossil preservation in the record and possible morphological responses of non-genetic origin – a feature well known in the animal group concerned (Fryer *et al.* 1983).

Evolutionary and environmental changes are interconnected in many ways, some possibly influenced by latitude. For example, tropical sequences of variable water balance and level change have led to at least local extinctions. This can be illustrated by the paucity or lack of endemic species in Lake Chad, once one of the largest of tropical lakes, compared to the refugia for endemics offered by the river systems of Zaïre and Amazon in the humid tropics. However, extensive past river-connections of the Chad basin, with consequent dispersal, are also involved.

Although extreme catastrophic events are clearly unfavourable, smaller variations in the hydrological environment can favour speciation. Thus new habitats may be created for colonization, or a single population may be split into several non-interbreeding derivatives by geographical or ecological barriers. In the big African rift lakes of Tanganyika, Malawi and Turkana, with predominantly atmospheric control of water balance (Chapter 2.2), level changes well in excess of 100 m have occurred. There have been corresponding changes in both littoral and pelagic regions, and subdivision of a lake at low level into separate basins. The latter would enhance opportunities for speciation in spatially separated regions (*allopatric* speciation), otherwise also possible on a locally varied shoreline; Fryer & Iles (1972) give examples for cichlid fishes in Lake Malawi.

Subdivision of Lake Victoria at low level has been a favoured influence for the origin of the species flock of cichlids there, but is not easily compatible with the recently suggested total drying up mentioned above.

Loss of species diversity after major environmental change can have origins other than by desiccation. It can constitute total extinction if the change impacts the whole distribution-range of a species or species-group, including a water-body with endemics. The outstanding recent example of the latter situation is Lake Victoria, where *introduction* of a fish (*Lates niloticus*) led to predation that has probably extinguished many – possibly hundreds – of endemic cichlids. On a smaller scale and more problematic is the apparent loss of *Orestias cuvieri* from Lake Titicaca after the introduction of the North American Rainbow Trout, *Onchorhynchus mykiss,* formerly *Salmo gairdneri* (Loubens 1992).

Local loss of diversity, but generally not overall extinction, has accompanied several other system changes summarized in Section 5.5b. World-wide, *eutrophication* usually involves the replacement of diverse biota of low abundance by those of fewer species at higher abundance. The end-products are widespread in the tropics, although the documentation of change is rarely available. Examples are seen in numerous enriched 'tanks' or reservoirs of India that bear dense blooms of *Microcystis* spp.; also in the Paranoá reservoir at Brasilia that was enriched by sewage and came to bear a dense phytoplankton dominated by the blue-green *Cylindrospermopsis raciborskii* (Branco & Senna 1994, 1996).

Salinization is another chemically based sequence that is marked by loss of diversity, with resulting biota of a few specialists that may be in high abundance (e.g., many soda lakes) or may not (e.g., Lake Abhé, Ethiopia). For some tropical salt lakes a dated reconstruction of historical change has been made from the sedimentary record – as for the diatom flora of Lake Abhé (Gasse 1977), showing loss of an original low-salinity (oligohaline) flora. Also chemically based, *pollution* involves a heterogeneous array of human inputs, some persistent, that usually lead to loss of biological diversity. The reverse is probably true of *reservoir creation and development*, with local transition from riverine to lacustrine biota. Here some extensive studies of time-sequences exist, as for the plankton of the South American Brokopondo Reservoir (van der Heide 1982), and the zoobenthos and fish fauna of the West African Volta Lake (see Section 5.2). Lastly, there is *harvest by man*, that chiefly influences the composition of fish communities. One pronounced example is the decline of the highly edible cichlid *Oreochromis esculentus* with gill-netting in Lake Victoria (Fig. 5.38).

In summary, prevalent levels of biological diversity represent a dynamic balance between gain and loss rates, on the long (evolutionary) and short (ecological) time scales.

5.5 Systems

The environmental basis of time-variability in tropical freshwaters has been interpreted in terms of three major factor-systems (regimes of radiation balance, water balance, wind), with interactions (Chapter 4.3). These interactions can modify cyclic responses of various frequencies, notably the diel and the annual. Examples of modification are discussed below on a broader and comparative basis, having regard to the aspects of amplitude and timing of successive phases and the interrelation of physical, chemical and biological cycles. A following section discusses individual sub-systems of associated and cross-linked time-changes that are traceable to certain major environmental events.

(a) Interactions of cyclic systems in time

Physical ⇌ physical interactions are fundamental, as the three major factor-systems are physical in nature. Cycles of thermal (density) stratification in lakes provide examples at diel and annual levels, as discussed in Chapter 4.2, 4.3. At both periods the divergence of temperature between surface and deeper levels often follows, with some lag, an increase of solar radiation. Conversely, convergence with surface cooling and vertical mixing usually occurs during periods of reduced radiation input. However, the coupling between cycles of radiation and stratification can be modified or even lost under the influence of other factors such as wind, air temperature and humidity. Cycles of these factors often occur on both the diel and annual scales, and their effects may either reinforce or oppose those of a radiation cycle. Their predominance is particularly likely in equatorial regions where the annual cycle of radiation input is of low amplitude. Reference may be made to the differing diel wind cycles of lakes George and Chad, the seasonal harmattan wind of West Africa, and the presumed uncoupling of the radiation and stratification cycles for the Ethiopian Bishoftu lakes under a seasonal regime of variable humidity and evaporation (Wood *et al.* 1976). Disturbance of seasonal stratification can also have an origin from inflowing floodwater. In two West African reservoir lakes, Volta in Ghana and Asejire in Nigeria, this and the harmattan wind influence combine to produce two seasonal events of salient change in the water-mass (Biswas 1972*a*; Egborge 1978).

Other physical–physical interactions follow from the influence of a primary cycle upon a dependent one. Thus in many tropical rivers (e.g., Amazon, Nile) and floodplains the annual cycle of water level and discharge is directly linked to a cycle of sediment load with reduced light penetration.

Physical $\rightleftharpoons$ chemical interactions generally follow the relationship of a primary (physical) with a secondary (chemical) cycle. Again the stratification cycles of lakes, diel or annual, provide clear examples. Various chemical accumulations below a thermal/density discontinuity, with depletion of O_2, necessarily accompany the stratified phase. Upward transfers from these accumulations can generate pulses of concentration in surface water that share the periodicity of the vertical mixing. The amplitude of concentration changes depends on many factors; these include the relative volumes of upper and lower layers, the input of organic matter available for decomposition and the duration of the stratification. The last is short within diel cycles and this typically limits the amplitude for N- and P-nutrients to levels below detection, although it was measurable in the very productive, warm, shallow waters of Lake George (Uganda) and Parakrama Samudra (Sri Lanka) (Section 5.1b). Respiration and decomposition, and photosynthesis, can induce very large diel changes of O_2 and CO_2 content in such productive waters (Section 5.1c). Thus two consequences of the diel radiation cycle determine dependent chemical cycles of these gases.

The physical phenomena of a variable hydrological balance are also typically cyclic (often annual) and induce dependent chemical cycles. Since, during changing flow, the ratio between solute release and volumetric dilution varies widely between chemical species, and hydrological paths of water flow can be multiple (cf. floodplains), the concentration–time relationships are also varied and usually skewed relative to the parent discharge–time relationship. Examples are provided by the Gambia River, West Africa (Lesack *et al.* 1984: see Fig. 5.5) and the Apure River, Venezuela (Saunders & Lewis 1989*a*). Other influences of water flow upon chemical characteristics are more indirect. High river flows are often associated with much turbidity that can eliminate the possibility of nutrient depletion by phytoplankton; longer water retention, as in reservoirs, can increase this possibility. Closed-basin lakes with relatively dilute inflows often show cyclic variation of salinity in relation to variable rainfall, on annual and longer time scales, with additional chemical modifications at very low water level (e.g., Lake Chilwa: Kalk *et al.* 1979).

Rarely, a physical cycle may be induced or modified by chemical factors. An example could be the variation in light absorption within a black-water river that follows variable seasonal inputs of dispersed humic substances. Deep solute accumulations also modify the variable extent of vertical mixing in meromictic (long-stratified) lakes. Lake Sonachi (Kenya) is a well-studied example (MacIntyre & Melack 1982), where over the period 1971–79 stability increased in years with more rainfall and consequent reduction of salinity and density in the upper layer.

Environmental ⇌ biological interactions on diel, lunar and annual cycles have already been abundantly illustrated for planktonic, benthic and fish communities. The ultimate cyclic controls are physical and lie in the regimes of radiation income, water balance and wind. Numerous secondary controlling cycles exist that include important chemical factors such as the concentrations of major ions, nutrients, dissolved O_2 and reduction products such as H_2S (Section 5.1). Excepting major ions, most such chemical cycles have a strong component of biological activity that is often mainly microbial, by bacteria and algae.

Coupling between biological and environmental cycles can arise in four main ways. An environmental trigger or cue may be recognized; a direct regulating influence may operate; a quantity transfer may induce depletion on one side and accumulation on the other; and the cycles may not relate directly but share relations with another master cycle.

At the *diel period*, biological cycles involve short-term physiology and behaviour, possibly with population redistribution but not significant amplitude of population size – unless, conceivably, there is a source of strong diel mortality or of diel recruitment. In unicellular organisms, especially algae, diel cycles of phased cell division are widely known, but examples in tropical freshwaters do not seem to have been investigated. The diel cycle of photosynthetic activity is clearly governed by that of radiant flux density (e.g., Fig. 3.13), but its time-course may be skewed relative to that of diel radiation by factors that include vertical redistribution of population (e.g., Talling 1957*a*) and photoinhibition behaviour (Chapter 3.1). The outstanding tropical study of photoinhibition is that of Vincent, Neale & Richerson (1984) and Neale & Richerson (1987) on the phytoplankton of Lake Titicaca. Here it is accentuated in near-surface water during the afternoon by the set-up of a diel temperature/density stratification that reduces cell dispersal. This feature is illustrated in Fig. 3.8, together with the depth-time distribution of a fluorescence change that is an index of photoinhibition. Thus two effects of the diel radiation cycle interact and partially reinforce.

Light-avoiding or photofuge behaviour plays a large part in the diel cycles of vertical distribution exhibited by some motile phytoflagellates, zooplankton and benthic invertebrates. Other physiological processes which regulate the volume of gas vesicles, together with reduced mixing during diel stratification, are responsible for the variable diel distribution cycles that are a feature of many blue-green algae in the tropical phytoplankton. Both daytime rise (Talling 1957*a*) or sinking (Ganf 1974*b*) may be predominant. In animals various combinations of behavioural and physiological diel cycles are widespread, as seen in the diel feeding cycles of zooplankton and a fish in Lake George (Fig. 5.54). In many examples the daily photoperiod has a direct effect, but the independent existence of endogenous and free-running circadian rhythms must also be recognized. Other indirect evocation, by environmental factor 'cues' with selective advantage, is of course important at longer, especially annual, time scales.

Lunar-biological cycles (e.g., of insect emergence, planktonic predation) are remarkable for the biological response to extremely low levels of irradiance, of the order 10^{-5} to 10^{-6} of sunlight.

Environmentally induced cycles of *annual period* involve more available time, as for extensive chemical transformations (input–biomass increment–decay), phenological succession (e.g., reproductive stages) and population dynamics. There is also opportunity for larger change of the inducing factors such as radiation, temperature and rainfall. The amplitude of many biological cycles is governed by the degree to which input and output processes operate continuously or are phased discontinuously in time.

Given such discontinuity, chemical transfers – as of nutrients in primary production – can potentially lead to environmental and biological stocks that are in antiphase cycles. One example is the depletion of soluble Si by diatoms in Lake Chad (Fig. 5.3); another, the depletion of soluble reactive phosphorus in the While Nile (Prowse & Talling 1958: see Fig. 3.21). However, a continuity of nutrient regeneration and resupply may often sustain a population pulse of abundance without a marked inverse cycle of nutrient stock.

Biological $\rightleftharpoons$ biological interactions are classically capable of generating linked cyclic patterns of species abundance in time through plant–herbivore, prey–predator, host–parasite and competing species relationships.

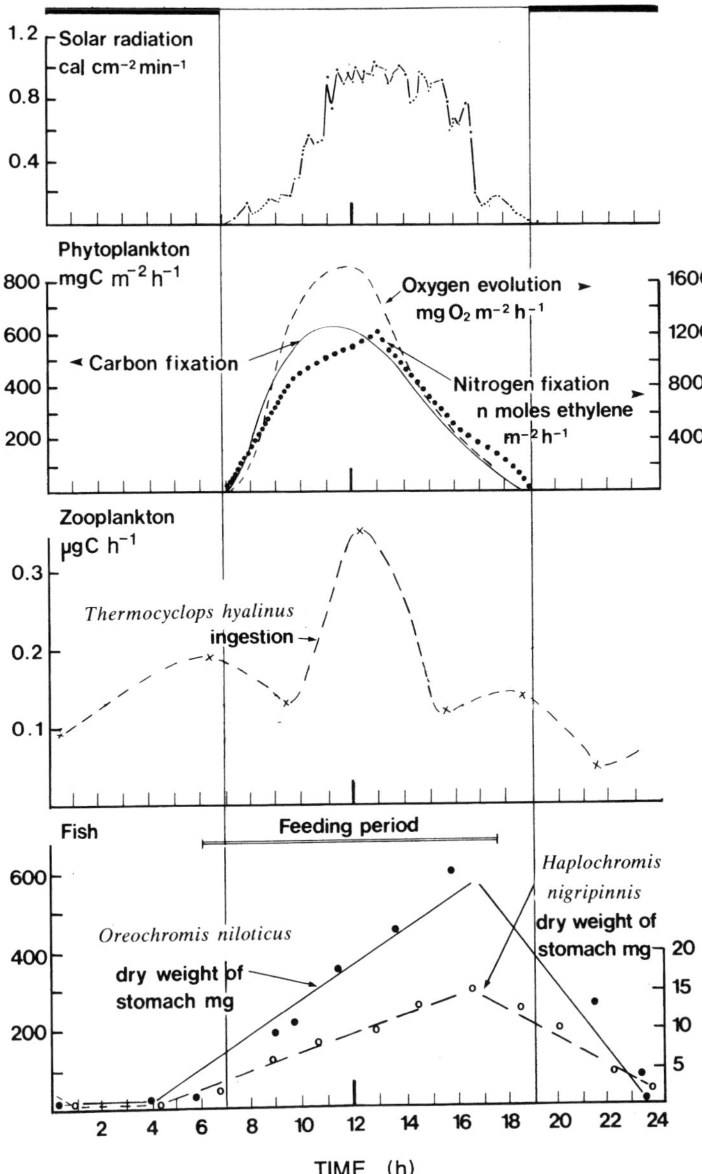

Fig. 5.54. Lake George, Uganda. Comparative examples of diel cycles, including solar radiation, fixation of C and N with evolution of O_2 by phytoplankton, and ingestion of food by a zooplankter and by two species of fish as stomach weight. From Burgis (1978).

Very few examples are established for tropical freshwaters. Thus, although there is much evidence (e.g., Lewis 1979; Saunders & Lewis 1988*a*) that predation by *Chaoborus* larvae can be important in the plankton, this activity rarely appears to be clearly shown as responsible for regular time-patterns of abundance. One possible instance is the late-season decline of Cladocera as *Chaoborus* larvae became abundant in a Venezuelan floodplain lake (Twombly & Lewis 1989). Predation by water-mites (Hydracarina) possibly may be responsible for a minimum of the dominant Cladocera in the rainy season at Gatún Lake, Panama (Gliwicz & Besiadka 1975). As already described (Chapter 4.4), variable predation by a pelagic sardine in Lake Cahora Bassa can mediate cyclic changes in the abundance of zooplankton that are controlled by visibility in moonlight and hence by the lunar cycle. Similarly, intense predation during the final low-level phase of a floodplain cycle will widen the amplitude of cyclic abundance of prey species, but the period of the cycle is set by the hydrology and not the biology. However, predator–prey fluctuations may exist in northern Lake Tanganyika, where the predator *Lates stappersii* and its prey *Stolothrissa tanganicae* appear to have alternate cyclical abundance with periods of 6–8 years (Coulter 1991*b*; Roest 1992).

It is common, world-wide, for successful expansions by an aquatic invader to be followed by some decline. Given dominance and high densities, this sequence is likely to induce related changes – negative or positive – among the accompanying populations. Examples are known from invasions of floating plants such as *Eichhornia crassipes*, *Salvinia molesta* and *Pistia stratiotes*. Outstanding invasions of lakes – notably Lake Victoria – in East Africa by the Nile Perch, *Lates niloticus*, have brought some prey species to presumed extinction. However, decline of the invader was progressively averted or delayed by switches to alternative prey, including (unstably) the young of *Lates* itself (Ogutu-Ohwayo 1990*a*).

(b) Systems of associated time-courses

Certain widespread events in tropical freshwaters have far-reaching effects upon components of environment and biota. They generate associations of individual, interacting time-courses, each association being related to one major generating event. Here eight such events are distinguished. Individual consequences have generally been considered in earlier sections; the present object is to give some overview of their association and interaction.

The diel radiation-pulse leads immediately to heat storage, raised surface-temperature and, often, to a temperature- and density-stratification (Chapter 4.2). This stratification is liable to incorporate – especially in productive waters – other components in dynamic flux, such as gases (O_2, CO_2: Section 5.1c) and sinking or buoyant particles that include phytoplankton (Section 5.2e). Simultaneously, photosynthesis operates with organic production, and, not infrequently, photoinhibition; contents of O_2 increase in the upper layers and those of CO_2 decrease. Nitrogen-fixation may also be activated, given the presence of heterocystous blue-green algae (Section 5.3). Other light-reactions, often photofuge, affect the movements and vertical distribution of zooplankton, and possibly some motile phytoplankters (Section 5.2e, f), some littoral zoobenthos, and fishes (Section 5.2h). For the zoobenthos of streams this alters relative representation in the invertebrates carried in the flowing water 'drift' (Section 5.2g). Light-dependent predation on some zooplankters may also increase if the depth-distributions of predator and prey are compatible. Here there can be interaction with lunar cycles of predation intensity (Chapter 4.4, Section 5.2f). A day–night variation of food intake, and excretion, also exists for many herbivores, from copepods to hippopotamus (Section 5.2f, i).

The seasonal floodpulse embodies the runoff from heavy seasonal rainfall, carried at high-water-discharge and level along the course of a river (Chapter 4.3). The Amazon and the Orinoco are well-studied examples (Sioli 1984; Lewis *et al.* 1990, 1995; Junk 1997) and exemplify the consequence of seasonal overspill to create or enlarge water-bodies (e.g., *várzea*-lakes) on a surrounding floodplain. Junk *et al.* (1989) have emphasized the distinctive character of the floodpulse as a drastic yet seasonally stable event, giving rise to many biological adaptations, to intermingling of terrestrial and aquatic biota, and to extensive transfers of organic food materials of terrestrial origin. There are limited parallels with the flooding of terrestrial areas on **reservoir creation** (see below): one is the incidence of wood-boring mayfly larvae in the Amazonian floodplain (*Asthenopus curtus*) and in African man-made lakes (*Povilla adusta*). However, there are transitions to floodpulses that are carried largely within the river channel (e.g., the Blue Nile) and those which exert an on-lake influence by a terminal discharge (e.g., River Chari – Lake Chad; River Omo – Lake Turkana).

Floodplain lakes supplied mainly by lateral river overspill from a strong floodpulse (reviewed by Melack & Fisher 1990; Tundisi 1994; Junk 1997) vary in depth with season. Cycles of temperature/density stratification may coexist on the diel and seasonal time scales; the latter is eliminated at low water level, and sometimes broken during cold spells with adverse biological consequences, especially on fishes (Section 5.2h) if anoxic and H_2S-rich deeper water is involved. Entering river water and local runoff can introduce substantial quantities of plant nutrients, that are later depleted by denser phytoplankton and macrophytes that adjust to changing water level. However, nutrient regeneration within the lake can also be intense, as demonstrated for Lake Calado in Amazonia (Melack 1996: Chapter 3.2c). The same time-sequence is marked by a succession of zooplankton (Section 5.2f) and of zoobenthos (Section 5.2g) for which the O_2-conditions are critical.

Correlated and adapted with the floodpulse are many features of fish biology: development and maturation of gonads, lateral and longitudinal migrations, spawning, and feeding that may exploit terrestrial sources of living food as well as detritivory and brief episodes of intense aquatic predation (Section 5.2h).

Despite the opportunities of habitats and food resources that are created laterally, the floodpulse of the main channel typically blocks the development there of phytoplankton and zooplankton by virtue of turbid water and wash-out. This is conspicuously true of the Blue Nile (Rzóska *et al.* 1955; Talling and Rzóska 1967; Hammerton 1972, 1976).

Lewis *et al.* (1990) interpreted the floodplain system as an ecosystem complex, comprising individual linked ecosystems and transmission channels. They distinguished between these two components according to whether their chemical boundary fluxes for C, N and P (omitting those driven metabolically) are less or greater than their internal fluxes associated with primary and secondary organic production plus heterotrophic decomposition. The quantification was provided from studies on the Orinoco river system. The floodpulse has the role of a synchronizing, 'setpoint', event.

De-stratification marks the transition to increased vertical circulation in the water-column of a lake, and the loss of much pre-existing layering of physical, chemical and biological components. Of these, the vertical differences of density associated with temperature differences are crucial. De-stratification can occur with regularity on a diel or seasonal time scale, or irregularly at long or short intervals. Predisposing factors are a net loss of heat and increased wind stress at the water surface.

Complete de-stratification is typically a rather abrupt event, but often is less ecologically influential than the preceding and longer extension downwards of the upper mixed layer. This extension also occurs in some (meromictic) lakes that preserve some stratification indefinitely. It entrains deep water, usually with upward transfer of plant nutrients that can evoke increased production of phytoplankton in the illuminated surface zone (examples, Lake Victoria: Fig. 5.20; Lake Tanganyika: Coulter 1963, 1968; Lake Malawi: Patterson & Kachinjika 1993, 1995; Bootsma 1993*a*, *b*). Conversely, a depression of this production is also possible (example, Lake Lanao: Lewis 1974); an oft-advanced reason is the reduced exposure to light of deep-circulating algal cells when the ratio of euphotic zone to mixed zone is low. At few tropical sites, however, these alternative mechanisms have been examined in detail. They are likely to coexist in mixed communities of diatoms and blue-green algae; in these groups contrasts of buoyancy and sinking will influence the vertical transport of cells during vertical mixing (Fig. 5.18). Previous deep accumulations of diatoms (e.g., *Aulacoseira* spp.) may be transported into upper regions, as followed in detail in Lake Victoria (Fish 1957; Talling 1966, 1969, 1986).

As with the phytoplankton, the response of zooplankton population density to de-stratification can also be positive or negative (Section 5.2f, and survey by Hart 1985). One reason is likely to be direct feeding links to phytoplankton fractions of different response to mixing. The usually positive response of diatom species is here relevant. In some examples there is likely to be an effect of upward transfer of toxic material, such as H_2S, from an anoxic hypolimnion (e.g., Lake Valencia: Infante *et al.* 1979), with simultaneous depression of O_2 concentration in surface water. These toxic influences can also induce fish-kills (Section 5.2h). However, there may also be positive correlations between fish production and vertical mixing that are probably based upon the stimulation of planktonic production and feeding linkages that can include the benthos.

Eutrophication is centred upon the long-term increase of nutrient input to, and content in, water-bodies. Of tropical examples (surveyed by Thornton 1987*b*), few are well characterized. Lake McIlwaine (Lake Chivero) in Zimbabwe, an impoundment that was enriched by sewage loading (later diverted), is a notable example with long-term records (Marshall & Falconer 1973; Thornton 1982). Another is the elongate coastal lagoon of Ebrié in West Africa, with sewage loading and nutrient enrichment varying with distance from the city of Abidjan (Dufour & Lemasson 1985). A third is the similarly enriched and highly productive

Paranoá Reservoir at Brasilia (Branco & Senna 1994). There are also some much studied sites in the subtropics like Lake Mariut (Egypt) and Hartbeespoort Dam (South Africa). The nutrient elements usually involved are phosphorus and nitrogen, but others such as carbon, iron and sulphur are not excluded. There are few estimates of external loading per unit area of tropical lakes. One of the highest for large lakes – 3.31 g P m^{-2} yr^{-1} – was obtained by Lewis & Weibezahn (1983) for Lake Valencia and is more than ten times the estimated natural background before sewage input. Loading rates to Lake McIlwaine decreased 3–4 fold after the reduction of sewage input between 1967 and 1977 (Thornton 1982), but since then have risen again (Moyo 1997). The apparent predominance of phosphorus in the man-induced eutrophication of temperate lakes is not necessarily matched in the tropics where natural P:N concentration ratios are often higher (Chapter 3.2).

The biological utilization of the increased nutrient income or stock typically leads to altered species representation and increased densities of biomass per unit area or volume for various communities; consequently also to increased modifications of the water-medium by their chemical activity and organic products. Of these modifications, O_2 depletion – especially in deep water – has numerous further effects, chemical and biological, and its extent has been used as an index of trends to eutrophy. In the tropics, as elsewhere, it increases with gross organic pollution, but for the deeper water of small stratified lakes appears to be developed at lower levels of other indicators of eutrophy than would be expected from experience of temperate lakes. Thus there was, historically, a divergence of indications of eutrophy between the chemical and some biological observations of the 1928–9 German Sunda Expedition (Ruttner 1931*b*; Thienemann 1932), that might result from accelerated rates of organic decomposition and hence O_2 consumption in the tropical waters (Chapter 3.4). Further evidence for the temperature factor in deoxygenation has been obtained more recently by Townsend (1995) from an unproductive lake in the Australian tropics. Here hypolimnetic anoxia typically develops around the summer solstice at temperature values of 26–28 °C, for a duration of eight weeks or more. This would be eliminated if one assumed an operating temperature of < 13.5 °C (usual in the hypolimnia of temperate lakes) and a temperature-sensitivity of O_2 depletion rate expressed by a Q_{10} value (rate-increase factor following a temperature rise of 10 °C) of 2.0.

At the present time the most outstanding example of tropical eutrophication, that of Lake Victoria (Hecky 1993; Lehman 1996; Bugenyi &

Magumba 1996), remains enigmatic regarding its cause. Suspected possibilities are nutrient inputs induced by altered land-use, nutrient inputs from atmospheric precipitation and marginal swamps flooded after 1961, and 'top-down' effects from introduced changes in the fish populations. The last were discounted by Lehman (1996) for reasons of timing. The eutrophication is diverse in its apparent consequences. These include increased N (especially NO_3-N) concentrations, much denser phytoplankton of altered composition, higher photosynthetic production (Mugidde 1993), reduced transparency, increased deoxygenation at depth and altered elemental composition of the sediments. Although there has been a decline of diatoms (especially *Aulacoseira* spp.) in the plankton, surface water has become strongly depleted in soluble reactive Si. Some of these changes, expressed in depth-profiles of concentration, are shown in Fig. 5.55. There are also implications of the greater O_2 depletion for the distribution of the deeper zoobenthos and fish community (Hecky *et al.* 1994).

Some studies of nutrient-rich and productive tropical waters have invoked a past eutrophication not recorded directly in time. Natural examples include lakes with fluxes of P from volcanic regions, as in East Africa (Golterman 1973) and Lake Patzcuaro in Mexico (Planas & Moreau 1990). In the latter human influences by erosion and sewage now dominate the P-input (Chacón-Torres 1993*b*). Another instance of eutrophication by sewage and urban runoff is the Waigani Lake near Port Moresby in Papua New Guinea (Osborne 1991).

A cool phase in the annual cycle is quite pronounced towards the edges of the tropics (Chapter 4.3). Lewis (1987, 1995) illustrates the broad latitudinal dependence of minimum water temperature that has numerous environmental and biological consequences. Many are similar qualitatively, if not quantitatively, to effects of winter at higher latitudes. The seasonal cooling around the winter solstice (Figs. 4.4, 4.5) induces one annual phase of complete or extended vertical mixing (i.e., monomictic regime) in many lakes, with minimal vertical differences in temperature (Fig. 4.9) and density. This mixing can, in turn, produce chemical and biological responses as outlined above. More directly, lowered temperature is often correlated with the seasonal maxima of nitrate concentration, probably through the balance between nitrification and denitrification (Section 5.1). The temperature-dependent rates of growth, and – inversely – the duration time of developmental stages, are also modified. Good examples are available from the zooplankton, zoo-

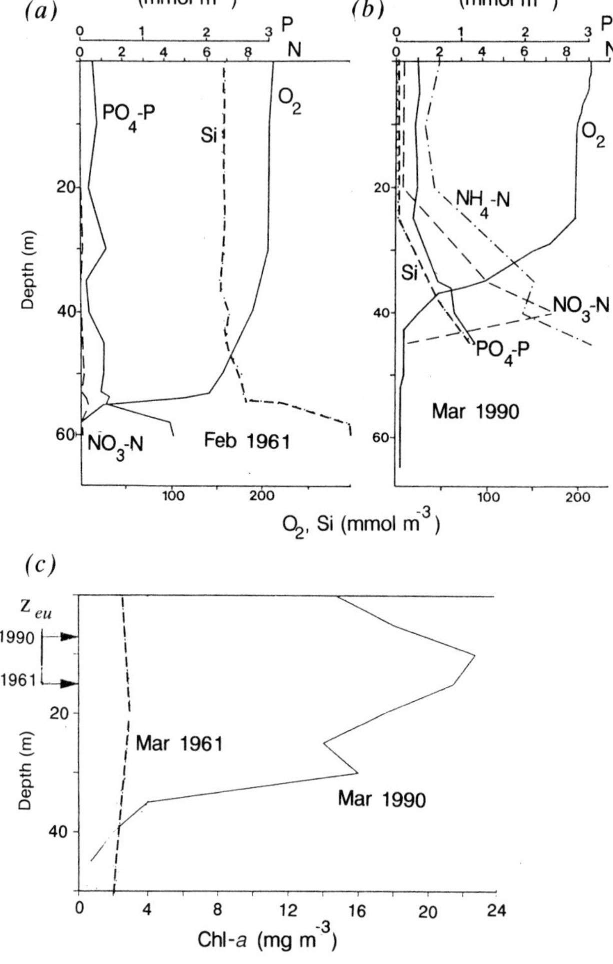

Fig. 5.55. Concentration-depth profiles measured in February–March from an offshore area of Lake Victoria, showing differences between 1960–1 and 1989–90 in (*a*, *b*) oxygen and various nutrients, (*c*) chl-*a*, March values. Change in euphotic depth (z_{eu}) is also indicated. From Hecky (1993).

benthos and fishes of Lake Chad (Section 5.2), where the seasonal depression of water temperature is ~10 °C (30 → 20 °C).

The cool phase often influences, usually by avoidance, the timing of critical phases in a life history. Examples include insect emergence (Section 5.2g) and the spawning season of fishes (Section 5.2h). Among algae, there may be a direct induction of some resting stages; for example, resting spores (akinetes) of planktonic blue-greens are uncommon in equatorial Africa but a feature of the cooler season in the Nile near Khartoum (Talling, unpublished).

Lowered temperature also affects the character of diel patterns of stratification during the cool phase, by virtue of a reduced sensitivity of water density to change in temperature. This was directly demonstrated (Talling 1957*a*) in comparative records made for the Jebel Aulia reservoir of the Nile under temperature ranges near 20 °C and 30 °C, respectively. Another influence lies in the greater opportunity for penetrative convection (Chapter 2.3).

More rarely, cool phases of brief duration can arrive irregularly as travelling disturbances from higher latitudes (e.g., 'friagems' of South America: Chapter 4.3).

A drying phase can sometimes dominate conditions in shallow tropical water-bodies, either within-years as during an extended dry season (Chapter 4.3b) or over a longer term with net water deficit (Chapter 4.5a). In rivers continuity of flow can be lost and the water-course reduced to isolated pools, often rich in life. One good example is the River Sokoto in Nigeria, studied by Holden & Green (1960) and Holden (1963); another, the River Dinder in Sudan (Rzóska 1976); yet another, streams in tropical northern Australia (Smith & Pearson 1987; Pearson 1994). Whether in river, lake or temporary pool, there is a time-sequence involving decreasing depth, increasing ionic concentration, sometimes salt precipitation and enforced concentration of larger aquatic organisms such as fish. The last condition often accentuates predation (e.g., Prejs & Prejs 1992).

River floodplain lakes are a special example, much studied in Amazonia, and already described in relation to hydrological stratification and plankton development (Chapter 4.3; Sections 5.2e, 5.2f), and fish biology (Section 5.2h). Fall of water level here relates mainly to river level rather than to a local dry season, but there can be an appreciable rise in total ionic concentration (Fig. 5.1).

Another special case is temporary rainpools (Chapter 4.3), with a short existence but an often rich, specialized fauna (Sections 5.2f, 5.2g; also review by Williams 1985). There are requirements for rapid development and a capacity to either survive desiccation by resistant stages or be re-introduced by oviposition from aerial adults bred elsewhere.

At the opposite extreme of time scale, long inter-annual periods of net water loss from lakes can induce large contractions in area and depth (Chapter 4.5) with salinity increase (Section 5.1), possible salt precipitation and modification of biological communities towards a more restricted composition with often abundant halophilic or salt-tolerant species. Vulnerable lakes are generally shallow and in semi-arid climates. They are especially numerous in sub-Saharan Africa, for which sequences and factors have been discussed comparatively by Talling (1992) and Dumont (1992). The outstanding and well-studied examples are lakes Chad (Carmouze, Durand & Lévêque 1983) and Chilwa (Kalk *et al.* 1979).

The decrease of level in Lake Chad has been associated with a number of environmental changes as well as reactions of the whole biota. The first observed changes involved water transparency, conductivity and ionic composition of the water, and were directly related to evaporation and depth decrease. Water transparency in this large lake was mainly determined by inorganic solids (clay) resuspended by wind-induced bottom stirring. Only at the end of the drying period did the phytoplankton take a significant part in light attenuation (Lemoalle 1979*a*; Carmouze, Chantraine & Lemoalle 1983). The Secchi transparency thus decreased with decreasing level until October 1973 in the eastern archipelago (Bol station) and until December 1975 in the northern basin (Kidjeria station). Afterwards, an extended marsh vegetation filtered new water inputs and reduced the wind fetch on small open water areas.

Reservoir creation involves much physical novelty, from which chemical and biological sequences develop in the long-term, and cycles of seasonal and lesser scales are modified. Water retention increases depth, extends area and reduces longitudinal (gravity-fed) currents to generally negligible values. It is usually indefinite and inter-annual, but if intra-annual its variation is generally influenced and often much increased by human regulation of the outflow as well as climatic seasonality. The direct replacement of within-year by long-term storage is uncommon, but occurred after 1964 on the Nile above Aswan with fundamental ecological changes (Entz 1976; Latif 1984). At the other extreme, there was little or no effect on Lake Victoria after the building of a dam that regulated its outflow. General surveys of this variety of ecological response are given by Baxter (1977) for reservoirs in general, by Petr (1978) for tropical reservoirs, and by McLachlan (1974), Davies (1980) and Obeng (1981) for African reservoirs or 'man-made lakes'.

Derived features of the river → lake conversion include the sedimentation of abiotic particulate material with consequent increase of light penetration; the build up – enabled mainly by longer retention – of a denser and qualitatively different phytoplankton (Section 5.2e); a corresponding increase of zooplankton; and possibilities for the development of populations of planktivorous fishes (Section 5.2h). Except in the shallowest reservoirs, greater depth and reduced turbulence promote the occurrence of temperature/density stratification, with accompanying chemical and biological modifications on the seasonal and diel time scales. Chemical modification includes a deep deoxygenation that is seasonal in many tropical reservoirs (Fig. 5.7), and continual in some (e.g., Lake Brokopondo: Fig. 5.6) often with accumulations of ammonium ions and H_2S. These features restrict colonization or redistribution of benthic animals (Section 5.2g). They are likely to be accentuated during the early years of a reservoir, when recent flooding incorporates terrestrial products (as during the floodpulse of floodplains) and the nutrient 'upsurge' (Fig. 5.12) favours a temporary phase of increased biological productivity that ranges from phytoplankton and macrophytes (Section 5.2c, 5.2e; Fig. 5.12c) to fishes (Section 5.2h).

A reservoir, like a natural lake, typically acts as a buffer which evens out over time the sharper fluctuations that might otherwise be transmitted down a river-system (Chapter 2.2, 2.3). If such fluctuations include a major seasonal floodpulse, the retained water-mass can be modified throughout the reservoir or in its higher reaches for features such as turbidity, ionic dilution and de-stratification. Stratification tends to be most persistent towards the dam that is usually distant from inflows and where the water is deeper. Examples of variable scale-effects in the absorption, over time and space, of floodpulse effects are well seen in the Nile reservoirs (small, at Roseires and Sennar: Hammerton 1972; large, above Aswan: Entz 1976; Latif 1984; Elewa 1985) and at Lake Kariba (Coche 1974; Lindmark 1997) where a chain of basins that are influenced consecutively can be distinguished (Fig. 4.19). At the Kainji Reservoir on the Niger an isotopic index – deuterium content – of the floodwater has been used (Zimmerman *et al.* 1976). This index is raised by evaporation in the 'internal delta' region upstream, with a seasonal variation that lags in the reservoir outflow over the inflow. It indicated variable degrees of horizontal mixing over the reservoir, the least being prevalent at low level.

On a very shallow scale, temporarily impounded and flooded rice fields are widespread in the tropics of South East Asia and have distinctive time-relations in their aquatic ecology (Whitton & Rother 1988; Ali 1990; Fernando 1993, 1995). Also common in some regions are fishponds that are discontinuously stocked and sometimes drained. A sequence of redevelopment after lime-sterilization and refilling has been followed with close-interval sampling for a pond in West Africa (Arfi *et al.* 1991; Arfi &

Guiral 1994; Guiral *et al.* 1994). This natural recolonization was first dominated by heterotrophic microbial assemblages of bacteria, flagellates and ciliates based on nutrient carry-over, later phytoplankton with nutrient depletion, and finally grazing with rotifer, copepod and chironomid consumers.

Invasive introductions alter the composition of animal or plant communities and often their environments as well. Floating plants include the conspicuous tropical invaders *Salvinia molesta*, *Eichhornia crassipes* and *Pistia stratiotes* (Section 5.2c) that produce, in dense stands, a closed canopy as a raft near the water surface. This excludes much light from the water-column below and reduces gaseous transfer, notably of O_2, at the air–water interface. Other chemical changes are likely to follow from the absorption of nutrients and the shedding of organic material, although for neither do they seem to be well documented. Dense, extending covers have often appeared in waters that had a high nutrient loading, as by the early nutrient 'upsurge' in recently created reservoirs like lakes Brokopondo (van Donselaar 1968), Kariba (Mitchell 1973; Marshall & Junor 1981), Cahora Bassa (Bond & Roberts 1978) and Volta (Obeng 1981) and in a polluted small water-body in Nigeria (Sharma & Sridhar 1981). Effects on other aquatic communities have been described for Lake Kariba (McLachlan 1969) and a dam basin on the White Nile (Abu-Gideiri & Yousif 1974). In the latter it was believed that plankton density had actually increased, with favourable consequences for planktivorous fishes. In Lake Kariba the early chironomid-dominated zoobenthos was suppressed under dense mats of *Salvinia molesta*. Fishing activity by man and other animals is impeded, as is navigation. The floating mats themselves harbour a distinctive fauna, in which changes with time are partly related to the growth-phase of the host plant. Petr (1968*a*) describes such changes for the invertebrates living on *Pistia* in Lake Volta. On the Upper White Nile after 1960 *Pistia* was largely replaced by invading *Eichhornia*, also a host for macro-invertebrates (Bailey & Litterick 1993) but ones that are qualitatively different from those of *Pistia* (Mitchell & Gopal 1991). A further invasion of Lake Victoria, probably via the Kagera River inflow, occurred in the 1980s with great expansion in the early 1990s.

The most ecologically influential animal introductions have been of fishes, although those of some larger Crustacea are also notable. They include the crayfish *Procambarus clarkii* that for some years almost eliminated submerged macrophytes from Lake Naivasha, Kenya (Harper 1992) – a lake that has seen a series of introductions of plants (*Salvinia*

(a)

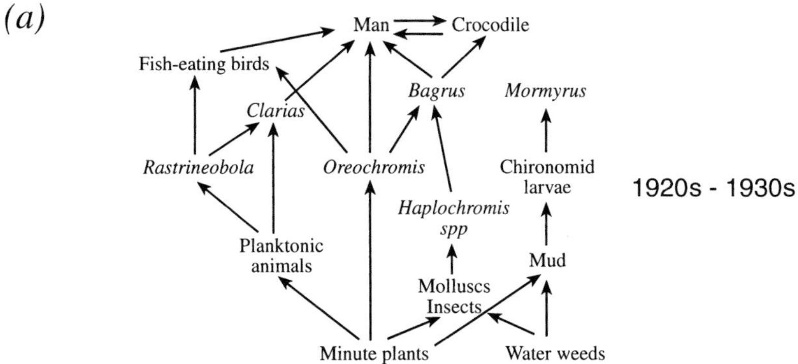

(b)

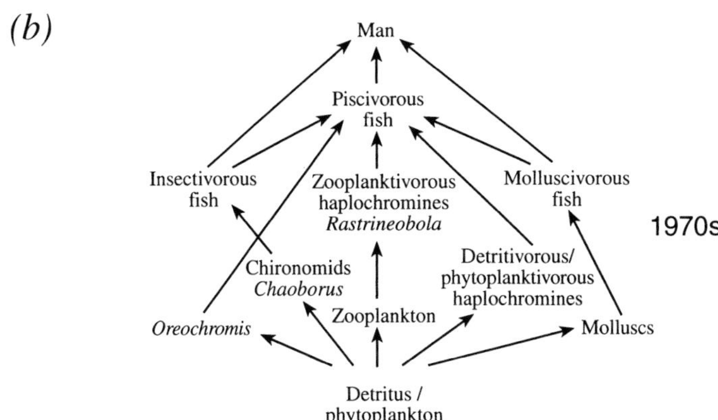

(c)

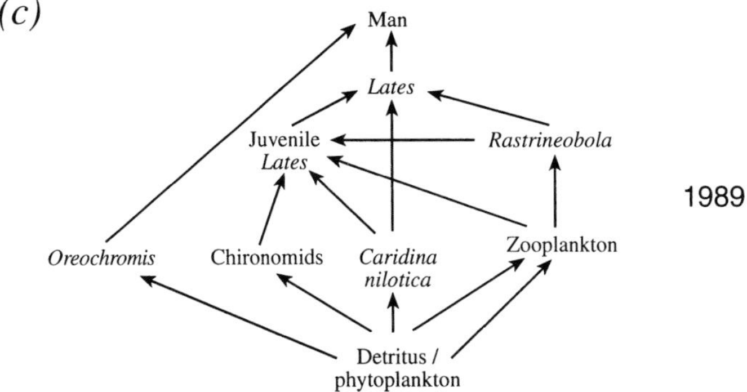

Fig. 5.56. Food webs of Lake Victoria, as interpreted from (*a*) observations of Worthington in the 1920s–1930s, and from those of Ligtvoet & Witte on sub-littoral (6–20 m) waters in (*b*) the 1970s and (*c*) 1989. From Lévêque (1995*a*).

molesta), invertebrates and fishes (Harper *et al.* 1990). In the nearby soda lake of Nakuru, an introduction about 1960 of the fish *Oreochromis alcalicus grahami* led to a dense population that supported a new diversity of fish-eating birds, dominated by the Great White Pelican (Vareschi 1979). Transfers of the clupeid (sardine) *Limnothrissa miodon* from Lake Tanganyika to Lake Kariba (Section 5.2h) led to a successful fishery (Fig. 5.39). There was also enhancement of predation upon zooplankton that greatly reduced *Chaoborus* and larger Crustacea (Marshall 1997) and, in a range-extension downstream at Lake Cahora Bassa, led to a lunar periodicity of Cladocera (Chapter 4.4; Section 5.2f) and apparently even a secondary impact on the concentration of suspended clay (Gliwicz 1986*b*).

At another size-extreme, predation by the introduced large piscivore *Lates niloticus* (Nile Perch) had very large effects in reducing or eliminating other, largely cichlid and endemic, fish stocks in Lake Victoria after *c.* 1980 (Section 5.2h). There have been wide repercussions on food webs in this lake (Fig. 5.56). It is, however, still conjectural whether there is another causal relation to the pronounced 'eutrophication' of the lake witnessed over the same period (Chapter 3.2; Section 5.4), by some form of 'top-down' effect that involved increase of phytoplankton.

The evidence advanced by Fryer (1991) strongly suggests that tropical freshwater communities are at least as susceptible to invasion as temperate ones.

6

Concluding: tropical distinctiveness

Previously considered from a diversity of viewpoints, evidence for the proposition of tropical distinctiveness may now be assessed.

Tropical conditions are related to environmental dynamics in several ways. First, purely local fluxes – as of solar energy – may determine conditions at a particular place. Second, variable circulation patterns between higher and lower latitudes may be involved. Examples are seen in global circulations in the atmosphere (e.g., Hadley cells) and oceans (e.g., Atlantic currents). Shorter-term incursions from higher latitudes appear in travelling polar fronts. Lastly, some tropical features are the legacy of changes in the remote past, such as continental drift and the evolution of regional faunas.

Distinctiveness can be sought in three main areas: in absolute magnitudes of environmental factors, in their time-variability and in the responses of biota. For all these, comparisons between tropical and non-tropical regions are implied. Some authors have based the comparison upon a few intensively studied sites. Examples include: the stability of stratification in some Indonesian and Austrian lakes (Ruttner 1938); environments and phytoplankton production in lakes Victoria and Windermere (Talling 1965*b*); zooplankton production in lakes George and Leven (Burgis & Walker 1972; Burgis 1974); production at multiple levels in lakes George, Leven and the River Thames (Burgis & Dunn 1978); temperature–time changes in an English and a Kenyan pond (Young 1975); seasonality of nutrients and phytoplankton production in an English and a Papuan lake (Osborne 1991); bacterial production in some Nicaraguan and Swedish lakes (Ahlgren *et al.* 1997); and the relative significances of N-, P- and light-limitation for phytoplankton in lakes Malawi and Superior (Guildford *et al.* in press). Here there is the possibility of relatively spurious differences of local origin.

Comparisons with a broader base are exemplified by the works of Kalff & Watson (1986) and Lewis (1990) on phytoplankton dynamics, abundance and composition, and of Dudgeon & Bretschko (1996) on the characteristics and invertebrate ecology of streams in South East Asia and Central Europe. The question of correlation with latitude in a very wide range of limnological characteristics has also been taken up. Lewis (1987, 1995) and Talling (1969, 1990, 1992) have shown strong correlations related to radiation income, and especially the annual minimum in this income. Kalff (1991), however, emphasized that latitude was a poor predictor of many lake characteristics, especially those expressing or dependent upon chemical concentrations.

Regarding absolute magnitudes, two correlations with latitude are indisputable. One is the maximum possible instantaneous flux of solar radiation, enhanced under a vertical sun with normal incidence. For many purposes, however, the daily totals of solar radiation are of more consequence. These are otherwise curtailed in the tropics by a relatively low upper limit to daylength and reach somewhat higher values in subtropical regions (e.g., Lewis 1995). The second latitudinal correlate is the earth-rotational (geostrophic) effect, the Coriolis force, that is minimal at the equator. Clear demonstration of its consequence for water motion in tropical lakes has yet to be made, although Lewis (1987, 1995) has given relative estimates for vertical mixing.

High temperature is by far the most familiar tropical correlate, that is even more pronounced in the depths of stratified lakes (hypolimnia) than in surface water. The connection must be qualified by remembering the universal depression with altitude (Fig. 4.6); also the variable importance of other terms in the energy budget that relates heat storage to energy income. The significance of high temperature for environmental processes and features is all-pervasive. Examples that have been treated include density stratification, potential evaporation and salinization, lakes with water budgets under atmospheric control, denitrification and lowered nitrate concentration, and aspects of chemical weathering including soil laterization.

Entry to the tropical zone at latitudes of 23°27′ N and S is generally in regions of low rainfall, but with exceptions in southern Brazil, northern India and the maritime areas of the Caribbean and South East Asia. The subtropical–tropical transition therefore includes a diversity of climates, in which the generally very seasonal and often scant rainfall affects time-variability in lakes and rivers. Here also the amplitude of annual temperature variation is considerable, although the highest amplitudes are typically found in the subtropics. These and other latitudinal gradients are further illustrated by Serruya & Pollingher (1983) and Lewis (1987, 1995).

Tropical freshwaters at high altitudes provide the unique combination of relatively low temperature with low amplitude of annual variation of solar radiation. The former appears to eliminate most faunistic and floristic features of typical tropical waters (e.g., Löffler 1964, 1968*a*, *b*; Carney *et al.* 1987; Pollingher & Berman 1991; Green 1995), but the latter leads to the tropical characteristic of a subordinate role of seasonal variability in radiation for primary production (Vincent *et al.* 1986).

A low annual amplitude of variation for any environmental factor or condition is likely, when subject to given absolute modifications, to increase the liability to irregular time-variability. Thus tropical time-variations of low amplitude might be inherently disposed to irregularity. This expectation was quantitatively modelled by Lewis (1987) for the stability of thermal stratification of a model lake at higher and lower latitudes, supposed subject to irregular episodic losses of 250 cal cm^{-2}. The relative stability of stratification in the equatorial lake was much more sensitive to decrease under such circumstances than was that of a lake at higher latitudes (>10°). At the fringes of the tropics, latitude range 20–25°, the difference in sensitivity from temperate lakes was slight.

Contrary to the expectation above, a number of tropical lakes of latitude <20° show relatively regular cycles of stratification over many years that nevertheless are of low amplitude (<6 °C). Examples include lakes Victoria (Talling 1966), Malawi (Talling 1969; Eccles 1974), Valencia (Lewis 1983*a*, 1984) and Titicaca (Dejoux & Iltis 1992). A circumstance contributing to annual regularity in many tropical areas is the relationship of seasonal climate to the movements of the inter-tropical convergence zone (ITCZ) linked to solar declination (cf. Fig. 4.11). This contrasts, for example, with the movements of travelling cyclones or depressions which contribute short-term climatic irregularity to maritime areas of both tropical and temperate regions. They are important, for example, for the stratification sequence in Lake Lanao, Philippines, that has been used as a model for tropical lakes (Lewis 1973).

In the tropics, it is probably not uncommon for a sequence of annual cycles of lake stratification to preserve a regular seasonal phasing of net gain and loss of heat, yet differ markedly from year to year in the vertical extent of seasonal mixing. Examples from Lake Pawlo, an Ethiopian crater lake and Lake Titicaca have led to emphasis (Wood *et al.* 1976;

Wood, Baxter & Prosser 1984; Vincent *et al.* 1985) on this source of year-to-year variability in tropical lakes.

Biologically, correlations that involve minima of solar radiation and temperature around the winter solstice are responsible for widespread restriction of growing season at high latitudes that disappears in the tropics. As a consequence, less wastage of resources over time in the tropics has been postulated by Lewis (1974). There, however, replacing restrictions are possible, notably of reduced seasonal water supply in rivers and shallow lakes of the less humid tropics. At intermediate latitudes a moderate seasonal depression of temperature often has consequences for reduced growth rate (e.g., Fig. 5.48). At the levels of temperature found in water-bodies at various latitudes, positive relationships to specific growth rate are predominant for most groups of poikilothermic organisms. In consequence tropical waters of low to moderate elevation provide many notable examples of high specific growth rate, high photosynthetic capacity and short stage-duration in a phased life history. Nevertheless an opposing temperature compensation is also possible (Bullock 1955), as is lethal temperature influence where the temperature-optima are low (e.g., the planktonic diatom *Asterionella formosa*, salmonid fishes).

If a prevalence of high developmental rates related to temperature regime exists in tropical freshwaters, there is potential for completing cycles of unusually short duration. The selective advantage of these is obvious in short-lived habitats such as rainpools or floodplain extensions. Related here may be the distinctive exploitation of the relatively short lunar cycles by some invertebrates. Diel (24-h) cycles are – as at higher latitudes – rich in behavioural linkages, but are too short for much involvement in population growth. Their tropical prominence is best illustrated by the linkage to cycles of density stratification and its correlates, influenced by the accentuated change of water density with temperature in warm waters. In the lake plankton, demographic population cycles do not seem to be recognizably biased towards responses of higher frequency compared to temperate zone behaviour. However, if environmental changes that are construed as interruptions are more frequent, a tropical sequence may include a larger number of species-successions, as proposed by Lewis (1978*a*) for phytoplankton in Lake Lanao. In tropical waters, as elsewhere, the complementary growth 'strategies' of *r*- and *K*-selected species are likely to exist and be correlated respectively with rapidly changing and more persistent population levels.

Other traits are inevitably influenced by the pervasive *biological* differentiation between temperate and tropical floras and – especially – faunas. One could instance the relative tropical success of the free-floating vegetation, of large predators such as reptiles, and the relative failure of freshwater amphipods and isopods. Persistent incursions from the higher latitudes – a biological analogue to travelling fronts of polar air – are, however, made by the more mobile of water birds, sometimes with remarkable crossings of subtropical desert like the Sahara (Moreau 1967, 1972).

In Chapter 4, cycles of factor complexes associated with solar radiation income, water balance (hydrology) and wind regime were recognized as the predominant sources of time-variability in tropical freshwaters. In these waters the hydrological factor-complex often played a major role on the annual time scale, within which the amplitude of factors more directly related to radiation was reduced. However, for latitudinal differentiation the fundamental distinction is between geostrophic (earth-rotational) factors and solar radiation–energy balance factors, with the set up of the ITCZ and tropical rainfall regimes as more indirect consequences of radiation balance. Wind regime, as a third environmental component, combines some directional influence that is geostrophic with generative influences related to pressure distribution patterns and traceable to the complex of solar radiation–energy balance. For time-variability this complex is, therefore, the main basis of tropical distinctiveness.

Appendix A

Sources of regional information

Comparison and generalization are ultimately founded upon information from individual tropical water-bodies and regional groupings. Useful sources of descriptive information are listed in Table A.1.

Table A.1. *Extended sources of regional information on tropical lakes and rivers*

A. General	Serruya & Pollingher (1983)	world-wide, with most subtropics
	Gopal & Wetzel (1995+)	country-based surveys
	Welcomme (1979, 1985)	floodplain rivers
B. Regional		
Central America:	Cole (1979)	general
	Zaret (1984)	lakes, esp. Gatún
South America:	Tundisi (1994)	general
Africa:	Beadle (1981)	general
	Symoens *et al.* (1981)	general
	Burgis & Symoens (1987)	shallow lakes, wetlands
	Talling (1992)	shallow lakes
	Dumont (1992)	shallow lakes
	Johnson & Odada (1996)	East African lakes
	John (1986)	West Africa
	Livingstone & Melack (1984)	subsaharan lakes
	Lévêque (1995*b*)	North West Africa, rivers and streams
	Harrison (1995)	North East Africa, rivers and streams
	Wood & Talling (1988)	Ethiopian lakes
	Allanson *et al.* (1990)	Southern Africa
South East Asia	Fernando (1984)	lakes and rivers
	Dudgeon (1995, in press)	rivers and streams
C. Individual rivers	Bishop (1973)	Sungai Gombak (Malaya)
	Rzóska (1976)	Nile
	Davies & Walker (1986)	Nile, Niger, Volta, Zaïre, Zambezi, Amazon, Mekong
	Grove (1985)	Niger, Volta, Senegal
	Sioli (1984)	Amazon
	Junk (1997)	Amazon
	Goulding *et al.* (1988)	Rio Negro
	Lewis *et al.* (1995)	Amazon, Orinoco
	Petr (1983)	Purari (Papua New Guinea)
	Dudgeon (1992)	Hong Kong streams
D. Individual lakes	Greenwood & Lund (1973)	George
	Balon & Coche (1974)	Kariba
	Moreau (1997)	Kariba
	Kalk *et al.* (1979)	Chilwa
	Thornton (1982)	McIlwaine = Chivero
	Hopson (1982)	Turkana
	Carmouze, Durand & Lévêque (1983)	Chad
	Latif (1984)	Nubia-Nasser
	Coulter (1991)	Tanganyika
	Coulter *et al.* (in press)	Tanganyika
	Crul (1993, 1995)	Victoria, Tanganyika, Malawi
	Dejoux & Iltis (1992)	Titicaca
	Menz (1995)	Malawi
	Tundisi & Saijo (1997)	Rio Doce Valley lakes

Appendix B

Name changes and synonymy

Information on a number of ecologically important species has been published under several taxonomic names for the same entity. Equivalences and preferred nomenclature are given in Table A.2.

Alternative names also exist for some much-studied water-bodies, as indicated.

Table A.2. *Name changes and synonymy*

(a) Organisms

Group	Preferred name	Alternative usage
Cyanophyta (blue-green algae, cyanoprokaryotes, cyanobacteria)	*Spirulina fusiformis* Voron. (in African tropics)	*Spirulina platensis* *Arthrospira platensis* *Arthrospira fusiformis*
	Cylindrospermopsis raciborskii (Wolosz.) Seen.	*Anabaenopsis raciborskii* *Cylindrospermopsis stagnale*
Bacillariophyta (diatoms)	*Aulacoseira* spp.	*Melosira* (most freshwater species)
Copepoda	*Thermocyclops hyalinus* Rehberg	*Thermocyclops crassus*
	Mesocyclops leuckarti group	*Mesocyclops leuckarti* (a non-tropical species)
	Lovenula africana Daday	*Paradiaptomus africanus*
'Cladocera' (a heterogeneous group: Fryer 1987) includes Anomopoda	*Ceriodaphnia rigaudi* Richard	*Ceriodaphnia cornuta*
	Moina micrura Kurz	*Moina dubia*
Cichlidae (cichlids)	*Oreochromis niloticus* (Linn.)	*Tilapia nilotica* *Sarotherodon niloticus*
	Oreochromis esculentus (Graham)	*Tilapia esculenta*
	Oreochromis mossambicus (Peters)	*Tilapia mossambica* *Sarotherodon mossambicus*
	Oreochromis alcalicus grahami (Boulenger)	*Tilapia grahami* *Sarotherodon alcalicus grahami*

(b) Water-bodies

Region	Preferred name	Alternative usage
Central America	Xolotlán	Managua
South America	Lobo Reservoir	Broa Reservoir
West-Central Africa	Zaïre River	Congo River
East Africa	Turkana	Rudolf
	Albert	Mobutu Sese Seko
Southern Africa	Malawi	Nyasa
	McIlwaine	Chivero

Appendix C
Quantities, units and conversion factors

The quantities used here have the fundamental dimensions of length (L), mass (M) and time (T). The corresponding fundamental units are metre (m), kilogram (kg) and second (s). These constitute the MKS system, as opposed to the older CGS system of centimetre-gram-second which is still often used or implied. Energy, with the unit Joule (J), is related to mass in several ways (kinetic, chemical, nuclear); here the thermochemical link as 'heat of combustion' is especially relevant.

Although chemical quantities are often given terms of mass ('weight') as in $\mathrm{g\,m^{-3} = mg\,dm^{-3} = mg\,l^{-1}}$, interpretation is generally easier in terms of *moles* (mol), as (mass in g)/(molecular weight), or (mass in g)/(atomic weight) for elements. For ions the former chemical measure of *equivalent* or its derivatives (meq, μeq) is often used; 1 eq = (mass in g)/(equivalent weight) = (mass in g)/(atomic or ion molecular weight ÷ ion charge). Thus for nitrate (NO_3^-), $\mathrm{1\,mol\,m^{-3} = 1\,mmol\,l^{-1} \equiv 1\,meq\,l^{-1}}$ $= 62\,\mathrm{mg}\,NO_3^-\,\mathrm{l^{-1}} = 14\,\mathrm{mg}\,NO_3\text{-N}\,\mathrm{l^{-1}}$. For calcium ($Ca^{2+}$) the corresponding series is $\mathrm{1\,mol\,m^{-3} = 1\,mmol\,l^{-1} \equiv 2\,meq\,l^{-1}} = 40.1\,\mathrm{mg}$ $Ca^{2+}\,\mathrm{l^{-1}}$.

In water-bodies 'stock' quantities are often expressed per unit volume (e.g., $\mathrm{g\,m^{-3}}$) or per unit area (e.g., $\mathrm{g\,m^{-2}}$). The linking dimension of depth (in m) may refer to the mean depth ($\bar{z}$) of the habitat, or to another depth interval such as the epilimnion.

Fluxes naturally incorporate the time dimension T^{-1}, but unfortunately the corresponding units widely used – s (fundamental), min, h, day, yr – are not in simple multiples. Fluxes can involve a flow quantity being expressed (or 'normalized') per unit stock quantity, giving a *specific rate*, e.g., of growth. If the stock shares the same measure (e.g., carbon) as the flux, then the specific rate has the dimension of T^{-1} only. Alternative usages exist, such as % $\mathrm{day^{-1}}$ and mg $\mathrm{mg^{-1}\,day^{-1}}$. A specific rate ($r$)

often expresses an exponential or logarithmic relationship between stock (S) and time (t), when its calculation from quantities S_1 and S_2 at times t_1, and t_2 involves a logarithmic transformation:

$$r = [\ln(S_2/S_1)]/(t_2 - t_1)$$

It can then be interpreted as an *instantaneous rate* (= $\mathrm{d}S/S.\mathrm{d}t$) over an element of time $\mathrm{d}t$. Its reciprocal $1/r$ is sometimes adopted as the 'turnover time', although this can have vague associations.

Interconversions between quantities and units often used are given in Table A.3. Some are precise equivalents, others are rough estimates from experience and subject to biological variation.

Table A.3. *Units, interconversions and some rough equivalences*

Quantity	Common units	Other measures
water volume	m^3	litre (l) = dm^3 [1 m^3 = 10^3 l]
energy	Joule (J)	calorie (cal) [1 cal = 4.18 J]
energy flux	$J\ m^{-2}\ s^{-1}$	× 1: $W\ m^{-2}$ = $kerg\ cm^{-2}\ s^{-1}$ × (60/4.18) × 10^{-3}: $cal\ cm^{-2}\ min^{-1}$
photon flux	$\mu mol\ m^{-2}\ s^{-1}$	approx. equivalence, PAR quantities: 1 $J\ m^{-2}\ s^{-1} \simeq 5\ \mu mol$ photons $m^{-2}\ s^{-1}$
chemical amount *chemical concn.*	mole (mol) $mol\ m^{-3}$ = $mmol\ l^{-1}$	gram (g), equivalent (eq) × Atomic Weight (element*) = $mg\ l^{-1}$ = $g\ m^{-3}$ × Molecular Weight (molecule†) = $mg\ l^{-1}$ = $g\ m^{-3}$ × charge (ion) = mequivalents (meq) l^{-1} *Na^+ = 23.0, K^+ = 39.0, Ca^{2+} = 40.1, Mg^{2+} = 24.8 C = 12.0, Cl^- = 35.5, S = 32.1, Si = 28.1, N = 14.0, P = 31.0 †O_2 = 32.0, CO_2 = 44.0, SiO_2 = 60.1
biomass index	g organic dry weight	rough equivalences of 1 g (see also Fig. 3.23.): ~5 g fresh weight (FW) ~0.45 g C ~20 mg chlorophyll *a* (phytoplankton) ~5 cm^3 (= 5 × $10^{12} \mu m^3$) cell volume (phytoplankton) ~4.5 kcal or 19 kJ

Appendix D
Symbols used in the text

For ready reference, the meanings of symbols used in this book are assembled below. Some multiple meanings have been unavoidable.

		example units
a_s	albedo, relative upward loss from water surface, of solar radiation by reflection and scattering	dimensionless
a_l	albedo of long-wave radiation	dimensionless
c_s	concentration of a gas in water at air-equilibrium	$\text{mol m}^{-3} = \text{mmol l}^{-1}$
c_w	concentration of a gas in water	$\text{mol m}^{-3} = \text{mmol l}^{-1}$
ε_a	efficiency factor: photosynthetic production per unit area as a fraction of that predicted in the absence of light-saturation	dimensionless
ε_i	efficiency factor: fractional light interception by a photosynthetic cover	dimensionless
ε_{max}	efficiency factor: maximum fractional efficiency of energy conversion in photosynthesis	dimensionless
ε_p	efficiency factor: fractional net penetration of light through the water surface	dimensionless
ε_r	efficiency factor: ratio of net to gross photosynthetic production	dimensionless
ε_s	efficiency factor: fraction of photo-synthetically available radiation (PAR) in the total solar radiant energy flux	dimensionless
ε_f	efficiency factor in filter feeding retention	dimensionless

Q'_{li}	energy flux by surface-penetrating long-wave radiation	$J\ m^{-2}\ s^{-1}$
Q_{lo}	energy flux by outgoing long-wave radiation	$J\ m^{-2}\ s^{-1}$
Q_r	energy flux by sum of radiation flux	$J\ m^{-2}\ s^{-1}$
Q_s	energy flux by solar radiation incident on water surface	$J\ m^{-2}\ s^{-1}$
Q'_s	energy flux by solar radiation penetrating water surface	$J\ m^{-2}\ s^{-1}$
R	rate of respiration per unit water volume; respiration loss	$mol\ m^{-3}\ h^{-1}$; mg
R'	rate of respiration per unit biomass	$\mu mol\ mg^{-1}\ h^{-1}$
R_i	Richardson Number in lake hydrodynamics; river input	dimensionless; $m^3\ day^{-1}$
R_o	river output	$m^3\ day^{-1}$
S	lake stability	$J\ m^{-2}$
T	period of seiche	h, day
U	wind velocity	$m\ s^{-1}$
V	volume of water-body	m^3
ΔV	change in water volume	m^3
W	Wedderburn Number in lake hydrodynamics	dimensionless
α	initial gradient of the relationship between photosynthetic rate (ϕ) and light flux density (I)	$mmol\ mg^{-1}(Jm^{-2})^{-1}$
ρ	density of water	$kg\ m^{-3}$
ρ_o	reference density of water ($\sim 10^3$)	$kg\ m^{-3}$
ρ_1	density of water in upper layer of two-layer model of a stratified lake	$kg\ m^{-3}$
ρ_2	the same in lower layer of two-layer model of a stratified lake	$kg\ m^{-3}$
σ	Stefan–Boltzmann constant governing emission of radiant energy	$J\ m^{-2}\ s^{-1}\ K^{-4}$
ϕ	rate of photosynthesis per unit biomass (or biomass index)	$mg\ mg^{-1}\ h^{-1}$
ϕ_{max}	light-saturated rate of photosynthesis per unit biomass	$mg\ mg^{-1}\ h^{-1}$
θ	temperature	°C, K
θ_t	annual range of temperature in near-surface water	°C
θ_z	range of temperature with depth in a lake	°C

References

Abiodun, A.A. & Adeniji, H.A. (1978). Movement of water columns in Lake Kainji. *Remote Sens. Env.*, **7**, 227–34.

Abu-Gideiri, Y.B. & Yousif, A.M. (1974). The influence of *Eichhornia crassipes* Solm. on planktonic development in the White Nile. *Arch. Hydrobiol.*, **74**, 463–7.

Adeniji, H.A. (1977). Preliminary study of the silica content and diatom abundance in Kainji Lake, Nigeria. In *Interactions between sediments and fresh water*, ed. H.L. Golterman, pp. 331–4. The Hague: Junk & PUDOC.

Adeniji, H.A. (1978). Diurnal vertical distribution of zooplankton during stratification in Kainji Lake, Nigeria. *Verh. int. Verein. Limnol.*, **20**, 1677–83.

Adeniji, H.A. (1981). Circadial vertical migration of zooplankton during stratification and its significance to fish distribution and abundance in Kainji Lake, Nigeria. *Verh. int. Verein. Limnol.*, **21**, 1021–4.

Ahlgren, I., Chacón, C., García, R., Mairena, I., Rivas, K. & Zelaya, A. (1997). Microbial activity in temperate and tropical lakes, a comparison between Swedish and Nicaraguan lakes. *Verh. int. Verein. Limnol.*, **26**, 429–34.

Akiyama, T., Kajumulo, A.A. & Olsen, S. (1977). Seasonal variations of plankton and physicochemical condition in Mwanza Gulf, Lake Victoria. *Bull. Freshwat. Fish. Res. Lab. Tokyo*, **27**, 49–61.

Albaret, J.J. & Ecoutin, J.M. (1990). Influence des saisons et des variations climatiques sur les peuplements de poissons d'une lagune tropicale en Afrique de l'Ouest. *Acta Oecol.*, **11**, 557–83.

Ali, A.B. (1990). Seasonal dynamics of microcrustacean and rotifer communities in Malaysian rice fields used for rice-fish farming. *Hydrobiologia*, **206**, 139–48.

Allanson, B.R. (1990). Physical processes and their biological impact. *Verh. int. Verein. Limnol.*, **24**, 112–16.

Allanson, B.R., Hart, R.C., O'Keeffe, J.H. & Robarts, R.D. (1990). *Inland waters of Southern Africa.* Monogr. Biologicae 64. Dordrecht: Kluwer.

Allison, E.H. (1996). Estimating fish production and biomass in the pelagic zone of Lake Malawi/Niassa: a comparison between acoustic observations and predictions based on biomass size-distribution theory. In *Stock assessment in freshwater fisheries*, ed. I.G. Cowx, pp. 224–42. Oxford: Fishing News Books, Blackwell.

Allison, E.H., Irvine, K., Thompson, A.B. & Ngatunga, B.P. (1996). Diets and food consumption rates of pelagic fish in Lake Malawi, Africa. *Freshwat. Biol.*, **35**, 489–515.

Allison, E.H., Patterson, G., Irvine, K., Thompson, A.B. & Menz, A. (1995). The pelagic ecosystem. In *The fishery potential and productivity of the pelagic zone of Lake Malawi/Niassa*, ed. A. Menz, pp. 351–67. UK: Natural Resources Institute, Overseas Development Administration.

Alvarez Cobelas, M. & Rojo, C. (1994). Spatial, seasonal and long-term variability of phytoplankton photosynthesis in lakes. *J. Plankton Res.*, **16**, 1691–716.

Amarasinghe, P.B., Vijverberg, J. & Boersma, M. (1997). Production biology of copepods and cladocerans in three South East Sri Lankan lowland reservoirs and its comparison to other tropical freshwater bodies. *Hydrobiologia*, **350**, 145–62.

Amon, R.M.W. & Benner, R. (1996). Photochemical and microbial consumption of dissolved organic carbon and dissolved oxygen in the Amazon River system. *Geochim. cosmochim. Acta*, **60**, 1783–92.

Andel, J. & Balek, J. (1971). Analysis of periodicity in hydrological sequences. *J. Hydrol.*, **14**, 66–82.

Anton, A., Kusnan, M., Yusoff, F.M. & Ong, E.S. (1996). Nutrient enrichment studies in a tropical reservoir: effect of N:P ratio on phytoplankton populations. In *Tropical limnology*, vol. 2, *Tropical lakes and reservoirs*, ed. K.H. Timotius & F. Göltenboth, pp. 179–85. Salatiga, Indonesia: Satya Wacana Christian University.

Apstein, C. (1907). Das Plankton im Colombo-See auf Ceylon. Sammelausbeute von A. Bogert, 1904–1905. *Zool. Jahrb. (Abt. System.)* **25**, 201–44.

Apstein, C. (1910). Das Plankton des Gregory-Sees auf Ceylon. Sammelausbeute von A. Bogert, 1904–1905. *Zool. Jahrb. (Abt. System.)* **29**, 661–80.

Araujo-Lima, C., Forsberg, C., Victoria, R. & Martinelli, L. (1986). Energy sources for detritivorous fishes in the Amazon. *Science, NY*, **234**, 1256–8.

Arcifa, M.S., Meschiatti, A.J. & Gomes, E.A.T. (1990). Thermal regime and stability of a tropical shallow reservoir: Lake Monte Alegre, Brazil, *Rev. Hydrobiol. trop.*, **23**, 271–81.

Arcifa-Zago, M.S. (1976). A preliminary investigation on the cyclomorphosis of *Daphnia gessneri* Herbst, 1967, in a Brazilian reservoir. *Bol. Zool. Univ., São Paulo*, **1**, 147–60.

Arcifa-Zago, M.S. (1978). Vertical migration of *Daphnia gessneri* Herbst (1967) in Americana Reservoir, State of São Paulo, Brazil. *Verh. int. Verein. Limnol.*, **20**, 1720–6.

Arfi, R. & Bouvy, M. (1995). Size, composition and distribution of particles related to wind-induced resuspension in a shallow tropical lagoon. *J. Plankton Res.*, **17**, 557–74.

Arfi, R. & Guiral, D. (1994). Chlorophyll budget in a productive tropical pond: algal production, sedimentation, and grazing by microzooplankton and rotifers. *Hydrobiologia*, **272**, 239–49.

Arfi, R., Guiral, D. & Torreton, J.P. (1991). Natural recolonization of a tropical pond: day to day variations in the photosynthetic parameters. *Aquat. Sci.*, **53**, 39–54.

Arunachalam, M., Madhusoodanan Nair, K.C., Vijverberg, J., Kortmulder, K. & Suriyanarayanan, H. (1991). Substrate selection and seasonal variation in densities of invertebrates in stream pools of a tropical river. *Hydrobiologia*, **213**, 141–8.

Ashton, P.J. (1979). Nitrogen fixation in a nitrogen-limited impoundment. *J. Wat. Poll. Control. Fed.*, **51**, 570–9.

Ashton, P.J. (1985*a*). Nitrogen transformations and the nitrogen budget of a hypertrophic impoundment (Hartbeespoort Dam, South Africa). *J. Limnol. Soc. Sth. Afr.*, **11**, 32–42.

Ashton, P.J. (1985*b*). Seasonality in Southern Hemisphere phytoplankton assemblages. *Hydrobiologia*, **125**, 179–90.

Ashton, P.J. & Schoeman, F.R. (1983). Limnological studies on the Pretoria Salt Pan, a hypersaline maar lake. I. Morphometric, physical and chemical features. *Hydrobiologia*, **99**, 61–73.

Ashton, P.J. & Schoeman, F.R. (1988). Thermal stratification and the stability of meromixis in the Pretoria Salt Pan, South Africa. *Hydrobiologia*, **158**, 253–65.

Bailey, R.G. (1972). Observations on the biology of *Nothobranchius guentheri* (Pfeffer) (Cyprinodontidae), an annual fish from the coastal region of East Africa. *Afr. J. trop. Hydrobiol. Fish.*, **2**, 33–43.

Bailey, R.G. & Litterick, M.R. (1993). The macroinvertebrate fauna of water hyacinth fringes in the Sudd swamps (River Nile, southern Sudan). *Hydrobiologia*, **250**, 97–103.

Baker, J.R. (1938). The relation between latitude and breeding seasons in birds. *Proc. Zool. Soc. Lond. Ser. A.*, **108**, 557–82.

Balek, J. (1977). *Hydrology and water resources in tropical Africa.* Amsterdam: Elsevier.

Balek, J. (1983). *Hydrology and water resources in tropical regions.* Amsterdam: Elsevier.

Balon, E.K. & Coche, A.G. (ed.) (1974). *Lake Kariba, a man-made tropical ecosystem in Central Africa.* Monogr. Biologicae 24. The Hague: Junk.

Banderas Tarabay, A., Gonzales Villela, R. & Espino, G.L. (1991). Limnological aspects of a high-mountain lake in Mexico. *Hydrobiologia*, **224**, 1–10.

Banse, K. & Mosher, S. (1980). Adult body mass and annual production/biomass relationships of field populations. *Ecol. Monogr.*, **50**, 355–79.

Barbieri, R. & Esteves, F.A. (1991). The chemical composition of some macrophyte species and implications for the metabolism of a tropical lacustrine ecosystem – Lobo Reservoir, São Paulo, Brazil. *Hydrobiologia*, **213**, 133–40.

Barbieri, R., Esteves, F.A. & Reid, J.W. (1984). Contribution of two aquatic macrophytes to the nutrient budget of Lobo Reservoir, São Paulo, Brazil. *Verh. int. Verein. Limnol.*, **22**, 1631–5.

Barbieri, S.M. & Godinho-Orlandi, M.J.L. (1989). Planktonic protozoa in a tropical reservoir: temporal variations in abundance and composition. *Rev. Hydrobiol. trop.*, **22**, 275–85.

Barbosa, F.A.R. & Tundisi, J.G. (1980). Primary production of phytoplankton and environmental characteristics of a shallow Quaternary lake at Eastern Brasil. *Arch. Hydrobiol.*, **90**, 139–61.

Barbosa, F.A.R. & Tundisi, J.G. (1989). Diel variations in a shallow tropical Brazilian lake. I. The influence of temperature variation on the distribution of dissolved oxygen and nutrients. *Arch. Hydrobiol.*, **116**, 333–49.

Barbosa, F.A.R., Tundisi, J.G. & Henry, R. (1989). Diel variations in a shallow tropical Brazilian lake. II. Primary production, photosynthetic efficiency and chlorophyll-*a* content. *Arch. Hydrobiol.*, **116**, 435–48.

Barel, C.D.N., Dorit, R., Greenwood, P.H., Fryer, G., Hughes, N., Jackson, P.B.N., Kawanabe, H., Lowe-McConnell, R.H., Nagoshi, M., Ribbinck,

A.J., Trewavas, E., Witte, F. & Yamaoka, K. (1985). Destruction of fisheries in Africa's lakes. *Nature, Lond.*, **315**, 19–20.

Barnes, J.R. & Shiozawa, D.K. (1985). Drift in Hawaian streams. *Verh. int. Verein. Limnol.*, **22**, 2119–24.

Barton, C.E., Solomon, D.K., Bowman, J.R., Cerling, T.E. & Sayer, M.D. (1987). Chloride budgets in transient lakes: Lakes Baringo, Naivasha, and Turkana. *Limnol. Oceanogr.*, **32**, 745–51.

Batanouny, K. & El-Fiky, A. (1975). The water hyacinth (*Eichhornia crassipes*) in the Nile system, Egypt. *Aquat. Bot.*, **1**, 243–52.

Bauer, K. (1983). Thermal stratification, mixis, and advective currents in the Parakrama Samudra Reservoir, Sri Lanka. In *Limnology of Parakrama Samudra, Sri Lanka*, ed. F. Schiemer, pp. 27–34. The Hague: Junk.

Baxter, R.M. (1977). Environmental effects of dams and impoundments. *Ann. Rev. Ecol. Syst.*, **8**, 253–83.

Baxter, R.M., Prosser, M.V., Talling, J.F. & Wood, R.B. (1965). Stratification in tropical African lakes at moderate altitudes (1500 to 2000 m). *Limnol. Oceanogr.*, **10**, 510–20.

Baxter, R.M., Wood, R.B. & Prosser, M.V. (1973). The probable occurrence of hydroxylamine in the water of an Ethiopian lake. *Limnol. Oceanogr.*, **18**, 470–2.

Bayley, P.B. (1988). Factors affecting growth rates of tropical floodplain fishes: seasonality and density-dependence. *Env. Biol. Fishes*, **21**, 127–42.

Beadle, L.C. (1932*a*). Scientific results of the Cambridge Expedition to the East African Lakes, 1930–1. 3. Observations on the bionomics of some East African swamps. *J. Linn. Soc. Zool.*, **38**, 135–55.

Beadle, L.C. (1932*b*). Scientific results of the Cambridge Expedition to the East African Lakes, 1930–1. 4. The waters of some East African Lakes in relation to their fauna and flora. *J. Linn. Soc. Zool.*, **38**, 157–211.

Beadle, L.C. (1981). *The inland waters of tropical Africa. An introduction to tropical limnology*, 2nd edn. London: Longman.

Beauchamp, R.S.A. (1953*a*). Hydrological data from Lake Nyasa. *J. Ecol.*, **41**, 226–39.

Beauchamp, R.S.A. (1953*b*). Sulphates in African inland waters. *Nature, Lond.*, **171**, 769–71.

Beauchamp, R.S.A. (1956). The electrical conductivity of the head waters of the White Nile. *Nature, Lond.*, **178**, 616–19.

Beauchamp, R.S.A. (1958). Utilizing the natural resources of Lake Victoria for the benefit of fisheries and agriculture. *Nature, Lond.*, **181**, 1634–6.

Beauchamp, R.S.A. (1964). The Rift Valley lakes of Africa. *Verh. int. Verein. Limnol.*, **15**, 91–9.

Begg, G.W. (1970). Limnological observations on Lake Kariba during 1967 with emphasis on some special features. *Limnol. Oceanogr.*, **15**, 776–88.

Begg, G.W. (1976). The relationship between the diurnal movements of some of the zooplankton and the sardine *Limnothrissa miodon* in Lake Kariba. *Limnol. Oceanogr.*, **21**, 529–39.

Belay, A. & Wood, R.B. (1984). Primary productivity of five Ethiopian Rift Valley lakes. *Verh. int. Verein. Limnol.*, **22**, 1187–92.

Bell, R.T., Erikson, R., Vammen, K., Vargas, M.H. & Zelaya, A. (1991). Heterotrophic bacterial production in Lake Xolotlán (Managua) during 1988–1989. *Hydrobiol. Bull., Amsterdam*, **25**, 145–9.

Bénech, V., Lemoalle, J. & Quensière, J. (1976). Mortalités de poissons et conditions de milieu dans le lac Tchad au cours d'une période de sécheresse. *Cah. ORSTOM sér. Hydrobiol.*, **10**, 119–30.

Bénech, V. & Quensière, J. (1983*a*). Migrations de poissons vers le lac Tchad à la décrue de la plaine inondée du Nord-Cameroun. II. Comportement et rythmes d'activité des principales espèces. *Rev. Hydrobiol. trop.*, **16**, 79–101.

Bénech, V. & Quensière, J. (1983*b*). Migrations de poissons vers le lac Tchad à la décrue de la plaine inondée du Nord-Cameroun. III. Variations annuelles en fonction de l'hydrologie. *Rev. Hydrobiol. trop.*, **16**, 287–316.

Bénech, V. & Quensière, J. (1985). Stratégies de reproduction des poissons du Tchad en période de Tchad Normal (1966–1971). *Rev. Hydrobiol. trop.*, **18**, 227–44.

Benke, A.C. (1993). Concepts and patterns of invertebrate production in running waters. *Verh. int. Verein. Limnol.*, **25**, 15–38.

Benner, R., Opsahl, S., Chin-Leo, G., Richey, J.E. & Forsberg, B.R. (1995). Bacterial carbon metabolism in the Amazon River system. *Limnol. Oceanogr.*, **40**, 1262–70.

Benson, L.J. & Pearson, R.G. (1987). Drift and upstream movement in Yuccabine Creek, a tropical Australian stream. *Hydrobiologia*, **153**, 225–39.

Berg, A. (1959). Analyse des conditions impropres au développement de la jacinthe d'eau *Eichhornia crassipes* (Mart.) Solms dans certaines rivières de la cuvette congolaise. *Bull. agric. Congo Belge*, **50**, 365–94.

Berg, A. (1961). Rôle écologique des eaux de la Cuvette congolaise sur la croissance de la jacinthe d'eau (*Eichhornia crassipes* (Mart.) Solms). *Mém. Acad. R. Sci. outre-mer, Sci. nat.* 8, n.s. 12(3).

Berg, H. (1995). Modelling of DDT dynamics in Lake Kariba, a tropical man-made lake, and its implications for the control of tsetse flies. *Ann. zool. fenn.*, **32**, 331–53.

Berg, H. & Kautsky, N. (1997). Persistent pollutants in the Lake Kariba ecosystem – a tropical man-made lake. In *African inland fisheries, aquaculture and the environment* ed. K. Remane, pp. 115–35. Oxford: Fishing News Books.

Berry, A.J. & Kadri, A.H. (1974). Reproduction in the Malayan freshwater cerithiacean gastropod *Melanoides tuberculata*. *J. Zool., Lond.*, **172**, 369–81.

Best, R.C. (1984). The aquatic mammals and reptiles of the Amazon. In *The Amazon*, ed. H. Sioli, pp. 371–412. Monogr. Biologicae 56. Dordrecht: Junk.

Binford, M.W. (1982). Ecological history of Lake Valencia, Venezuela: interpretation of animal microfossils and some chemical, physical and geological features. *Ecol. Monogr.*, **52**, 307–33.

Bishop, J.E. (1973). *Limnology of a small Malayan river, Sungai Gombak.* Monogr. Biologicae 22. The Hague: Junk.

Biswas, S. (1969). The Volta Lake: some ecological observations on the phytoplankton. *Verh. int. Verein. Limnol.*, **17**, 259–72.

Biswas, S. (1972*a*). Distribution of phytoplankton during the early development of Volta Lake (1964–1968). *Hydrobiologia*, **40**, 201–7.

Biswas, S. (1972*b*). Ecology of phytoplankton of the Volta lake. *Hydrobiologia*, **39**, 277–88.

Biswas, S. (1975). Phytoplankton in Volta Lake, Ghana, during 1964–1973. *Verh. int. Verein. Limnol.*, **19**, 1928–34.

Biswas, S. (1978). Observations on phytoplankton and primary productivity in Volta Lake, Ghana. *Verh. int. Verein. Limnol.*, **20**, 1672–6.

Blache, J. & Goossens, J. (1954). Monographie piscicole d'une zone de pêche au Cambodge. *Cybium, Paris*, **8**, 1–49.

Blake, B.F. (1977). Lake Kainji, Nigeria, a summary of the changes within the fish population since the impoundment of the Niger in 1968. *Hydrobiologia*, **53**, 131–7.

Blažka, P., Backiel, T. & Taub, F.R. (1980). Trophic relationships and efficiencies. In *The functioning of freshwater ecosystems*, ed. E.D. Le Cren & R.H. Lowe-McConnell, pp. 393–410. International Biological Programme 22. Cambridge: Cambridge Univ. Press.

Blomfield, C., Brown, G. & Catt, J.A. (1970). The distribution of sulphur in the mud of Lake Victoria. *Plant Soil*, **33**, 479–81.

Boland, K.T. & Griffiths, D.J. (1995). Water column stability as a major determinant of shifts in phytoplankton dominance: evidence from two tropical lakes in northern Australia. In *Tropical limnology*, vol. 2, *Tropical lakes and reservoirs*, ed. K.H. Timotius & F. Göltenboth, pp. 113–22. Salatiga, Indonesia: Satya Wacana Christian University. Reprinted 1996 in *Perspectives in tropical limnology*, ed. F. Schiemer & K.T. Boland, pp. 89–99. Amsterdam: SPB Academic Publishing.

Boland, K.T. & Imberger, J. (1994). An analysis of dynamic stability in two inland lakes of northern Australia. *Mitt. int. Verein. Limnol.*, **24**, 337–44.

Bond, W.J. & Roberts, M.G. (1978). The colonization of Cabora Bassa, Moçambique, a new man-made lake, by floating aquatic macrophytes. *Hydrobiologia*, **60**, 243–59.

Bonell, M., Hufschmidt, M.M. & Gladwell, J.S. (1993). *Hydrology and water management in the humid tropics*. Cambridge: Cambridge Univ. Press.

Booth, W.E. & Choy, S.C. (1995). Phototrophic bacteria in the Benutan Reservoir, Brunei Darussalam. In *Tropical limnology*, vol. 2, *Tropical lakes and reservoirs*, ed. K.H. Timotius & F. Göltenboth, pp. 171–7. Salatiga, Indonesia: Satya Wacana Chrisitian University.

Bootsma, H.A. (1993*a*). *Algal dynamics in an African Great Lake, and their relation to hydrographic and meteorological conditions*. Ph.D. thesis, University of Manitoba.

Bootsma, H.A. (1993*b*). Spatio-temporal variation of phytoplankton biomass in Lake Malawi, Central Africa. *Verh. int. Verein. Limnol.*, **25**, 882–6.

Bootsma, H.A., Bootsma, M.J. & Hecky, R.E. (1996). The chemical composition of precipitation and its significance to the nutrient budget of Lake Malawi. In *The limnology, climatology and paleoclimatology of the East African lakes*, ed. T.C. Johnson & E.O. Odada, pp. 251–65. Amsterdam: Gordon & Breach.

Bootsma, H.A. & Hecky, R.E. (1993). Conservation of the African Great Lakes: a limnological perspective. *Cons. Biol.*, **7**, 1–13.

Bootsma, H.A., Hecky, R.E., Hesslein, R.H. & Turner, G.F. (1996). Food partitioning among Lake Malawi nearshore fishes as revealed by stable isotope analyses. *Ecology*, **77**, 1286–90.

Böttger, K. (1975). Produktionsbiologische Studien an dem zentralafrikanischen Bergbach Kalengo. *Arch. Hydrobiol.*, **75**, 1–31.

Bouvy, M., Arfi, R. & Guiral, D. (1994). Short-term variations of seston characteristics in a shallow tropical lagoon: effect of wind-induced resuspension. *J. aquat. Ecol.*, **28**, 433–40.

Bowen, S.H. (1976). Mechanism for digestion of detrital bacteria by the cichlid fish *Sarotherodon mossambicus* (Peters). *Nature, Lond.*, **260**, 137–8.

Bowen, S.H. (1979). Determinants of the chemical composition of periphytic detrital aggregate in a tropical lake (Lake Valencia, Venezuela). *Arch. Hydrobiol.*, **87**, 166–77.

Bowen, S.H. (1980). Detrital nonprotein amino acids are the key to rapid growth of *Tilapia* in Lake Valencia, Venezuela. *Science, NY*, **207**, 1216–18.

Bowen, S.H. (1981). Digestion and assimilation of periphytic detrital aggregate by *Tilapia mossambica*. *Trans. Amer. Fish. Soc.*, **110**, 239–45.

Bowen, S.H. (1984). Detritivory in neotropical fish communities. In *Evolutionary ecology of neotropical freshwater fishes*, ed. T.M. Zaret, pp. 59–66. The Hague: Junk.

Bowmaker, A.P. (1963). Cormorant predation on two central African lakes. *Ostrich*, **34**, 2–26.

Bowmaker, A.P. (1976). The physicochemical limnology of the Mwenda River mouth, Lake Kariba. *Arch. Hydrobiol.*, **77**, 66–108.

Bozniak, E.G., Schanen, N.S., Parker, B.C. & Keenan, C.M. (1969). Limnological features of a tropical meromictic lake. *Hydrobiologia*, **34**, 524–32.

Bradbury, J.P. (1971). Paleolimnology of Lake Texcoco, Mexico. Evidence from diatoms. *Limnol. Oceanogr.*, **16**, 180–200.

Bradbury, J.P., Leyden, B., Salgado-Labouriau, M., Lewis, W.M., Schubert, C., Binford, M.W., Frey, D.G., Whitehead, D.R. & Weizebahn, F.H. (1981). Late Quaternary environmental history of Lake Valencia, Venezuela. *Science, NY*, **214**, 1299–305.

Bradley, W.H. (1966). Tropical lakes, copropel, and oil shale. *Bull. geol. Soc. Amer.*, **77**, 1333–8.

Branco, C.W.C. & Senna, P.A.C. (1994). Factors influencing the development of *Cylindrospermopsis raciborskii* and *Microcystis aeruginosa* in the Paranoá Reservoir, Brasilia, Brazil. *Arch. Hydrobiol., Suppl.* **105** (*Algological Studies* 75), 85–96.

Branco, C.W.C. & Senna, P.A.C. (1996). Phytoplankton composition, community structure and seasonal changes in a tropical reservoir (Paranoá Reservoir, Brazil). *Arch. Hydrobiol. Suppl.* **114**, 69–84.

Brandorff, G.O. & Andrade, E.R. (1978). The relationship between the water level of the Amazon River and the fate of the zooplankton population in Lago Jacaretinga, a *várzea* lake in the Central Amazon. *Stud. Neotr. Fauna Envir.*, **13**, 63–70.

Branstrator, D.K., Ndawula, L.M. & Lehman, J.T. (1996). Zooplankton dynamics in Lake Victoria. In *The limnology, climatology and paleoclimatology of the East African lakes*, ed. T.C. Johnson & E.O. Odada, pp. 337–55. Amsterdam: Gordon & Breach.

Breen, C.M. & Stormanns, C.H. (1991). Observations on the growth and production of *Cyperus papyrus* L. in a sub-tropical swamp. *Verh. int. Verein. Limnol.*, **24**, 2722–5.

Bright, G.R. (1982). Secondary benthic production in a tropical island stream. *Limnol. Oceanogr.*, **27**, 472–80.

Brinkmann, W.L.F. (1986). Particulate and dissolved materials in the Rio Negro-Amazon basin. In *Sediments and water interactions*, ed. P.G. Sly, pp. 3–12. New York: Springer-Verlag.

Brinkmann, W.L.F. & Santos, U. de M. (1973). Heavy fish-kill in unpolluted floodplain lakes of Central Amazon, Brasil. *Biol. Conserv.*, **5**, 146–7.

Brinkmann, W.L.F. & Santos, U. de M. (1974). The emission of biogenic hydrogen sulfide from amazonian floodplain lakes. *Tellus*, **26**, 261–7.

Brook, A.J. & Rzóska, J. (1954). The influence of the Gebel Aulyia Dam on the development of Nile plankton. *J. anim. Ecol.*, **23**, 101–14.

Brown, D.S. (1994). *Freshwater snails of Africa and their medical importance.* Chapter II. *Life cycles and populations*, 2nd edn. London: Taylor & Francis.

Brown, L.H. & Root, A. (1971). The breeding behaviour of the lesser flamingo. *Ibis*, **113**, 147–72.

Brown, T.E., Morley, A.W. & Koontz, D.V. (1985). The limnology of a naturally acidic tropical water system in Australia. II. Dry season characteristics. *Verh. int. Verein. Limnol.*, **22**, 2131–5.

Buddington, R.K. (1979). Digestion of an aquatic macrophyte by *Tilapia zillii*. *J. Fish. Biol.*, **15**, 449–56.

Bugenyi, F.W.B. & Magumba, K.M. (1996). The present physicochemical ecology of Lake Victoria, Uganda. In *The limnology, climatology and paleoclimatology of the East African lakes*, ed. T.C. Johnson & E.O. Odada, pp. 141–54. Amsterdam: Gordon & Breach.

Bullock, A. (1993). Perspectives on hydrology and water resource management of natural freshwater wetlands and lakes in the humid tropics. In *Hydrology and water management in the humid tropics*, ed. M. Bonell, M.H. Hufschmidt & J.S. Gladwell, pp. 273–300. UNESCO. Cambridge: Cambridge Univ. Press.

Bullock, T.A. (1955). Compensation for temperature in the metabolism and activity of poikilotherms. *Biol. Rev.*, **30**, 311–42.

Burgis, M.J. (1970). The effect of temperature on the development time of eggs of *Thermocyclops sp.*, a tropical cyclopoid from Lake George, Uganda. *Limnol. Oceanogr.*, **15**, 742–7.

Burgis, M.J. (1971). The ecology and production of copepods, particularly *Thermocyclops hyalinus*, in the tropical Lake George, Uganda. *Freshwat. Biol.* **1**, 169–92.

Burgis, M.J. (1973). Observations on the Cladocera of Lake George, Uganda. *J. Zool., Lond.*, **170**, 339–49.

Burgis, M.J. (1974). Revised estimates for the biomass and production of zooplankton in Lake George, Uganda. *Freshwat. Biol.* **4**, 535–41.

Burgis, M.J. (1978). Case studies of lake ecosystems at different latitudes: the tropics. The Lake George ecosystem. *Verh. int. Verein. Limnol.*, **20**, 1139–52.

Burgis, M.J. (1984). An estimate of zooplankton biomass for Lake Tanganyika. *Verh. int. Verein. Limnol.*, **22**, 1199–203.

Burgis, M.J. (1986). Food chain efficiency in the open water of Lake Tanganyika. *Bull. Séances Acad. Roy. Sci. Outre-Mer*, **30**, 282–4.

Burgis, M.J., Darlington, J.P.E.C., Dunn, I.G., Ganf, G.G., Gwahaba, J.J. & McGowan, L.M. (1973). The biomass and distribution of organisms in Lake George, Uganda. *Proc. R. Soc. (B)*, **184**, 271–98.

Burgis, M.J. & Dunn, I.G. (1978). Production in three contrasting ecosystems. In *Ecology of freshwater fish production*, ed. S.D. Gerking, pp. 137–58. Oxford: Blackwell.

Burgis, M.J., Mavuti, K.M., Moreau, J. & Moreau, I. (1987). The central plateau. In *African wetlands and shallow water bodies. Directory*, ed. M.J. Burgis & J.J. Symoens, pp. 359–88. Paris: ORSTOM.

Burgis, M.J. & Symoens, J.J. (1987). *African wetlands and shallow water bodies. Directory.* Paris: ORSTOM.

Burgis, M.J. & Walker, A.F. (1972). A preliminary comparison of the zooplankton in a tropical and a temperate lake (Lake George, Uganda and Loch Leven, Scotland). *Verh. int. Verein. Limnol.*, **18**, 647–55.

Buschkiel, A.L. (1936). Periodizitäten und Biozyklen im tropischen Süsswasser. *Arch. Hydrobiol. Suppl. (Tropische Binnengewässer)* **14**, 506–11.

Cadée, G.C. (1978). Primary production and chlorophyll in the Zaïre river, estuary and plume. *Neth. J. Sea Res.*, **12**, 368–81.

Calder, E.A. (1959). Nitrogen fixation in a Uganda swamp soil. *Nature, Lond.*, **184**, 746.

Cantrell, M.A. (1988). Effect of lake level fluctuations on the habitats of benthic invertebrates in a shallow tropical lake. *Hydrobiologia*, **158**, 125–31.

Carbonnel, J.P. & Guiscafré, J. (1965). *Grand Lac du Cambodge. Sédimentologie et hydrologie. 1962–63.* Paris: Ministère des Affaires Etrangères.

Carignan, R., Neiff, J.J. & Planas, D. (1994). Limitation of water hyacinth by nitrogen in subtropical lakes of the Paraná floodplain (Argentina). *Limnol. Oceanogr.*, **39**, 439–43.

Carignan, R. & Planas, D. (1994). Recognition of nutrient and light limitation in turbid mixed layers: three approaches compared in the Paraná floodplain (Argentina). *Limnol. Oceanogr.*, **39**, 580–96.

Carmouze, J.P. (1983). Hydrochemical regulation of the lake. In *Lake Chad. Ecology and productivity of a shallow tropical ecosystem*, ed. J.P. Carmouze, J.R. Durand & C. Lévêque, pp. 95–123. Monogr. Biologicae 53. The Hague: Junk.

Carmouze, J.P. (1992). The energy balance. In *Lake Titicaca*, ed. C. Dejoux & A. Iltis, pp. 149–60. Monogr. Biologicae 68. Dordrecht: Kluwer.

Carmouze, J.P., Aquize, E., Arze, C. & Quintanilla, J.J. (1983). Le bilan énergétique du lac Titicaca. *Rev. Hydrobiol. trop.*, **16**, 135–44.

Carmouze, J.P., Arze, C. & Quintanilla, J. (1981). Régulation hydrochimique du lac Titicaca et l'hydrochimie de ses tributaires. *Rev. Hydrobiol. trop.*, **14**, 329–48.

Carmouze, J.P., Arze, C. & Quintanilla, J. (1982). Hydrochemical regulation of the lake and water chemistry of its inflow rivers. In *Lake Titicaca*, ed. C. Dejoux & A. Iltis, pp. 98–112. Monogr. Biologicae 68. Dordrecht: Kluwer.

Carmouze, J.P., Chantraine, J.M. & Lemoalle, J. (1983). Physical and chemical characteristics of the waters. In *Lake Chad. Ecology and productivity of a shallow tropical ecosystem*, ed. J.P. Carmouze, J.R. Durand & C. Lévêque, pp. 65–94. Monogr. Biologicae 53. The Hague: Junk.

Carmouze, J.P., Durand, J.R. & C. Lévêque, C. (ed.) (1983). *Lake Chad. Ecology and productivity of a shallow tropical ecosystem.* Monogr. Biologicae 53. The Hague: Junk.

Carmouze, J.P., Fotius, G. & Lévêque, C. (1978). Influence qualitative des macrophytes sur la régulation hydrochimique du lac Tchad. *Cah. ORSTOM sér. Hydrobiol.*, **12**, 65–9.

Carmouze, J.P., Golterman, H.L. & Pedro, G. (1976). The neoformation of sediments in Lake Chad. Their influence on the salinity control. In *Interaction between sediments and freshwater*, ed. H.L. Golterman, pp. 33–9. The Hague: Junk & PUDOC.

Carmouze, J.P. & Lemoalle, J. (1983). The lacustrine environment. In *Lake Chad. Ecology and productivity of a shallow tropical ecosystem*, ed. J.P. Carmouze, J.R. Durand & C. Lévêque, pp. 27–63. Monogr. Biologicae 53. The Hague: Junk.

Carney, H.J. (1984). Productivity, population growth and physiological responses to nutrient enrichments by phytoplankton of Lake Titicaca, Peru – Bolivia. *Verh. int. Verein. Limnol.*, **22**, 1253–7.

Carney, H.J., Richerson, P.J. & Eloranta, P. (1987). Lake Titicaca (Peru/Bolivia) phytoplankton: species composition and structural comparison with other tropical and temperate lakes. *Arch. Hydrobiol.*, **110**, 365–85.

Carter, G.S. (1934). Results of the Cambridge Expedition to British Guiana, 1933. The fresh waters of the rain-forest areas of British Guiana. *J. Linn. Soc. Zool.*, **39**, 147–93.

Carter, G.S. & Beadle, L.C. (1930). The fauna of the swamps of the Paraguayan Chaco in relation to its environment. I. Physico-chemical nature of the environment. *J. Linn. Soc. Zool.*, **37**, 205–58.

Carvajal-Chitty, H.I. (1993). Some notes about the Intermediate Disturbance Hypothesis and its effects on the phytoplankton of the middle Orinoco river. *Hydrobiologia*, **249**, 117–24.

Carvalho, M.L. (1984). Influence of predation by fish and water turbidity on a *Daphnia gessneri* population in an Amazonian floodplain. *Hydrobiologia*, **113**, 243–7.

Caulton, M.S. (1977). A quantitative assessment of the daily ingestion of *Panicum repens* L. by *Tilapia rendalli* Boulenger (Cichlidae) in Lake Kariba. *Trans. Rhodesia Sci. Ass.*, **58**, 38–42.

Caulton, M.S. (1978). The importance of habitat temperature for growth in the tropical cichlid *Tilapia rendalli* Boulenger. *J. Fish. Biol.*, **13**, 99–112.

Caumette, P., Pagano, M. & Saint-Jean, L. (1983). Répartition verticale du phytoplancton, des bactéries et du zooplancton dans un milieu stratifié en Baie de Biétri (Lagune Ebrié, Côte-d'Ivoire). Relations trophiques. *Hydrobiologia*, **106**, 135–48.

Chacón-Torres, A. (1993*a*). Lake Patzcuaro, Mexico: effects of turbidity in a tropical high altitude lake. *Trop. freshwat. Biol.*, **3**, 251–72.

Chacón-Torres, A. (1993*b*). Lake Patzcuaro, Mexico: watershed and water quality deterioration in a tropical high-altitude Latin American lake. *Lake Reservoir Mgmt.*, **8**, 37–47.

Chale, F.M.M. (1987). Plant biomass and nutrient levels of a tropical macrophyte (*Cyperus papyrus* L.) receiving domestic wastewater. *Hydrobiol. Bull., Amsterdam*, **21**, 167–70.

Chapman, D.W. & van Well, P. (1978). Growth and mortality of *Stolothrissa tanganicae*. *Trans. Am. Fish. Soc.*, **107**, 26–35.

Chapman, L.J. & Chapman, C.A. (1993). Fish populations in tropical floodplain pools: a re-evaluation of Holden's data on the River Sokoto. *Ecol. freshwat. Fish.*, **2**, 23–30.

Chapman, L.J. & Kramer, D.L. (1991). The consequences of flooding for the dispersal and fate of poeciliid fish in an intermittent tropical stream. *Oecologia*, **87**, 299–306.

Chatfield, C. (1984). *The analysis of time series. An introduction.* London: Chapman & Hall.

Christensen, V. & Pauly, D. (eds.) (1993). *Trophic models of aquatic ecosystems.* ICLARM Conf. Proc. 26.

Chutter, F.M. (1985). Seasonality/aseasonality: chairman's summary. *Hydrobiologia*, **125**, 191–4.

Cisneros, R.O. & Mangas, E.I. (1991). Zooplankton studies in a tropical lake (Lake Xolotlán, Nicaragua). *Verh. int. Verein. Limnol.*, **24**, 1167–70.

Cloudsley-Thompson, J.L. (1964). Diurnal rhythm of activity in the Nile crocodile. *Animal Behav.*, **12**, 98–100.

Coche, A.G. (1974). Limnological study of a tropical reservoir, part I. In *Lake Kariba, a man-made tropical ecosystem in Central Africa*, ed. E.K. Balon & A.G. Coche, pp. 1–248. Monogr. Biologicae 24. The Hague: Junk.

Coe, M.J. (1966). The biology of *Tilapia grahami* in Lake Magadi, Kenya. *Acta tropica*, **23**, 146–77.

Cole, J.J., Carago, N.F., Kling, G.W. & Kratz, T.K. (1994). Carbon dioxide supersaturation in the surface waters of lakes. *Science, NY*, **265**, 1568–70.

Collot, P., Koriyama, F. & Garcia, E. (1983). Répartitions, biomasses et productions des macrophytes du lac Titicaca. *Rev. Hydrobiol. trop.*, **16**, 241–61.

Compère, P. & Iltis, A. (1983). The phytoplankton. In *Lake Chad. Ecology and productivity of a shallow tropical ecosystem*, ed. J.P. Carmouze, J.R. Durand & C. Lévêque, pp. 145–97. Monogr. Biologicae 68. The Hague: Junk.

Corbet, P.S. (1958). Lunar periodicity of aquatic insects in Lake Victoria. *Nature, Lond.*, **182**, 330–1.

Corbet, P.S. (1961). The food of non-cichlid fishes in the Lake Victoria basin, with remarks on their evolution and adaptation to lacustrine conditions. *Proc. Zool. Soc. Lond.*, **136**, 1–101.

Corbet, P.S. (1964). Temporal patterns of emergence in aquatic insects. *Can. Ent.*, **96**, 264–79.

Costa-Moreira, A.L. & Carmouze, J.P. (1991). La lagune de Saquarema (Brésil): hydroclimat, seston et éléments biogéniques au cours d'un cycle annuel. *Rev. Hydrobiol. trop.*, **24**, 13–23.

Cott, H.B. (1954). The status of the Nile crocodile in Uganda. *Uganda J.*, **18**, 1–12.

Cott, H.B. (1963). Scientific results of an enquiry into the ecology and economic status of the Nile Crocodile (*Crocodilus niloticus*) in Uganda and Northern Rhodesia. *Trans. Zool. Soc. Lond.*, **29**, 211–346.

Coulter, G.W. (1963). Hydrological changes in relation to biological production in southern Lake Tanganyika. *Limnol. Oceanogr.*, **8**, 463–77.

Coulter, G.W. (1968). Hydrological processes and primary production in Lake Tanganyika. *Proc. Conf. Great Lakes Res.*, **11**, 609–26.

Coulter, G.W. (1970). Population changes within a group of fish species in Lake Tanganyika following their exploitation. *J. Fish Biol.*, **2**, 329–53.

Coulter, G.W. (1981). Biomass, production and potential yield of the Lake Tanganyika pelagic fish community. *Trans. Amer. Fish. Soc.*, **110**, 325–35.

Coulter, G.W. (1988). Seasonal hydrodynamic cycles in Lake Tanganyika. *Verh. int. Verein. Limnol.*, **23**, 86–9.

Coulter, G.W. (ed.) (1991*a*). *Lake Tanganyika and its life*. Oxford: Natural History Museum Publications, Oxford Univ. Press.

Coulter, G.W. (1991*b*). Pelagic fish. In *Lake Tanganyika and its life*, ed. G.W. Coulter, pp. 111–138. Oxford: Oxford Univ. Press.

Coulter, G.W. (1991*c*). Zoogeography, affinities and evolution with special regard to the fish. In *Lake Tanganyika and its life*, ed. G.W. Coulter, pp. 275–305. Oxford: Oxford Univ. Press.

Coulter, G.W. (1994). Speciation and fluctuating environments, with reference to ancient East African lakes. *Arch. Hydrobiol. Beih., Ergebn. Limnol.*, **44**, 127–37.

Coulter, G.W., Roest, F.C. & Lindqvist, O. (eds.) (in press). Limnology and fisheries in Lake Tanganyika. *Mitt. int. Verein. Limnol.*, **26**.

Coulter, G.W. & Spigel, R.H. (1991). Hydrodynamics. In *Lake Tanganyika and its life*, ed. G.W. Coulter, pp. 49–75. Oxford: Oxford Univ. Press.

Craig, J.F. (1992). Human-induced changes in the composition of fish communities in the African Great Lakes. *Rev. Fish Biol. Fisheries*, **2**, 93–124.

Cressa, A.C. & Lewis, W.H. (1986). Ecological energetics of *Chaoborus* in a tropical lake. *Oecologia*, **70**, 326–31.

Crill, P.M., Bartlett, K.B., Wilson, J.O., Sebacher, D.I., Harriss, R.C., Melack, J.M., MacIntyre, S., Lesack, L. & Smith-Morrill, L. (1988). Tropospheric methane from an Amazonian floodplain lake. *J. geophys. Res.*, **93**, 1564–70.

Cronberg, G. (1997). Phytoplankton in Lake Kariba 1986–1990. In *Advances in the ecology of Lake Kariba*, ed. J. Moreau, pp. 66–101. Harare: Univ. of Zimbabwe Publications.

Cronberg, G., Gieske, A., Martins, E., Prince Nengu, J. & Stenström, I.M. (1996). Major ion chemistry, plankton and bacterial assemblages of the Jao/Boro River, Okavango Delta, Botswana: the swamps and flood plains. *Arch. Hydrobiol. Suppl.* **107**, 287–334.

Crul, R.C.M. (1993). *Monographs of the African Great Lakes. Limnology and hydrology of Lake Victoria.* UNESCO/IHP Project M-51.

Crul, R.C.M. (1995). *Monographs of the African Great Lakes*, part 1, *Lake Tanganyika*; part 2, *Lake Malawi.* UNESCO/IHP Project M-51.

Crutzen, P.J. & Andreae, M.O. (1990). Biomass burning in the tropics: impact on atmospheric chemistry and biogeochemical cycles. *Science, NY*, **250**, 1669–78.

Cruz, V. & Salazar, P. (1989). Biomasa y producción primaria del perifiton en una sabane inundable de Venezuela. *Rev. Hydrobiol. trop.*, **22**, 213–22.

Darlington, J.P.E.C. (1977). Temporal and spatial variation in the benthic invertebrate fauna of Lake George, Uganda. *J. Zool. Lond.*, **181**, 95–111.

Darragi, F. & Tardy, Y. (1987). Authigenic trioctahedral smectites controlling pH, alkalinity, silica and magnesium concentrations in alkaline lakes. *Chem. Geol.*, **63**, 59–72.

Dávalos, L.O., Lind, O.T. & Doyle, R.D. (1989). Evaluation of phytoplankton-limiting factors in Lake Chapala, Mexico: turbidity and the spatial and temporal variation in algal assay response. *Lake Reservoir Mgmt.*, **5**, 99–104.

Davies, B.R. (1980). Stream regulation in Africa: a review. In *The ecology of regulated streams*, ed. J.W. Ward & J.A. Stanford, pp. 113–42. New York: Plenum.

Davies, B.R. & Hart, R.C. (1981). Invertebrates. In *The ecology and utilization of African inland waters*, ed. J.J. Symoens, M. Burgis & J.J. Gaudet, pp. 51–68. United Nations Environmental Programme, Reports & Proceedings Series 1.

Davies, B.R. & Walker, K.F. (1986). *The ecology of river systems.* Dordrecht: Junk.

Day, J.A. (1993). The major ion chemistry of some southern African saline systems. *Hydrobiologia*, **267**, 37–59.

Deacon, E.L. (1969). Physical processes near the surface of the earth. In *General climatology*, ed. H. Flohn, pp. 46. In *World survey of climatology*, vol. 2, ed. H.E. Landsberg. Amsterdam: Elsevier.

Deb, D. (1995). Scale-dependence of food web structures: tropical ponds as paradigm. *Oikos*, **72**, 245–62.

Deevey, E.S. (1955). Limnological studies in Guatemala and El Salvador. *Verh. int. Verein. Limnol.*, **12**, 278–83.

Deevey, E.S. (1957). Limnological studies in Middle America. *Trans. Conn. Acad. Arts Sci.*, **39**, 213–328.

Degnbol, P. & Mapila, S. (1985). Limnological observations on the pelagic zone of Lake Malawi from 1978 to 1981. In *Fishery expansion project. Malawi.*

Biological studies on the pelagic ecosystem of Lake Malawi, pp. 5–47. MLW/75/019 Tech. Rep. Rome: FAO.

Dejoux, C. (1971). Recherches sur le cycle de développement de *Chironomus pulcher*. *Can. Ent.*, **103**, 465–70.

Dejoux, C. (1983*a*). The fauna associated with the aquatic vegetation. In *Lake Chad. Ecology and productivity of a shallow tropical ecosystem*, ed. J.P. Carmouze, J.R. Durand & C. Lévêque, pp. 273–92. Monogr. Biologicae 53. The Hague: Junk.

Dejoux, C. (1983*b*). The impact of birds on the lacustrine ecosystem. In *Lake Chad. Ecology and productivity of a shallow tropical ecosystem*, ed. J.P. Carmouze, J.R. Durand & C. Lévêque, pp. 519–25. Monogr. Biologicae 53. The Hague: Junk.

Dejoux, C. & Iltis, A. (ed.) (1992). *Lake Titicaca. Synthesis of present limnological knowledge*. Monogr. Biologicae 68. Dordrecht: Kluwer.

de Lima, W.C., Marins, M. de A. & Tundisi, J.G. (1983). Influence of wind on the standing stock of *Melosira italica* (Ehr.) Kütz. *Revta bras. Biol.*, **43**, 317–20.

Delincé, G. (1992). *The ecology of the fish pond ecosystem with special reference to Africa*. Dordrecht: Kluwer.

de Merona, B., Hecht, T. & Moreau, J. (1988). Growth of African fishes. In *Biology and ecology of African freshwater fishes*, ed. C. Lévêque, M.N. Bruton & G.W. Ssentongo, pp. 191–219. Trav. Docum. 216, Paris: ORSTOM.

Denny, P. (1972). Lakes of South-Western Uganda. Physical and chemical studies in Lake Bunyoni. *Freshwat. Biol.*, **2**, 143–58.

Denny, P. (ed.) (1985). *The ecology and management of African wetland vegetation*. Dordrecht: Junk.

Devol, A.H., Dos Santos, A., Forsberg, B.R. & Zaret, T.M. (1984). Nutrient addition experiments in Lago Jacaretinga, Central Amazon, Brazil. 2. The effect of humic and fulvic acids. *Hydrobiologia*, **109**, 97–103.

Devol, A.H., Forsberg, B.R., Richey, J.E. & Pimentel, T.A. (1995). Seasonal variations in chemical distributions in the Amazon River: a multi-year time series. *Global Biogeochem. Cycles*, **9**, 307–28.

Devol, A.H., Richey, J.E., Quay, P. & Martinelli, L. (1987). The role of gas exchange in the inorganic carbon, oxygen, and ^{222}Rn budgets of the Amazon River. *Limnol.Oceanogr.*, **32**, 235–48.

Devol, A.H., Zaret, T.M. & Forsberg, B.R. (1984). Sedimentary organic matter diagenesis and its relation to the carbon budget of tropical Amazon floodplain lakes. *Verh. int. Verein. Limnol.*, **22**, 1299–304.

Dhonneur, G. (1985). *Traité de météorologie tropicale*. Paris: Direction de la Météorologie.

Dobesch, H. (1983). Energy and water budget of a tropical man-made lake. *Limnology of Parakrama Samudra – Sri Lanka*, ed. F. Schiemer, pp. 19–26. The Hague: Junk.

Dokulil, M., Bauer, K. & Silva, I. (1983). An assessment of the phytoplankton biomass and primary productivity of Parakrama Samudra, a shallow man-made lake in Sri Lanka. In *Limnology of Parakrama Samudra – Sri Lanka*, ed. F. Schiemer, pp. 49–76. The Hague: Junk.

Domingos, P. & Carmouze, J.P. (1995). Influence des intrusions de masses d'air polaires sur le phytoplancton et le métabolisme d'une lagune tropicale. *Rev. Hydrobiol. trop.*, **26**, 257–67.

Downing, J.A. & Plante, C. (1993). Production of fish populations in lakes. *Can. J. Fish. aquat. Sci.*, **50**, 110–20.

Downing, J.A., Plante, C. & Lalonde, S. (1990). Fish production correlated with primary productivity, not the morphoedaphic index. *Can. J. Fish. aquat. Sci.*, **47**, 1929–36.

Downing, J.A. & Rigler, F.H. (1984). *A manual on methods for the assessment of secondary productivity in fresh waters*, 2nd edn. Oxford: Blackwell.

Doyle, R.D. & Fisher, T.R. (1994). Nitrogen fixation by periphyton and plankton on the Amazon floodplain at Lake Calado. *Biogeochemistry*, **26**, 41–6.

Drago, E.C. (1989). Morphological and hydrological characteristics of the floodplain ponds of the Middle Paraná River. *Rev. Hydrobiol. trop.*, **22**, 183–90.

Drayton, R.S. (1984). Variations in the level of Lake Malawi. *Hydrol. Sci. J.*, **29**, 1–12.

Dubois, J.Th. (1958). Composition chimique des affluents du nord du lac Tanganika. *Bull. Séances Acad. Roy. Belg., Sci. Colon.*, **4**, 1226–37.

Dudgeon, D. (1983). An investigation of the drift of aquatic insects in Tai Po Kau Forest Stream, New Territories, Hong Kong. *Arch. Hydrobiol.*, **96**, 434–47.

Dudgeon, D. (1984). Seasonal and long-term changes in the hydrobiology of the Lam Tsuan River, New Territories, Hong Kong with special reference to benthic macroinvertebrate distribution and abundance. *Arch. Hydrobiol. Suppl.* **69**, 55–129.

Dudgeon, D. (1985). The population dynamics of some freshwater carideans (Crustacea: Decapoda) in Hong Kong, with special reference to *Neocaridina serrata* (Atyidae). *Hydrobiologia*, **120**, 141–9.

Dudgeon, D. (1990). Seasonal dynamics of invertebrate drift in a Hong Kong stream. *J. Zool., Lond.*, **222**, 187–96.

Dudgeon, D. (1992). *Patterns and processes in stream ecology. A synoptic review of Hong Kong running waters.* Die Binnengewässer 29.

Dudgeon, D. (1993). The effects of spate-induced disturbance, predation and environmental complexity on macroinvertebrates in a tropical stream. *Freshwat. Biol.*, **30**, 189–98.

Dudgeon, D. (1995). The ecology of rivers and streams in tropical Asia. In *River and stream ecosystems*, ed. C.E. Cushing, K.W. Cummins & G.W. Minshall, pp. 615–57. Amsterdam: Elsevier.

Dudgeon, D. (1996). The life history, secondary production and microdistribution of *Ephemera* spp. (Ephemeroptera: Ephemerida) in a tropical forest stream. *Arch. Hydrobiol*, **135**, 473–83.

Dudgeon, D. (in press). *Tropical Asian streams: zoobenthos and ecology*. Hong Kong: Hong Kong Univ. Press.

Dudgeon, D. & Bretschko, G. (1995). Land–water interactions and stream ecology: comparison of tropical Asia and temperate Europe. In *Tropical limnology*, vol. 1, ed. K.H. Timotius & F. Göltenboth, pp. 69–108. Salatiga, Indonesia: Satya Wacana Christian University.

Dudgeon, D. & Bretschko, G. (1996). Allochthonous inputs and land–water interactions in seasonal streams: tropical Asia and temperate Europe. In *Perspectives in tropical limnology*, ed. F. Schiemer & K.T. Boland, pp. 161–79. Amsterdam: SPB Academic Publishing.

Dufour, P. (1982). Modèles semi-empiriques de la production phytoplanctonique en milieu lagunaire tropical (Côte-d'Ivoire). *Acta Oecologica, Oecol. Gener.*, **3**, 223–39.

Dufour, P., Cremoux, J.L. & Slepoukha, M. (1981). Contrôle nutritif de la biomasse du seston dans une lagune tropicale de Côte-d'Ivoire. I. Etude méthodologique et premiers résultats. *J. exp. mar. Biol. Ecol.*, **51**, 247–67.

Dufour, P. & Durand, J.R. (1982). La production végétale des lagunes de Côte-d'Ivoire. *Rev. Hydrobiol. trop.*, **15**, 209–30.

Dufour, P., Kouassi, A.M. & Lanusse, A. (1994). Les pollutions. In *Environment et ressources aquatiques de Côte-d'Ivoire*. II. *Les milieux lagunaires*, ed. J.R. Durand, P. Dufour, D. Guiral & S.G. Zabi, pp. 309–33. Paris: ORSTOM.

Dufour, P. & Lemasson, L. (1985). Le régime nutritif de la lagune tropicale Ebrié (Côte-d'Ivoire). *Oceanogr. trop.*, **20**, 41–69.

Dufour, P., Lemasson, L. & Cremoux, J.L. (1981). Contrôle nutritif de la biomasse du seston dans une lagune tropicale de Côte-d'Ivoire. II. Variations géographiques et saisonnières. *J. exp. mar. Biol. Ecol.*, **51**, 269–84.

Dufour, P. & Slepoukha, M. (1981). Etude de la fertilité d'une lagune tropicale au moyen de tests biologiques sur populations phytoplanctoniques naturelles. *Rev. Hydrobiol. trop.*, **14**, 103–14.

Dumont, H.J. (1992). The regulation of plant and animal species and communities in African shallow lakes and wetlands. *Rev. Hydrobiol. trop.*, **25**, 303–46.

Duncan, A. (1983). The influence of temperature upon the duration of embryonic development of tropical *Brachionus* species (Rotifera). In *Limnology of Parakrama Samudra – Sri Lanka*, ed. F. Schiemer, pp. 107–15. Monogr. Biologicae 53. The Hague: Junk.

Duncan, A. (1984). Assessment of factors influencing the composition, body size and turnover rate of zooplankton in Parakrama Samudra, an irrigation reservoir in Sri Lanka. *Hydrobiologia*, **113**, 201–15.

Dunn, I.G. (1967). Diurnal fluctuations of physicochemical conditions in a shallow tropical pond. *Limnol. Oceanogr.*, **12**, 151–4.

Dunn, I.G. (1970). Recovery of a tropical pond zooplankton community after destruction by algal bloom. *Limnol. Oceanogr.*, **15**, 373–9.

Dunne, T. (1978). Rates of chemical denudation of silicate rocks in tropical catchments. *Nature, Lond.*, **274**, 244–6.

Durand, J.R. & Chantraine, J.M. (1982). L'environnement climatique des lagunes ivoiriennes. *Rev. Hydrobiol. trop.*, **15**, 85–114.

Durve, V.S. & Rao, P.S. (1987). Seasonal variation in primary productivity and its interrelationships with chlorophyll in the Lake Jaisamand (Rajasthan, India). *Acta Hydrochim. Hydrobiol.*, **15**, 379–87.

Dyer, T.G.J. (1979). Pseudo-periodicities in lake level changes. In *Lake Chilwa*, ed. M. Kalk, A.J. McLachlan & C. Howard-Williams, pp. 47–9. Monogr. Biologicae 53. The Hague: Junk.

Eccles, D.H. (1962). An internal wave in Lake Nyasa and its probable significance in the nutrient cycle. *Nature, Lond.*, **194**, 832–3.

Eccles, D.H. (1974). An outline of the physical limnology of Lake Malawi (Lake Nyasa). *Limnol. Oceanogr.*, **19**, 730–42.

Eccles, D.H. (1976). Mass mortalities of *Lates* (Pisces: Centropomidae) in Lake Albert. *J. Limnol. Soc. Sth. Afr.*, **2**, 7–10.

Edmond, J.M., Palmer, M.R., Measures, C.I., Grant, B. & Stallard, R.F. (1995). The fluvial geochemistry and denudation rate of the Guayana Shield in Venezuela, Colombia and Brazil. *Geochim. cosmochim. Acta*, **59**, 301–25.

Edmond, J.M., Stallard, R.F., Craig, H., Craig, V., Weiss, R.F. & Coulter, G.W. (1993). Nutrient chemistry of the water column: Lake Tanganyika. *Limnol. Oceanogr.*, **38**, 725–38.

Egborge, A.B.M. (1974). The seasonal variation and distribution of phytoplankton in the River Oshun, Nigeria. *Freshwat. Biol.*, **4**, 177–91.

Egborge, A.B.M. (1978). Seasonal variations in the density of a small West African Lake. *Hydrobiologia*, **61**, 195–203.

Egborge, A.B.M. (1979). The effect of impoundment on the phytoplankton of the River Oshun, Nigeria. *Nova Hedwigia*, **31**, 407–18.

Egborge, A.B.M. & Ogbekene, L. (1986). Cyclomorphosis in *Keratella tropica* (Apstein) of Lake Asejire, Nigeria. *Hydrobiologia*, **135**, 179–91.

Elewa, S.A. (1985). Effect of flood water on the salt content of Aswan High Dam Reservoir. *Hydrobiologia*, **128**, 249–54.

Ellery, K., Ellery, W.N., Rogers, K.H. & Walker, B.H. (1990). Formation, colonization and fate of floating sudds in the Maunachira river system of the Okavango Delta, Botswana, *Aquat. Bot.*, **38**, 315–29.

Elouard, J.M. & Forge, P. (1978). Emergence et activité de vol nocturne de quelques espèces d'Ephéméroptères de Côte-d'Ivoire. *Rev. Hydrobiol. trop.*, **7**, 187–95.

Elouard, J.M. & Lévêque, C. (1977). Rythme nycthéméral de dérive des insectes et des poissons dans les rivières de Côte-d'Ivoire. *Cah. ORSTOM sér. Hydrobiol.*, **11**, 179–83.

Engle, D.L. & Melack, J.M. (1990). Floating meadow epiphyton: biological and chemical features of epiphytic material in an Amazon floodplain lake. *Freshwat. Biol.*, **23**, 474–94.

Engle, D.L. & Melack, J.M. (1993). Consequences of riverine flooding for seston and the periphyton of floating meadows in an Amazon floodplain lake. *Limnol. Oceanogr.*, **38**, 1500–20.

Engle, D.L. & Sarnelle, O. (1990). Algal use of sedimentary phosphorus from an Amazon floodplain lake: implications for total phosphorus analysis in turbid waters. *Limnol. Oceanogr.*, **35**, 483–90.

Enikeff, M.G. (1939). Le transport de sels dissous par le Niger en 1938. *C. R. Acad. Sci. (Paris)*, **209**, 229–31.

Entz, B. (1976). Lake Nasser and Lake Nubia. In *The Nile, biology of an ancient river*, ed. J. Rzóska, pp. 271–98. Monogr. Biologicae 29. The Hague: Junk.

Entz, B. (1978). Sedimentation processes above the Aswan High Dam in Lake Nasser – Nubia (Egypt – Sudan). *Verh. int. Verein. Limnol.*, **20**, 1667–71.

Eppley, R.W. (1972). Temperature and phytoplankton growth in the sea. *Fish. Bull. (USA)*, **70**, 1063–86.

Erikson, R., Hooker, E. & Meija, M. (1991*a*). The dynamics of photosynthetic activity in Lake Xolotlán (Nicaragua). *Verh. int. Verein. Limnol.*, **24**, 1163–6.

Erikson, R., Hooker, E. & Meija, M. (1991*b*). Underwater light penetration, phytoplankton biomass and photosynthetic activity in Lake Xolotlán (Managua). *Hydrobiol. Bull., Amsterdam*, **25**, 137–44.

Erikson, R., Pum, M., Vammen, K., Cruz, A., Ruiz, M. & Zamora, H. (1997). Nutrient availability and the stability of phytoplankton biomass and production in Lake Xolotlán (Lake Managua, Nicaragua). *Limnologia*, **27**, 157–64.

Ertel, J.R., Hedges, J.I., Devol, A.H. & Richey, J.E. (1986). Dissolved humic substances of the Amazon River system. *Limnol. Oceanogr.*, **31**, 739–54.

Esteves, F.A. & Barbieri, R. (1983). Dry weight and chemical changes during decomposition of tropical macrophytes in Lobo Reservoir – São Paulo, Brazil. *Aquat. Bot.*, **16**, 285–95.

Eugster, H.P. & Jones, B.F. (1979). Behavior of major solutes during closed-basin brine evolution. *Am. J. Sci.*, **279**, 609–31.

Eugster, H.P. & Maglione, G. (1979). Brines and evaporites of the Lake Chad basin, Africa. *Geochim. cosmochim. Acta*, **43**, 973–81.

Evans, J.H. (1961). Growth of Lake Victoria phytoplankton in enriched cultures. *Nature, Lond.*, **189**, 417.

Evans, J.H. (1997). Spatial and seasonal distribution of phytoplankton in an African Rift Valley Lake (Lake Albert, Uganda–Zaïre). *Hydrobiologia*, **354**, 1–16.

Farrell, T.P., Finlayson, C.M. & Griffiths, D.J. (1979). Studies of the hydrobiology of a tropical lake in northwestern Queensland. I. Seasonal changes in chemical characteristics. *Austr. J. mar. freshwat. Res.*, **30**, 579–95.

Fatimah, M.Y., Mohsin, A.K.M. & Kamal, A.S.M. (1984). Phytoplankton composition and productivity of a shallow tropical lake. *Pertanika*, **7**, 101–13.

Faure, H. & Gac, J.Y. (1981). Climate variability, solar cycle, stratospheric and atmospheric circulation. *Pangea (France)*, 15–16, 17–28.

Fee, E.J. (1969). A numerical model for the estimation of photosynthetic production, integrated over time and depth, in natural waters. *Limnol. Oceanogr.*, **14**, 906–11.

Fee, E.J. (1973*a*). A numerical model for determining integral primary production and its application to Lake Michigan. *J. Fish. Res. Bd. Can.*, **30**, 1447–68.

Fee, E.J. (1973*b*). Modelling primary production in water bodies; a numerical approach that allows vertical inhomogeneities. *J. Fish Res. Bd. Can.*, **30**, 1469–73.

Fenchel, T. (1974). Intrinsic rate of natural increase: the relationship with body size. *Oecologia*, **14**, 317–26.

Ferguson, A.J.D. (1982). Studies on the zooplankton of Lake Turkana. In *Lake Turkana. A report on the findings of the Lake Turkana project, 1972–1975*, ed. A.J. Hopson, pp. 163–245. London: Overseas Development Administration.

Ferguson, A.J.D. & Harbott, B.J. (1982). Geographical, physical and chemical aspects of Lake Turkana. In *Lake Turkana. A report on the findings of the Lake Turkana Project 1972–1975*, ed. A.J. Hopson, pp. 1–107. London: Overseas Development Administration.

Fernandes, V.O. & Esteves, F.A. (1996). Temporal variation of dry weight, organic matter, chlorophyll *a* + phaeopigments and organic carbon of the periphyton on leaves of *Typha dominguensis*. *Arch. Hydrobiol. Suppl.* **114**, 85–98.

Fernando, C.H. (1980*a*). The freshwater zooplankton of Sri Lanka with a discussion of tropical freshwater zooplankton composition. *Int. Rev. ges. Hydrobiol.* **651**, 85–125.

Fernando, C.H. (1980*b*). The species and size composition of tropical freshwater zooplankton with special reference to the Oriental Region (South East Asia). *Int. Rev. ges. Hydrobiol.* **651**, 411–26.

Fernando, C.H. (1984). Reservoirs and lakes of Southeast Asia (Oriental Region). In *Ecosystems of the world.* 23. *Lakes and reservoirs.* ed. F.B. Taub, pp. 411–46. Amsterdam: Elsevier.

Fernando, C.H. (1991). Impacts of fish introductions in tropical Asia and America. *Can. J. Fish. aquat. Sci.*, **48**, (Suppl. 1), 24–32.

Fernando, C.H. (1993). Rice field ecology and fish culture – an overview. *Hydrobiologia*, **259**, 91–113.

Fernando, C.H. (1995). Rice fields are aquatic, semi-aquatic, terrestrial and agricultural: a complex and questionable limnology. In *Tropical limnology*, vol. 1, ed. K.H. Timotius & F. Göltenboth, pp. 121–48. Salatiga, Indonesia: Satya Wacana Christian University.

Fernando, C.H. & Holcík, J. (1991*a*). Some impacts of fish introductions into tropical freshwaters. In *Ecology of biological invasion in the tropics*, ed. P.S. Ramakrishnan, pp. 103–29. New Delhi: International Scientific Publications.

Fernando, C.H. & Holcík, J. (1991*b*). Fish in reservoirs. *Int. Revue ges. Hydrobiol.*, **76**, 146–67.

Finlay, B.J., Curds, C.R., Bamforth, S.S. & Bafort, J.M. (1987). Ciliated Protozoa and other micro-organisms from two African soda lakes (Lake Nakuru and Lake Simbi, Kenya). *Arch. Protistenk.*, **133**, 81–91.

Finlayson, C.M., Farrell, T.P. & Griffiths, D.J. (1980). Studies of a tropical lake in north-western Queensland. 2. Seasonal changes in thermal and dissolved oxygen characteristics. *Aust. J. mar. freshwat. Res.*, **31**, 589–96.

Finlayson, C.M., Farrell, T.P. & Griffiths, D.J. (1984). The hydrobiology of five man-made lakes in south-western Queensland. *Proc. R. Soc. Queensland*, **95**, 29–40.

Fish, G.R. (1955). The food of Tilapia in East Africa. *Uganda J.*, **19**, 85–9.

Fish, G.R. (1956). Chemical factors limiting growth of phytoplankton in Lake Victoria. *East Afr. Agric. J.*, **21**, 152–8.

Fish, G.R. (1957). A seiche movement and its effect on the hydrology of Lake Victoria. *Fish. Publs. Colon. Office, Lond.*, **10**, 1–68.

Fisher, T.R. (1979). Plankton and primary production in aquatic systems of the central Amazon basin. *J. comp. Biochem. Physiol.*, **621**, 31–8.

Fisher, T.R., Doyle, R.D. & Peele, E.R. (1988). Size-fractionated uptake and regeneration of ammonium and phosphate in a tropical lake. *Verh. int. Verein. Limnol.*, **23**, 637–41.

Fisher, T.R. & Lean, D.R.S. (1992). Interpretation of radiophosphate dynamics in lake waters. *Can. J. Fish. aquat. Sci.*, **49**, 252–8.

Fisher, T.R., Lesack, L.F.W. & Smith, L.K. (1991). Input, recycling and export of N and P on the Amazon floodplain at Lake Calado. In *Phosphorus cycles in terrestrial and aquatic ecosystems. Regional workshop 3. South and Central America*, ed. H. Tiessen, D. Lopes-Hernandez & I.H. Salcedo, pp. 34–53. Saskatoon, Canada: Sakatchewan Inst. Petrology.

Fisher, T.R., Melack, J.M., Robertson, B., Hardy, E. & Alves, L.F. (1983). Vertical distribution of zooplankton and physico-chemical conditions during a 24 hour period in an Amazon floodplain lake (Lago Calado, Brazil). *Acta Amazonica*, **13**, 475–87.

Fisher, T.R., Morrissey, K.M., Carlson, P.R., Alves, L.F. & Melack, J.M. (1988). Nitrate and ammonium uptake in an Amazon River floodplain lake. *J. Plankton Res.*, **10**, 7–29.

Fisher, T.R. & Parsley, P.E. (1979). Amazon lakes: water storage and nutrient stripping by algae. *Limnol. Oceanogr.*, **24**, 547–53.

Flecker, A.S. & Feifarek, B. (1994). Disturbance and the temporal variability of invertebrate assemblages in two Andean streams. *Freshwat. Biol.*, **31**, 131–42.

Fleet, A.J., Kelts, K. & Talbot, M.R. (ed.) (1988). *Lacustrine petroleum source rocks*. Oxford: Blackwell.

Flores, J.F. & Balagot, V.F. (1981). Climate of the Philippines. In *Climates of southern and western Asia*, ed. H. Arakawa, pp. 159–213. *World survey of climatology*, vol. 8, ed. H.E. Landsberg. Amsterdam: Elsevier.

Fontaine, B. (1991). Variations pluviométriques et connexions climatiques: l'exemple des aires de mousson indienne et ouest-africaine. *Sécheresse*, **2**, 259–64.

Forsberg, B.R. (1984). Nutrient processing in Amazon floodplain lakes. *Verh. int. Verein. Limnol.*, **22**, 1294–8.

Forsberg, B.R., Araujo-Lima, C., Padovani, C.R., Fernandez, J., Martinelli, L.A. & Victoria, R. (1995). Carbon flow in Amazon food webs. International Association of Limnology, 26th Congress, São Paulo. Abstracts.

Forsberg, B.R., Devol, A.H., Richey, J.E., Martinelli, L.A. & Dos Santos, H. (1988). Factors controlling nutrient concentrations in Amazon floodplain lakes. *Limnol. Oceanogr.*, **33**, 41–56.

Forsberg, B.R., Pimentel, T.P. & Nobre, A.D. (1991). Photosynthetic parameters for phytoplankton in Amazon floodplain lakes. April–May 1987. *Verh. int. Verein. Limnol.*, **24**, 1188–91.

France, R.L. (1992). Climatic governance of the latitudinal cline in seasonality of freshwater phytoplankton production. *Int. J. Biometeorol.*, **36**, 243–4.

Frécaut, R. (1982). *Eléments d'hydrologie et de dynamique fluviales. I. Hydrologie et dynamique fluviale des régions chaudes et humides des basses latitudes.* Nancy: Univ. Nancy.

Freeth, S.J. & Kay, R.L.F. (1987). The Lake Nyos gas disaster. *Nature, Lond.*, **325**, 104–5.

Freitas, E.A.C. & Godhino-Orlandi, M.J.L. (1991). Distribution of bacteria in the sediment of an oxbow tropical lake (Lagoa do Infernão, SP, Brazil). *Hydrobiologia*, **211**, 33–41.

Freson, R.E. (1972). Aspect de la limnochemie et de la production primaire au lac de la Lubumbashi. *Verh. int. Verein. Limnol.*, **18**, 661–5.

Fritsch, J.M. (1992). *Les effets du défrichement de la forêt amazonienne et de la mise en culture sur l'hydrologie de petits bassins versants: opération ECEREX en Guyane Française.* Paris: ORSTOM.

Froehlich, C.G. & Arcifa, M.S. (1984). An oligomictic man-made lake in southeastern Brazil. *Verh. int. Verein. Limnol.*, **22**, 1620–4.

Fryer, G. (1959). Lunar rhythm of emergence, differential behaviour of the sexes, and other phenomena in the African midge, *Chironomus brevibucca. Bull. ent. Res.*, **50**, 1–8.

Fryer, G. (1965). Predation and its effects on migration and speciation in African fishes: a comment . . . with further comments by P.H. Greenwood and a reply by P.B.N Jackson. *Proc. Zool. Soc. Lond.*, **144**, 301–22.

Fryer, G. (1969). Speciation and adaptive radiation in African lakes. *Verh. int. Verein. Limnol.*, **17**, 303–22.

Fryer, G. (1973). The Lake Victoria fisheries: some facts and fallacies. *Biol. Conserv.*, **3**, 304–8.

Fryer, G. (1987). Morphology and the classification of the so-called Cladocera. *Hydrobiologia*, **145**, 19–28.

Fryer, G. (1991). Biological invasions in the tropics: hypotheses versus reality. In *Ecology of biological invasion in the tropics*, ed. P.S. Ramakrishnan, pp. 87–101. New Delhi: International Scientific Publications.

Fryer, G. (1997). Biological implications of a suggested late Pleistocene desiccation of Lake Victoria. *Hydrobiologia*, **354**, 177–82.

Fryer, G., Greenwood, P.H. & Peake, J.F. (1983). Punctuated equilibria, morphological stasis and the palaeontological documentation of speciation: a biological appraisal of a case history in an African lake. *Biol. J. Linnean Soc.*, **20**, 195–205.

Fryer, G. & Iles, T.D. (1972). *The cichlid fishes of the Great Lakes of Africa.* Edinburgh: Oliver & Boyd.

Fryer, G. & Talling, J.F. (1986). Africa: the FBA connection. *Rep. Freshwat. biol. Ass.*, **54**, 97–122.

Fukuhara, H., Torres, G.E. & Monteiro, S.M.C. (1997). Emergence ecology of chaoborids in Lake Dom Helvécio. In *Limnological studies on the Rio Doce Valley Lakes, Brazil*, ed. J.G. Tundisi & Y. Saijo, pp. 353–8. São Carlos: Brazilian Academy of Sciences.

Fülleborn, F. (1900). Über Untersuchungen im Nyassa-See und in den Seen im nördlichen Nyassa-land. *Verh. Ges. Erdk. Berl.* **28**, 332–8.

Furch, K. (1982). Jahreszeitliche chemische Veränderungen in einen Várzea-See des mittleren Amazonas (Lago Calado, Brasilien). *Arch. Hydrobiol.*, **95**, 47–67.

Furch, K. (1984). Seasonal variation of the major cation content of the *várzea* lake Lago Camaleão, middle Amazon, Brazil, in 1981 and 1982. *Verh. int. Verein. Limnol.*, **22**, 1288–93.

Furch, K. & Junk, W.J. (1992). Nutrient dynamics of submersed decomposing Amazonian herbaceous plant species *Paspalum fasciculatum* and *Echinochloa polystachya*. *Rev. Hydrobiol. trop.*, **25**, 75–85.

Furch, K. & Junk, W.J. (1993). Seasonal nutrient dynamics in an Amazonian floodplain lake. *Arch. Hydrobiol.*, **128**, 277–85.

Furch, K., Junk, W.J., Dieterich, J. & Kochert, N. (1983). Seasonal variation in the major cation (Na, K, Mg and Ca) content of the water of Lago Camaleão, an Amazonian floodplain-lake near Manaus, Brazil. *Amazoniana*, **8**, 75–89.

Furtado, J.I. & Mori, S. (ed.) (1982). *Tasek Bera. The ecology of a freshwater swamp*. Monogr. Biologicae 47. The Hague: Junk.

Furtado, J.I. & Verghese, S. (1981). Nutrient turnover in a freshwater inundated forested swamp, the Tasek Bera, Malaysia. *Verh. int. Verein. Limnol.*, **21**, 1200–6.

Gac, J.Y. (1980). *Géochimie du bassin du lac Tchad. Bilan de l'altération de l'érosion et de la sédimentation*. Trav. docum. 123. Paris: ORSTOM.

Gac, J.Y., Cogels, F.X. & Vincke, P.P. (1987). Lac de Guiers. In *African wetlands and shallow water bodies*. Directory, ed. M.J. Burgis & J.J. Symoens, pp. 204–13. Paris: ORSTOM.

Games, I. & Moreau, J. (1997). The feeding ecology of two Nile crocodile populations in the Zambezi valley. In *Advances in the ecology of Lake Kariba*, ed. J. Moreau, pp. 183–95. Zimbabwe: Univ. of Zimbabwe Publications.

Ganapati, S.V. & Sreenivasan, A. (1970). Energy flow in natural aquatic ecosystems in India. *Arch. Hydrobiol.*, **66**, 458–98.

Ganf, G.G. (1969). *Physiological and ecological aspects of the phytoplankton of Lake George, Uganda*. Ph.D thesis, University of Lancaster.

Ganf, G.G. (1972). The regulation of net primary production in Lake George, Uganda, East Africa. In *Productivity problems of freshwaters*, ed. Z. Kajak & A. Hillbricht-Ilkowska, pp. 693–708. Kraków: Polish Scientific Publishers.

Ganf, G.G. (1974*a*). Rates of oxygen uptake by the planktonic community of a shallow equatorial lake (Lake George, Uganda). *Oecologia*, **15**, 17–32.

Ganf, G.G. (1974*b*). Phytoplankton biomass and distribution in a shallow eutrophic lake (Lake George, Uganda). *Oecologia*, **16**, 9–29.

Ganf, G.G. (1974*c*). Incident solar irradiance and underwater light penetration as factors controlling the chlorophyll *a* content of a shallow equatorial lake (Lake George, Uganda). *J. Ecol.*, **62**, 593–609.

Ganf, G.G. (1974*d*). Diurnal mixing and the vertical distribution of phytoplankton in a shallow equatorial lake (Lake George, Uganda). *J. Ecol.*, **62**, 611–29.

Ganf, G.G. (1975). Photosynthetic production and irradiance-photosynthesis relationships of the phytoplankton from a shallow equatorial lake (Lake George, Uganda). *Oecologia*, **18**, 165–83.

Ganf, G.G. & Blažka, P. (1974). Oxygen uptake, ammonia and phosphate excretion by zooplankton of a shallow equatorial lake (Lake George, Uganda). *Limnol. Oceanogr.*, **19**, 313–26.

Ganf, G.G. & Horne, A.J. (1975). Diurnal stratification, photosynthesis and nitrogen-fixation in a shallow equatorial lake (Lake George, Uganda). *Freshwat. Biol.*, **5**, 13–39.

Ganf, G.G. & Milburn, T.R. (1971). A conductimetric method for the determination of total inorganic and particulate organic carbon fractions in freshwater. *Arch. Hydrobiol.*, **69**, 1–13.

Ganf, G.G. & Viner, A.B. (1973). Ecological stability in a shallow equatorial lake (Lake George, Uganda). *Proc. R. Soc. (B)*, **184**, 321–46.

Garrod, D.J. (1963). An estimation of the mortality rates in a population of *Tilapia esculenta* Graham (Pisces, Cichlidae) in Lake Victoria, East Africa. *J. Fish. Res. Bd. Can.*, **20**, 195–227.

Gasse, F. (1977). Evolution of an intertropical African lake from 70 000 BP to present: lake Abhé (Ethiopia and T.F.A.I.). *Nature, Lond.*, **265**, 42–5.

Gasse, F. (1987). Diatoms for reconstructing palaeoenvironments and palaeohydrology in tropical semi-arid zones. Examples of some lakes from Niger since 1200 BP. *Hydrobiologia*, **154**, 127–63.

Gasse, F., Rognon, P. & Street, F.A. (1980). Quaternary history of the Afar and Ethiopan Rift lakes. In *The Sahara and the Nile*, ed. M.A.J. Williams & H. Faure, pp. 361–400. Rotterdam: Balkema.

Gasse, F. & Street, F.A. (1978). Late Quaternary lake-level fluctuations and environments of the northern Rift Valley and Afar region (Ethiopia and Djibouti). *Palaeogeogr. Palaeoclimat. Palaeoecol.*, **24**, 279–325.

Gasse, F., Talling, J.F. & Kilham, P. (1983). Diatom assemblages in East Africa: classification, distribution and ecology. *Rev. Hydrobiol. trop.*, **16**, 3–34.

Gaudet, J.J. (1973). Growth of a floating aquatic weed, *Salvinia* under standard conditions. *Hydrobiologia*, **41**, 77–106.

Gaudet, J.J. (1975). Mineral concentrations in papyrus in various African swamps. *J. Ecol.*, **63**, 483–91.

Gaudet, J.J. (1976). Nutrient relationships in the detritus of a tropical swamp. *Arch. Hydrobiol.*, **78**, 213–39.

Gaudet, J.J. (1977*a*). Natural drawdown on Lake Naivasha, Kenya, and the formation of papyrus swamps. *Aquat. Bot.*, **3**, 1–47.

Gaudet, J.J. (1977*b*). Uptake, accumulation and loss of nutrients by papyrus in tropical swamps. *Ecology*, **58**, 415–22.

Gaudet, J.J. (1979*a*). Aquatic weeds in African man-made lakes. *Pest Articles News Summaries (PANS)*, **25**, 279–86.

Gaudet, J.J. (1979*b*). Seasonal changes in nutrients in a tropical swamp: North Swamp, Lake Naivasha, Kenya. *J. Ecol.*, **67**, 953–81.

Gaudet, J.J. & Melack, J.M. (1981). Major ion chemistry in a tropical African lake basin. *Freshwat. Biol.*, **11**, 309–33.

Gaudet, J.J. & Muthuri, F.M. (1981*a*). Nutrient regeneration in shallow tropical lake water. *Verh. int. Verein. Limnol.*, **21**, 725–9.

Gaudet, J.J. & Muthuri, F.M. (1981*b*). Nutrient relationships in shallow water in an African lake, Lake Naivasha. *Oecologia*, **49**, 109–18.

Gay, P.A. (1960). Ecological studies of *Eichhornia crassipes* Solms in the Sudan. I. Analysis of spread in the Nile. *J. Ecol.*, **48**, 183–91.

Gebre-Mariam, Z. & Taylor, W.D. (1989*a*). Seasonality and spatial variation in abundance, biomass and activity of heterotrophic bacterioplankton in relation to some biotic and abiotic variables in an Ethiopian rift-valley lake (Awassa). *Freshwat. Biol.*, **22**, 355–68.

Gebre-Mariam, Z. & Taylor, W.D. (1989*b*). Heterotrophic bacterioplankton production and grazing mortality rates in an Ethiopian rift-valley lake (Awassa). *Freshwat. Biol.*, **22**, 369–81.

Gessner, F. (1956). Das Plankton des Lago Maracaibo. *Ergebn. deutschen limnologischen Venezuela-Expedition* 1952, **1**, 67–92.

Gessner, F. & Hammer, L. (1967). Limnologische Untersuchungen an Seen der Venezolanischen Hochlanden. *Int. Rev. ges. Hydrobiol.*, **52**, 301–20.

Gianesella-Galvão, S.M.F. (1985). Primary production in ten reservoirs in southern Brazil. *Hydrobiologia*, **122**, 81–8.

Gliwicz, Z.M. (1976*a*). Stratification of kinetic origin and its biological consequences in a neotropical man-made lake. *Ekol. pol.*, **24**, 197–209.

Gliwicz, Z.M. (1976*b*). Plankton photosynthetic activity and its regulation in two neotropical man-made lakes. *Polsk. Arch. Hydrobiol.*, **23**, 61–93.

Gliwicz, Z.M. (1986*a*). A lunar cycle in zooplankton. *Ecology*, **67**, 883–7.

Gliwicz, Z.M. (1986*b*). Suspended clay concentration controlled by filter-feeding zooplankton in a tropical reservoir. *Nature, Lond.*, **323**, 330–2.

Gliwicz, Z.M. & Besiadka, E. (1975). Pelagic water mites (Hydracarina) and their effect on the plankton community in a neotropical man-made lake. *Arch. Hydrobiol.*, **76**, 65–88.

Goldschmidt, T. & Witte, F. (1990). Reproductive strategies of zooplanktivorous haplochromine cichlids (Pisces) from Lake Victoria before the Nile perch boom. *Oikos*, **58**, 356–68.

Goldschmidt, T., Witte, F. & de Visser, J. (1990). Ecological segregation in zooplanktivorous haplochromine species (Pisces: Cichlidae) from Lake Victoria. *Oikos*, **58**, 343–55.

Goldschmidt, T., Witte, F. & Wanink, J. (1993). Cascading effects of the introduced Nile Perch on the detritivorous/phytoplanktivorous species in the sublittoral areas of Lake Victoria. *Conserv. Biol.*, **7**, 686–700.

Golterman, H.L. (1971). The determination of mineralization losses in correlation with the estimation of net primary production with the oxygen method and chemical inhibitors. *Freshwat. Biol.*, **1**, 249–56.

Golterman, H.L. (1973). Natural phosphate sources in relation to phosphate budgets: a contribution to the understanding of eutrophication. *Wat. Res.*, **7**, 3–17.

Gonfiantini, R., Zuppi, G.M., Eccles, P.H. & Ferro, W. (1979). Isotope investigation of Lake Malawi. In *Isotopes in lake studies*, ed. International Atomic Energy Agency, pp. 195–208. Vienna: International Atomic Energy Agency.

Gonzáles, E., Paolini, J. & Infante, A. (1991). Water chemistry, physical features and primary production of phytoplankton in a tropical blackwater reservoir (Embalse de Gusi, Venezuela). *Verh. int. Verein. Limnol.*, **24**, 1477–81.

Gopal, B. (1987). *Water hyacinth. Biology, ecology and development.* Amsterdam: Elsevier.

Gopal, B. & Wetzel, R.W. (1995). *Limnology in developing countries.* New Delhi: International Association for Limnology/International Scientific Publications.

Gophen, M., Ochumba, P.B.O., Pollingher, U. & Kaufman, L.S. (1993). Nile Perch (*Lates niloticus*) invasion in Lake Victoria (East Africa). *Verh. int. Verein. Limnol.*, **25**, 856–9.

Gophen, M., Ochumba, P.B.O. & Kaufman, L.S. (1995). Some aspects of perturbation in the structure and biodiversity of the ecosystem of Lake Victoria (East Africa). *Aquat. Living Res.*, **8**, 27–41.

Gordon Colón, E. & Velásquez, J. (1989). Variaciones estacionales de la biomassa de *Eleocharis interstincta* (Valley) R. & S. (Cyperaceae) en la laguna El Burro (Guárico, Venezuela). *Rev. Hydrobiol. trop.*, **22**, 201–12.

Goulding, M. (1980). *The fishes and the forest*. Berkeley: Univ. California Press.

Goulding, M., Carvalho, M.L. & Ferreira, E.G. (1988). *Rio Negro, rich life in poor water*. The Hague: SPB Academic Publishing.

Gras, R., Iltis, A. & Saint-Jean, L. (1971). Biologie des Crustacés du lac Tchad. II. Régime alimentaire des Entomostracés planctoniques. *Cah. ORSTOM sér. Hydrobiol.*, **5**, 285–96.

Gras, R. & Saint-Jean, L. (1969). Biologie des Crustacés du lac Tchad. I. Durée de développement embryonnaire et post embryonnaire: premiers résultats. *Cah. ORSTOM sér. Hydrobiol.*, **3**, 43–60.

Gras, R. & Saint-Jean, L. (1976). Durée du développement embryonnaire chez quelques espèces de Cladocères et de Copépodes du lac Tchad. *Cah. ORSTOM sér. Hydrobiol.*, **10**, 233–54.

Gras, L. & Saint-Jean, L. (1978). Durée et caractéristiques du développement juvénile de quelques Cladocères du lac Tchad. *Cah. ORSTOM sér. Hydrobiol.*, **12**, 119–36.

Gras, R. & Saint-Jean, L. (1981). Durée de développement juvénile de quelques copépodes planctoniques du lac Tchad. *Rev. Hydrobiol. trop.*, **14**, 39–51.

Gras, R. & Saint-Jean, L. (1983). Production du zooplancton du lac Tchad. *Rev. Hydrobiol. trop.*, **16**, 57–77.

Green, J. (1967). The distribution and variation of *Daphnia lumholtzi* (Crustacea: Cladocera) in relation to fish predation in Lake Albert, East Africa. *J. Zool., Lond.*, **151**, 181–97.

Green, J. (1976). Changes in the zooplankton of lakes Mutanda, Bunyoni and Mulehe (Uganda). *Freshwat. Biol.*, **6**, 433–6.

Green, J. (1995). Altitudinal distribution of tropical planktonic Cladocera. *Hydrobiologia*, **307**, 75–84.

Green, J., Corbet, S.A., Watts, E. & Lan, O.B. (1976). Ecological studies on Indonesian lakes. Overturn and restratification of Ranu Lamongan. *J. Zool., Lond.*, **180**, 315–54.

Greenwood, P.H. & Lund, J.W.G. (1973). A discussion on the biology of an equatorial lake: Lake George, Uganda. *Proc. R. Soc. (B)*, **184**, 227–346.

Griffiths, J.F. (1972*a*) Semi-arid zones. In *Climates of Africa*, ed. J.F. Griffiths, pp. 193–219. *World survey of climatology*, vol. 10, ed. H.E. Landsberg. Amsterdam: Elsevier.

Griffiths, J.F. (1972*b*). Eastern Africa. In *Climates of Africa*, ed. J.F. Griffiths, pp. 313–47. *World survey of climatology*, vol. 10, ed. H.E. Landsberg. Amsterdam: Elsevier.

Griffiths, D.J. & Faithful, J.W. (1996). Effects of the sediment load of a tropical north-Australian river on water column characteristics in the receiving impoundment. *Arch. Hydrobiol. Suppl.* **113**, 147–57.

Grillet, M.E. & Barrera, R. (1997). Spatial and temporal abundance, substrate partitioning and species co-occurrence in a guild of Neotropical blackflies (Diptera – Simuliidae). *Hydrobiologia*, **345**, 197–208.

Grobbelaar, J.U. (1983). Availability to algae of N and P adsorbed on suspended solids in turbid waters of the Amazon River. *Arch. Hydrobiol.*, **96**, 302–16.

Grobbelaar, J.U. (1985). Phytoplankton productivity in turbid waters. *J. Plankton Res.*, **7**, 653–63.

Grove, A.T. (1972). The dissolved and solid load carried by some West African rivers: Senegal, Niger, Benue and Shari. *J. Hydrol.*, **16**, 277–300.

Grove, A.T. (1985). The environmental setting. In *The Niger and its neighbours*, ed. A.T. Grove, pp. 3–19. Rotterdam: Balkema.

Guildford, S.J., Bootsma, H.A., Fee, E.J., Hecky, R.E. & Patterson, G (in press). Phytoplankton nutrient status and mean water column light intensity in Lakes Malawi and Superior. In *Great lakes of the world*, ed. M. Munawar & R. E. Hecky. Burlington: Aquatic Ecosystem Health Management Society.

Guiral, D., Arfi, R., Bouvy, M., Pagano, M. & Saint-Jean, L. (1994). Ecological organization and succession during natural recolonization of a tropical pond. *Hydrobiologia*, **294**, 229–42.

Gunatilaka, A. (1983). Phosphorus and phosphatase dynamics in Parakrama Samudra based on diurnal observations. In *Limnology of Parakrama Samudra – Sri Lanka*, ed. F. Schiemer, pp. 35–47. The Hague: Junk.

Gunatilaka, A. (1984). Observations on phosphorus dynamics and orthophosphate turnover in a tropical lake – Parakrama Samudra, Sri Lanka. *Verh. int. Verein. Limnol.*, **22**, 1567–71.

Gunatilaka, A. & Senaratna, C. (1981). Parakrama Samudra (Sri Lanka) project, a study of a tropical lake ecosystem. II. Chemical environment with special reference to nutrients. *Verh. int. Verein. Limnol.*, **21**, 994–1000.

Gupta, M.K., Shrivastava, P. & Singhal, P.K. (1996). Decomposition of young water hyacinth leaves in lake water. *Hydrobiologia*, **335**, 33–41.

Gwahaba, J. (1978). The biology of cichlid fishes (Teleostei) in an equatorial lake (Lake George, Uganda). *Arch. Hydrobiol.*, **83**, 538–51.

Haberyan, K.A. (1987). Fossil diatoms and the paleolimnology of Lake Rukwa, Tanzania. *Freshwat. Biol.*, **17**, 429–36.

Haberyan, K.A. (1990). The misrepresentation of the planktonic diatom assemblage in traps and sediments: southern Lake Malawi, Africa. *J. Paleolimnol.*, **3**, 35–44.

Haberyan, K.A. & Mhone, O.K. (1991). Algal communities near Cape Maclear, southern Lake Malawi, Africa. *Hydrobiologia*, **215**, 175–88.

Habib, O.A. & Aruga, Y. (1988). Changes of the distribution of phytoplankton chlorophyll *a* in the main channel of the High Dam Lake, Egypt. *J. Tokyo Univ. Fisheries*, **75**, 343–52.

Habib, O.A., Ioriya, T. & Aruga, Y. (1987). The distribution of chlorophyll *a* as an index of primary productivity in Khor el Ramla of the High Dam Lake, Egypt. *J. Tokyo Univ. Fisheries*, **74**, 145–57.

Halat, K.M. & Lehman, J.T. (1996). Temperature-dependent energetics of *Chaoborus* populations: hypothesis for anomalous distributions in the great lakes of East Africa. *Hydrobiologia*, **330**, 31–6.

Halfman, J.D. (1993). Water column characteristics from modern CTD data, Lake Malawi, Africa. *J. Grt Lakes Res.*, **19**, 512–20.

Halfman, J.D., Johnson, T.C. & Finney, B.P. (1994). New AMS dates, stratigraphic correlations and decadal climatic cycles for the past 4 ka at Lake Turkana, Kenya. *Palaeogeogr. Palaeoclimat. Palaeoecol.*, **111**, 83–98.

Hall, J.B. & Okali, D.V.V. (1974). Phenology and productivity of *Pistia stratiotes* on the Volta Lake, Ghana. *J. appl. Ecol.*, **11**, 709–25.

Hamilton, S.K. & Lewis, W.M. (1987). Causes of seasonality in the chemistry of a lake on the Orinoco River floodplain, Venezuela. *Limnol. Oceanogr.*, **32**, 1277–90.

Hamilton, S.K. & Lewis, W.M. (1990). Basin morphology in relation to chemical and ecological characteristics of lakes on the Orinoco River floodplain, Venezuela. *Arch. Hydrobiol.*, **119**, 393–425.

Hamilton, S.K., Lewis, W.M. & Sippel, S.J. (1992). Energy sources for aquatic animals in the Orinoco River floodplain: evidence from stable isotopes. *Oecologia*, **89**, 324–30.

Hamilton, S.K., Sippel, S.J., Calheiros, D.F. & Melack, J.M. (1997). An anoxic event and other biogeochemical effects of the Pantanal wetland on the Paraguay River. *Limnol. Oceanogr.*, **42**, 257–72.

Hamilton, S.K., Sippel, S.J., Lewis, W.M. & Saunders, J.F. (1990). Zooplankton abundance and evidence for its reduction by macrophyte mats in two Orinoco floodplain lakes. *J. Plankton Res.*, **12**, 345–63.

Hamilton, S.K., Sippel, S.J. & Melack, J.M. (1995). Oxygen depletion and carbon dioxide and methane production in waters of the Pantanal wetland of Brazil. *Biogeochemistry*, **30**, 115–41.

Hammer, L. (1965). Photosynthese und Primärproduktion im Rio Negro. *Int. Rev. ges. Hydrobiol.*, **50**, 335–9.

Hammerton, D. (1972). The Nile River – a case history. In *River ecology and man*, ed. R.T. Oglesby, C.A. Carlson & J.A. McCann, pp. 171–214. New York & London: Academic Press.

Hammerton, D. (1976). The Blue Nile in the plains. In *The Nile, biology of an ancient river*, ed. J. Rzóska, pp. 243–56. Monogr. Biologicae 29. The Hague: Junk.

Haney, J.F. & Trout, M.A. (1985). Size selective grazing by zooplankton in Lake Titicaca. *Arch. Hydrobiol. Beih., Ergebn. Limnol.*, **21**, 147–60.

Haniffa, M.A. (1977). Secondary productivity and energy flow in a tropical pond. *Hydrobiologia*, **59**, 49–65.

Haniffa, M.A. & Pandian, T.J. (1978). Morphometry, primary productivity and energy flow in a tropical pond. *Hydrobiologia*, **59**, 23–48.

Hanna, N.S. & Schiemer, F. (1993). The seasonality of zooplanktivorous fish in an African reservoir (Gebel Aulia Reservoir, White Nile, Sudan). *Hydrobiologia*, **250**, 173–85, 187–99.

Harbott, B.J. (1976). Preliminary observations on the feeding of *Tilapia nilotica* in Lake Rudolf. *Afr. J. trop. Hydrobiol. Fish.*, **4**, 27–37.

Harbott, B.J. (1982). Studies on algal dynamics and primary productivity in Lake Turkana. In *Lake Turkana. A report on the findings of the Lake Turkana Project 1972–1975*, ed. A.J. Hopson, pp. 108–61. London: Overseas Development Administration.

Harding, D. (1963). Studies on the hydrology of Lake Nyasa and associated rivers. In *Report on the survey of northern Lake Nyasa 1954–55*, ed. P.B.N. Jackson, T.D. Iles, D. Harding & G. Fryer, pp. 8–44. Zomba, Nyasaland: Government Printer.

Harding, D. (1964). Hydrology and fisheries in Lake Kariba. *Verh. int. Verein. Limnol.*, **15**, 139–49.

Hardy, E.R. (1993). Changes in species composition of Cladocera and food availability in a floodplain lake, Lago Jacaretinga, Central Amazon. *Amazoniana*, **12**, 155–68.

Hardy, E.R., Robertson, B. & Koste, E. (1984). About the relationship between the zooplankton and fluctuating water levels of Lago Camaleão, a Central Amazonian *várzea* lake. *Amazoniana*, **9**, 43–52.

Hare, L. & Carter, J.C.H. (1984). Diel and seasonal physico-chemical fluctuations in a small natural West African lake. *Freshwat. Biol.*, **14**, 597–610.

Hare, L. & Carter, J.C.H. (1986). The benthos of a natural West African lake, with emphasis on the diel migrations and lunar and seasonal periodicities of the *Chaoborus* populations (Diptera, Chaoboridae). *Freshwat. Biol.*, **16**, 759–80.

Hare, L. & Carter, J.C.H. (1987). Zooplankton populations and the diets of three *Chaoborus* species (Diptera, Chaoboridae) in a tropical lake. *Freshwat. Biol.*, **17**, 275–90.

Harikumar, S., Padmanabhan, K.G., John, P.A. & Kortmülder, K. (1994). Dry-season spawning in a cyprinid fish of southern India. *Env. Biol. Fishes*, **39**, 129–36.

Harper, D. (1984). Recent changes in the ecology of Lake Naivasha, Kenya. *Verh. int. Verein. Limnol.*, **22**, 1193–7.

Harper, D. (ed.) (1987). *Studies on the Lake Naivasha ecosystem, 1982–4*. Final Report to the Kenya Government, May 1987. Leicester: Univ. Leicester.

Harper, D. (1992). The ecological relationships of aquatic plants at Lake Naivasha, Kenya. *Hydrobiologia*, **232**, 65–71.

Harper, D.M., Mavuti, K.M. & Muchiri, S.M. (1990). Ecology and management of Lake Naivasha, Kenya, in relation to climatic change, alien species introductions and agricultural development. *Env. Conserv.*, **17**, 328–36.

Harper, D.M., Philips, G., Chilvers, A., Kitaka, N. & Mavuti, K. (1993). Eutrophication prognosis for Lake Naivasha, Kenya. *Verh. int. Verein. Limnol.*, **25**, 861–5.

Harris, G.P. (1986). *Phytoplankton ecology. Structure, function and fluctuation.* London: Chapman & Hall.

Harrison, A.D. (1966). Recolonization of a Rhodesian stream after drought. *Arch. Hydrobiol.*, **62**, 405–21.

Harrison, A.D. (1995). Northeastern Africa rivers and streams. In *River and stream ecosystems*, ed. C.E. Cushing, K.W. Cummings & G.W. Minshall, pp. 507–17. Ecosystems of the world 22. Amsterdam: Elsevier.

Hart, R.C. (1980). Embryonic duration, and post-embryonic growth rates of the tropical freshwater shrimp *Caridina nilotica* (Decapoda: Atyidae) under laboratory and experimental field conditions. *Freshwat. Biol.*, **10**, 297–315.

Hart, R.C. (1981). Population dynamics and production of the tropical freshwater shrimp *Caridina nilotica* (Decapoda: Atyidae) in the littoral of Lake Sibaya. *Freshwat. Biol.*, **11**, 531–47.

Hart, R.C. (1985). Seasonality of aquatic invertebrates in low-latitude and southern hemisphere inland waters. *Hydrobiologia,* **125**, 151–78.

Hart, R.C. & Allanson, B.R. (1981). Energy requirements of the tropical freshwater shrimp *Caridina nilotica* (Decapoda: Atyidae). *Verh. int. Verein. Limnol.*, **21**, 1597–602.

Hart, R.C., Irvine, K. & Waya, R. (1995). Experimental studies on food dependency of development times and reproductive effort (fecundity and egg size) of *Tropodiaptomus cunningtoni* in relation to its natural distribution in Lake Malawi. *Arch. Hydrobiol.*, **133**, 23–47.

Hartland-Rowe, R. (1955). Lunar rhythm in the emergence of an ephemeropteran. *Nature, Lond.*, **176**, 657.

Hartland-Rowe, R. (1958). The biology of a tropical mayfly *Povilla adusta* Navas (Ephemeroptera, Polymitarcidae) with special reference to the lunar rhythm of emergence. *Rev. Zool. Bot. afr.*, **58**, 185–202.

Hartman, E., Asbury, C. & Coler, R. (1981). Seasonal variation in primary productivity in Lake Tapacura, a tropical reservoir. *J. Freshwat. Ecol.*, **1**, 203–13.

Hastenrath, S., Wu, M.C. & Chu, P.S. (1984). Towards the monitoring and prediction of north-east Brazil droughts. *Quart. J.R. met. Soc.*, **110**, 411–25.

Hawkins, P.R. (1985). Thermal and chemical stratification and mixing in a small tropical reservoir, Solomon Dam, Australia. *Freshwat. Biol.*, **15**, 193–503.

Hawkins, P.R. (1988). The zooplankton of a small tropical reservoir (Solomon Dam, North Queensland). *Hydrobiologia*, **157**, 105–18.

Hawkins, P.R. & Griffiths, D.J. (1986). Light attenuation in a small tropical reservoir (Solomon Dam, North Queensland): seasonal changes and the effects of artificial aeration. *Aust. J. mar. freshwat. Res.*, **37**, 199–208.

Hawkins, P.R. & Griffiths, D.J. (1993). Artificial destratification of a small tropical reservoir: effects upon the phytoplankton. *Hydrobiologia*, **254**, 169–81.

Haworth, E.Y. (1977). The sediments of Lake George (Uganda). V. The diatom assemblages in relation to the ecological history. *Arch. Hydrobiol.*, **80**, 200–15.

Heckman, C.W. (1994). The seasonal succession of biotic communities in wetlands of the tropical wet-and-dry climatic zone. I. Physical and chemical causes and biological effects in the Pantanal of Mato Grosso, Brazil. *Int. Rev. ges. Hydrobiol.*, **79**, 397–421.

Hecky, R.E. (1984). African lakes and their trophic efficiencies: a temporal perspective. In *Trophic interactions within aquatic ecosystems*, ed. D.G. Meyers & J.R. Strickler, pp. 405–48. Amer. Assoc. Adv. Sci. Selected Symposium 85.

Hecky, R.E. (1991). The pelagic ecosystem. In *Lake Tanganyika and its life*, ed. G.W. Coulter, pp. 90–110. Oxford: Oxford Univ. Press.

Hecky, R.E. (1993). The eutrophication of Lake Victoria. *Verh. int. Verein. Limnol.*, **25**, 39–48.

Hecky, R.E., Bootsma, H.A., Mugidde, R. & Bugenyi, F.W.B. (1996). Phosphorus pumps, nitrogen sinks and silicon drains: plumbing nutrients in the African Great Lakes. In *The limnology, climatology and paleoclimatology of the East African lakes*, ed. T.C. Johnson & E.O. Odada, pp. 205–24. Amsterdam: Gordon & Breach.

Hecky, R.E., Bugenyi, F.W.B., Ochumba, P., Talling, J.F., Mugidde, R., Gophen, M. & Kaufman, L. (1994). Deoxygenation of the deep water of Lake Victoria, East Africa, *Limnol. Oceanogr.*, **39**, 1476–81.

Hecky, R.E., Campbell, P. & Hendzel, L.L. (1993). The stoichiometry of carbon, nitrogen and phosphorus in particulate matter of lakes and oceans. *Limnol. Oceanogr.*, **38**, 709–24.

Hecky, R.E. & Fee, E.J. (1981). Primary production and rates of algal growth in Lake Tanganyika. *Limnol. Oceanogr.*, **26**, 532–47.

Hecky, R.E., Fee, E.J., Kling, H.J. & Ruud, J.M.W. (1981). The relationship between primary production and fish production in Lake Tanganyika. *Trans. Amer. Fish Soc.*, **110**, 336–45.

Hecky, R.E. & Hesslein, R.H. (1995). Contributions of benthic algae to lake food webs as revealed by stable isotope analysis. *J. North Amer. benthol. Soc.*, **14**, 631–53.

Hecky, R.E. & Kilham, P. (1973). Diatoms in alkaline saline lakes: ecology and geochemical implications. *Limnol. Oceanogr.*, **18**, 53–71.

Hecky, R.E. & Kilham, P. (1988). Nutrient limitation of phytoplankton in freshwater and marine environments: a review of recent evidence on the effects of enrichment. *Limnol. Oceanogr.*, **33**, 796–822.

Hecky, R.E. & Kling, J.J. (1981). The phytoplankton and protozooplankton of the euphotic zone of Lake Tanganyika: species composition, biomass, chlorophyll content, and spatio-temporal distribution. *Limnol. Oceanogr.*, **26**, 548–64.

Hecky, R.E. & Kling, H.J. (1987). Phytoplankton ecology of the great lakes in the rift valleys of Central Africa. *Arch. Hydrobiol. Beih., Ergebn. Limnol.*, **25**, 197–228.

Hecky, R.E., Spigel, R.H. & Coulter, G.W. (1991). The nutrient regime. In *Lake Tanganyika and its life*, ed. G.W. Coulter, pp. 76–89. Oxford: Oxford Univ. Press.

Hedges, J.I., Clark, W.A., Quay, P.D., Richey, J.E., Devol, A.H. & Santos, U. de M. (1986). Compositions and fluxes of particulate organic material in the Amazon River. *Limnol. Oceanogr.*, **31**, 717–38.

Henry, R. (1993). Thermal regime and stability of Jurumirin Reservoir (Paranapanema River, São Paulo, Brazil). *Int. Rev. ges. Hydrobiol.*, **78**, 501–11.

Henry, R. & Barbosa, F.A.R. (1989). Thermal structure, heat content and stability of two lakes in the National Park of Rio Doce Valley (Minas Gerais, Brazil). *Hydrobiologia*, **171**, 189–99.

Henry, R., Hino, K. Gentil, J.G. & Tundisi, J.G. (1985*a*). Primary production and effects of enrichment with nitrate and phosphate on phytoplankton in the Barra Bonita Reservoir (State of São Paulo, Brazil). *Int. Rev. ges. Hydrobiol.*, **70**, 561–73.

Henry, R., Hino, K., Tundisi, J.G. & Ribeiro, J.S.B. (1985*b*). Responses of phytoplankton in Lake Jacaretinga to enrichment with nitrogen and phosphorus in concentrations similar to those of the River Solimões (Amazon, Brazil). *Arch. Hydrobiol.*, **103**, 453–77.

Henry, R. & Tundisi, J.G. (1982). Evidence of limitation by molybdenum and nitrogen on the growth of the phytoplankton community of the Lobo Reservoir (São Paulo, Brazil). *Rev. Hydrobiol. trop.*, **15**, 201–8.

Henry, R. & Tundisi, J.G. (1983). Responses of the phytoplankton community of a tropical reservoir (São Paulo, Brazil) to the enrichment with nitrate, phosphate and EDTA. *Int. Rev. ges. Hydrobiol.*, **68**, 853–62.

Henry, R., Tundisi, J.G. & Curi, P.R. (1984). Effects of phosphorus and nitrogen enrichment on the phytoplankton in a tropical reservoir (Lobo Reservoir, Brazil). *Hydrobiologia*, **118**, 177–85.

Henry, R., Tundisi, J.G. & Rodrigues de Ibañez, M.S. (1997). Enrichment experiments and their effects on phytoplankton (biomass and primary productivity). In *Limnological studies on the Rio Doce Valley Lakes, Brazil*, ed. J.G. Tundisi & Y. Saijo, pp. 243–63. São Carlos: Brazilian Academy of Sciences.

Hesse, P.R. (1958*a*). The distribution of sulphur in the muds, water and vegetation of Lake Victoria. *Hydrobiologia*, **11**, 29–39.

Hesse, P.R. (1958*b*). Fixation of sulphur in the muds of Lake Victoria. *Hydrobiologia*, **11**, 171–81.

Hiez, G. & Dubreuil, P. (1984). *Les régimes hydrologiques en Guyane Française*. Paris: ORSTOM.

Ho, S.C. (1976). Periphyton production in a tropical lowland stream polluted by inorganic sediments and organic wastes. *Arch. Hydrobiol.*, **77**, 458–74.

Hodgkiss, I.J. (1974). Studies on Plover Cove Reservoir, Hong Kong: composition and distribution of phytoplankton and its relationship to environmental factors. *Freshwat. Biol.*, **4**, 111–26.

Hofer, F. & Schiemer, R. (1983). Feeding ecology, assimilation efficiencies and energetics of two herbivorous fish: *Sarotherodon (Tilapia) mossambicus* (Peters) and *Puntius filamentosus* (Cuv. et Val.). In *Limnology of Parakrama Samudra, Sri Lanka*, ed. F. Schiemer, pp. 155–64. The Hague: Junk.

Holden, M.J. (1963). The populations of fish in the dry season pools of the River Sokoto. *Fish. Publs. Colon. Office, Lond.*, **19**, 1–58.

Holden, M.J. & Green, J. (1960). The hydrology and plankton of the River Sokoto. *J. anim. Ecol.*, **29**, 65–84.

Holsinger, E.C.T. (1955). The distribution and periodicity of the phytoplankton of three Ceylon lakes. *Hydrobiologia*, **7**, 25–35.

Hooker, E.L., Hernandez, S., Chow, N. & Vargas, L. (1991). Phytoplankton studies in a tropical lake (Lake Xolotlán, Nicaragua). *Verh. int. Verein. Limnol.*, **24**, 1158–62.

Hopson, A.J. (1968). Winter scale rings in *Lates niloticus* (Pisces, Centropomidae) from Lake Chad. *Nature, Lond.*, 208, 1013–14.

Hopson, A.J. (1972). *A study of the Nile Perch in Lake Chad.* Overseas Res. Publ., London no. 19. UK: Overseas Development Administration.

Hopson, A.J. (1972). *Breeding and growth in two populations of* Alestes baremose (*Joannis*) (*Pisces, Characidae*) *from the northern basin of Lake Chad.* Overseas Res. Publ., London no. 20. UK: Overseas Development Administration.

Hopson, A.J. (ed.) (1982). *Lake Turkana. A report on the findings of the Lake Turkana project, 1972–75*, 6 vols. UK: Overseas Development Administration.

Horne, A.J. & Viner, A.B. (1971). Nitrogen fixation and its significance in tropical Lake George, Uganda. *Nature, Lond.*, **232**, 417–18.

Howard-Williams, C. (1972). Limnological studies in an African swamp: seasonal and spatial changes in the swamps of Lake Chilwa, Malawi. *Arch. Hydrobiol.*, **70**, 379–91.

Howard-Williams, C. (1975). Vegetation changes in a shallow African lake: response of the vegetation to a recent dry period. *Hydrobiologia*, **47**, 381–97.

Howard-Williams, C. (1979*a*). The aquatic environment: II. Chemical and physical characteristics of the Lake Chilwa swamps. In *Lake Chilwa*, ed. M. Kalk, A.J. McLachlan & C. Howard-Williams, pp. 79–245. Monogr. Biologicae 35. The Hague: Junk.

Howard-Williams, C. (1979*b*). The distribution of aquatic macrophytes in Lake Chilwa: annual and long-term environmental fluctuations. In *Lake Chilwa*, ed. M. Kalk, A.J. McLachlan & C. Howard-Williams, pp. 105–22. Monogr. Biologicae 35. The Hague: Junk.

Howard-Williams, C. (1979*c*). Interactions between swamp and lake. In *Lake Chilwa*, ed. M. Kalk, A.J. McLachlan & C. Howard-Williams, pp. 231–45. Monogr. Biologicae 35. The Hague: Junk.

Howard-Williams, C. & Davies, B.R. (1979). The rates of dry matter and nutrient loss from decomposing *Potamogeton pectinatus* in a brackish south-temperate coastal lake. *Freshwat. Biol.*, **9**, 13–21.

Howard-Williams, C. & Gaudet, J.J. (1985). The structure and functioning of African swamps. In *The ecology and management of African wetland vegetation*, ed. P. Denny, pp. 153–75. Dordrecht: Junk.

Howard-Williams, C. & Howard-Williams, W. (1978). Nutrient leaching from the swamp vegetation of Lake Chilwa, a shallow African lake. *Aquat. Bot.*, **4**, 257–67.

Howard-Williams, C. & Junk, W.J. (1976). The decomposition of aquatic macrophytes in the floating meadows of a central Amazonian *várzea* lake. *Biogeographica*, **7**, 115–23.

Howard-Williams, C. & Lenton, G.M. (1975). The role of the littoral zone in the functioning of a shallow tropical lake ecosystem. *Freshwat. Biol.*, **5**, 445–59.

Hughes, N.F. (1986). Changes in the feeding biology of the Nile perch, *Lates niloticus* (L.) (Pisces: Centropomidae), in Lake Victoria, East Africa, since its introduction in 1960 and its impact on the native fish community in the Nyanza Gulf. *J. Fish Biol.*, **29**, 541–8.

Hulme, M. (1992). Rainfall changes in Africa: 1931–1960 to 1961–1990. *Int. J. Climatology*, **12**, 685–99.

Hurst, H.E. (1952). *The Nile*. London: Constable.

Hurst, H.E. & Philips, P. (1938). *The Nile Basin. The hydrology of the Lake Plateau and Bahr el Jebel.* Physical Dept. Paper no. 35. Cairo: Ministry of Public Works.

Hussainy, S.U. (1967). Studies on the limnology and primary production of a tropical lake. *Hydrobiologia*, **30**, 335–52.

Hustler, K. (1997). The ecology of fish eating birds and their impact on the inshore fisheries of Lake Kariba. In *Advances in the ecology of Lake Kariba*, ed. J. Moreau, pp. 196–218. Zimbabwe: Univ. of Zimbabwe Publications.

Hustler, K. & Marshall, B.E. (1990). Population dynamics of two small cichlid fish species in a tropical man-made lake (Lake Kariba). *Hydrobiologia*, **190**, 253–62.

Hutton, J.M. (1987). Growth and feeding ecology of the Nile crocodile *Crocodylus niloticus* at Ngezi, Zimbabwe. *J. anim. Ecol.*, **56**, 25–38.

Hynes, J.D. (1975*a*). Annual cycles of macro-invertebrates of a river in southern Ghana. *Freshwat. Biol.*, **5**, 71–83.

Hynes, J.D. (1975*b*). Downstream drift of invertebrates in a river in southern Ghana. *Freshwat. Biol.*, **5**, 515–32.

Idso, S.B. (1973). On the concept of lake stability. *Limnol. Oceanogr.*, **18**, 681–3.

Ignatow, M., Mbahinzireki, G. & Lehman, J.T. (1996). Secondary production and energetics of the shrimp *Caridina nilotica* in Lake Victoria, East Africa: model development and application. *Hydrobiologia*, **332**, 175–81.

Ikusima, I. (1978). Primary production and population ecology of the aquatic sedge *Lepironia articulata* in a tropical swamp, Tasek Bera, Malaysia. *Aquat. Bot.*, **4**, 269–80.

Ikusima, I. (1982). Production estimates based on oxygen dynamics in a closed water column. In *Tasek Bera. The ecology of a freshwater swamp*, ed. J.I. Furtado & S. Mori, pp. 269–77. Monogr. Biologicae 47. The Hague: Junk.

Ikusima, I., Hino, K. & Tundisi, J.G. (1983). Daily oxygen budgets in a submerged plant stand in Broa Reservoir, Southern Brazil. *Jap. J. Limnol.*, **44**, 304–10.

Iltis, A. (1968). Tolérance de salinité de *Spirulina platensis* (Gom.) Geitl. (Cyanophyta) dans les mares natronées du Kanem (Tchad). *Cah. ORSTOM sér. Hydrobiol.*, **2**, 119–25.

Iltis, A. (1982). Peuplements algaux des rivières de Côte-d'Ivoire. 3. Etude du périphyton. *Rev. Hydrobiol. trop.*, **15**, 303–12.

Iltis, A. & Lemoalle, J. (1983). The aquatic vegetation of Lake Chad. In *Lake Chad. Ecology and productivity of a shallow tropical ecosystem*, ed. J.P.

Carmouze, J.R. Durand & C. Lévêque, pp. 125–43. Monogr. Biologicae 53. The Hague: Junk.

Iltis, A. & Riou-Duwat, S. (1971). Variations saisonnières du peuplement en Rotifères des eaux natronées du Kanem (Tchad). *Cah. ORSTOM sér. Hydrobiol.*, **5**, 101–12.

Iltis, A. & Saint-Jean, L. (1983). Trophic relations between the phytoplankton and the zooplankton. In *Lake Chad. Ecology and productivity of a shallow tropical ecosystem*, ed. J.P. Carmouze, J.R. Durand & C. Lévêque, pp. 483–8. Monogr. Biologicae 53. The Hague: Junk.

Imberger, J. (1985). Thermal characteristics of standing waters: an illustration of dynamic processes. *Hydrobiologia*, **125**, 7–29.

Imberger, J. & Patterson, J.C. (1990). Physical limnology. *Adv. appl. Mechanics*, **27**, 303–475.

Imevbore, A.M.A. & Boszormenyi, Z. (1975). Preliminary measurements of photosynthetic productivity in Lake Kainji. In *The ecology of Lake Kainji*, ed. A.M.A. Imevbore & O.S. Adegoke, pp. 136–45. Ife: Univ. of Ife Press.

Infante, A. (1982). Annual variations in abundance of zooplankton in Lake Valencia (Venezuela). *Arch. Hydrobiol.*, **93**, 194–208.

Infante, A., Infante, O., Marquez, M., Lewis, W.M. & Weibezahn, F.H. (1979). Conditions leading to mass mortality of fish and zooplankton in Lake Valencia, Venezuela. *Acta Cient. Venez.*, **3**, 67–73.

Infante, A. & Riehl, W. (1984). The effect of Cyanophyta upon zooplankton in a eutrophic tropical lake (Lake Valencia, Venezuela). *Hydrobiologia*, **113**, 293–8.

Irvine, K. (1995*a*). Standing biomass, production, spatial and temporal distributions of the crustacean zooplankton. In *The fishery potential and productivity of the pelagic zone of Lake Malawi/Niassa*, ed. A. Menz, pp. 85–108. UK: Natural Resources Institute, Overseas Development Administration.

Irvine, K. (1995*b*). Ecology of the lakefly *Chaoborus edulis*. In *The fishery potential and productivity of the pelagic zone of Lake Malawi/Niassa*, ed. A. Menz, pp. 109–40. UK: Natural Resources Institute, Overseas Development Administration.

Irvine, K. & Waya, R. (1993). Predatory behaviour of the cyclopoid copepod *Mesocyclops aequatorialis aequatorialis* in Lake Malawi, a deep tropical lake. *Verh. int. Verein. Limnol.*, **25**, 877–81.

Irvine, K., Waya, R. & Hart, R.C. (1995). Additional studies into the ecology of *Tropodiaptomus cunningtoni*. In *The fishery potential and productivity of the pelagic zone of Lake Malawi/Niassa*, ed. A. Menz, pp. 141–58. UK: Natural Resources Institute, Overseas Development Administration.

Jana, B.B. (1974). Diurnal rhythm of plankton in a tropical freshwater pond in Santiniketan, India. *Ekol. pol.*, **22**, 287–94.

Jannasch, H.W. (1975). Methane oxidation in Lake Kivu (Central Africa). *Limnol. Oceanogr.*, **20**, 860–4.

Jarvis, M.J.F., Mitchell, D.S. & Thornton, J.A. (1982). Aquatic macrophytes and *Eichhornia crassipes*. In *Lake McIlwaine, the eutrophication and recovery of a tropical African lake*, ed. J.A. Thornton, pp. 137–44. Monogr. Biologicae 49. The Hague: Junk.

Jenkin, P.M. (1936). Reports on the Percy Sladen Expedition to some Rift Valley lakes in Kenya in 1929. VII. Summary of the ecological results, with special reference to the alkaline lakes. *Ann. Mag. Nat. Hist.*, ser. 10, **18**, 133–81.

Jenkin, P.M. (1957). The filter-feeding and food of flamingoes (Phoenicoptera). *Phil. Trans. Roy. Soc. (B)*, **240**, 401–93.

Jewson, D.H., Khondker, M., Rahaman, M.H. & Lowry, S. (1993). Auxosporulation of the freshwater diatom *Aulacoseira herzogii* in Lake Banani, Bangladesh. *Diatom Res.*, **8**, 403–18.

John, D.M. (1986). The inland waters of tropical West Africa. *Arch. Hydrobiol. Beih., Ergebn. Limnol.*, **23**, 1–244.

John, D.M., Obeng-Asamoa, E.K. & Appler, H.N. (1981). Periphyton in the Volta Lake. II. Seasonal changes on wooden blocks with depth. *Hydrobiologia*, **76**, 207–15.

Johnson, T.C. & Odada, E.O. (ed.) (1996). *The limnology, climatology and paleoclimatology of the East African lakes.* Amsterdam: Gordon & Breach.

Johnson, T.C., Scholz, C.A., Talbot, M.R., Kelts, K., Ricketts, R.D., Nagobi, G., Beuning, K., Ssemmanda, I. & McGill, J.W. (1996). Late Pleistocene desiccation of Lake Victoria and rapid evolution of cichlid fishes. *Science, NY*, **273**, 1091–3.

Jones, M.B. (1986). Photosynthesis in wetlands. In *Photosynthesis in contrasting environments*, ed. N.R. Baker & S.P. Long, pp. 103–38. The Hague: Elsevier.

Jones, M.B. (1987). The photosynthetic characteristics of papyrus in a tropical swamp. *Oecologia*, **71**, 355–9.

Jones, M.B. (1988). Photosynthetic responses of C_3 and C_4 wetland species in a tropical swamp. *J. Ecol.*, **76**, 253–62.

Jones, M.B. & Milburn, T.R. (1978). Photosynthesis in papyrus (*Cyperus papyrus* L.). *Photosynthetica*, **12**, 197–9.

Jones, M.B. & Muthuri, F.M. (1985). The canopy structure and microclimate of papyrus (*Cyperus papyrus*) swamps. *J. Ecol.*, **73**, 481–91.

Junk, W. (1970). Investigations in the ecology and production-biology of the 'floating meadows' (Paspalo-Echinochloetum) on the middle Amazon. I. The floating vegetation and its ecology. *Amazoniana*, **2**, 449–95.

Junk, W.J. (1980). Die Bedeutung der Wasserstandsschwankungen für die Ökologie von Überschwemmungsgebieten, dargestellt an der Várzea des mittleren Amazonas. *Amazoniana*, **7**, 19–29.

Junk, W.J. (1982). Amazonian floodplains: their ecology, present and potential use. *Rev. Hydrobiol. trop.*, **15**, 285–302.

Junk, W.J. (1984). Ecology of the *várzea*, floodplain of Amazonian white-water rivers. In *The Amazon*, ed. H. Sioli, pp. 215–43. Monogr. Biologicae 56. Dordrecht: Junk.

Junk, W.J. (ed.) (1997). *The Central Amazon floodplain. Ecology of a pulsating system.* New York: Springer.

Junk, W.J. Bayley, P.B. & Sparks, R.E. (1989). The flood pulse concept in river-floodplain systems. In *Proceedings of the international large lakes symposium (LARS)*, ed. D.P. Dodge, *Can. Spec. Publ. Fish. Aquat. Sci.*, **106**, 110–27.

Junk, W.J. & Furch, K. (1991). Nutrient dynamics in Amazonian floodplains: decomposition of herbaceous plants in aquatic and terrestrial environments. *Verh. int. Verein. Limnol.*, **24**, 2080–4.

Junk, W.J. & Howard-Williams, C. (1984). Ecology of aquatic macrophytes in Amazonia. In *The Amazon*, ed. H. Sioli, pp. 269–93. Monogr. Biologicae 56. Dordrecht: Junk.

Junk, W.J. & Piedade, M.T.F. (1993). Biomass and primary production of herbaceous plant communities in the Amazon floodplain. *Hydrobiologia*, **63**, 155–62.

Kalff, J. (1983). Phosphorus limitation in some tropical African lakes. *Hydrobiologia*, **100**, 101–12.

Kalff, J. (1991). The utility of latitude and other environmental factors as predictors of nutrients, biomass and production in lakes worldwide: problems and alternatives. *Verh. int. Verein. Limnol.*, **24**, 1235–9.

Kalff, J. & Brumelis, D. (1993). Nutrient loading, wind speed and phytoplankton in a tropical African lake. *Verh. int. Verein. Limnol.*, **25**, 860.

Kalff, J. & Watson, S. (1986). Phytoplankton and its dynamics in two tropical lakes: a tropical and temperate zone comparison. *Hydrobiologia*, **138**, 161–76.

Kalk, M. (1979*a*). Zooplankton in Lake Chilwa: adaptations to changes. In *Lake Chilwa. Studies of change in a tropical ecosystem*, ed. M. Kalk, A.J. McLachlan & C. Howard-Williams, pp. 123–41. Monogr. Biologicae 35. The Hague: Junk.

Kalk, M. (1979*b*). Zooplankton in a quasi-stable phase in an endorheic lake (Lake Chilwa, Malawi). *Hydrobiologia*, **66**, 7–15.

Kalk, M., McLachlan, A.J. & Howard-Williams, C. (ed.) (1979). *Lake Chilwa. Studies of change in a tropical ecosystem*. Monogr. Biologicae 35. The Hague: Junk.

Kalk, M. & Schulten-Senden, C.M. (1977). Zooplankton in a tropical endorheic lake (Lake Chilwa, Malawi) during drying and recovery phases. *J. Limnol. Soc. Sth. Afr.*, **3**, 1–7.

Kannan, V. & Job, S.V. (1980*a*). Seasonal and diurnal changes of chlorophyll in a tropical fresh water impoundment. *Hydrobiologia*, **69**, 267–71.

Kannan, V. & Job, S.V. (1980*b*). Diurnal depth-wise and seasonal changes of physico-chemical factors in Sathiar reservoir. *Hydrobiologia*, **70**, 103–17.

Kannan, V. & Job, S.V. (1980*c*). Diurnal seasonal and vertical study of primary production in Sathiar Reservoir. *Hydrobiologia*, **70**, 171–8.

Karlman, S.G. (1973). *Kainji Lake Research Project, Nigeria. Pelagic primary production in Kainji Lake*. FL: DP/NIR/24. Tech. Rep. 3. Rome: FAO.

Karlman, S.G. (1982). The annual flood regime as a regulatory mechanism for phytoplankton production in Kainji Lake, Nigeria. *Hydrobiologia*, **86**, 93–7.

Kaufman, L.S. (1992). Catastrophic changes in species rich freshwater ecosystems: the lessons of Lake Victoria. *BioScience*, **42**: 846–58.

Kautsky, N. & Kiibus, M. (1997). Biomass, ecology and production of benthic fauna in Lake Kariba. In *Advances in the ecology of Lake Kariba*, ed. J. Moreau, pp. 162–82. Harare: Univ. of Zimbabwe Publications.

Kebede, E. (1997). Response of *Spirulina platensis (= Arthrospira fusiformis)* from Lake Chitu, Ethiopia, to salinity stress from sodium salts. *J. appl. Phycol.*, **9**, 551–8.

Kebede, E. & Ahlgren, G. (1996). Optimum growth conditions and light utilization efficiency of *Spirulina platensis (= Arthrospira fusiformis)* (Cyanophyta) from Lake Chitu, Ethiopia. *Hydrobiologia*, **332**, 99–109.

Kebede, E. & Belay, A. (1994). Species composition and phytoplankton biomass in a tropical African lake (Lake Awasa, Ethiopia). *Hydrobiologia*, **288**, 13–32.

Kebede, E., Mariam, Z.G. & Ahlgren, I. (1994). The Ethiopian Rift Valley lakes: chemical characteristics of a salinity–alkalinity series. *Hydrobiologia*, **288**, 1–12.

Kendall, R.L. (1969). An ecological history of the Lake Victoria basin. *Ecol. Monogr.*, **39**, 121–76.

Kern, J. & Darwich, A. (1997). Nitrogen turnover in the *várzea*. In *The central Amazon floodplain. Ecology of a pulsing system*, ed. W.J. Junk, pp. 119–35. Berlin: Springer.

Kern, J., Darwich, A., Furch, K. & Junk, W.J. (1996). Seasonal denitrification in flooded and exposed sediments from the Amazon floodplain at Lago Camaleão. *Microb. Ecol.*, **32**, 47–57.

Kern, J., Darwich, A. & Junk, W.J. (1998). The contribution of gaseous nitrogen flux in the nitrogen budget on the Amazon floodplain at Lago Camaleão. *Verh. int. Verein. Limnol.*, **26**, 926-8.

Khondker, M. & Kabir, M.A. (1995). Phytoplankton primary production in a mesotrophic pond in subtropical Bangladesh. *Hydrobiologia*, **304**, 39–47.

Khondker, M. & Parveen, L. (1993). Daily rate of primary productivity in hypertrophic Dhanmondi Lake. In *Hypertrophic and polluted freshwater ecosystems: ecological bases for water-resource management*, ed. M.M. Tilzer & M. Khondker, pp. 181–91. Bangladesh: Department of Botany, University of Dhaka.

Kifle, D. & Belay, A. (1990). Seasonal variation in phytoplankton primary production in relation to light and nutrients in Lake Awasa, Ethiopia. *Hydrobiologia*, **196**, 217–27.

Kiibus, M. & Kautsky, N. (1996). Respiration, nutrient excretion and filtration rate of tropical freshwater mussels and their contribution to production and energy flow in Lake Kariba, Zimbabwe. *Hydrobiologia*, **331**, 25–32.

Kilham, P. (1981). Pelagic bacteria: extreme abundances in African saline lakes. *Naturwissenschaften*, **67**, 380–1.

Kilham, P. (1982). The effect of hippopotamuses on the potassium and phosphate ion concentrations in an African lake. *Am. Midl. Nat.*, **108**, 202–5.

Kilham, P. (1990*a*). Ecology of *Melosira* species in the Great Lakes of Africa. In *Large lakes: ecological structure and function*, ed. M.M. Tilzer & C. Serruya, pp. 414–27. Berlin: Springer.

Kilham, P. (1990*b*). Mechanisms controlling the chemical composition of lakes and rivers: data from Africa. *Limnol. Oceanogr.*, **35**, 80–3.

Kilham, P. & Kilham, S.S. (1989). Endless summer: internal loading processes dominate nutrient cycling in tropical lakes. *Freshwat. Biol.*, **23**, 379–89.

Kilham, P., Kilham, S.S. & Hecky, R.E. (1987). Hypothesized resource relationships among African planktonic diatoms. *Limnol. Oceanogr.*, **31**, 1169–81.

Kilham, S.S. (1990). Relationship of phytoplankton and nutrients to stoichiometric measures. In *Large lakes: ecological structure and function*, ed. M.M. Tilzer & C. Serruya, pp. 403–13. Berlin: Springer.

King, R. & Lee, R.E. (1974). The effect of river flooding on the Mwenda River mouth, Lake Kariba, Rhodesia. *Arch. Hydrobiol.*, **74**, 32–8.

King, R.D. & Thomas, D.P. (1985). Environmental conditions and phytoplankton in the Mwenda River, a small intermittent river flowing into Lake Kariba. *Hydrobiologia*, **126**, 81–9.

Kite, G.W. (1981). Recent changes in level of Lake Victoria. *Hydrol. Sci. Bull.*, **9**, 233–43.

Kite, G.W. (1982). Analysis of Lake Victoria levels. *Hydrol. Sci. Bull.*, **27**, 99–110.

Kittel, T. & Richerson, P.J. (1978). The heat budget of a large tropical lake, Lake Titicaca (Peru–Bolivia). *Verh. int. Verein. Limnol.*, **20**, 1203–9.

Kizito, Y.S. & Nauwerck, A. (1995). Temporal and vertical distribution of planktonic rotifers in a meromictic crater lake, Lake Nyahirya (Western Uganda). *Hydrobiologia*, **313**, 303–12.

Kizito, Y.S. & Nauwerck, A. (1996). The distribution of planktonic rotifers in Lake Nkuruba, Western Uganda. *Limnologia*, **26**, 263–73.

Kling, G.W. (1982). Seasonal mixing and catastrophic degassing in tropical lakes, Cameroon, West Africa. *Science, NY*, **237**, 1022–4.

Kling, G.W. (1988). Comparative transparency, depth of mixing and stability of stratification in lakes of Cameroon, West Africa. *Limnol. Oceanogr.*, **33**, 27–40.

Kling, G.W., Clark, M.A., Compton, H.R., Devine, J.D., Evans, W.C., Humphrey, A.M., Koenigsberg, E.J., Lockwood, J.P., Tuttle, M.L. & Wagner, G.N. (1987). The August 1986 Lake Nyos disaster. *Science, NY*, **236**, 169–70.

Kling, G.W., Evans, W.C. & Tuttle, M.L. (1991). A comparative view of Lakes Nyos and Monoun, Cameroon, West Africa. *Verh. int. Verein. Limnol.*, **24**, 1102–5.

Kling, G.W., Tuttle, M.L. & Evans, W.C. (1989). The evolution of thermal structure and water chemistry in Lake Nyos. *J. Volcanol. Geotherm. Res.*, **39**, 151–65.

Knoppers, B., Kjerve, B. & Carmouze, J.P. (1991). Trophic and hydrodynamic turnover time in six choked coastal lagoons of Brazil. *Biogeochemistry*, **14**, 149–66.

Kramer, D.L. (1978). Reproductive seasonality in the fishes of a tropical stream. *Ecology*, **59**, 976–85.

Kundu, G. & Jana, B.B. (1994). Influence of an atypical summer on the primary productivity of phytoplankton in two tropical ponds. *Int. Rev. ges. Hydrobiol.*, **79**, 249–57.

Kutzbach, J.E. & Street-Perrott, F.A. (1985). Milankovitch forcing of fluctuations in the level of tropical lakes from 18 to 0 kyr BP. *Nature, Lond.*, **317**, 130–4.

Lacaux, J.P., Loemba-Ndembi, B., Lefeivre, B., Cros, B. & Delmas, R. (1992). Biogenic emissions and biomass influences on the chemistry of fogwater and stratiform precipitations in the African equatorial forest. *Atmos. Environ.*, **26A**, 541–51.

Lacaux, J.P., Servant, J. & Baudet, J.G.R. (1987). Acid rain in the tropical forests of the Ivory Coast. *Atmos. Environ.*, **21**, 2643–7.

Laiz, O., Quintana, I., Blomqvist, P. & Broberg, A. (1993*a*). Limnology of Cuban reservoirs. I. Lebrije. *Trop. freshwat. Biol.*, **3**, 371–96.

Laiz, O., Quintana, I., Blomqvist, P., Broberg, A. & Infante, A. (1993*b*). Limnology of Cuban reservoirs. II. Higuanojo. *Acta Cient. Venezolana*, **44**, 297–306.

Laiz, O., Quintana, I., Blomqvist, P., Broberg, A. & Infante, A. (1993*c*). Limnology of Cuban reservoirs. IV. Tuinicu. *Trop. freshwat. Biol.*, **3**, 452–71.

Laiz, O., Quintana, I., Blomqvist, P., Broberg, A. & Infante, A. (1994). Comparative limnology of four Cuban reservoirs. *Int. Rev. ges. Hydrobiol.*, **79**, 27–45.

Lakshminarayana, S.S. (1965*a*). Studies of the phytoplankton of the River Ganges, Varanasi, India. I. The physico-chemical characteristics of the River Ganges. *Hydrobiologia*, **25**, 119–37.

Lakshminarayana, S.S. (1965*b*). Studies of the phytoplankton of the River Ganges, Varanasi, India. II. The seasonal growth and succession of the plankton algae in the River Ganges. *Hydrobiologia*, **25**, 138–67.

Lamb, H.F., Gasse, F., Benkaddour, A., El Hamouti, N., van der Kaars, S., Perkins, W.T., Pearce, N.J. & Roberts, C.N. (1995). Relation between century-scale Holocene arid intervals in tropical and temperate zones. *Nature, Lond.*, **373**, 134–7.

Lancaster, N. (1979). The physical environment of Lake Chilwa. In *Lake Chilwa: studies of change in a tropical ecosystem*, ed. M.J. Kalk, A.J. McLachlan & C. Howard-Williams, pp. 17–40. Monogr. Biologicae 35. The Hague: Junk.

Landsberg, H.E. (ed.) (1972–81). *World survey of climatology*, vols. 9, 10, 12. Amsterdam: Elsevier.

Latif, A.F.A. (1984). Lake Nasser – the new man-made lake in Egypt (with reference to Lake Nubia). In *Ecosystems of the world. 23. Lakes and reservoirs*, ed. F.B. Taub, pp. 411–46. Amsterdam: Elsevier.

Lauzanne, L. (1978). Etude quantitative de l'alimentation de *Sarotherodon galilaeus* (Pisces, Cichlidae) du lac Tchad. *Cah. ORSTOM sér. Hydrobiol.*, **12**, 71–81.

Lauzanne, L. (1983). Trophic relations of fishes in Lake Chad. In *Lake Chad. Ecology and productivity of a shallow tropical ecosystem*, ed. J.P. Carmouze, J.R. Durand & C. Lévêque, pp. 489–518. Monogr. Biologicae 53. The Hague: Junk.

Laws, R.M. & Clough, G. (1966). Observations on reproduction in the hippopotamus (*Hippopotamus amphibius* Linn.). *Symp. Zool. Soc. Lond.*, **15**, 117–40.

Lazzaro, X. (1981). Biomasses, peuplements phytoplanctoniques et production primaire du lac Titicaca. *Rev. Hydrobiol. trop.*, **14**, 349–80.

Lazzaro, X. (1987). A review of planktivorous fishes: their evolution, feeding behaviours, selectivities and impacts. *Hydrobiologia*, **146**, 97–167.

Leatherland, J.F., Farbridge, K.J. & Boujard, T. (1992). Lunar and semi-lunar rhythms in fishes. In *Rhythms in fishes*, ed. M.A. Ali, pp. 83–107. London & New York: Plenum.

Leguizamon, M., Hammerly, J. Maine, M.A., Sune, N. & Pizarro, M.J. (1992). Decomposition and nutrient liberation rates of plant material in the Parana Medio river (Argentina). *Hydrobiologia*, **230**, 157–64.

Lehman, J.T. (1996). Pelagic food webs of the East African Great Lakes. In *The limnology, climatology and paleoclimatology of the East African lakes*, ed. T.C. Johnson & E.O. Odada, Amsterdam: Gordon & Breach.

Lehman, J.T. & Branstrator, D.K. (1993). Effects of nutrients and grazing on the phytoplankton of Lake Victoria. *Verh. int. Verein. Limnol.*, **25**, 850–5.

Lehman, J.T. & Branstrator, D.K. (1994). Nutrient dynamics and turnover rates of phosphate and sulfate in Lake Victoria, East Africa. *Limnol. Oceanogr.*, **39**, 227–33.

Lehmusluoto, P., Machbub, B., Terangna, N., Achmed, F., Boer, L., Setiadji, B., Brahmana, S.S. & Priadi, B. (1995). Major lakes and reservoirs in Indonesia, an overview. In *Tropical limnology*, vol. 1, ed. K.H. Timotius & F. Göltenboth, pp. 11–28. Salatiga, Indonesia: Satya Wacana Christian University.

Lelek, A. (1973). Sequence of changes in fish populations of the new tropical man-made lake, Kainji, Nigeria, West Africa. *Arch. Hydrobiol.*, **71**, 381–420.

Lemasson, L. & Pagès, J. (1982). Apports de phosphore et d'azote par la pluie en zone tropicale (Côte-d'Ivoire). *Rev. Hydrobiol. trop.*, **15**, 9–14.

Lemasson, L., Pagès, J. & Cremoux, J.L. (1982). Echanges d'éléments nutritifs dissous entre l'eau et le sédiment dans une lagune tropicale saumâtre. *Océanogr. trop.*, **17**, 45–58.

Lemma, B. (1994). Changes in the limnological behaviour of a tropical African explosion crater lake: Lake Hora-Kilole, Ethiopia. *Limnologica*, **24**, 57–70.

Lemoalle, J. (1969). Premières données sur la production primaire dans la région de Bol (avril-octobre 1968) (lac Tchad). *Cah. ORSTOM sér. Hydrobiol.*, **3**, 107–19.

Lemoalle, J. (1973*a*). L'énergie lumineuse et l'activité photosynthétique du phytoplancton dans le lac Tchad. *Cah. ORSTOM sér. Hydrobiol.*, **7**, 95–116.

Lemoalle, J. (1973*b*). Azote et phosphore dans les eaux de pluies à Fort-Lamy (1970). *Cah. ORSTOM sér. Hydrol.*, **9**, 61–3.

Lemoalle, J. (1975). L'activité photosynthétique du phytoplancton en relation avec le niveau des eaux du lac Tchad (Afrique). *Verh. int. Verein. Limnol.*, **19**, 1398–1403.

Lemoalle, J. (1978). Relations silice-diatomées dans le lac Tchad. *Cah. ORSTOM, sér. Hydrobiol.*, **12**, 137–41.

Lemoalle, J. (1979*a*). *Biomasse et production phytoplanctoniques du lac Tchad (1968–1976). Relations avec les conditions du milieu.* Paris: ORSTOM.

Lemoalle, J. (1979*b*). Application des données landsat à l'estimation de la production du phytoplancton dans le lac Tchad. *Cah. ORSTOM sér. Hydrobiol.*, **13**, 35–46.

Lemoalle, J. (1981*a*). Photosynthetic production and phytoplankton in the euphotic zone of some African and temperate lakes. *Rev. Hydrobiol. trop.*, **11**, 31–7.

Lemoalle, J. (1981*b*). Photosynthetic activity. In *The ecology and utilization of African inland waters*, ed. J.J. Symoens, M. Burgis & J.J. Gaudet, pp. 45–50. Nairobi: United Nations Environmental Programme, Reports & Proceedings Series.

Lemoalle, J. (1983). Phytoplankton production. In *Lake Chad. Ecology and productivity of a shallow tropical ecosystem*, ed. J.P. Carmouze, J.R. Durand & C. Lévêque, pp. 357–84. Monogr. Biologicae 53. The Hague: Junk.

Lemoalle, J. (1991). Eléments d'hydrologie du lac Tchad au cours d'une période de sécheresse (1973–1989). *FAO Fisheries Report* 445, 54–61.

Lenz, P.H., Melack, J.M., Robertson, B. & Hardy, E. (1986). Ammonium and phosphate regeneration by the zooplankton of an Amazon floodplain lake. *Freshwat. Biol.*, **16**, 821–30.

Leroux, M. (1983). *Le climat de l'Afrique tropicale*. Paris: Editions Champion.

Lesack, L.F.W. (1993*a*). Export of nutrients and major ionic solutes from a rain forest catchment in the Central Amazon Basin. *Wat. Resour. Res.*, **29**, 743–58.

Lesack, L.F.W. (1993*b*). Water balance and hydrologic characteristics of a rain forest catchment in the Central Amazon Basin. *Wat. Resour. Res.*, **29**, 759–73.

Lesack, L.F.W. (1995). Seepage exchange in an Amazon floodplain lake. *Limnol. Oceanogr.*, **40**, 598–609.

Lesack, L.F.W., Hecky, R.E. & Melack, J.M. (1984). Transport of carbon, nitrogen, phosphorus and major solutes in the Gambia River, West Africa. *Limnol. Oceanogr.*, **29**, 816–30.

Lesack, L.F.W. & Melack, J.M. (1991). The deposition, composition and potential sources of major ionic solutes in rain of the Central Amazon Basin. *Wat. Resour. Res.*, **27**, 2953–77.

Lesack, L.F.W. & Melack, J.M. (1995). Flooding hydrology and mixture dynamics of lake water derived from multiple sources in an Amazon floodplain lake. *Wat. Resour. Res.*, **31**, 329–45.

Lévêque, C. (1973*a*). Dynamique des peuplements, biologie, et estimation de la production des mollusques benthiques du lac Tchad. *Cah. ORSTOM sér. Hydrobiol.*, **7**, 117–47.

Lévêque, C. (1973*b*). Bilans énergétiques des populations naturelles de mollusques benthiques du lac Tchad. *Cah. ORSTOM sér. Hydrobiol.*, **7**, 151–65.

Lévêque, C. (1995*a*). Role and consequences of fish diversity in the functioning of African freshwater ecosystems: a review. *Aquat. Living Resour.*, **8**, 59–78.

Lévêque, C. (1995*b*). River and stream ecosystems of northwestern Africa. In *Ecosystems of the world. 22. River and stream ecosystems*, ed. C.E. Cushing, K.W. Cummins & G.W. Minshall, pp. 519–36. Amsterdam: Elsevier.

Lévêque, C. (1997). *Biodiversity dynamics and conservation. The freshwater fishes of Africa.* Cambridge: Cambridge Univ. Press.

Lévêque, C., Bruton, M.N. & Ssentongo, G.W. (1988). *Biology and ecology of African freshwater fishes.* Trav. et Docum. 216. Paris: ORSTOM.

Lévêque, C., Dejoux, C. & Lauzanne, L. (1983). The benthic fauna: ecology, biomass and communities. In *Lake Chad. Ecology and productivity of a shallow tropical ecosystem*, ed. J.P. Carmouze, J.R. Durand & C. Lévêque, pp. 233–72. Monogr. Biologicae 53. The Hague: Junk.

Lévêque, C., Durand, J.R. & Ecoutin, J.M. (1977). Relations entre le rapport P/B et la longévité des organismes. *Cah. ORSTOM sér. Hydrobiol.*, **11**, 17–32.

Lévêque, C. & Saint-Jean, L. (1983). Secondary production (zooplankton and benthos). In *Lake Chad. Ecology and productivity of a shallow tropical ecosystem*, ed. J.P. Carmouze, J.R. Durand & C. Lévêque, pp. 385–424. Monogr. Biologicae 53. The Hague: Junk.

Levine, S.N. & Lewis, W.M. (1984). Diel variation of nitrogen fixation in Lake Valencia, Venezuela. *Limnol. Oceanogr.*, **29**, 887–93.

Levine, S.N. & Lewis, W.M. (1986). A numerical model of nitrogen fixation and its application to Lake Valencia. *Freshwat. Biol.*, **17**, 265–74.

Levring, T. & Fish, G.R. (1956). The penetration of light into certain East African lake waters. *Oikos*, **7**, 98–109.

Lewis, D.J. (1956). Chironomidae as a pest in the northern Sudan. *Acta tropica*, **13**, 142–58.

Lewis, D.S.C. (1974). The effects of the formation of Lake Kainji upon the indigenous fish populations. *Hydrobiologia*, **45**, 281–301.

Lewis, D.S.C. (1981). Preliminary comparisons between the ecology of the haplochromine cichlid fishes of Lake Victoria and Lake Malawi. *Neth. J. Zool.*, **31**, 746–61.

Lewis, D.[S.C.] (ed.) (1988). *Predator–prey relationships, population dynamics and fisheries productivities of large African lakes.* FAO/CIFA Occas. Pap. 15.

Lewis, W.M. (1973). The thermal regime of Lake Lanao (Philippines) and its theoretical implications for tropical lakes. *Limnol. Oceanogr.*, **18**, 200–17.

Lewis, W.M. (1974). Primary production in the plankton community of a tropical lake. *Ecol. Monogr.*, **44**, 377–409.

Lewis, W.M. (1975). Distribution and feeding habits of a tropical *Chaoborus* population. *Verh. int. Verein. Limnol.*, **19**, 3106–19.

Lewis, W.M. (1977). Feeding selectivity of a tropical *Chaoborus* population. *Freshwat. Biol.*, **7**, 311–25.

Lewis, W.M. (1978*a*). Dynamics and succession of the phytoplankton in a tropical lake: Lake Lanao, Philippines. *J. Ecol.*, **66**, 849–80.

Lewis, W.M. (1978*b*). Analysis of succession in a tropical phytoplankton community and a new measure of succession rate. *Am. Nat.*, **112**, 401–14.

Lewis, W.M. (1979). *Zooplankton community analysis. Studies on a tropical system*. New York: Springer.
Lewis, W.M. (1981). Precipitation chemistry and nutrient loading by precipitation in a tropical watershed (Venezuela). *Wat. Resour. Res.*, **17**, 169–81.
Lewis, W.M. (1982). Vertical eddy diffusivities in a large tropical lake. *Limnol. Oceanogr.*, **27**, 161–3.
Lewis, W.M. (1983*a*). Temperature, heat and mixing in Lake Valencia, Venezuela. *Limnol. Oceanogr.*, **28**, 273–86.
Lewis, W.M. (1983*b*). Interception of atmospheric fixed nitrogen: an explanation of scum formation in nutrient-stressed blue-green algae. *J. Phycol.*, **19**, 534–6.
Lewis, W.M. (1983*c*). Water budget of Lake Valencia, Venezuela. *Acta Cient. Venezolana*, **34**, 248–51.
Lewis, W.M. (1984). A five-year record of temperature, mixing and stability for a tropical lake (Lake Valencia, Venezuela). *Arch. Hydrobiol.*, **99**, 340–6.
Lewis, W.M. (1985). Protozoan abundance in the plankton of two tropical lakes. *Arch. Hydrobiol.*, **104**, 337–43.
Lewis, W.M. (1986*a*). Phytoplankton succession in Lake Valencia, Venezuela. *Hydrobiologia*, **138**, 189–203.
Lewis, W.M. (1986*b*). Nitrogen and phosphorus runoff losses from a nutrient-poor tropical moist forest. *Ecology*, **67**, 1275–82.
Lewis, W.M. (1987). Tropical limnology. *Ann. Rev. Ecol. Syst.*, **18**, 158–84.
Lewis, W.M. (1988). Primary production in the Orinoco river. *Ecology*, **69**, 679–92.
Lewis, W.M. (1990). Comparisons of phytoplankton biomass in temperate and tropical lakes. *Limnol. Oceanogr.*, **35**, 1838–45.
Lewis, W.M. (1995). Tropical lakes: how latitude makes a difference. In *Tropical limnology*, vol. 1, ed. K.H. Timotius & F. Göltenboth, pp. 29–44. Salatiga, Indonesia: Satya Wacana Christian University. Reprinted 1996 in *Perspectives in tropical limnology*, ed. F. Schiemer & K.T. Boland, pp. 43–64. Amsterdam: SPB Academic Publishing.
Lewis, W.M., Frost, T. & Morris, D. (1986). Studies of planktonic bacteria in Lake Valencia, Venezuela. *Arch. Hydrobiol.*, **106**, 289–305.
Lewis, W.M., Hamilton, S.K., Jones, S.L. & Runnels, D.D. (1987). Major ion chemistry, weathering and element yields for the Caura River drainage, Venezuela. *Biogeochemistry*, **4**, 159–81.
Lewis, W.M., Hamilton, S.K. & Saunders, J.F. (1995). Rivers of northern South America. In *Ecosystems of the world. 22. River and stream ecosystems*, ed. C.E. Cushing, K.W. Cummins & G.W. Minshall, pp. 219–56. Amsterdam: Elsevier.
Lewis, W.M. & Levine, S.H. (1984). The light response of nitrogen fixation in Lake Valencia, Venezuela. *Limnol. Oceanogr.*, **29**, 894–900.
Lewis, W.M. & Saunders, J.F. (1989). Concentration and transport of dissolved and suspended substances in the Orinoco River. *Biogeochemistry*, **7**, 203–40.
Lewis, W.M. & Weibezahn, F.H. (1976). Chemistry, energy flow, and community structure in some Venezuelan fresh waters. *Arch. Hydrobiol. Suppl.* **50**, 145–207.
Lewis, W.M. & Weibezahn, F.H. (1981*a*). Acid rain and major seasonal variation of hydrogen ion loading in a tropical watershed. *Acta Cient. Venezolana*, **32**, 236–8.
Lewis, W.M. & Weibezahn, F.H. (1981*b*). Chemistry of a 7.5-m sediment core from Lake Valencia, Venezuela. *Limnol. Oceanogr.*, **26**, 907–24.

Lewis, W.M. & Weibezahn, F.H. (1983). Phosphorus and nitrogen loading of Lake Valencia, Venezuela. *Acta Cient. Venezolana*, **34**, 345–9.

Lewis, W.M., Weibezahn, F.H., Saunders, J.F. & Hamilton, S.K. (1990). The Orinoco River as an ecological system. *Interciencia*, **15**, 346–57.

Lim, R.P. & Furtado, J.I. (1975). Population changes in the aquatic fauna inhabiting the bladderwort, *Utricularia flexuosa* Vahl., in a tropical swamp, Tasek Bera, Malaysia. *Verh. int. Verein. Limnol.*, **19**, 1390–7.

Limón, J.G., Lind, O.T., Vodopich, D.S., Doyle, R. & Trotter, B.G. (1989). Long- and short-term variation in the physical and chemical limnology of a large, shallow, turbid tropical lake (Lake Chapala, Mexico). *Arch. Hydrobiol. Suppl.* **83**, 57–81.

Limpadanai, D. & Brahamanonda, P. (1978). Salinity intrusion into Lake Songkla, a lagoonal lake of Southern Thailand. *Verh. int. Verein. Limnol.* **20**, 1111–15.

Linacre, E.T. (1969). Empirical relationships involving the global radiation intensity and ambient temperature at various latitudes and altitudes. *Arch. Met. Geoph. Biokl. ser. B*, **17**, 1–20.

Lind, O.T. & Dávalos-Lind, L. (1991). Association of turbidity and organic carbon with bacterial abundance and cell size in a large, turbid, tropical lake. *Limnol. Oceanogr.*, **36**, 1200–8.

Lind, O.T., Doyle, R., Vodopich, D.S., Trotter, B.G., Limón, J.G. & Dávalos-Lind, L. (1992). Clay turbidity: regulation of phytoplankton production in a large, nutrient-rich tropical lake. *Limnol. Oceanogr.*, **37**, 549–65.

Lindell, M. & Edling, H. (1996). Influence of light on bacterioplankton in a tropical lake. *Hydrobiologia*, **323**, 67–73.

Lindmark, G. (1976). *The Lago do Paranoá Restoration Project. Bioassay – field and laboratory experiments.* Report PAHO/WHO, 76/PW/BRA/2000.

Lindmark, G. (1997). Sediment characteristics in relation to nutrients distribution in littoral and pelagic waters in Lake Kariba. In *Advances in the ecology of Lake Kariba*, ed. J. Moreau, pp. 11–57. Zimbabwe: Univ. of Zimbabwe Publications.

Linn, I.J. & Campbell, K.L.I. (1992). Interactions between white-breasted cormorants *Phalacocorax carbo* (Aves: Phalacocoracidae) and the fisheries of Lake Malawi. *J. appl. Ecol.*, **29**, 619–34.

List, M.J. (1951). Smithsonian meteorological tables, 6th edn. *Smithsonian Misc. Collect.* **114**, 411–47.

Liti, L., Källqvist, T. & Lien, L. (1991). Limnological aspects of Lake Turkana, Kenya. *Verh. int. Verein. Limnol.*, **24**, 1108–11.

Little, E.C.S. (1966). The invasion of man-made lakes by plants. In *Man-made lakes*, ed. R.H. McConnell, pp. 75–86. London: Academic Press.

Livingstone, D.A. (1975). Late Quaternary climate changes in Africa. *Ann. Rev. Ecol. Syst.*, **6**, 249–80.

Livingstone, D.A. & Melack, J.M. (1984). Some lakes of subsaharan Africa. In *Ecosystems of the world. 23. Lakes and reservoirs*, ed. F.B. Taub, pp. 467–97. Amsterdam: Elsevier.

Löffler, H. (1964). The limnology of tropical high-mountain-lakes. *Verh. int. Verein Limnol.*, **15**, 176–93.

Löffler, H. (1968*a*). Die Hochgebirgsseen Ostafrikas. *Hochgebirgsforschung*, **1**, 1–68.

Löffler, H. (1968*b*). Tropical high mountain lakes. *Colloquium Geographicum*, **9**, 57–76.

Loubens, G. (1974). Quelques aspects de la biologie des *Lates niloticus* du Tchad. *Cah. ORSTOM sér. Hydrobiol.*, **8**, 3–21.

Loubens, G. (1992). Introduced species. I. *Salmo gairdneri* (Rainbow Trout). In *Lake Titicaca. A synthesis of limnological knowledge*, ed. C. Dejoux & A. Iltis, pp. 420–6. Monogr. Biologicae 68. Dordrecht: Kluwer.

Loubens, G., Lauzanne, L. & Le Guennec, B. (1992). Les milieux aquatiques de la région de Trinidad (Béni, Amazonie bolivienne). *Rev. Hydrobiol. trop.*, **25**, 3–21.

Loubens, G. & Sarmiento, J. (1985). Observations sur les poissons de la partie bolivienne du lac Titicaca. 2. *Orestias agassii*, Valenciennes 1846 (Pisces, Cyprinodontidae). *Rev. Hydrobiol. trop.*, **18**, 159–71.

Lowe-McConnell, R.H. (1956). Observations on the biology of *Tilapia* (Pisces-Cichlidae) in Lake Victoria, East Africa. *E. Afr. Fish. Res. Org. Suppl. Publ.* no. 1, 1–72.

Lowe-McConnell, R.H. (1964). The fishes of the Rupununi savanna district of British Guiana, South America. I. Ecological groupings of fish species and effects of the seasonal cycle on the fish. *J. Linn. Soc. Zool.*, **45**, 103–44.

Lowe-McConnell, R.H. (1975). *Fish communities in tropical freshwaters. Their distribution, ecology and evolution.* London: Longman.

Lowe-McConnell, R.H. (1979). Ecological aspects of seasonality in fishes of tropical waters. In *Fish phenology*, ed. P.J. Miller, *Symp. Zool. Soc. Lond.*, **44**, 219–41.

Lowe-McConnell, R.H. (1987) *Ecological studies in tropical fish communities.* Cambridge: Cambridge Univ. Press.

MacDonald, W.W. (1956). Observations on the biology of Chaoborids and Chironomids in Lake Victoria and on the feeding habits of the 'Elephant-snout fish' (*Mormyrus kannume* Forsk.) *J. anim. Ecol.*, **25**, 36–53.

Machena, C. (1997). The organization and production of the submerged macrophyte communities in Lake Kariba. In *Advances in the ecology of Lake Kariba*, ed. J. Moreau, pp. 139–61. Zimbabwe: Univ. of Zimbabwe Publications.

Machena, C. & Fair, P. (1986). Comparison of fish yields from prediction models between lakes Tanganyika and Kariba. *Hydrobiologia*, **137**, 29–32.

Machena, C. & Kautsky, N. (1988). A quantitative diving survey of benthic vegetation and fauna in Lake Kariba, a tropical man-made lake. *Freshwat. Biol.*, **19**, 1–14.

Machena, C., Kautsky, N., & Lindmark, G. (1990). Growth and production of *Lagarosiphon ilicifolius* in Lake Kariba – a man-made tropical lake. *Aquat. Bot.*, **37**, 1–15.

Machena, C., Kolding, J. & Sanyanga, R.A. (1993). A preliminary assessment of the trophic structure of Lake Kariba, Africa. In *Trophic models of aquatic ecosystems*, ed. V. Christensen & D. Pauly, pp. 130–7. ICLARM Conf. Proc. 26.

MacIntyre, S. (1981). *Stratification and mixing in shallow tropical African lakes.* Ph.D. thesis, Duke Univ.

MacIntyre, S. (1984). Current fluctuations in the surface waters of small lakes. In *Gas transfer at water surfaces*, ed. W. Brutsaert & G.H. Jirka, pp. 125–31. Dordrecht: Riedel.

MacIntyre, S. & Melack, J.M. (1982). Meromixis in an equatorial African soda lake. *Limnol. Oceanogr.*, **27**, 595–609.

MacIntyre, S. & Melack, J.M. (1988). Frequency and depth of vertical mixing in an Amazon floodplain lake (L. Calado, Brazil). *Verh. int. Verein. Limnol.*, **23**, 80–5.

MacIntyre, S. & Melack, J.M. (1995). Vertical and horizontal transport in lakes: linking littoral, benthic and pelagic habitats. *J. North Amer. benthol. Soc.*, **14**, 599–615.

Maglione, G. & Maglione, M.H. (1972). Étude thermodynamique de quelques silicates néoformés en milieu confiné carbonaté sodique: conséquences géochimiques. *Sciences Géologiques Bull.*, **25**, 231–50.

Mahon, R. & Balon, E.K. (1977). Fish production in Lake Kariba, reconsidered. *Env. Biol. Fishes*, **1**, 215–18.

Maley, J. (1981). Etudes palynologiques dans le bassin du Tchad et paléoclimatologie de l'Afrique nord-tropicale de 30 000 ans à l'époque actuelle. Trav. Doc. 129. Paris: ORSTOM.

Marchant, R. (1982). Seasonal variation in the macroinvertebrate fauna of billabongs along Magela Creek, Northern Territory. *Austr. J. mar. freshwat. Res.*, **33**, 329–42.

Marchant, R. & Yule, C.M. (1996). A method for estimating larval life spans of aseasonal aquatic insects from streams on Bougainville Island, Papua New Guinea. *Freshwat. Biol.*, **35**, 101–7.

Mariazzi, A.A., Romero, M.C., Nakanishi, M. & Conzonno, V.H. (1983). Influence of temperature on the photosynthesis rate in the Embalse del Rio Tercero (Cordoba Province, Argentina). *Limnobios*, **2**, 419–29.

Marshall, A.J. & Roberts, J.D. (1959). The breeding biology of equatorial vertebrates: reproduction of cormorants (Phalacocoracidae) at latitude 0°20′ N. *Proc. Zool. Soc. Lond.*, **132**, 617–25.

Marshall, B.E. (1978). Aspects of the ecology of benthic fauna in Lake McIlwaine, Rhodesia. *Freshwat. Biol.*, **8**, 241–9.

Marshall, B.E. (1982*a*). The bottom fauna of Lake McIlwaine. In *Lake McIlwaine*, ed. J.A. Thornton, pp. 144–55. Monogr. Biologicae 49. The Hague: Junk.

Marshall, B.E. (1982*b*). The influence of river flow on pelagic sardine catches in L. Kariba. *J. Fish. Biol.*, **20**, 465–9.

Marshall, B.E. (1987). Growth and mortality of the introduced Lake Tanganyika clupeid, *Limnothrissa miodon*, in Lake Kariba. *J. Fish Biol.*, **31**, 603–15.

Marshall, B.E. (1988). Seasonal and annual variations in the abundance of pelagic sardines in Lake Kariba, with special reference to the effects of drought. *Arch. Hydrobiol.*, **112**, 399–409.

Marshall, B.E. (1995). Changes in the benthic fauna of Lake Chivero, Zimbabwe, over thirty years. *Sth. Afr. J. aquat. Sci.*, **21**, 22–8.

Marshall, B.E. (1997). A review of zooplankton ecology in Lake Kariba. In *Advances in the ecology of Lake Kariba*, ed. J. Moreau, pp. 102–19. Harare: Univ. of Zimbabwe Publications.

Marshall, B.E. & Falconer, A.C. (1973). Eutrophication of a tropical African impoundment (Lake McIlwaine, Rhodesia). *Hydrobiologia*, **43**, 109–23.

Marshall, B.E. & Junor, F.J.R. (1981). The decline of *Salvinia molesta* on Lake Kariba. *Hydrobiologia*, **83**, 477–84.

Martens, K., Goddeeris, B. & Coulter, G. (1994). Speciation in ancient lakes. *Arch. Hydrobiol. Beih., Ergebn. Limnol.*, **44**, 1–508.

Martens, K. & Tudorancea, C. (1991). Seasonality and spatial distribution of the ostracods of Lake Zwai, Ethiopia (Crustacea: Ostracoda). *Freshwat. Biol.*, **25**, 233–41.

Marzolf, G.R. & Saunders, G.W. (1984). Patterns of diel oxygen change in ponds of tropical India. *Verh. int. Verein. Limnol.*, **22**, 1722–6.

Masundire, H.M. (1991). Seasonal variation in size of *Bosmina longirostris* O.F. Müller in tropical Lake Kariba, Zimbabwe. *Verh. int. Verein. Limnol.*, **24**, 1455–9.

Masundire, H.M. (1994). Seasonal trends in zooplankton densities in Sanyati basin, Lake Kariba: multivariate analysis. *Hydrobiologia*, **272**, 211–30.

Masundire, H.M. (1997). Spatial and temporal variations in the composition and density of crustacean plankton in the five basins of Lake Kariba, Zambia-Zimbabwe. *J. Plankton Res.*, **19**, 43–62.

Mathooko, J.M. (1996). Rainbow trout (*Oncorhynchus mykiss* Walbaum) as a potential natural 'drift sampler' in a tropical lotic ecosystem. *Limnologia*, **26**, 245–54.

Mathooko, J.M. & Mavuti, K.M. (1992). Composition and seasonality of benthic invertebrates, and drift in the Naro Moru River, Kenya. *Hydrobiologia*, **232**, 47–56.

Mathooko, J.M. & Mavuti, K.M. (1994). Factors influencing drift transport and concentration in a second-order high altitude tropical river in central Kenya. *Afr. J. Ecol.*, **32**, 39–49.

Matsumura-Tundisi, T. & Okano, W.Y. (1983). Seasonal fluctuations of Copepod populations in lake Dom Helvécio (Parque Florestal, Rio Doce, Minas Gerais, Brazil). *Rev. Hydrobiol. trop.*, **16**, 35–9.

Matsumura-Tundisi, T., Okano, W. & Tundisi, J.G. (1997). Vertical migration of copepod populations in the tropical monomictic Lake Dom Helvécio. In *Limnological studies on the Rio Doce Valley Lakes, Brazil*, ed. J.G. Tundisi & Y. Saijo, pp. 297–307. São Carlos: Brazilian Academy of Sciences.

Matsumura-Tundisi, T., Tundisi, J.G., Saggio, A., Oliveira Neto, A.L. & Espindola, E.G. (1991). Limnology of Samuel Reservoir (Brazil, Rondônia) in the filling phase. *Verh. int. Verein. Limnol.*, **24**, 1482–8.

Matsumura-Tundisi, T., Tundisi, J.G. & Tavares, L.S. (1984). Diel migration and vertical distribution of Cladocera in Lake D. Helvécio (Minas Gerais, Brazil). *Hydrobiologia*, **113**, 299–306.

Mavuti, K.M. (1990). Ecology and role of zooplankton in the fishery of Lake Naivasha. *Hydrobiologia*, **208**, 131–40.

Mavuti, K.M. (1992). Diel vertical distribution of zooplankton in Lake Naivasha, Kenya. *Hydrobiologia*, **232**, 31–41.

Mavuti, K.M. (1994). Durations of development and production estimates by two crustacean zooplankton species *Thermocyclops oblongatus* Sars (Copepoda) and *Diaphanosoma excisum* Sars (Cladocera), in Lake Naivasha, Kenya. *Hydrobiologia*, **272**, 185–200.

Mavuti, K.M. & Litterick, M.R. (1981). Species composition and distribution of zooplankton in a tropical lake, Lake Naivasha, Kenya. *Arch. Hydrobiol.*, **93**, 52–8.

Mavuti, K., Moreau, J., Munyandorero, J. & Plisnier, P.D. (1996). Analysis of trophic relationships in two shallow equatorial lakes Lake Naivasha (Kenya) and Ihema (Rwanda) using a multispecies trophic model. *Hydrobiologia*, **321**, 89–100.

McClain, M.E. & Richey, J.E. (1996). Regional-scale linkages of terrestrial and lotic ecosystems in the Amazon basin: a conceptual model for organic matter. *Arch. Hydrobiol. Suppl.* **113**, 111–25.

McElravy, E.P., Wolda, H. & Resh, V.H. (1982). Seasonality and annual variability of caddisfly adults (Trichoptera) in a non-seasonal tropical environment. *Arch. Hydrobiol.*, **94**, 302–17.

McGowan, L.M. (1974). Ecological studies on *Chaoborus* (Diptera, Chaoboridae) in Lake George, Uganda. *Freshwat. Biol.*, **4**, 483–505.

McLachlan, A.J. (1969). The effect of aquatic macrophytes on the variety and abundance of benthic fauna in a newly created lake in the tropics (Lake Kariba). *Arch. Hydrobiol.*, **66**, 212–16.

McLachlan, A.J. (1970*a*). Some effects of annual fluctuations in water level on the larval chironomid communities of Lake Kariba. *J. anim. Ecol.*, **39**, 79–90.

McLachlan, A.J. (1970*b*). Submerged trees as a substrate for benthic fauna in the newly created Lake Kariba, Central Africa. *J. appl. Ecol.*, **7**, 253–66.

McLachlan, A.J. (1974). Development of some lake ecosystems in tropical Africa, with special reference to the invertebrates. *Biol. Rev.*, **49**, 365–97.

McLachlan, A.J. (1975). The role of aquatic macrophytes in the recovery of the benthic fauna of a tropical lake after a dry phase. *Limnol. Oceanogr.*, **20**, 54–63.

McLachlan, A.J. (1977). The changing role of terrestrial and autochthonous organic matter in newly flooded lakes. *Hydrobiologia*, **54**, 215–17.

McLachlan, A.J. (1979). Decline and recovery of the benthic invertebrate communities. In *Lake Chilwa*, ed. M. Kalk, A.J. McLachlan & C. Howard-Williams, pp. 143–60. Monogr. Biologicae 35. The Hague: Junk.

McLachlan, A.J. (1983). Life history tactics of rain-pool dwellers. *J. anim. Ecol.*, **52**, 545–61.

McLachlan, A.J. & Cantrell, M.A. (1980). Survival strategies in tropical rain pools. *Oecologia*, **47**, 344–51.

McLachlan, S.M. (1971). The rate of nutrient release from grass and dung following immersion in lake water. *Hydrobiologia*, **37**, 521–30.

Megard, R.O., Combs, W.S., Smith, P.D. & Knoll, A.S. (1979). Attenuation of light and daily integral rates of photosynthesis attained by planktonic algae. *Limnol. Oceanogr.*, **24**, 1038–50.

Melack, J.M. (1976). Primary productivity and fish yields in tropical lakes. *Trans. Am. Fish. Soc.*, **105**, 575–80.

Melack, J.M. (1978). Morphometric, physical and chemical features of the volcanic crater lakes of western Uganda. *Arch. Hydrobiol.*, **84**, 430–53.

Melack, J.M. (1979*a*). Temporal variability of phytoplankton in tropical lakes. *Oecologica*, **44**, 1–7.

Melack, J.M. (1979*b*). Photosynthesis and growth of *Spirulina platensis* (Cyanophyta) in an equatorial lake (Lake Simbi, Kenya). *Limnol. Oceanogr.*, **24**, 753–60.

Melack, J.M. (1979*c*). Photosynthetic rates in four tropical African fresh waters. *Freshwat. Biol.*, **9**, 555–71.

Melack, J.M. (1980). An initial measurement of photosynthetic productivity in Lake Tanganyika. *Hydrobiologia*, **72**, 243–7.

Melack, J.M. (1981). Photosynthetic activity of phytoplankton in tropical African soda lakes. *Hydrobiologia*, **81**, 71–85.

Melack, J.M. (1982). Photosynthetic activity and respiration in an equatorial African soda lake. *Freshwat. Biol.*, **12**, 381–400.

Melack, J.M. (1988). Primary producer dynamics associated with evaporative concentration in a shallow, equatorial soda lake (Lake Elmenteita, Kenya). *Hydrobiologia*, **158**, 1–14.

Melack, J.M. (1996). Recent developments in tropical limnology. *Verh. int. Verein. Limnol.*, **26**, 211–17.

Melack, J.M. & Fisher, T.R. (1983). Diel oxygen variations and their ecological implications in Amazon floodplain lakes. *Arch. Hydrobiol.*, **98**, 422–42.

Melack, J.M. & Fisher, T.R. (1988). Denitrification and nitrogen fixation in an Amazon floodplain lake. *Verh. int. Verein. Limnol.*, **23**, 2232–6.

Melack, J.M. & Fisher, T.R. (1990). Comparative limnology of tropical floodplain lakes with an emphasis on the central Amazon. *Acta Limnol. Brasil.*, **3**, 1–48.

Melack, J.M. & Kilham, P. (1974). Photosynthetic rates of phytoplankton in East African alkaline, saline lakes. *Limnol. Oceanogr.*, **19**, 743–55.

Melack, J.M., Kilham, P. & Fisher, T.R. (1982). Response of phytoplankton to experimental fertilization with ammonium and phosphate in an African soda lake. *Oecologia*, **52**, 321–6.

Melack, J.M. & MacIntyre, S. (1992). Phosphorus concentrations, supply and limitation in tropical African lakes and rivers. In *Phosphorus cycles in terrestrial and aquatic ecosystems. Africa*, ed. H. Tiessen & E. Frossard, pp. 1–18. Canada: Saskatchewan Inst. of Pedology and SCOPE.

Mengestou, S. & Fernando, C.H. (1991*a*). Seasonality and abundance of some dominant crustacean zooplankton in Lake Awasa, a tropical rift valley lake in Ethiopia. *Hydrobiologia*, **226**, 137–52.

Mengestou, S. & Fernando, C.H. (1991*b*). Biomass and production of the major dominant crustacean zooplankton in a tropical Rift Valley lake, Awasa, Ethiopia. *J. Plankton Res.*, **13**, 831–51.

Mengestou, S., Green, J. & Fernando, C.H. (1991). Species composition, distribution and seasonal dynamics of Rotifera in a Rift Valley lake in Ethiopia (Lake Awasa). *Hydrobiologia*, **209**, 203–14.

Menz, A. (ed.) (1995). *The fishery potential and productivity of the pelagic zone of Lake Malawi/Niassa.* UK: National Resources Institute, Overseas Development Administration.

Mepham, S. (1987). Southern Africa. In *African wetlands and shallow water bodies. Directory*, ed. M.J. Burgis & J.J. Symoens, pp. 457–594. Paris: ORSTOM.

Meyer, A., Montero, C. & Spreinat, A. (1994). Evolutionary history of the cichlid fish species flocks of the East African great lakes inferred from molecular phylogenetic data. *Arch. Hydrobiol. Beih., Ergebn. Limnol.*, **44**, 407–23.

Meyer, A., Montero, C.M. & Spreinat, A. (1996). Molecular phylogenetic inferences about the evolutionary history of East African cichlid fish radiations. In *The limnology, climatology and paleoclimatology of the East African lakes*, ed. T.C. Johnson & E.O. Odada, pp. 303–23. Amsterdam: Gordon & Breach.

Michael, R.G. & Anselm, V.M. (1979). Role of nannoplankton in primary productivity studies in tropical ponds. *Verh. int. Verein. Limnol.*, **20**, 2196–201.

Miller, M.C., Kannan, M. & Colinvaux, P.A. (1984). Limnology and primary productivity of Andean and Amazonian tropical lakes of Ecuador. *Verh. int. Verein. Limnol.*, **22**, 1264–70.

Mitamura, O., Saijo, Y. & Hino, K. (1995). Cycling of urea associated with photosynthetic activity of phytoplankton in the euphotic zone of tropical lakes, Brazil. *Jap. J. Limnol.*, **56**, 95–105.

Mitamura, O., Saijo, Y. & Hino, K. (1997). Cycling of urea associated with photosynthetic activity of phytoplankton in the euphotic layer in Lakes Dom

Helvécio, Jacaré and Carioca. In *Limnological studies on the Rio Doce Valley Lakes, Brazil*, ed. J.G. Tundisi & Y. Saijo, pp. 129–39. São Carlos: Brazilian Academy of Sciences.

Mitamura, O., Saijo, Y., Hino, K. & Barbosa, F.A.R. (1995). The significance of regenerated nitrogen for phytoplankton productivity in the Rio Doce Valley Lakes, Brazil. *Arch. Hydrobiol.*, **134**, 179–94.

Mitchell, D.S. (1969). The ecology of vascular hydrophytes on Lake Kariba. *Hydrobiologia*, **34**, 448–64.

Mitchell, D.S. (1973). Supply of plant nutrient materials in Lake Kariba. In *Man-made lakes: their problems and environmental effects*, ed. W.C. Ackermann, G.F. White & E.B. Worthington, pp. 165–9. Washington, DC: Amer. Geophys. Union.

Mitchell, D.S. & Gopal, B. (1991). Invasion of tropical freshwaters by alien aquatic plants. In *Ecology of biological invasion in the tropics*, ed. P.S. Ramakrishnan, pp. 139–54. New Delhi: International Scientific Publishers.

Mitchell, D.S. & Rogers, K.H. (1985). Seasonality/aseasonality of aquatic macrophytes in Southern Hemisphere inland waters. *Hydrobiologia*, **125**, 137–50.

Mitchell, D.S. & Tur, N.M. (1975). The rate of growth of *Salvinia molesta* (*S. auriculata* auct.) in laboratory and natural conditions. *J. appl. Ecol.*, **12**, 213–25.

Moghraby, A.I. (1977). A study on diapause of zooplankton in a tropical river – the Blue Nile. *Freshwat. Biol.*, **7**, 207–12.

Moghraby, A.I. & Adam, M.E. (1984). Ring formation and annual growth in *Corbicula consobrina* Caillaud, 1827 (Bivalvia, Corbiculidae). *Hydrobiologia*, **110**, 219–25.

Monnin, C. & Schott, J. (1984). Determination of the solubility products of sodium carbonate minerals and an application to trona deposition in Lake Magadi (Kenya). *Geochim. cosmochim. Acta*, **48**, 571–81.

Monteith, J.L. (1972). Solar radiation and productivity in tropical ecosystems. *J. appl. Ecol.*, **9**, 747–66.

Monteith, J.L. (1973). *Principles of environmental physics.* London: Edward Arnold.

Montenegro-Guillén, S. (1991). Limnology of Lake Xolotlán, Nicaragua: an overview. *Verh. int. Verein. Limnol.*, **24**, 1155–7.

Montenegro-Guillén, S. (1993). A note on the eolic action as an ecological factor upon Lake Xolotlán (Nicaragua). *Verh. int. Verein. Limnol.*, **25**, 894–6.

Moreau, J. (1982). Le lac Ihotry, lac plat, hypersalé (Madagascar). Ecologie et peuplement piscicole. *Rev. Hydrobiol. trop.*, **15**, 71–80.

Moreau, J. (ed.) (1997). *Advances in the ecology of Lake Kariba.* Harare: Univ. of Zimbabwe Publications.

Moreau, J. (in press). The adaptation and impact on trophic relationships of introduced cichlids in SE Asian lakes and reservoirs. In *Fish and fisheries of lakes and reservoirs in Southeast Asia and Africa*, ed. W.L.T. van Densen & M.J. Morris. Otley, UK: Westbury Publishing.

Moreau, J., Arrignon, J. & Jubb, R.A. (1988). Introduction of foreign fishes in African inland waters. In *Biology and ecology of African freshwater fishes*, ed. C. Lévêque, M.N. Bruton & G.W. Ssentongo, pp. 395–425. Trav. et Docum. 216. Paris: ORSTOM.

Moreau, J., Cronberg, G., Games, I., Hustler, K., Kautsky, N., Kiibus, M., Machena, C. & Marshall, B. (1997). Biomass flows in Lake Kariba, towards an ecosystem approach. In *Advances in the ecology of Lake Kariba*, ed. J. Moreau, pp. 219–30. Harare: Univ. of Zimbabwe Publications.

Moreau, R.E. (1967). Water birds over the Sahara, *Ibis*, **109**, 232–59.

Moreau, R.E. (1972). The Palaearctic-African bird migration systems. London & New York: Academic Press.

Morgan, N.C., Backiel, T., Bretschko, G., Duncan, A., Hillbricht-Ilkowska, A., Kajak, Z., Kitchell, J.F., Larsson, P., Lévêque, C., Nauwerck, A., Schiemer, F. & Thorpe, J.E. (1980). Secondary production. In *The functioning of freshwater ecosystems*, ed. E.D. Le Cren & R.H. Lowe-McConnell, pp. 247–340. International Biological Programme 22. Cambridge: Cambridge Univ. Press.

Moriarty, D.J.W. (1973). The physiology of digestion of blue-green algae in the cichlid fish, *Tilapia nilotica*. *J. Zool., Lond.*, **171**, 25–39.

Moriarty, D.J.W., Darlington, J.P.E.C., Dunn, I.G., Moriarty, C.M. & Tevlin, M.P. (1973). Feeding and grazing in Lake George, Uganda. *Proc. R. Soc. (B)*, **184**, 299–319.

Moriarty, C.M. & Moriarty, D.J.W. (1973*a*). Quantitative estimation of the daily ingestion of phytoplankton by *Tilapia nilotica* and *Haplochromis nigripinnis* in Lake George, Uganda. *J. Zool., Lond.*, **171**, 15–23.

Moriarty, D.J.W. & Moriarty, C.M. (1973*b*). The assimiliation of carbon from phytoplankton by two herbivorous fishes: *Tilapia nilotica* and *Haplochromis nigripinnis*. *J. Zool., Lond.*, **171**, 41–5.

Morley, A.W., Brown, T.E. & Koontz, D.V. (1985). The limnology of a naturally acidic tropical water system in Australia. I. General description and wet season characteristics. *Verh. int. Verein. Limnol.*, **22**, 2125–30.

Morrissey, K.M. & Fisher, T.R. (1988). Regeneration and uptake of ammonium by plankton in an Amazon floodplain lake. *J. Plankton Res.*, **10**, 31–48.

Moss, B. (1969). Limitation of algal growth in some Central African waters. *Limnol. Oceanogr.*, **14**, 591–601.

Moss, B. & Moss, J. (1969). Aspects of the limnology of an endorheic African lake (Lake Chilwa, Malawi). *Ecology*, **50**, 109–18.

Moukolo, N., Bricquet, J.P. & Biyedi, J. (1990). Bilans et variations des exportations de matières sur le Congo à Brazzaville de janvier 1987 à décembre 1988. *Hydrol. Continentale*, **5**, 41–52.

Moyo, N.A.G. (1997). *Lake Chivero: a polluted lake*. Harare: IUCN & University of Zimbabwe Publications.

Moyo, S. (1991). Cyanobacterial nitrogen fixation in Lake Kariba, Zimbabwe. *Verh. int. Verein. Limnol*, **24**, 1123–7.

Moyo, S. (1997). Contribution of nitrogen fixation to the nitrogen budget of Lake Kariba. In *Advances in the ecology of Lake Kariba*, ed. J. Moreau, pp. 58–65. Harare: Univ. of Zimbabwe Publications.

Mtada, O.S.M. (1986). Thermal stratification in a tropical African reservoir (the Guma Dam, Sierra Leone). *Arch. Hydrobiol.*, **107**, 183–96.

Mugidde, R. (1993). The increase in phytoplankton primary productivity and biomass in Lake Victoria (Uganda). *Verh. int. Verein. Limnol.*, **25**, 846–9.

Mühlhauser, H.A., Hrepic, N., Mladinic, P., Montecino, V. & Cabrera, S. (1995). Water quality and limnological features of a high altitude Andean Lake, Chungará, in northern Chile. *Rev. Chilean Hist. Nat.*, **68**, 341–9.

Mukankomeje, R., Plisnier, P.D., Descy, J.P. & Massault, L. (1993). Lake Muzahi, Rwanda: limnological features and phytoplankton production. *Hydrobiologia*, **257**, 107–20.

Munawar, M. & Talling, J.F. (ed.) (1986). Seasonality of freshwater phytoplankton. A global perspective. *Hydrobiologia*, **138**, 1–236.

Mwaura, F.B. & Widdowson, D. (1992). Nitrogenase activity in the papyrus swamps of Lake Naivasha, Kenya. *Hydrobiologia*, **232**, 23–30.

Nakai, K., Yanagisawa, Y., Sato, T., Nimura, Y. & Gashagaza, M.M. (1990). Lunar synchronization of spawning in cichlid fishes of the tribe Lamprologini in Lake Tanganyika. *J. Fish Biol.*, **37**, 589–98.

Nakamoto, N., Marins, M. de A. & Tundisi, J.G. (1976). Synchronous growth of a freshwater diatom *Melosira italica* under natural environment. *Oecologia*, **23**, 179–84.

Nduku, W.K. & Robarts, R.D. (1977). The effect of catchment geochemistry and geomorphology on the productivity of a tropical African montane lake (Little Connemara Dam no. 3, Rhodesia). *Freshwat. Biol.*, **7**, 19–30.

Neale, P.J. & Richerson, P.J. (1987). Photoinhibition and the diurnal variation of phytoplankton photosynthesis. I. Development of a photosynthesis-irradiance model from studies of *in situ* responses. *J. Plankton Res.*, **9**, 167–93.

Newbold, J.D., Sweeney, B.W., Jackson, J.K. & Kaplan, L.A. (1995). Concentrations and export of solutes from six mountain streams in northwestern Costa Rica. *J. North Amer. benthol. Soc.*, **14**, 21–37.

Newell, B.S. (1960). The hydrology of Lake Victoria. *Hydrobiologia*, **15**, 363–83.

Newman, F.C. (1976). Temperature steps in Lake Kivu: a bottom heated saline lake. *J. phys. Oceanogr.*, **6**, 157–63.

Nicholson, S.E. & Entekhabi, D.I. (1986). The quasi-periodic behaviour of rainfall variability in Africa and its relationship to the Southern Oscillation. *Arch. Met. Geoph. Biokl. ser. A,* **34**, 311–48.

Nicholson, S.E., King, J. & Hoopingarner, J. (1988). *Atlas of African rainfall and its interannual variability*. Tallahassee, USA: Dept of Meteorology, Florida State University.

Nieuwohlt, S. (1981). The climates of continental Southeast Asia. In *Climates of southern and western Asia*, ed. K. Takahashi & H. Arakawa, pp. 1–66. *World survey of climatology*, vol. 9, ed. H.E. Landsberg. Elsevier: Amsterdam.

Nogrady, T. (1983). Succession of planktonic rotifer populations in some lakes of the Eastern Rift Valley. *Hydrobiologia*, **98**, 45–54.

Nolte, U., de Oliveira, M.J. & Stur, E. (1997). Seasonal, discharge-driven patterns of mayfly assemblages in an intermittent Neotropical stream. *Freshwat. Biol.*, **37**, 333–45.

Northcote, T.G., Arcifa, M.S. & Mouro, K.A. (1990). An experimental study of the effects of fish zooplanktivory on the phytoplankton of a Brazilian reservoir. *Hydrobiologia*, **19**, 31–45.

Nwadiaro, C.S. & Oji, E.O. (1986). Phytoplankton productivity and chlorophyll-*a* concentration of Oguta Lake in southeastern Nigeria. *Hydrobiol. Bull., Amsterdam*, **19**, 123–31.

Obeid, M. (ed.) (1975). *Aquatic weeds in the Sudan with special reference to water hyacinth*. Khartoum: National Council for Research.

Obeng, L.E. (1981). Man's impact on tropical rivers. In *Perspectives in running water ecology*, ed. M.A. Lock & D.D. Williams, pp. 265–88. New York: Plenum.

Obeng-Asamoa, E.K., John, D.M. & Appler, H.N. (1980). Periphyton in the Volta Lake. I. Seasonal changes on the trunks of flooded trees. *Hydrobiologia*, **76**, 191–200.

Ochumba, P.B.O. (1987). Periodic massive fish kills in the Kenyan part of Lake Victoria. *Water Qual. Bull.*, **12**, 119–22, 130.

Ochumba, P.B.O. (1990). Massive fish kills within the Nyanza Gulf of Lake Victoria, Kenya. *Hydrobiologia*, **208**, 93–9.

Ochumba, P.B.O. & Kibaara, D.I. (1988). An instance of thermal instability in Lake Simbi, Kenya. *Hydrobiologia*, **158**, 247–52.

Ochumba, P.B.O. & Kibaara, D.I. (1989). Observations on blue-green algal blooms in the open waters of Lake Victoria, Kenya. *Afr. J. Ecol.*, **271**, 23–34.

Odinetz-Collart, O. (1987). La pêche crevettière de *Macrobrachium amazonicum* (Palaeomonideae) dans le Bas-Tocantins, après la fermeture du barrage de Tucurui (Brésil). *Rev. Hydrobiol. trop.*, **20**, 131–44.

Ogutu-Ohwayo, R. (1988). Reproductive potential of the Nile perch, *Lates niloticus* L. and the establishment of the species in Lakes Kyoga and Victoria (East Africa). *Hydrobiologia*, **162**, 193–200.

Ogutu-Ohwayo, R. (1990*a*). Changes in the prey ingested and the variations in the Nile perch and other fish stocks of Lake Kyoga and the northern waters of Lake Victoria (Uganda). *J. Fish Biol.*, **37**, 55–63.

Ogutu-Ohwayo, R. (1990*b*). The decline of the native fishes of lakes Victoria and Kyoga (East Africa) and the impact of introduced species, especially the Nile perch, *Lates niloticus*, and the Nile tilapia, *Oreochroms niloticus*. *Env. Biol. Fishes*, **27**, 81–96.

Ogutu-Ohwayo, R. (1992). The purpose, costs and benefits of fish introductions: with specific reference to the Great Lakes of Africa. *Mitt. int. Verein. Limnol.*, **23**, 37–44.

Ogutu-Ohwayo, R. & Hecky, R.E. (1991). Fish introductions in Africa and some of their implications. *Can. J. Fish. aquat. Sci.*, **48** (Suppl. 1), 8–12.

Okali, D.U.U. & Hall, J.B. (1974). Die-back of *Pistia stratiotes* on Volta Lake, Ghana. *Nature, Lond.*, **248**, 452–3.

Olah, J., Sinha, V.R.P., Ayyappan, S., Purushothaman, C.S. & Radheysyam, S. (1987). Detritus associated respiration during macrophyte decomposition. *Arch. Hydrobiol.*, **111**, 309–15.

Oldfield, F., Appleby, P.G. & Thompson, R. (1980). Palaeoecological studies of lakes in the highlands of Papua New Guinea. I. The chronology of sedimentation. *J. Ecol.*, **68**, 457–77.

Olivry, J.C., Bricquet, J.P. & Mahé, G. (1993). Vers un appauvrissement durable des ressources en eau de l'Afrique humide?. In *Hydrology of warm humid regions*, ed. J.S. Gladwell, pp. 67–78. Assoc. Internat. Hydrol. Sci. Publication 216.

Olivry, J.C., Chouret, A., Vuillaume, G., Lemoalle, J. & Bricquet, J.P. (1996). Hydrologie du lac Tchad. Paris: ORSTOM.

Osborne, P.L. (1991). Seasonality in nutrients and phytoplankton production in two shallow lakes: Waigani lake, Papua New Guinea, and Barton Broad, Norfolk, England. *Int. Rev. ges. Hydrobiol.*, **76**, 105–12.

Osborne, P.L., Kyle, J.H. & Abramski, M.S. (1987). Effects of seasonal water level changes on the chemical and biological limnology of Lake Murray, Papua New Guinea. *Aust. J. mar. freshwat. Res.*, **38**, 397–408.

Osborne, P.L. & Polunin, N.V.C. (1986). From swamp to lake: recent changes in a lowland tropical swamp. *J. Ecol.*, **74**, 197–210.

Osborne, P.L. & Totome, R.G. (1992). Influences of oligomixis on the water and sediment chemistry of Lake Kutubu, Papua New Guinea. *Arch. Hydrobiol.*, **124**, 427–49.

Owen, R.B. & Crossley, R. (1992). Spatial and temporal distribution of diatoms in sediments of Lake Malawi, Central Africa, and ecological implications. *J. Paleolimnol.*, **7**, 55–71.

Owen, R.B., Crossley, R., Johnson, J.C., Tweddle, D., Kornfield, I., Davison, S., Eccles, D.H. & Engstrom, D.E. (1990). Major low levels of Lake Malawi and their implications for speciation rates in cichlid fishes. *Proc. R. Soc. (B)*, **240**, 519–53.

Padgett, D.E. (1976). Leaf decomposition by fungi in a tropical rainforest stream. *Biotropica*, **8**, 166–78.

Pagès, J. & Citeau, J. (1990). Rainfall and salinity of a Sahelian estuary between 1927 and 1987. *J. Hydrol.*, **113**, 325–41.

Pagès, J. & Debenay, J.P. (1987). Evolution saisonnière de la salinité de la Casamance. *Rev. Hydrobiol. trop.*, **20**, 203–17.

Pagès, J., Debenay, J.P. & Lebrusq, J.Y. (1987). L'environnement estuarien de la Casamance. *Rev. Hydrobiol. trop.*, **20**, 191–202.

Pagès, J., Lemasson, L. & Dufour, P. (1979). Eléments nutritifs et production primaire dans les lagunes de Côte-d'Ivoire. Cycle annuel. *Arch. Sci. Centre Rech. Océanogr. Abidjan,* **5**, 1–60.

Pagès, J., Lemasson, L. & Dufour, P. (1981). Primary production measurement in a brackish tropical lagoon. Effect of light, as studied at some stations by the ^{14}C method. *Rev. Hydrobiol. trop.,* **14**, 3–15.

Paolini, J.E. (1991). Organic carbon in the Orinoco River. *Verh. int. Verein. Limnol.*, **24**, 2077–9.

Paolini, J. (1994). Dissolved organic carbon in a tropical blackwater reservoir, Guri Lake (Venezuela). *Verh. int. Verein. Limnol.*, **25**, 1291–4.

Patterson, G. & Kachinjika, O. (1993). Effect of wind-induced mixing on the vertical distribution of nutrients and phytoplankton in Lake Malawi. *Verh. int. Verein. Limnol.*, **25**, 872–6.

Patterson, G. & Kachinjika, O. (1995). Limnology and phytoplankton ecology. In *Scientific Report of the UK/SADC Pelagic Fisheries Assessment Project*, ed. A. Menz, pp. 1–67, UK: Natural Resources Institute, Overseas Development Agency.

Patterson, G. & Wilson, K.K. (1995). The influence of the diel climatic cycle on the depth-time distribution of phytoplankton and photosynthesis in a shallow equatorial lake (Lake Baringo, Kenya). *Hydrobiologia*, **304**, 1–8.

Patterson, G., Wooster, M.J. & Sear, C.B. (1998). Satellite-derived surface temperatures and the interpretation of the 3-dimensional structure of Lake Malawi, Africa: the presence of a profile-bound density current and the persistence of thermal stratification. *Verh. int. Verein. Limnol.,* **26**, 252–5.

Paugy, D. (1994). Ecologie des poissons tropicaux d'un cours d'eau temporaire (Baoulé, haut bassin du Sénégal au Mali): adaptation au milieu et plasticité du régime alimentaire. *Rev. Hydrobiol. trop.*, **27**, 157–72.

Pauly, D. & Christensen, V. (1995). Primary production required to sustain global fisheries. *Nature, Lond.*, **374**, 255–7.

Pawley, A.L. & Alfaro, R. (1984). Zooplankton in the Lake Titicaca ecosystem: the importance of regenerated nitrogen to phytoplankton productivity. *Verh. int. Verein. Limnol.*, **22**, 1258–63.

Pawley, A.L. & Richerson, P.J. (1992). Temporal and spatial variation of zooplankton in Lago Grande. In *Lake Titicaca*, ed. C. Dejoux & A. Iltis, pp. 276–88. Dordrecht: Kluwer.

Payne, A.I. (1971). An experiment on the culture of *Tilapia esculenta* Graham and *Tilapia zillii* (Gervais) in fish ponds. *J. Fish Biol.*, **3**, 325–40.

Payne, A.I. (1986). *The ecology of tropical lakes and rivers*. Chichester: Wiley.

Pearson, R.G. (1994). Limnology in the northeast tropics of Australia, the wettest part of the driest continent. *Mitt. int. Verein. Limnol.*, **24**, 155–63.

Pearson, R.G., Tobin, R.K., Smith, R.E.W. & Benson, L.J. (1989). Standing crop and processing of rainforest litter in a tropical Australian stream. *Arch. Hydrobiol.*, **45**, 481–98.

Pedrozo, F. & Bonetto, C. (1987). Nitrogen and phosphorus transport in the Bermejo River (South America). *Rev. Hydrobiol. trop.*, **20**, 91–9.

Penman, H.L. (1956). Evaporation from Lake Volta. In *The Volta River Project*, pp. 37–47. London: HMSO.

Pennycuik, C.J. & Bartholomew, G.A. (1973). Energy budget of the lesser flamingo (*Phoeniconaias minor* Geoffrey). *East Afr. Wildl. J.*, **11**, 199–207.

Pereira, A. (1994). *Contribution à l'étude de la qualité des eaux des retenues amazoniennes: application de la modélisation mathématique à la retenue de Tucurui (Brésil)*. Thèse, Ecole Nat. Ponts et Chaussées, Paris.

Pereira, A., Tassin, B. & Joergensen, S.E. (1994). A model for decomposition of the (drowned) vegetation in an Amazonian reservoir. *Ecol. Model.*, **75–76**, 447–58.

Pérez Eiriz, M. & Pubillones, M.A. (1976*a*). Influencia de la luz sobre la intensidad de la fotosintèsis en la columna de agua en un embalse de la Sierra del Rosario. *Acad. Cienc. Cuba Ser. Forestal* no. 26.

Pérez Eiriz, M. & Pubillones, M.A. (1976*b*). Dinamica diaria de la producción primaria del fitoplancton on un embalse de la Sierra del Rosario, Provincia de Pinar del Rio, Cuba. *Acad. Cienc. Cuba Ser. Forestal* no. 28.

Pérez Eiriz, M. & Pubillones, M.A. (1977). Light regulated intensities of photosynthesis in the water layers of the Reservoir Sierra del Rozario, Cuba. *Inf. Byull. biol. vnutr. Vod.*, **35**, 20–4. [In Russian.]

Pérez Eiriz, M., Romanenko, V.I., Kudriatsev, V.M. & Pubillones, M.A. (1980). Particularidades del proceso de producción primaria de materia organica por la fotosíntesis del fitoplancton en los embalses de Cuba. *Acad. Cienc. Cuba Ser. Forestal* no. 127.

Peters, R.H. (1983). *The ecological implications of body size*. Cambridge: Cambridge Univ. Press.

Peters, R.H. & MacIntyre, S. (1976). Orthophosphate turnover in East African lakes. *Oecologia*, **25**, 313–19.

Petersen, R.C. (1984). Detritus decomposition in endogenous and exogenous rivers of a tropical wetland. *Verh. int. Verein Limnol.*, **22**, 1926–31.

Petr, T. (1967). Fish population changes in the Volta Lake in Ghana during its first sixteen months. *Hydrobiologia*, **30**, 193–220.

Petr, T. (1968*a*). Population changes in aquatic invertebrates living on two water plants in a tropical man-made lake. *Hydrobiologia*, **32**, 449–85.

Petr, T. (1968*b*). The establishment of lacustrine fish population in the Volta Lake in Ghana during 1964–1966. *Bull. Inst. fr. Afr. noire* 30 ser. A, **1**, 257–69.

Petr, T. (1970). Macroinvertebrates of flooded trees in the man-made Volta Lake (Ghana) with special reference to the burrowing mayfly *Povilla adusta* Navas. *Hydrologia*, **36**, 373–98.

Petr, T. (1972). Benthic fauna of a tropical man-made lake (Volta Lake, Ghana 1965–1968). *Arch. Hydrobiol.*, **70**, 484–533.

Petr, T. (1974). Dynamics of benthic invertebrates in a tropical man-made lake (Volta Lake 1964–1968). Standing crop and bathymetric distribution. *Arch. Hydrobiol.*, **73**, 245–65.

Petr, T. (1975). On some factors associated with the initial high fish catches in new African man-made lakes. *Arch. Hydrobiol.*, **75**, 32–49.

Petr, T. (1978). Tropical man-made lakes – their ecological impact. *Arch. Hydrobiol.*, **81**, 368–85.

Petr, T. (ed.) (1983). *The Purari – tropical environment of a high rainfall river basin.* Monogr. Biologicae 51. The Hague: Junk.

Piedade, M.T.F., Junk, W.J. & Long, S.P. (1991). The productivity of the C_4 grass *Echinochloa polystachya* on the Amazon floodplain. *Ecology*, **72**, 1456–63.

Piedade, M.T.F., Junk, W.J. & Long, S.P. (1997). Nutrient dynamics of the highly productive C_4 macrophyte *Echinochloa polystachya* on the Amazon floodplain. *Funct. Ecol.*, **11**, 60–5.

Piedade, M.T.F., Long, S.P. & Junk, W.J. (1994). Leaf and canopy CO_2 uptake of a stand of *Echinochloa polystachya* on the Central Amazon floodplain. *Oecologia*, **97**, 193–201.

Piet, G.J., Pet, J.S. & Guruge, W.A.H.P. (1994). Niche partioning of indigenous riverine carps and the exotic lacustrine tilapia (*Oreochromis mossambicus*) in a reservoir in S.E. Sri Lanka. *Verh. int. Verein. Limnol.*, **25**, 2183–7.

Piet, G.J., Vijverberg, K. & van Densen, W.L.T. (in press). Foodweb structure of a Sri Lankan reservoir. In *Fish and fisheries in Southeast Asia and Africa*, ed. W.L.T. van Densen & M.J. Morris. Otley, UK: Westbury Publishing.

Pike, J.G. (1965). The sunspot/lake level relationship and the control of Lake Nyasa. *J. Instn. Wat. Engrs.*, **19**, 221–30.

Pilskaln, C.H. & Johnson, T.C. (1991). Seasonal signals in Lake Malawi sediments. *Limnol. Oceanogr.*, **36**, 544–57.

Piper, B.S., Plinston, D.T. & Sutcliffe, J.V. (1986). The water balance of Lake Victoria. *Hydrol. Sci. J.*, **31**, 25–37.

Pitcher, T.J. & Hart, P.J.B. (ed.) (1995). *The impact of species changes in African lakes.* London: Chapman & Hall.

Piyasiri, S. (1985). Dependence of food on growth and development of two freshwater tropical and temperate calanoid species. *Verh. int. Verein. Limnol.*, **22**, 3185–9.

Planas, D. & Moreau, G. (1990). Natural eutrophication in a warm volcanic lake. *Verh. int. Verein. Limnol.*, **24**, 554–9.

Plasencia, J.M. & Květ, J. (1993). Production dynamics of *Typha domingensis* (Pers.) Kunth populations in Cuba. *J. aquat. Pl. Mgmt.*, **31**, 240–3.

Platt, T. & Denman, K. (1975). Spectral analysis in ecology. *Ann. Rev. Ecol. Syst.*, **6**, 189–210.

Plisnier, P.D., Langenberg, V., Mwape, L., Chitamwebwa, D. & Coenen, E. (1995). Limnological cycles in Lake Tanganyika. Internal waves and hypothesis of pulsed production. In *Symposium on Lake Tanganyika Research, Sept. 11–15, 1995*, ed. H. Molsa, pp. 33. Finland: Kuopio. [Abstracts.]

Pollingher, U. (1986). Phytoplankton periodicity in a subtropical lake (Lake Kinneret, Israel). *Hydrobiologia*, **138**, 127–38.

Pollingher, U. & Berman, T. (1991). Phytoplankton composition and activity in lakes of the warm belt. *Verh. int. Verein. Limnol.*, **24**, 1230–84.

Polunin, N.V.C. (1984). The decomposition of emergent macrophytes in freshwater. *Adv. Ecol. Res.*, **14**, 115–66.

Pourriot, R. & Rougier, C. (1975). Dynamique d'une population expérimentale de *Brachionus dimidiatus* (Bryce) (Rotifère) en fonction de la nourriture et de la température. *Annals Limnol.*, **11**, 125–43.

Pouyaud, B. (1986). *Contribution a l'évaluation de l'évaporation de nappes d'eau libre en climat tropical sec.* Coll. Etudes et thèses. Paris: ORSTOM.

Pouyaud, B. (1987*a*). L'évaporation des nappes d'eau libre. L'exemple du lac de Bam au Burkina-Faso. 1. Échelles décadaire et mensuelle. *Hydrol. continentale*, **2**, 29–46.

Pouyaud, B. (1987*b*). L'évaporation des nappes d'eau libre. L'exemple du lac de Bam au Burkina-Faso. 2. Échelles journalières et infra-journalières. *Hydrol. continentale*, **2**, 127–49.

Powell, T., Kirkish, M.H., Neale, P.J. & Richerson, P.J. (1984). The diurnal cycle of stratification in Lake Titicaca: eddy diffusion. *Verh. int. Verein. Limnol.*, **22**, 1237–43.

Prejs, A. & Prejs, K. (1987). Feeding of tropical freshwater fishes: seasonality in resource availability and resource use. *Oecologia*, **71**, 397–404.

Prejs, K. & Prejs, A. (1992). Importance of predation in regulating density of meio- and macrofauna in seasonal tropical waters. *Hydrobiologia*, **242**, 77–86.

Pringle, C.M., Trisks, F.J. & Browder, G. (1990). Spatial variation in basic chemistry of streams draining a volcanic landscape on Costa Rica's Caribbean slope. *Hydrobiologia*, **206**, 73–85.

Pringle, C.M., Paaby-Hansen, P., Vaux, P.D. & Goldman, C.R. (1986). *In situ* nutrient assays of periphyton growth in a lowland Costa Rican stream. *Hydrobiologia*, **134**, 207–13.

Prosser, M.V. (1987). The biological effects of variation in water quality. In *The Jonglei Canal*, ed. P. Howell, M. Lock & S. Cobb, pp. 126–45. Cambridge: Cambridge Univ. Press.

Prowse, G.A. (1964). Some limnological problems in tropical fish ponds. *Verh. int. Verein. Limnol.*, **15**, 480–4.

Prowse, G.A. (1972). Some observations on primary and fish production in experimental fish ponds in Malacca, Malaysia. In *Productivity problems of freshwaters*, eds. Z. Kajak & A. Hillbricht-Ilkowska, pp. 555–61. Warszawa-Kraków: Polish Scientific Publishers.

Prowse, G.A. & Talling, J.F. (1958). The seasonal growth and succession of plankton algae in the White Nile. *Limnol. Oceanogr.*, **3**, 223–38.

Quay, P.D., Wilbur, D.O., Richey, J.E., Devol, A.H., Benner, R. & Forsberg, B.R. (1995). The ^{18}O: ^{16}O of dissolved oxygen in rivers and lakes in the Amazon Basin: determining the ratio of respiration to photosynthesis rates in freshwaters. *Limnol. Oceanogr.*, **40**, 718–29.

Rai, H. (1979). Microbiology of Central Amazon lakes. *Amazoniana*, **6**, 583–99.

Rai, H. & Hill, G. (1981). Bacterial biodynamics in Lago Tupé, a Central Amazonian black water 'Ria Lake'. *Arch. Hydrobiol. Suppl.*, **58**, 420–68.

Rai, H. & Hill, G. (1982*a*). Establishing the pattern of heterotrophic bacterial activity in three Central Amazonian lakes. *Hydrobiologia*, **86**, 121–6.

Rai, H. & Hill, G. (1982*b*). On the nature of the ecological cycle of Lago Janauari: a Central Amazon riávárzea lake. *Trop. Ecol.*, **23**, 1–50.

Rai, H. & Hill, G. (1984). Microbiology of Amazonian waters. In *The Amazon*, ed. H. Sioli, pp. 413–41. Monogr. Biologicae 56. Dordrecht: Junk.

Ramage, C.S. (1971). *Monsoon meteorology*. New York: Academic Press.

Ramberg, L. (1987). Phytoplankton succession in the Sanyati Basin, Lake Kariba. *Hydrobiologia*, **153**, 193–202.

Rao, K.N. (1981). The climate of the Indian subcontinent. In *Climates of southern and western Asia*, ed. K. Takahashi & H. Arakawa, pp 47–182. *World survey of climatology*, vol. 9, ed. H.E. Landsberg. Amsterdam: Elsevier.

Ratisbona, L.R. (1976). Climate of Brazil. In *Climates of Central and South America*, ed. W. Schwerdfeger, pp. 219–93. *World survey of climatology*, vol. 12, ed. H.E. Landsberg. Amsterdam: Elsevier.

Redfield, A.C. & Doe, A.E. (1964). Lake Maracaibo. *Verh. int. Verein. Limnol.*, **15**, 100–111.

Reiss, F. (1976). Die Benthoszoozönosen Zentralamazonischer Várzeaseen und ihre Anpassungen an die jahresperiodischen Wasserstandsschwankungen. *Biogeographica*, **7**, 125–35. [English translation, 1977, *Geo-Eco-Trop.*, **1**, 65–75.]

Reynolds, C.S., Tundisi, J.G. & Hino, K. (1983). Observations on a metalimnetic *Lyngbya* population in a stably stratified tropical lake (Lagoa Carioca, Eastern Brazil). *Arch. Hydrobiol.*, **97**, 7–17.

Reynolds, J.D. (1974). Biology of the small pelagic fishes in the new Volta Lake in Ghana. 3. Sex and reproduction. *Hydrobiologia*, **45**, 489–508.

Richardson, J.L. & Dussinger, R.A. (1986). Paleolimnology of mid-elevation lakes in the Kenya Rift Valley. *Hydrobiologia*, **143**, 167–74.

Richardson, J.L. & Jin, L.T. (1975). Algal productivity of natural and artificially enriched fresh waters in Malaya. *Verh. int. Verein. Limnol.*, **19**, 1383–9.

Richardson, J.L. & Richardson, A.E. (1971). History of an African rift lake and its climatic implications. *Ecol. Monogr.*, **42**, 499–534.

Richerson, P.J. (1992). The thermal stratification regime in Lake Titicaca. In *Lake Titicaca. A synthesis of limnological knowledge*, ed. C. Dejoux & A. Iltis, pp. 120–30. Monogr. Biologicae 68. Dordrecht: Kluwer.

Richerson, P.J. & Carney, H.J. (1988). Patterns of temporal variation in Lake Titicaca, a high altitude tropical lake. 2. Succession rate and diversity of the phytoplankton. *Verh. int. Verein. Limnol.*, **23**, 734–8.

Richerson, P.J., Neale, P.J., Alfaro Tapio, R., Carney, H.J., Lazzaro, X., Vincent, W. & Wurtsbaugh, W. (1992). Patterns of planktonic primary production and algal biomass. In *Lake Titicaca. A synthesis of limnological knowledge*, ed. C. Dejoux & A. Iltis, pp. 196–222. Monogr. Biologicae 68. Dordrecht: Kluwer.

Richerson, P.J., Neale, P.J., Wurtsbaugh, W., Alfaro, T.R. & Vincent, W. (1986). Patterns of temporal variation in Lake Titicaca, a high altitude tropical lake. I. Background, physical and chemical processes, and primary production. *Hydrobiologia*, **138**, 205–20.

Richerson, P.J., Widmer, C. & Kittel, T. (1977). *The limnology of Lake Titicaca (Peru–Bolivia), a large high altitude tropical lake.* Univ. Calif. Davis, Inst. Ecology Publ. 14.

Richey, J.E. (1981). Particulate and dissolved carbon in the Amazon River: preliminary annual budget. *Verh. int. Verein. Limnol.*, **21**, 914–17.

Richey, J.E., Brock, J.T., Naimaw, R.J., Wissmar, R.C. & Stallard, R.F. (1980). Organic carbon: oxidation and transport in the Amazon River. *Science, NY*, **207**, 1348–51.

Richey, J.E., Devol, A.H., Wofsy, S.C., Victoria, R. & Riberio, M.N.G. (1988). Biogenic gases and oxidation and reduction of carbon in Amazon River and floodplain waters. *Limnol. Oceanogr.*, **33**, 551–61.

Richey, J.E., Hedges, J.I., Devol, A.H. & Quay, P.D. (1990). Biogeochemistry of carbon in the Amazon River. *Limnol. Oceanogr.*, **35**, 352–71.

Richey, J.E., Nobre, C. & Deser, C. (1989). Amazon river discharge and climate variability: 1903 to 1985. *Science, NY*, **246**, 101–3.

Riehl, H. (1979). *Climate and weather in the tropics*. New York: Academic Press.

Rigler, F.H. & Downing, J.A. (1984). The calculation of secondary productivity. In *A manual on methods for the assessment of secondary productivity in fresh waters*, 2nd edn., ed. J.A. Downing & F.H. Rigler, pp. 19–58. Oxford: Blackwell.

Rippey, B. & Wood, R.B. (1985). Trends in major ion composition of five Bishoftu crater lakes. *Sinet: Ethiop. J. Sci.*, **8**, 9–28.

Robarts, R.D. (1979). Underwater light penetration, chlorophyll *a* and primary production in a tropical African lake (Lake McIlwaine, Rhodesia). *Arch. Hydrobiol.*, **86**, 423–44.

Robarts, R.D. (1987). Decomposition in freshwater ecosystems. *J. Limnol. Soc. Sth. Afr.*, **12**, 72–89.

Robarts, R.D. & Southall, G.C. (1975). Algal bioassays of two tropical Rhodesian reservoirs. *Acta Hydrochim. Hydrobiol.*, **3**, 369–77.

Robarts, R.D. & Southall, G.C. (1977). Nutrient limitation of phytoplankton growth in seven tropical man-made lakes, with special reference to Lake McIlwaine, Rhodesia. *Arch. Hydrobiol.*, **79**, 1–35.

Robarts, R.D. & Ward, P.R.B. (1978). Vertical diffusion and nutrient transport in a tropical lake (Lake McIlwaine, Rhodesia). *Hydrobiologia*, **59**, 213–21.

Robinson, A.H. & Robinson, P.K. (1971). Seasonal distribution of zooplankton in the northern basin of Lake Chad. *J. Zool., Lond.*, **163**, 25–61.

Rocha, O., Matsumura-Tundisi, T. & Tundisi, J.G. (1982). Seasonal fluctuation of *Argyrodiaptomus furcatus* (Sars) population in Lobo (Broa) Reservoir, São Carlos, S.P., Brasil. *Trop. Ecol.*, **23**, 134–50.

Roche, M.A., Bourges, J., Cortes, J. & Mattos, R. (1992). Climatology and hydrology of the Lake Titicaca basin. In *Lake Titicaca. A synthesis of limnological knowledge*, ed. C. Dejoux & A. Iltis, pp. 63–88. Monogr. Biologicae 68. The Hague: Kluwer.

Rodhe, H. & Herrera, R. (ed.) (1988). *Acidification in tropical countries*. SCOPE Report 36. Chichester: Wiley.

Roest, F.C. (1992). The pelagic fisheries resources of Lake Tanganyika. *Mitt. int. Verein. Limnol.*, **23**, 11–15.

Rogers, K.H. & Breen, C.M. (1980). Growth and reproduction of *Potamogeton crispus* in a South African lake. *J. Ecol.*, **68**, 561–71.

Romanenko, V.I., Perez Eiriz, M., Kudryavchev, V.M. & Avrora Pubillones, M. (1979). Intensity of photosynthesis of phytoplankton in reservoirs of Cuba. *Trudy Inst. Biol. vnutr. Vod.*, **37**(40), 21–59. [In Russian.]

Ronchail, J. (1989). Advections polaires en Bolivie: mise en évidence et caractérisation des effets climatiques. *Hydrol. Continentale*, **4**, 49–56.

Room, P.M. & Thomas, P.A. (1986*a*). Population growth of the floating weed *Salvinia molesta*: field observations and a global model based on temperature and nitrogen. *J. appl. Ecol.*, **23**, 1013–28.

Room, P.M. & Thomas, P.A. (1986*b*). Nitrogen, phosphorus and potassium in *Salvinia molesta* Mitchell in the field: effects of weather, insect damage, fertilizers and age. *Aquat. Bot*, **24**, 213–32.

Rossiter, A. (1991). Lunar spawning synchroneity in a freshwater fish. *Naturwissenschaften*, **78**, 182–4.

Rudd, J.W.M. (1980). Methane oxidation in Lake Tanganyika (East Africa). *Limnol. Oceanogr.*, **25**, 958–63.

Ruttner, F. (1931*a*). Die Schichtung in tropischen Seen. *Verh. int. Verein. Limnol.*, **5**, 44–67.

Ruttner, F. (1931*b*). Hydrographische und hydrochemische Beobachtungen auf Java, Sumatra und Bali. *Arch. Hydrobiol. Suppl.* **8** (*Tropische Binnengewässer* 1), 197–454.

Ruttner, F. (1938). Stabilität und Umschichtung in tropischen und temperierten Seen. *Arch. Hydrobiol. Suppl.* 15 (*Tropische Binnengewässer* 7), 178–86.

Ruttner, F. (1943). Beobachtungen über die tägliche Vertikalwanderung des Planktons in tropischen Seen. *Arch. Hydrobiol.*, **40**, 474–92.

Ruttner, F. (1952). Planktonstudien der Deutschen Limnologischen Sunda-Expedition. *Arch. Hydrobiol. Suppl.* **21** (*Tropische Binnengewässer* 10), 1–274.

Rzóska, J. (1956). On the variability and status of the cladocera *Ceriodaphnia cornuta* and *Ceriodaphnia rigaudi*. *Ann. Mag. Nat. Hist. ser. 12*, **9**, 505–10.

Rzóska, J. (1958). Observations on tropical rainpools and general remarks on temporary waters. *Hydrobiologia*, **17**, 265–86.

Rzóska, J. (1964). Mass outbreaks of insects in the Sudanese Nile basin. *Verh. int. Verein. Limnol.*, **15**, 194–200.

Rzóska, J. (1968). Observations on zooplankton distribution in a tropical river dam-basin (Gebel Aulia, White Nile, Sudan). *J. anim. Ecol.*, **37**, 185–98.

Rzóska, J. (1974). The Upper Nile swamps: a tropical wetland study. *Freshwat. Biol.*, **4**, 1–30.

Rzóska, J. (ed.) (1976). *The Nile, biology of an ancient river*. Monogr. Biologicae 29. The Hague: Junk.

Rzóska, J. (1984). Temporary and other waters. In *Key Environments – The Sahara*, ed. J. Cloudsley-Thompson, pp. 105–14. Oxford: Pergamon Press.

Rzóska, J., Brook, A.J. & Prowse, G.A. (1953). Seasonal plankton development in the White and Blue Nile near Khartoum. *Verh. int. Verein. Limnol.*, **12**, 327–34.

Saha, L.C. & Pandit, B. (1987). Comparative primary production in one pond and river system at Bhagalpur, India. *Limnologica*, **18**, 313–16.

Saijo, Y., Mitamura, O. & Barbosa, F.A.R. (1991). Chemical studies on sediments in the Rio Doce valley lakes, Brazil. *Verh. int. Verein. Limnol.*, **24**, 1192–6.

Saint-Jean, L. (1983). The zooplankton. In *Lake Chad. Ecology and productivity of a shallow tropical ecosystem*, ed. J.P. Carmouze, J.R. Durand & C. Lévêque, pp. 199–232. Monogr. Biologicae 53. The Hague: Junk.

Saint-Jean, L. & Bonou, C. (1994). Growth, production, and demography of *Moina micrura* in brackish tropical fishponds (Layo, Ivory Coast). *Hydrobiologia*, **272**, 125–46.

Sale, P.J.M., Orr, P.T., Shell, G.S. & Erskine, D.J.C. (1985). Photosynthesis and growth rates in *Salvinia molesta* and *Eichhornia crassipes*. *J. appl. Ecol.*, **22**, 125–37.

Salonen, K. & Sarvala, J. (1994). First results of phytoplankton primary production studies. *Lake Tanganyika Res. Newslett.*, **10**, 10–11.

Santos, U. de M. (1973). Beobachtungen über Wasserbewegungen, chemische Schichtung und Fischwanderungen in Várzea-Seen am mittleren Solimões (Amazonas). *Oecologia*, **13**, 239–46.

Sarvala, J. & Salonen, K. (1995). *Preliminary experiments on phytoplankton production ecology in Lake Tanganyika*. FAO/FINNIDA Research for the management of the fisheries on Lake Tanganyika. Bujumbura: FAO/FINNIDA.

Saunders, J.F. (1980). Diel patterns of reproduction in rotifer populations from a tropical lake. *Freshwat. Biol.*, **10**, 35–9.

Saunders, J.F. & Lewis, W.M. (1988*a*). Dynamics and control mechanisms in a tropical zooplankton community (Lake Valencia, Venezuela). *Ecol. Monogr.*, **58**, 337–53.

Saunders, J.F. & Lewis, W.M. (1988*b*). Composition and seasonality of the zooplankton community of Lake Valencia, Venezuela. *J. Plankton Res.*, **16**, 957–85.

Saunders, J.F. & W.M. Lewis, W.M. (1988*c*). Zooplankton abundance in the Caura River, Venezuela. *Biotropica*, **20**, 206–14.

Saunders, J.F. & Lewis, W.M. (1988*d*). Transport of phosphorus, nitrogen and carbon by the Apure River, Venezuela. *Biogeochemistry*, **5**, 323–42.

Saunders, J.F. & Lewis, W.M. (1988*e*). Zooplankton abundance and transport in a tropical white-water river. *Hydrobiologia*, **162**, 147–55.

Saunders, J.F. & Lewis, W.M. (1989*a*). Transport of major solutes and the relationship between concentrations and discharge in the Apure River, Venezuela. *Biogeochemistry*, **8**, 101–13.

Saunders, J.F. & Lewis, W.M. (1989*b*). Zooplankton abundance in the lower Orinoco river, Venezuela. *Limnol. Oceanogr.*, **34**, 397–409.

Schiemer, F. (ed.) (1983). *Limnology of Parakrama Samudra, Sri Lanka. A case study of an ancient man-made lake in the tropics.* The Hague: Junk.

Schlesinger, W.H. & Melack, J.M. (1981). Transport of organic carbon in world rivers. *Tellus*, **33**, 172–87.

Schmidle, W. (1902). Das Chloro- und Cyanophyceenplankton des Nyassa und einiger anderer innerafrikanischer Seen. *Bot. Jb.*, **33** (1), 1–33.

Schmidt, G.W. (1970). Numbers of bacteria and algae and their interrelations in some Amazonian waters. *Amazoniana*, **2**, 393–400.

Schmidt, G.W. (1972). Seasonal changes in water chemistry of a tropical lake (Lago do Castanho, Amazonia, South America). *Verh. int. Verein. Limnol.*, **18**, 613–21.

Schmidt, G.W. (1973*a*). Primary production of phytoplankton in the three types of Amazonian waters. II. The limnology of a tropical flood-plain in central Amazonia (Lago do Castanho). *Amazoniana*, **4**, 139–203.

Schmidt, G.W. (1973*b*). Primary production of phytoplankton in the three types of Amazonian waters. III. Primary productivity of phytoplankton in a tropical flood-plain lake of central Amazonia, Lago do Castanho, Amazonas, Brazil. *Amazoniana*, **4**, 379–404.

Schmidt, G.W. (1976). Primary production of phytoplankton in the three types of Amazonian waters. IV. On the primary productivity of phytoplankton in a bay of the lower Rio Negro (Amazonas, Brazil). *Amazoniana*, **5**, 517–28.

Schmidt, G.W. (1982). Primary production of phytoplankton in the three types of Amazonian waters. V. Some investigations on the phytoplankton and its primary productivity in the clear water of the lower Rio Tapajóz (Pará, Brazil). *Amazoniana*, **7**, 335–48.

Sen, N.S., Kapoor, V.K. & Gopalkrishna, G. (1990). Seasonal growth of *Eichhornia crassipes* (Mart.) and its possible impact on the primary productivity and fishery structure in a tropical reservoir. *Acta Hydrochim. Hydrobiol.*, **18**, 307–23.

Sene, K.J., Gash, J.H.C. & McNeil, D.D. (1991). Evaporation from a tropical lake: comparison of theory with direct measurements. *J. Hydrol.*, **127**, 193–217.

Serruya, C. & Pollingher, U. (1983). *Lakes of the warm belt*. Cambridge: Cambridge Univ. Press.

Servais, F. (1957). *Etude théorique des oscillations libres (seiches) du lac Tanganika*. Result. sci. Explor. hydrobiol. Lac Tanganika, (1946–1947), **2** (3).

Servant, M. & Servant, S. (1983). Paleolimnology of an upper quaternary endorheic lake in the Chad basin. In *Lake Chad. Ecology and productivity of a shallow tropical ecosystem*, ed. J.P. Carmouze, J.R. Durand & C. Lévêque, pp. 11–26. Monogr. Biologicae 53. The Hague: Junk.

Setaro, F.V. & Melack, J.M. (1984). Responses of phytoplankton to experimental nutrient enrichment in an Amazon floodplain lake. *Limnol. Oceanogr.*, **29**, 972–84.

Sharma, B.M. & Sridhar, M.K.C. (1981). The productivity of *Pistia stratiotes* L. in a eutrophic lake. *Envir. Pollut.*, **24**, 277–89.

Sharma, K.P. & Pradham, V.N. (1983). Study on growth and biomass of underground organs of *Typha angustata* Bory & Chaub. *Hydrobiologia*, **99**, 89–93.

Silva, E.I.L. & Davies, R.W. (1986). Primary productivity and related parameters in three different types of inland waters in Sri Lanka. *Hydrobiologia*, **137**, 239–49.

Silva, E.I.L. & Davies, R.W. (1987). The seasonality of monsoonal primary productivity in Sri Lanka. *Hydrobiologia*, **150**, 165–75.

Sinada, F. & Abdel Karim, A.G. (1984*a*). Physical and chemical characteristics of the Blue Nile and the White Nile at Khartoum. *Hydrobiologia*, **110**, 21–32.

Sinada, F. & Abdel Karim, A.G. (1984*b*). A quantitative study of the phytoplankton in the Blue and White Niles at Khartoum. *Hydrobiologia*, **110**, 47–55.

Sinada, F. & Abdel Karim, A.G. (1984*c*). Primary production and respiration of the phytoplankton in the Blue and White Niles at Khartoum. *Hydrobiologia*, **110**, 57–9.

Singh, N.K., Munshi, J.S.D. & Bilgrami, K.S. (1983). A quantitative study of phytoplankton of the River Ganges at Bhagalpur, India. *Polsk. Arch. Hydrobiol.*, **30**, 81–7.

Sioli, H. (ed.) (1984). *The Amazon. Limnology and landscape ecology of a mighty tropical river and its basin*. Monogr. Biologicae 56. Dordrecht: Junk.

Smith, R.E.W. & Pearson, R.G. (1987). The macroinvertebrate communities of temporary pools in an intermittent stream in tropical Queensland. *Hydrobiologia*, **150**, 45–67.

Sommer, U. & Gliwicz, Z.M. (1986). Long range vertical migration of *Volvox* in tropical Lake Cahora Bassa (Mozambique). *Limnol. Oceanogr.*, **31**, 650–3.

Spigel, R.H. & Coulter, G.W. (1996). Comparison of hydrology and physical limnology of the East African Great Lakes: Tanganyika, Malawi, Victoria, Kivu and Turkana (with reference to some North American Great Lakes). In *The limnology, climatology and paleoclimatology of the East African lakes*, ed. T.C. Johnson & E.O. Odada, pp. 103–39. Amsterdam: Gordon & Breach.

Sreenivasan, A. (1964). The limnology, primary production and fish production in a tropical pond. *Limnol. Oceanogr.*, **9**, 391–6.

Sreenivasan, A. (1965). Limnology of tropical impoundments. III. Limnology and productivity of Amaravathy Reservoir (Madras State), India. *Hydrobiologia*, **26**, 501–16.

Sreenivasan, A. (1974). Limnological features of a tropical impoundment, Bhavanisagar Reservoir (Tamil Nadu), India. *Int. Rev. ges. Hydrobiol.*, **59**, 327–42.

Ssentongo, G.W. (1988). Population structure and dynamics. In *Biology and ecology of African freshwater fishes*, ed. C. Lévêque, M.N. Bruton & G.W. Ssentongo, pp. 363–77. Trav. et Docum. 216. Paris: ORSTOM.

Stager, J.C., Cumming, B. & Meeker, L. (1997). A high-resolution 11 400-yr diatom record from Lake Victoria, East Africa. *Quaternary Res.*, **47**, 81–9.

Stallard, R.F. (1985). River chemistry, geology, geomorphology, and soils in the Amazon and Orinoco basins. In *The chemistry of weathering*, ed. J.I. Drever, pp. 293–316. Dordrecht: Reidel.

Stallard, R.F. & Edmond, J.M. (1981). Geochemistry of the Amazon. I. Precipitation chemistry and the marine contribution to dissolved load at the time of peak discharge. *J. geophys. Res.*, **86**, 9844–58.

Stallard, R.F. & Edmond, J.M. (1983). Geochemistry of the Amazon. 2. The influence of geology and weathering environment on the dissolved load. *J. geophys. Res.*, **88**, 9671–88.

Stallard, R.F. & Edmond, J.M. (1987). Geochemistry of the Amazon. 3. Weathering, chemistry and limits to dissolved inputs. *J. geophys. Res.*, **92**, 8293–302.

Starling, F.L.R.M. & Rocha, A.J.A. (1990). Experimental study of the impacts of planktivorous fishes on plankton community and eutrophication of a tropical Brazilian reservoir. *Hydrobiologia*, **200/201**, 581–91.

Statzner, B. (1976). Die Köcherfliegen-Emergenz (Trichoptera, Insecta) aus dem zentralafrikanischen Bergbach Kalengo. *Arch. Hydrobiol.*, **78**, 102–37.

Statzner, B., Dejoux, C. & Elouard, J.M. (1984). Field experiments on the relationship between drift and benthic densities of aquatic insects in tropical streams (Ivory Coast). I. *Rev. Hydrobiol. trop.*, **17**, 319–34.

Statzner, B., Dejoux, C. & Elouard, J.M. (1985*a*). Field experiments of the relationship between drift and benthic densities of aquatic insects in tropical streams (Ivory Coast). 2. *Cheumatopsyche falcifera* (Trichoptera, Hydropsychidae). *J. anim. Ecol.*, **55**, 93–110.

Statzner, B., Elouard, J.M. & Dejoux, C. (1985*b*). Field experiments on the relationship between drift and benthic densities of aquatic insects in tropical streams (Ivory Coast). 3. Trichoptera. *Freshwat. Biol.*, **17**, 391–404.

Steinitz-Kannan, M., Colinvaux, P.A. & Kannan, R. (1983). Limnological studies in Ecuador. I. A survey of chemical and physical properties of Ecuadorian lakes. *Arch. Hydrobiol. Suppl.* **65**, 61–105.

Stout, R.J. (1982). Effects of a harsh environment on the life history patterns of two species of tropical aquatic hemiptera (family: Naucoridae). *Ecology*, **63**, 75–83.

Stout, R.J. (1989). Effects of condensed tannins on leaf processing in mid-latitude and tropical streams: a theoretical approach. *Can. J. Fish. aquat. Sci.*, **46**, 1097–106.

Street, F.A. (1981). The relative importance of climate and local hydrogeological factors in influencing lake-level fluctuations. *Palaeo-ecology of Africa*, **12**, 137–58.

Street, F.A. & Grove, A.T. (1979). Global maps of lake-level fluctuations since 30 000 yr BP. *Quaternary Res.*, **12**, 83–118.

Street-Perrott, F.A. & Perrott, R.A. (1990). Abrupt climate fluctuations in the tropics: the influence of Atlantic Ocean circulation. *Nature, Lond.*, **343**, 607–12.

Sugunan, V.V. (1980). Seasonal fluctuations of plankton of Nagarjunasagar reservoir A.P., India. *J. Inland Fish. Soc., India*, **12** (1), 79–91.

Sutcliffe, J.V. (1987). The use of historical records in flood frequency analysis. *J. Hydrol.*, **96**, 159–71.

Sutcliffe, J.V. (1988). The influence of Lake Victoria: climatic change, and variation in river flows. In *The Jonglei Canal*, ed. P. Howell, M. Lock & S. Cobb, pp. 87–99. Cambridge: Cambridge Univ. Press.

Sutcliffe, J.V. & Parks, Y.P. (1989). Comparative water balances of some selected African wetlands. *Hydrol. Sci. J.*, **34**, 49–62.

Symoens, J.J. (1956). Sur la formation de 'fleurs d'eau' à Cyanophycées (*Anabaena flos-aquae*) dans le bassin Nord du lac Tanganika. *Bull. Acad. Sci. Colon. n.s.*, **2**, 414–19.

Symoens, J.J., Burgis, M. & Gaudet, J.J. (ed.) (1981). *The ecology and utilization of African inland waters.* United Nations Environmental Programme, Reports & Proceedings Ser. 1.

Tait, R.D., Shiel, R.J. & Koste, W. (1984). Structure and dynamics of zooplankton communities, Alligator Rivers Region, N.T., Australia. *Hydrobiologia*, **113**, 1–13.

Takamura, K. (1988). The first measurement of the primary production of epilithic algae in Lake Tanganyika. *Physiol. Ecol., Japan*, **25**, 1–7.

Talling, J.F. (1957*a*). Diurnal changes of stratification and photosynthesis in some tropical African waters. *Proc. R. Soc. (B)*, **147**, 57–83.

Talling, J.F. (1957*b*). Some observations on the stratification of Lake Victoria. *Limnol. Oceanogr.*, **3**, 213–21.

Talling, J.F. (1957*c*). The longitudinal succession of water characteristics in the White Nile. *Hydrobiologia*, **11**, 73–89.

Talling, J.F. (1957*d*). The phytoplankton population as a compound photosynthetic system. *New Phytol.*, **56**, 133–49.

Talling, J.F. (1963). Origin of stratification in an African Rift lake. *Limnol. Oceanogr.*, **8**, 68–78.

Talling, J.F. (1965*a*). The photosynthetic activity of phytoplankton in East African lakes. *Int. Rev. ges. Hydrobiol.*, **50**, 1–32.

Talling, J.F. (1965*b*). Comparative problems of phytoplankton production and photosynthetic activity in a tropical and a temperate lake. *Mem. Ist. Ital. Hydrobiol.*, **18**, Suppl., 399–424.

Talling, J.F. (1966). The annual cycle of stratification and phytoplankton growth in Lake Victoria (East Africa). *Int. Rev. ges. Hydrobiol.*, **51**, 545–621.

Talling, J.F. (1969). The incidence of vertical mixing, and some biological and chemical consequences, in tropical African lakes. *Verh. int. Verein. Limnol.*, **17**, 998–1012.

Talling, J.F. (1970). Generalized and specialized features of phytoplankton as a form of photosynthetic cover. In *Prediction and measurement of photosynthetic productivity*, pp. 431–45. Wageningen: Centre for Agricultural Publishing and Documentation.

Talling, J.F. (1976). Water characteristics. In *The Nile, biology of an ancient river*, ed. J. Rzóska, pp. 357–84. Monogr. Biologicae 29. The Hague: Junk.

Talling, J.F. (1982). Utilization of solar radiation by phytoplankton. In *Trends in photobiology*, ed. C. Helene, M. Charlier, T. Montenay-Garestier & G. Laustriat, pp. 619–31. New York: Plenum.

Talling, J.F. (1986). The seasonality of phytoplankton in African lakes. *Hydrobiologia*, **138**, 139–60.

Talling, J.F. (1987). The phytoplankton of Lake Victoria (East Africa). *Arch. Hydrobiol. Beih., Ergebn. Limnol.*, **25**, 229–56.

Talling, J.F. (1990). Diel and seasonal energy transfer, storage and stratification in African reservoirs and lakes. *Arch. Hydrobiol. Beih., Ergebn. Limnol.*, **33** 651–60.

Talling, J.F. (1992). Environmental regulation in African shallow lakes and wetlands. *Rev. Hydrobiol. trop.*, **25**, 87–144.

Talling, J.F. (1995*a*). Tropical limnology and the Sunda expedition. In *Tropical limnology*, vol. I, ed. K.H. Timotius & F. Göltenboth, pp. 3–10. Salatiga, Indonesia: Satya Wacana Christian University. Reprinted 1996 in *Perspectives in tropical limnology*, ed. F. Schiemer & K.T. Boland, pp. 19–26. Amsterdam: SPB Academic Publishing.

Talling, J.F. (1995*b*). Phytoplankton as an increasingly compound photosynthetic system: an historical perspective. *Hydrobiologia*, **315**, 9–14.

Talling, J.F. & Rzóska. J. (1967). The development of plankton in relation to hydrological regime in the Blue Nile. *J. Ecol.*, **55**, 637–62.

Talling, J.F. & Talling, I.B. (1965). The chemical composition of African lake waters. *Int. Rev. ges. Hydrobiol.*, **50**, 421–63.

Talling, J.F., Wood, R.B., Prosser, M.V. & Baxter, R.M. (1973). The upper limit of photosynthetic productivity by phytoplankton: evidence from Ethiopian soda lakes. *Freshwat. Biol.*, **3**, 53–76.

Taylor, M. & Aquize, E. (1984). A climatological energy budget of Lake Titicaca (Peru/Bolivia). *Verh. int. Verein. Limnol.*, **22**, 1246–51.

Thienemann, A. (1932). Tropische Seen und Seetypenlehre. *Arch. Hydrobiol. Suppl.* **9**, 205–31.

Thomas, J.D. & Radcliffe, P.J. (1973). Observations on the limnology and primary production of a small man-made lake in the West African savanna. *Freshwat. Biol.*, **3**, 573–612.

Thompson, A.B., Allison, E.H., Ngatunga, B.P. & Bulirani, A. (1995). First growth and feeding biology. In *The fishery potential and productivity of the pelagic zone of Lake Malawi/Niassa*, ed. A. Menz, pp. 279–306. UK: Natural Resources Institute, Overseas Development Administration.

Thompson, K., Shewry, P.R. & Woolhouse, H.W. (1979). Papyrus swamp development in the Upemba Basin, Zaïre: studies of population structure in *Cyperus papyrus* stands. *Bot. J. Linnean Soc.*, **78**, 299–316.

Thornton, I.W.B. (1957). Faunal succession in umbels of *Cyperus papyrus* L. on the upper White Nile. *Proc. R. ent. Soc. Lond. (A)*, **32**, 119–31.

Thornton, J.A. (ed.) (1982). *Lake McIlwaine. The eutrophication and recovery of a tropical African man-made lake.* Monogr. Biologicae 49. The Hague: Junk.

Thornton, J.A. (1987*a*). Nutrients in African lake ecosystems: do we know all? *J. Limnol. Soc. Sth. Afr.*, **12**, 6–21.

Thornton, J.A. (1987*b*). Aspects of eutrophication management in tropical/subtropical regions. *J. Limnol. Soc. Sth. Afr.*, **13**, 25–43.

Tiercelin, J.J., Pflumio, C., Costrec, M., Boulègue, J., Gente, P., Rolet, J., Coussement, C., Stetter, K.O., Huber, R., Buku, S. & Mifundu, W. (1993). Hydrothermal vents in Lake Tanganyika, East African Rift System. *Geology*, **21**, 499–502.

Tjönneland, A. (1962). The nocturnal flight activity and the lunar rhythm of emergence in the African midge, *Conochironomus acutistilus* (Freeman). *Contrib. Fac. Sci., Univ. College Addis Ababa, Ser. C (Zool.)* no. 4, 1–21.

Toews, D.R. & Griffin, J.S. (1979). Empirical estimates of potential fish yields for the Lake Bangweulu, Zambia, Central Africa. *Trans. Amer. Fish. Soc.*, **108**, 241–52.

Torréton, J.P., Bouvy, M. & Arfi, R. (1994). Diel fluctuations of bacterial abundance and productivity in a shallow eutrophic tropical lagoon. *Arch. Hydrobiol.*, **131**, 79–92.

Townsend, S.A. (1994). The occurrence of natural fish kills, and their causes, in the Darwin-Katherine-Jabiru region of northern Australia. *Mitt. int. Verein. Limnol.*, **24**, 197–205.

Townsend, S.A. (1995). Metalimnetic and hypolimnetic deoxygenation in an Australian tropical reservoir of low trophic status. In *Tropical limnology*, vol. 2, *Tropical lakes and reservoirs*, ed. K.H. Timotius & F. Göltenboth, pp. 255–64. Salatiga, Indonesia: Satya Wacana Christian Univ. Reprinted 1996 in *Perspectives in tropical limnology*, ed. F. Schiemer & K.T. Boland, pp. 151–60. Amsterdam: SPB Academic Publishing.

Townsend, S.A. (1998). The influence of retention time and wind exposure on stratification and mixing in two tropical Australian lakes. *Arch. Hydrobiol.*, **41**, 333–71.

Townsend, S.A., Boland, K.T. & Luong-Van, J.T. (1997). Wet and dry season heat loss in two tropical Australian reservoirs. *Arch. Hydrobiol.*, **139**, 51–68.

Townsend, S.A., Luong-Van, J.T. & Boland, K.T. (1996). Retention time as a primary determinant of colour and light attenuation in two tropical Australian reservoirs. *Freshwat. Biol.*, **36**, 57–69.

Tudorancea, C., Baxter, R.M. & Fernando, C.H. (1989). A comparative limnological study of zoobenthic associations in lakes of the Ethiopian Rift Valley. *Arch. Hydrobiol. Suppl.* **83**, 121–74.

Tuite, C.H. (1979). Population size, distribution and biomass density of the lesser flamingo in the Eastern Rift Valley, 1974–76. *J. appl. Ecol.*, **16**, 765–75.

Tuite, C.H. (1981). Standing crop densities and distribution of *Spirulina* and benthic diatoms in East African alkaline saline lakes. *Freshwat. Biol.*, **11**, 345–60.

Tundisi, J.G. (1983). A review of basic ecological processes interacting with production and standing crop of phytoplankton in lakes and reservoirs in Brazil. *Hydrobiologia*, **100**, 223–43.

Tundisi, J.G. (1994). Tropical South America: present and perspectives. In *Limnology now: a paradigm of planetary problems*, ed. R. Margalef, pp. 353–424. Amsterdam: Elsevier.

Tundisi, J.G., Forsberg, B.R., Devol, A.H., Zaret, T.M., Tundisi, T.M., Dos Santos, A., Ribeiro, J.. & Hardy, E.R. (1984). Mixing patterns in Amazon lakes. *Hydrobiologia*, **108**, 3–15.

Tundisi, J.G., Gentil, J.G. & Dirickson, M.C. (1978). Seasonal cycle of primary production of nanno and microphytoplankton in a shallow tropical reservoir. *Rev. Brasil. Bot.*, **1**, 35–9.

Tundisi, J.G., Matsumura-Tundisi, T., Pontes, M.C.F. & Gentil, J.G. (1981). Limnological studies at Quaternary Lakes in Eastern Brazil. I. Primary production of phytoplankton and ecological factors at Lake D. Helvécio. *Rev. Brasil. Bot.*, **4**, 5–14.

Tundisi, J.G. & Saijo, Y. (ed.) (1997). *Limnological studies on the Rio Doce Valley Lakes, Brazil.* São Carlos: Brazilian Academy of Sciences.

Tundisi, J.G., Saijo, Y., Henry, R. & Nakamoto, N. (1997). Primary productivity, phytoplankton biomass and light photosynthesis responses in four lakes. In

Limnological studies on the Rio Doce Valley Lakes, Brazil., ed. J.G. Tundisi & Y. Saijo, pp. 199–225. São Carlos: Brazilian Academy of Sciences.

Turcotte, P. & Harper, P.P. (1982). Drift patterns in a high Andean stream. *Hydrobiologia*, **89**, 141–51.

Turner, B.F., Gardner, L.R., Sharp, W.E. & Blood, E.R. (1996). The geochemistry of Lake Bosumtwi, a hydrologically closed basin in the humid zone of tropical Ghana. *Limnol. Oceanogr.*, **41**, 1415–24.

Twombly, S. (1983). Seasonal and short term fluctuations in zooplankton abundance in tropical lake Malawi. *Limnol. Oceanogr.*, **28**, 1214–24.

Twombly, S. & Lewis, W.H. (1987). Zooplankton abundance and species composition in Laguna La Orsinera, a Venezuelan floodplain lake. *Arch. Hydrobiol. Suppl.* **79**, 87–107.

Twombly, S. & Lewis, W.H. (1989). Factors regulating cladoceran dynamics in a Venezuelan floodplain lake. *J. Plankton Res.*, **11**, 317–33.

Unger, P.A. & Lewis, W.M. (1991). Population ecology of the pelagic fish, *Xenomelaniris venezuelae* (Atherinidae), in tropical lake Valencia. *Ecology*, **72**, 440–56.

Vammen, K., Erikson, R., Vargas, M.H. & Bell, R. (1991). Heterotrophic activity and bacterial growth in a tropical lake (Lake Xolotlán, Managua, Nicaragua). *Verh. int. Verein. Limnol.*, **24**, 1171–3.

van der Heide, J. (1973). Plankton development during the first years of inundation of the Van Blommestein (Brokopondo) Reservoir in Suriname, S. America. *Verh. int. Verein. Limnol.*, **18**, 1784–1791.

van der Heide, J. (1978). Stability of diurnal stratification in the forming Brokopondo Reservoir in Suriname, S. America. *Verh. int. Verein. Limnol.*, **201**, 1702–9.

van der Heide, J. (1982). *Lake Brokopondo. Filling phase limnology of a man-made lake in the humid tropics.* Thesis, Free Univ. of Amsterdam.

van Donselaar, J. (1968). Water and marsh plants in the artificial Brokopondo Lake (Suriname) during the first three years of its existence. *Acta Bot. Neerland.*, **17**, 183–96.

van Someren, V.D. (1950). The 'winter check' on trout scales in East Africa. *Nature, Lond.*, **165**, 473–4.

van Someren, V.D. (1962). The migration of fish in a small Kenya river. *Rev. Zool. Bot. afr.*, **66**, 375–93.

Vareschi, E. (1978). The ecology of Lake Nakuru, Kenya. I. Abundance and feeding of the lesser flamingo. *Oecologia*, **32**, 11–35.

Vareschi, E. (1979). The ecology of Lake Nakuru (Kenya). II. Biomass and spatial distribution of fish. *Oecologia*, **37**, 321–35.

Vareschi, E. (1982). The ecology of Lake Nakuru (Kenya). III. Abiotic factors and primary production. *Oecologia*, **55**, 81–101.

Vareschi, E. & Jacobs, J. (1984). The ecology of Lake Nakuru (Kenya). V. Production and consumption of consumer organisms. *Oecologia*, **61**, 83–98.

Vareschi, E. & Jacobs, J. (1985). The ecology of Lake Nakuru. VI. Synopsis of production and energy flow. *Oecologia*, **65**, 412–24.

Vareschi, E. & Vareschi, A. (1984). The ecology of Lake Nakuru (Kenya). IV. Biomass and distribution of consumer organisms. *Oecologia*, **60**, 70–82.

Vásquez, E. (1992). Temperature and dissolved oxygen in lakes of the lower Orinoco River floodplain. *Rev. Hydrobiol. trop.*, **25**, 23–33.

Vincent, W.F. (1992). The daily pattern of nitrogen uptake by phytoplankton in dynamic mixed layer environments. *Hydrobiologia*, **238**, 37–52.

Vincent, W.F., Neale, P.J. & Richerson, P.J. (1984). Photoinhibition: algal responses to bright light during diel stratification and mixing in a tropical alpine lake. *J. Phycol.*, **20**, 201–11.

Vincent, W.F., Vincent, C.L., Downes, M.T. & Richerson, P.J. (1985). Nitrate cycling in Lake Titicaca (Peru–Bolivia): the effects of high-altitude and tropicality. *Freshwat. Biol.*, **15**, 31–42.

Vincent, W.F., Wurtsbaugh, W., Neale, P.J. & Richerson, P.J. (1986). Polymixis and algal production in a tropical lake: latitudinal effects on the seasonality of photosynthesis. *Freshwat. Biol.*, **16**, 781–803.

Vincent, W.F., Wurtsbaugh, W.A., Vincent, C.L. & Richerson, P.J. (1984). Seasonal dynamics of nutrient limitation in a tropical high-altitude lake (Lake Titicaca, Peru–Bolivia): application of physiological bioassays. *Limnol. Oceanogr.*, **29**, 540–52.

Viner, A.B. (1973). Responses of a mixed phytoplankton population to nutrient enrichments of ammonia and phosphate, and some associated ecological implications. *Proc. R. Soc. (B)*, **183**, 351–70.

Viner, A.B. (1975*a*). The supply of minerals to tropical rivers and lakes (Uganda). In *Coupling of land and water systems*, ed. A.D. Hasler, pp. 227–61. New York: Springer.

Viner, A.B. (1975*b*). The sediments of Lake George (Uganda). I. Redox potentials, oxygen consumption and carbon dioxide output. *Arch. Hydrobiol.*, **76**, 181–97.

Viner, A.B. (1975*c*). The sediments of Lake George, Uganda. II. Release of ammonium and phosphate from an undisturbed mud surface. *Arch. Hydrobiol.*, **76**, 368–78.

Viner, A.B. (1975*d*). The sediments of Lake George (Uganda). III. The uptake of phosphate. *Arch. Hydrobiol.*, **76**, 393–410.

Viner, A.B. (1975*e*). Non-biological factors affecting phosphate recycling in the water of a tropical eutrophic lake. *Verh. int. Verein. Limnol.*, **19**, 1404–15.

Viner, A.B. (1977*a*). The sediments of Lake George (Uganda). IV. Vertical distribution of chemical features in relation to ecological history and nutrient recycling. *Arch. Hydrobiol.*, **80**, 40–69.

Viner, A.B. (1977*b*). The influence of sediments upon nutrient exchanges in tropical lakes. In *Interactions between sediments and fresh water*, ed. H.L. Golterman, pp. 210–15. Wageningen: PUDOC Publ.

Viner, A.B. (1977*c*). Relationships of nitrogen and phosphorus to a tropical phytoplankton population. *Hydrobiologia*, **52**, 185–96.

Viner, A.B. (1982*a*). A quantitative assessment of the nutrient phosphate transported by particles in a tropical river. *Rev. Hydrobiol. trop.*, **15**, 3–8.

Viner, A.B. (1982*b*). Nitrogen fixation and denitrification in sediments of two Kenyan lakes. *Biotropica*, **14**, 91–8.

Viner, A.B. & Smith, I.R. (1973). Geographical, historical and physical aspects of Lake George. *Proc. R. Soc. (B)*, **184**, 235–70.

von Damm, K.L. & Edmond, J.M. (1984). Reverse weathering in the closed-basin lakes of the Ethiopian rift. *Am. J. Sci.*, **284**, 835–62.

von Herzen, R.P. & Vacquier, V. (1967). Terrestrial heat flow in Lake Malawi, Africa. *J. geophys. Res.*, **72**, 4221–6.

Vyverman, W. & Tyler, P. (1995). Fine-layer zonation and short-term changes of microbial communities in two coastal meromictic lakes (Madang Province, Papua New Guinea). *Arch. Hydrobiol.*, **132**, 385–406.

Waichman, A.V. (1996). Autotrophic carbon sources for heterotrophic bacterioplankton in a floodplain lake of Central Amazon. *Hydrobiologia*, **341**, 27–36.

Walker, I. & Ferreira, M. J. de N. (1985). On the population dynamics and ecology of the shrimp species (Crustacea, Decapoda, Natantia) in the Central Amazonian river Tarumã Mirim. *Oecologia*, **66**, 264–70.

Walker, I. Henderson, P.A. & Sterry, P. (1991). On the patterns of biomass transfer in the benthic fauna of an Amazonian blackwater river, as evidenced by P-32 label experiment. *Hydrobiologia*, **215**, 153–62.

Walker, T. & Tyler, P. (1984). Tropical Australia, a dynamic limnological environment. *Verh. int. Verein. Limnol.*, **22**, 1727–34.

Walling, D.E. (1984). The sediment yields of African rivers. In *Challenges in African hydrology and water resources*, ed. D.E. Walling, S.S.D. Foster & P. Wurzel, pp. 265–83. Wallingford: IAHS Press.

Ward, P.R.B. (1979). Seiches, tides and wind set-up on Lake Kariba. *Limnol. Oceanogr.*, **24**, 151–7.

Waters, T.F. (1977). Secondary production in inland waters. *Adv. Ecol. Res.*, **10**, 91–164.

Weers, E.T. & Zaret, T.M. (1975). Grazing effects on nanoplankton in Gatun Lake, Panama. *Verh. int. Verein. Limnol.*, **19**, 1480–3.

Weir, J.S. (1968). Seasonal variation in alkalinity in pans in Central Africa. *Hydrobiologia*, **32**, 69–80.

Welcomme, R.L. (1970). Studies of the effects of abnormally high water levels on the ecology of fish in certain shallow regions of Lake Victoria. *J. Zool., Lond.*, **160**, 405–36.

Welcomme, R.L. (1979). *Fisheries ecology of floodplain rivers*. London: Longman.

Welcomme, R.L. (1985). *River fisheries*. FAO Fish. Tech. Pap. no. 262.

Westlake, D.F. (1975). Primary production of freshwater macrophytes. In *Photosynthesis and productivity in different environments*, ed. J.P. Cooper, pp. 189–206. Cambridge: Cambridge Univ. Press.

Whitton, B.A. & Rother, J.A. (1988). Diel changes in the environment of a deepwater rice-field in Bangladesh. *Verh. int. Verein. Limnol.*, **23**, 1074–9.

Williams, N.J. & Goldman, C.R. (1975). Succession rates in lake phytoplankton communities. *Verh. int. Verein. Limnol.*, **19**, 808–11.

Williams, W.D. (1985). Biotic adaptations in temporary lentic waters, with special reference to those in semi-arid and arid regions. *Hydrobiologia*, **125**, 85–110.

Williamson, P.G. (1981). Paleontological documentation of speciation in Cenozoic molluscs from Turkana Basin. *Nature, Lond.*, **293**, 437–43.

Wirrmann, D., Ybert, J.P. & Mourguiart, P. (1992). A 20 000 years paleohydrological record from Lake Titicaca. In *Lake Titicaca. A synthesis of limnological knowledge*, ed. C. Dejoux & A. Iltis, pp. 40–8. Monogr. Biologicae 68. Dordrecht: Kluwer.

Wissmar, R.C., Richey, J.E., Stallard, R.F. & Edmond, J.M. (1981). Plankton metabolism and carbon processes in the Amazon River, its tributaries, and floodplain waters, Peru–Brazil, May–June 1977. *Ecology*, **62**, 1622–33.

Witte, F., Goldschmidt, T., Wanink, J., van Oijen, M., Goudswaard, K., Witte-Maas, E. & Bouton, N. (1992). The destruction of an endemic species flock: quantitative data on the decline of the haplochromine cichlids of Lake Victoria. *Envir. Biol. Fishes*, **34**, 1–28.

Wodajo, K. & Belay, A. (1984). Species composition and seasonal abundance of zooplankton in two Ethiopian Rift Valley lakes – Lakes Abiata and Langano. *Hydrobiologia*, **113**, 129–36.

Wolda, H. & Flowers, R.W. (1985). Seasonality and diversity of mayfly adults (Ephemeroptera) in a 'nonseasonal' tropical environment. *Biotropica*, **17**, 330–5.

Wood, K.G., Kannan, V. & Saunders, G.W. (1984). Photosynthesis in South India ponds: CO_2 variations. *Verh. int. Verein. Limnol.*, **22**, 1717–21.

Wood, R.B., Baxter, R.M. & Prosser, M.V. (1984). Seasonal and comparative aspects of chemical stratification in some tropical crater lakes, Ethiopia. *Freshwat. Biol.*, **14**, 551–73.

Wood, R.B., Prosser, M.V. & Baxter, R.M. (1976). The seasonal pattern of thermal characteristics of four of the Bishoftu crater lakes, Ethiopia. *Freshwat. Biol.*, **6**, 519–30.

Wood, R.B., Prosser, M.V. & Baxter, R.M. (1979). Optical characteristics of the Rift Valley lakes, Ethiopia. *Sinet: Ethiop. J. Sci.*, **1**, 73–85.

Wood, R.B. & Talling, J.F. (1988). Chemical and algal relationships in a salinity series of Ethiopian inland waters. *Hydrobiologia*, **158**, 29–67.

Wooster, M.J., Sear, C.B., Patterson, G. & Haigh, J. (1994). Tropical lake surface temperature from locally received NOAA-II AVHRR data – comparison with *in situ* measurements. *Int. J. Remote Sensing*, **15**, 183–9.

Worthington, E.B. (1930). Observations on the temperature, hydrogen-ion concentration, and other physical conditions of the Victoria and Albert Nyanzas. *Int. Rev. ges Hydrobiol.*, **24**, 328–57.

Worthington, E.B. (1931). Vertical movements of freshwater macroplankton. *Int. Rev. ges. Hydrobiol.*, **25**, 394–436.

Worthington, E.B. & Ricardo, C.K. (1936). Scientific results of the Cambridge expedition to the East African lakes, 1930–1. 17. The vertical distribution and movements of the plankton in lakes Rudolf, Naivasha, Edward and Bunyoni. *J. Linn. Soc. Zool.*, **40**, 33–69.

Wüest, A., Piepke, G. & Halfman, J.D. (1996). Combined effects of dissolved solids and temperature on the density stratification of Lake Malawi. In *The limnology, climatology and paleoclimatology of the East African Lakes*, ed. T.C. Johnson & E.O. Odada, pp. 183–202. Amsterdam: Gordon & Breach.

Wurtsbaugh, W.A., Vincent, W.F., Alfaro Tapia, R., Vincent, C.L. & Richerson, P.J. (1985). Nutrient limitation of algal growth and nitrogen fixation in a tropical alpine lake, Lake Titicaca (Peru, Bolivia). *Freshwat. Biol.*, **15**, 185–95.

Wurtsbaugh, W.A., Vincent, W.F., Vincent, C.L., Carney, H.J., Richerson, P.J. & Alfaro Tapia, R. (1992). Nutrients and nutrient limitation of phytoplankton. In *Lake Titicaca. A synthesis of limnological knowledge*, ed. C. Dejoux & A. Iltis, pp. 147–60. Monogr. Biologicae 68. Dordrecht: Kluwer.

Wyngaard, G.A., Elmore, J.L. & Cowell, B.C. (1982). Dynamics of a subtropical plankton community, with emphasis on the copepod *Mesocyclops edox*. *Hydrobiologia*, **89**, 39–48.

Ybert, J.P. (1992). Ancient lake environments as deduced from pollen analysis. In *Lake Titicaca. A synthesis of limnological knowledge*, ed. C. Dejoux & A. Iltis, pp. 49–62. Monogr. Biologicae 68. Dordrecht: Kluwer.

Young, J.O. (1975). Seasonal and diurnal changes in the water temperature of a temperate pond (England) and a tropical pond (Kenya). *Hydrobiologia*, **47**, 513–26.

Yule, C.M. (1995). The ecology of an aseasonal tropical river on Bougainville Island, Papua New Guinea. In *Tropical Limnology*, vol. 3, ed. K.H. Timotius & F. Göltenboth, pp. 1–14. Salatiga, Indonesia: Satya Wacana Christian

Univ. Reprinted 1996 in *Perspectives in tropical limnology*, ed. F. Schiemer & K.T. Boland, pp. 239–54. Amsterdam: SPB Academic Publishing.

Yule, C.M. & Pearson, R.G. (1996). Aseasonality of benthic invertebrates in a tropical stream on Bougainville Island, Papua New Guinea. *Arch. Hydrobiol.*, **137**, 95–117.

Yuretich, R.F. & Cerling, T.E. (1983). Hydrogeochemistry of Lake Turkana, Kenya: mass balance and mineral reactions in an alkaline lake. *Geochim. cosmochim. Acta*, **47**, 1099–109.

Zafar, A.R. (1986). Seasonality of phytoplankton in some South Indian lakes. *Hydrobiologia*, **138**, 177–87.

Zaret, T. (1969). Predation-balanced polymorphism of *Ceriodaphnia cornuta* Sars. *Limnol. Oceanogr.*, **14**, 301–3.

Zaret, T. (1972*a*). Predators, invisible prey and the nature of polymorphism in the Cladocera (Class Crustacea). *Limnol. Oceanogr.*, **17**, 171–84.

Zaret, T. (1972*b*). Predator–prey interaction in a tropical lacustrine ecosystem. *Ecology*, **53**, 248–57.

Zaret, T.M. (1984). Central American limnology and Gatún Lake, Panama. In *Ecosystems of the World*. 23. *Lakes and reservoirs*, ed. F.B. Taub, pp. 447–65. Amsterdam: Elsevier.

Zaret, T.M., Devol, A.D. & Dos Santos, A. (1981). Nutrient addition experiments in Lago Jacaretinga, Central Amazon Basin, Brazil. *Verh. int. Verein. Limnol.*, **21**, 721–4.

Zaret, T. & Paine, R.T. (1973). Species introduction in a tropical lake. *Science, NY*, **218**, 444–5.

Zaret, T.M. & Suffern, J.S. (1976). Vertical migration in zooplankton as a predator avoidance mechanism. *Limnol. Oceanogr.*, **21**, 804–13.

Zimmerman, U., Baumann, U., Imevbore, A.M.A., Henderson, F. & Adeniji, H.A. (1976). Study of the mixing pattern of Lake Kainji using stable isotopes. *Catena*, **3**, 63–76.

Zinabu, G.M. & Taylor, W.D. (1997). Bacteria–chlorophyll relationships in Ethiopian lakes of varying salinity: are soda lakes different? *J. Plankton Res.*, **19**, 647–54.

Index to water-bodies

Floodplains

Lakes and reservoirs (R)

Rivers

Streams

Swamps

General index